# ELEMENTARY
# EXCITATIONS
# IN SOLIDS

# ELEMENTARY EXCITATIONS IN SOLIDS

The Cortina Lectures, July 1966, and selected
lectures from the Conference on Localized
Excitations, Milan, July 25-26, 1966

Edited by

## A. A. Maradudin

*Department of Physics*
*University of California, Irvine*

and

## G. F. Nardelli

*Gruppo Nazionale Struttura della Materia, C.N.R.*
*and*
*Physics Institute*
*University of Milan*

Springer Science+Business Media New York 1969

Library of Congress Catalog Card Number 68-26772

ISBN 978-1-4899-5534-0      ISBN 978-1-4899-5532-6 (eBook)
DOI 10.1007/978-1-4899-5532-6

© 1969 Springer Science+Business Media New York
Originally published by Plenum Press, New York in 1969.
Softcover reprint of the hardcover 1st edition 1969

# Preface

The idea for an Advanced Study Institute devoted to elementary excitations in solids grew out of a conversation between Eli Burstein and Gianfranco Nardelli in Burstein's office at the University of Pennsylvania in December, 1965. From the start, it was intended that the Institute be tutorial in nature, rather than an arena for the presentation of the latest results in this area of physics. The selection of topics reflected this aim, and the lecturers, a distinguished and far flung group of physicists, implemented it, as is evident from their lectures, collected in this book.

To provide an opportunity for the specialists gathered at Cortina d'Ampezzo to describe the results of recent investigations in the area of localized excitations in solids, a two-day conference on this topic was organized at the University of Milan by Gianfranco Nardelli on July 25–26, 1966. Many of the invited and contributed talks presented at this conference are also included in the present volume.

The editors of this volume would like to express their appreciation to the NATO Science Affairs Division for its support of the Advanced Study Institute; to the Italian Council for Research, Province of Milan; and to Eugenio Depaolini Delvecchio; for their support of the Conference on Localized Excitations, Milan. Finally, they would like to thank the lecturers at both the Advanced Study Institute and the Localized Excitations Conference for the care with which they prepared and presented their talks. Through their efforts, the aims of these meetings were realized.

A. A. MARADUDIN

G. F. NARDELLI

# Contributors

G. Baldini
  *Istituto di Fisica, Università Degli Studi, Milan, Italy*

M. Balkanski
  *Laboratoire de Physique, Ecole Normale Supérieure, Paris, France*

E. Burstein
  *Physics Department and Laboratory for the Research on the Structure of Matter, University of Pennsylvania, Philadelphia, Pennsylvania*

J. Callaway
  *Department of Physics, Louisiana State University, Baton Rouge, Louisiana*

G. Chiarotti
  *Department of Physics, University of Rome, Rome, Italy*

M. F. Collins
  *A.E.R.E., Harwell, Berkshire, England, United Kingdom*

R. J. Elliot
  *Department of Physics, University of Oxford, Oxford, England*

J. J. Hopfield
  *Department of Physics, Princeton University, Princeton, New Jersey*

I. P. Ipatova
  *A. F. Ioffe Physico-Technical Institute, Leningrad, USSR*

A. A. Klochikhin
  *A. F. Ioffe Physico-Technical Institute, Leningrad, USSR*

W. Ludwig
  *Institute of Theoretical Physics, University of Giessen, Germany*

A. A. Maradudin
  *University of California, Irvine, California*

G. F. Nardelli
  *Institute of Physics, University of Milan, Milan, Italy, and Gruppo Nazionale Struttura della Materia del Consiglio Nazionale delle Recerche, Milan, Italy*

R. O. Pohl
*Laboratory of Atomic and Solid State Physics, Cornell University, Ithaca, New York*

P. Resibois
*Institute of Physics, Université Libre, Brussels, Belgium*

A. J. Sievers
*Laboratory of Atomic and Solid State Physics, Cornell University, Ithaca, New York*

R. F. Wallis
*University of California, Irvine, California*

# Contents

# Introduction to Elementary Excitations in Solids

## A. A. Maradudin*

*Department of Physics*
*University of California*
*Irvine, California*

## I. INTRODUCTION

Nature presents the solid-state physicist with a vast range of physical phenomena of inordinate complexity which require understanding and explanation. However, an understanding of the thermal, mechanical, electrical, magnetic, and optical properties of solids requires that we study properties of systems of many strongly interacting particles. In these circumstances it is understandable, even natural, that solid state theorists have sought approximations which render these systems tractable for study. That a qualitative, and in many cases quantitative, understanding of many properties of the many-body systems of interest in solid state physics has been achieved, can in a large measure be attributed to the following circumstances ([1]). For many purposes the total binding energy of the ground state of a given system is not an important physical quantity. What is important is the behavior of the lowest excited states relative to the ground state, when the response of the system to generalized external forces, such as electric and magnetic fields or temperature, is sought. For example, you are all familiar with the fact that the low-temperature specific heat of an insulating crystal and the deviation of the spontaneous magnetization of a ferromagnetic crystal from its saturation value at low temperatures have their origin in the thermal excitation of small numbers of elastic waves or spin waves, respectively. The fact that it is the low-lying excited states of a many-body system which play the central role in determining the response of the system to external stimuli would be of comparatively little usefulness if it were not for the fact that these states in most cases of physical interest can be obtained by carrying out a principal axis transformation of the Hamiltonian of the system which,

*This research was supported in part by the Air Force Office of Scientific Research, Office of Aerospace Research, U.S. Air Force, under AFOSR Grant No. 1080–66.

1

in a zero-order of approximation, re-expresses it as the sum of the Hamiltonians of a collection of independent, i.e., noninteracting, entities. The energies required to raise each of these independent entities to its lowest exicted state comprise the energy spectrum of the low-lying excited states of the system.

For each of the systems to be studied at this Institute, namely, the atomic vibrations in a crystal, interacting spins in a ferromagnetic crystal, the interacting electron gas, and the low-lying excited states of an insulating crystal, the principal axis transformation which diagonalizes the corresponding Hamiltonian in zero order expresses it as the sum of the Hamiltonians of individual, independent, harmonic oscillators. The energy levels of an harmonic oscillator are equally spaced, with a spacing which is $\hbar$ times the frequency of the oscillator. The quantum of energy required to excite the oscillator from any level to the next higher level, therefore, is independent of the particular levels being considered, and for each of the systems being considered has been given a special name. For the lattice vibrations it is a *phonon*; for the interacting spin system it is a *magnon*; for the interacting electron gas it is a *plasmon*; and for the low-lying excited states of an insulator it is an *exciton*.

If the system under consideration has perfect periodicity, such as a crystal which is free from defects, or has perfect translational invariance, such as the interacting electron gas, there exists a unique relation between the energy of a low-lying excited state and its momentum, or wavevector. (However, it should not be thought that periodicity or translational invariance is necessary for the existence of elementary excitations. It is not, and we will see an example of this later when we come to discuss phonons.)

We can then give a tentative definition of an elementary excitation of momentum $\hbar\mathbf{k}$ as that harmonic oscillator creation operator which when applied to the ground state of a system creates the lowest excited state of momentum $\hbar\mathbf{k}$. However, in any real system it is not at all uncommon to find several different kinds of elementary excitations present at the same time. Thus in a magnetic crystal we can have both phonons and magnons, for example, present at the same time. This means that we should sharpen our definition of elementary excitations somewhat. We accordingly define an elementary excitation of momentum $\hbar\mathbf{k}$ as that harmonic oscillator creation operator which when applied to the ground state of a system creates the lowest excited state of a particular type of momentum $\hbar\mathbf{k}$.

Of course, it is only in a zero order of approximation that the Hamiltonian of a system of interacting particles can be diagonalized by a principal axis transformation, except, perhaps, in the case of rather trivial systems. The principal axis transformation which effects this diagonalization at the same time introduces interactions between the noninteracting entities of the zero-order calculation. However, and this is the point of this discussion, if the transformation has been carried out cleverly, and most of the interactions or the most important interactions between the particles of the original Hamiltonian have been absorbed into the zero-order Hamiltonian, the interactions between the zero-order entities are small, and in many cases can be treated satisfactorily by perturbation theory. This is the case, for example, for phonons in an anharmonic crystal. The mere diagonalization of a system's Hamiltonian by some truncation of it does not automatically lead to a set of zero-order elementary excitations which interact weakly through the residual forces omitted by the truncation. We will see an example where this is not the case in our discussion of phonons. The trick is to diagonalize the Hamiltonian approximately in such a fashion that the residual interactions are truly small. Thus, as has been emphasized by Anderson ($^2$), the concept of elementary excitations is a way of linearizing the equations of the system about the true ground state, rather than about some independent particle approximation.

If this has been achieved, the effects of the interactions on the zero-order elementary excitations lead to true elementary excitations of the system of two kinds: quasi-particles and collective excitations.

Quasi-particles are elementary excitations which pass continuously into the zero-order elementary excitations when the interactions between the latter are turned off. In general, they no longer possess a unique relation between their energies and their wavevectors. That is, quasi-particles of a given wavevector $\mathbf{k}$ have a distribution of energies which is peaked about a value which is usually somewhat displaced from the energy of the zero-order excitation of the same wavevector. The width of this energy distribution is related by the uncertainty principle to the lifetime of this excitation. This lifetime can be regarded as the time during which a zero-order elementary excitation introduced into the interacting system retains a recognizable identity before it decays through its interactions with the other zero-order excitations into a less recognizable combination of these excitations. In general it is the quasi-particles of lowest energy which have the longest lifetimes, i.e., most closely resemble the zero-order elementary excitations into which they go in the absence of interactions. Therefore, as long as there are sufficiently few elementary excitations that only the lowest lying levels are populated, the quasi-particles can be regarded as independent of each other to a very good approximation. In fact, few here means few with respect to the total rumber of particles in the system, so that even when the absolute number of elementary excitations in very large, they can still be regarded as noninteracting.

However, unlike the situation which prevails for the rigorously independent elementary excitations of the zero-order diagonalization of the Hamiltonian, which for the calculation of many properties of the system can be regarded as comprising an ideal gas of fermions or bosons to which the standard formulas of statistical mechanics apply, quasi-particles cannot be so regarded in general when it comes to the calculation of properties of the system, such as, e.g., the thermodynamic functions. For example, in calculating the specific heat of an anharmonic crystal, one obtains an incorrect result if he substitutes the quasi-phonon frequencies into the expression for the specific heat of an harmonic oscillator of the frequency. On the other hand, the correct result, at least to the lowest orders of perturbation theory, for the entropy is obtained by this kind of procedure ([3]). The explanation of this result lies in the fact that in calculating the entropy we are really only counting states of the system, and these are not altered in number by the introduction of interactions among the zero-order elementary excitations. Similarly, it is found that the entropy of a system of fermions which interact through short-range forces is given correctly if the quasi-particle energies are used in the expression for the entropy of a system of noninteracting fermions [see, e.g. Ambegaokar ([4])].

These results have a certain intuitive appeal because they mean, for example, that thermodynamic functions of many-particle systems may be written as the sums of contributions from independent entities whose numbers at least in the case of phonons and magnons, equal the number of degrees of freedom in these systems, thereby assuring that the extensive nature of these functions is manifest.

In contrast with quasi-particles, which have their direct analogs in the noninteracting elementary excitations of the zero-order diagonalization of the Hamiltonian, collective excitations have no such counterparts among the zero-order elementary excitations. Instead, they are linear superpositions of the latter with a suitable choice of phases brought about by the interactions among themselves, so that they describe an organized or correlated motion of these formerly independent excitations. The best known example of such a collective excitation is the plasmon, a collective excitation in an interacting electron gas, of which I shall have more to say below. A second collective excitation is zero sound in a system of fermions interacting with short-range repulsive forces. Unlike plasmons, however, zero sound has not been observed experimentally as yet, and I will have no more to say about it here.

Finally, we remark that although many of the interesting properties of solids do not require knowledge of the total binding energy of the ground state of the solid for their understanding, but only of the elementary excitations above the ground state, it is interesting that experimental studies of elementary excitations can provide some information about the ground state. A notable example of this is found in the experimental determination of phonon dispersion curves by neutron spectrometry. Analysis of the experimental curves yields values for the second derivatives of the interatomic potential function evaluated at the equilibrium separations of the first, second, ... neighbor atoms in the crystal. While knowledge of the second derivatives of the interatomic potential function is not sufficient for a calculation of the binding energy of the ground state, it serves to put severe constraints on phenomenological interatomic potential functions used in such calculations.

In this talk I cannot hope to cover all aspects of elementary excitations and their interactions in solids: the large number of different kinds of elementary excitations, the many properties which each of these excitations possesses, and the variety of the interactions in which they participate, with other excitations of the same type or with excitations of other types, render such an encyclopedic discussion beyond the scope of the most ambitious summer school, let alone of a single talk [see, e.g., Pines ([5])].

I will be satisfied if I can achieve a much more modest goal, which is to preface the lectures presented here by considering in turn the four kinds of elementary excitations to which this summer school is devoted, viz., phonons, magnons, electrons and plasmons, and excitons; to indicate the kind of principal axis transformation which in zero order gives rise to these excitations; to point out certain general features of each of these excitations and the reasons for them; and, finally, to point out the similarities and differences among these excitations. The modifications of the results which I will describe, which are brought about by the interactions among the independent excitations of the zero-order theory by the terms in the system Hamiltonian which have been neglected in this approximation, various consequences of these interactions, and experimental techniques for studying elementary excitations, which in many cases are consequences of the interactions between different kinds of elementary excitations, comprise the subject matter of the lectures and seminars which will be given during the remainder of this summer school.

## II. LATTICE VIBRATIONS AND PHONONS

The existence of principal axis transformations which diagonalize the Hamiltonian of a system of coupled harmonic oscillators was well known from the work of Bernoulli([6]) and Lagrange([7]) on the vibrations of linear chains of particles, long before the development of quantum mechanics. It is therefore probably not surprising that it was in the context of lattice dynamics that elementary excitations were first introduced into solid state physics. I refer, of course, to the work of Debye in 1912([8]) in which he calculated the frequencies of the long-wavelength vibration modes of a crystal and combined them with Planck's quantum theory of the harmonic oscillator to obtain a correct explanation for the observed temperature dependence of the specific heats of crystals at low temperatures. Subsequent theoretical studies of the harmonic oscillator on the basis of quantum mechanics showed that the quantum mechanical and the classical equations of motion are identical, so that many caluclations of the dynamical properties of crystals obtained prior to 1926, such as the crystalline specific heat, required no modification to bring them into conformity with the results of purely quantum mechanical calculations.

Because the harmonic oscillator is one of the very few systems for which an exact quantum mechanical solution exists, it is perhaps not surprising that all of the principal axis transformations which have been employed in the study of many-body systems other than crystalline vibrations transform the Hamiltonians of these systems in zero order into the Hamiltonian of an assembly of noninteracting harmonic oscillators. Because of this, I will discuss the quantization of atomic vibrations in crystals in some detail in this talk, and will then apply the results of this discussion more or less directly to show how the elementary excitations called magnons, plasmons, and excitons arise in a system of interacting spins, in the interacting electron gas, and in insulating crystals, respectively.

The Hamiltonian which describes the atomic vibrations of an arbitrary crystal can be written

$$H = \sum_{l\kappa\alpha} \frac{P_\alpha^2(l\kappa)}{2M_{l\kappa}} + \frac{1}{2} \sum_{l\kappa\alpha} \sum_{l'\kappa'\beta} \Phi_{\alpha\beta}(l\kappa; l'\kappa') U_\alpha(l\kappa) U_\beta(l'\kappa')$$

$$+ \sum_{n=3}^{\infty} \frac{1}{n!} \sum_{l_1\kappa_1\alpha_1} \cdots \sum_{l_n\kappa_n\alpha_n} \Phi_{\alpha_1\cdots\alpha_n}(l_1\kappa_1; \cdots; l_n\kappa_n) U_{\alpha_1}(l_1\kappa_1) \cdots U_{\alpha_n}(l_n\kappa_n) \quad (2.1)$$

where $P_\alpha(lk)$ and $U_\alpha(lk)$ are the $\alpha$-Cartesian components of the momentum and the displacement from equilibrium of the $\kappa$th atom in the $l$th primitive unit cell of the crystal. $M_{lk}$ is mass of the atom $(lk)$ and the coefficients $\Phi_{\alpha_1\cdots\alpha_n}$ $(l_1\kappa_1; \ldots; l_n\kappa_n)$ are the so-called atomic force constants. They are the coefficients in the Taylor series expansion of the crystal potential energy in powers of the atomic dispacements:

$$\Phi_{\alpha_1\cdots\alpha_n}(l_1\kappa_1; \cdots l_n\kappa_n) = \frac{\partial^n\Phi}{\partial U_{\alpha_1}(l\kappa) \cdots \partial U_{\alpha_n}(l_n\kappa_n)}\Bigg)_{\{U_\alpha(l\kappa)\}=0} \quad (2.2)$$

The first two terms on the right-hand side of equation (2.1) give the crystal Hamiltonian in the harmonic approximation, that is, in the approximation in which the restoring force on an atom when it is displaced from its equilibrium position is a linear function of its displacement and those of all the other atoms in the crystal. The remaining terms in the Hamiltonian by definition comprise the anharmonic terms in the crystal potential energy.

It was recognized a long time ago that if the anharmonic terms are neglected in the crystal potential energy, the resulting Hamiltonian

$$H_0 = \sum_{l\kappa\alpha} \frac{P_\alpha^2(l\kappa)}{2M_{l\kappa}} + \frac{1}{2} \sum_{l\kappa\alpha} \sum_{l'\kappa'\beta} \Phi_{\alpha\beta}(l\kappa; l'\kappa')U_\alpha(l\kappa)U_\beta(l'\kappa') \qquad (2.3)$$

can be transformed into a new Hamiltonian which is the Hamiltonian of an assembly of independent harmonic oscillators. It was subsequently realized that this result is a general one, and can be stated as a mathematical theorem: it is always possible to find a transformation which simultaneously diagonalizes two positive-definite quadratic forms. That the kinetic energy is a positive-definite quadratic form is apparent from its representation as a sum of squares with positive coefficients. The positive-definite character of the quadratic form representing the potential energy must be postulated. If this is not the case, the frequencies of vibration of some of the independent oscillators become imaginary. The amplitude of vibration of an oscillator for which this happens will erupt exponentially with time into the past or into the future, implying instability of the crystal.

Because of the possibility of obtaining exact solutions of either the classical equations of motion obtained from the Hamiltonian $H_0$ or of the Schrodinger equation corresponding to this Hamiltonian, the theory of lattice dynamics has been based on the harmonic Hamiltonian $H_0$, with the anharmonic terms on the right-hand side of equation (2.1) regarded as a perturbation on $H_0$. The digonalization of $H_0$ can be carried out if we first consider the equations of motion of the crystal (either classical or quantum mechanical: they are the same)

$$M_{l\kappa}\ddot{U}_\alpha(l\kappa) = -\sum_{l'\kappa'\beta} \Phi_{\alpha\beta}(l\kappa; l'\kappa')U_\beta(l'\kappa') \qquad (2.4)$$

If we assume solutions of the form

$$U_\alpha(l\kappa) = \frac{B_\alpha(l\kappa)}{(M_{l\kappa})^{1/2}}e^{-i\omega t} \qquad (2.5)$$

the equations satisfied by the time-independent amplitudes $\{B_\alpha(l\kappa)\}$ are

$$\sum_{l'\kappa'\beta} D_{\alpha\beta}(l\kappa; l'\kappa')B_\beta(l'\kappa') = \omega^2 B_\alpha(l\kappa) \qquad (2.6)$$

where

$$D_{\alpha\beta}(l\kappa; l'\kappa') = \frac{\Phi_{\alpha\beta}(l\kappa; l'\kappa')}{(M_{l\kappa}M_{l'\kappa'})^{1/2}} \qquad (2.7)$$

If there are $N$ primitive unit cells in the crystal, and $\tau$ atoms in a primitive unit cell, the matrix $\mathbf{D}$ defined by equation (2.7) is a real, symmetric, $3\tau N \times 3\tau N$ matrix. The eigenvalue equation (2.6), therefore, has $3\tau N$ solutions, and we label these solutions by an index $s$ which runs from 1 to $3\tau N$:

$$\sum_{l'\kappa'\beta} D_{\alpha\beta}(l\kappa; l'\kappa') B_\beta^{(s)}(l'\kappa') = \omega_s^2 B_\alpha^{(s)}(l\kappa) \tag{2.8}$$

Because $\mathbf{D}$ is a real matrix, we can choose the eigenvectors $\{B_\alpha^{(s)}(l\kappa)\}$ to be real, and it is convenient to do so. In addition, the symmetry of $\mathbf{D}$ ensures that the eigenvectors are, or can be made to be, mutually orthogonal. With no loss of generality we can normalize these eigenvectors to unity, so that they satisfy the orthonormality and closure conditions

$$\sum_{l\kappa\alpha} B_\alpha^{(s)}(l\kappa) B_\alpha^{(s')}(l\kappa) = \delta_{ss'} \tag{2.9a}$$

$$\sum_s B_\alpha^{(s)}(l\kappa) B_\beta^{(s)}(l'\kappa') = \delta_{ll'}\delta_{\kappa\kappa'}\delta_{\alpha\beta} \tag{2.9b}$$

The displacement pattern described by the eigenvector $B_\alpha^{(s)}(l\kappa)$ is called the $s$th normal mode of vibration of the crystal. In this mode all of the atoms oscillate with the same frequency $\omega_s$.

Having obtained the atomic displacements in each of the normal modes, and the corresponding normal-mode frequencies, we can use these results to make a change of variables in the Hamiltonian (2.3) which simplifies it. We replace the operators $\{U_\alpha(l\kappa)\}$ and $\{P_\alpha(l\kappa)\}$ by new operators $\{b_s\}$ and $\{b_s^\dagger\}$ through the relations

$$U_\alpha(l\kappa) = \left(\frac{\hbar}{2M_{l\kappa}}\right)^{1/2} \sum_s \frac{B_\alpha^{(s)}(l\kappa)}{(\omega_s)^{1/2}}(b_s + b_s^\dagger) \tag{2.10a}$$

$$P_\alpha(l\kappa) = \frac{1}{i}\left(\frac{\hbar M_{l\kappa}}{2}\right)^{1/2} \sum_s (\omega_s)^{1/2} B_\alpha^s(l\kappa)(b_s - b_s^\dagger) \tag{2.10b}$$

The inverse transformations are readily found to be

$$b_s = \left(\frac{1}{2\hbar}\right)^{1/2} \sum_{l\kappa\alpha} B_\alpha^{(s)}(lk)\left\{(M_{l\kappa}\omega_s)^{1/2}U_\alpha(l\kappa) + \frac{iP_\alpha(l\kappa)}{(M_{l\kappa}\omega_s)^{1/2}}\right\} \tag{2.11a}$$

$$b_s^\dagger = \left(\frac{1}{2\hbar}\right)^{1/2} \sum_{l\kappa\alpha} B_\alpha^{(s)}(l\kappa)\left\{(M_{l\kappa}\omega_s)^{1/2}U_\alpha(l\kappa) - \frac{iP_\alpha(l\kappa)}{(M_{l\kappa}\omega_s)^{1/2}}\right\} \tag{2.11b}$$

With the aid of these transformations and the commutation rules

$$[U_\alpha(l\kappa), P_\beta(l'\kappa')] = i\hbar\delta_{ll'}\delta_{\kappa\kappa'}\delta_{\alpha\beta}$$
$$[U_\alpha(l\kappa), U_\beta(l'\kappa')] = [P_\alpha(l\kappa), P_\beta(l'\kappa')] = 0 \tag{2.12}$$

it is found that the new operators satisfy the simple commutation rules

$$[b_s, b_{s'}^\dagger] = \delta_{ss'}$$
$$[b_s, b_{s'}] = [b_s^\dagger, b_{s'}^\dagger] = 0 \tag{2.13}$$

and that the Hamiltonian $H_0$ takes the form

$$H_0 = \sum_{s=1}^{3\tau N} \hbar\omega_s\left(b_s^\dagger b_s + \frac{1}{2}\right) \tag{2.14}$$

From the commutation rules (2.13) it follows directly that the operators $b_s^\dagger b_s$ and $b_{s'}^\dagger b_{s'}$ commute regardless of whether $s = s'$ or not. Consequently, we can find a representation in which all of the operators

$$H_s = \hbar\omega_s(b_s^\dagger b_s + \tfrac{1}{2}) \qquad s = 1, 2, \cdots, 3\tau N \tag{2.15}$$

are simultaneously diagonal. In the representation in which the operator $b_s^\dagger b_s$ is diagonal, it is found that the eigenvalues of this operator are non-negative integers ([9]):

$$b_s^\dagger b_s |n_s\rangle = n_s |n_s\rangle \qquad n_s = 0, 1, 2, \cdots \tag{2.16}$$

It follows from this result and the commutation rules (2.13) that the non-zero matrix elements of the operators $b_s^\dagger$ and $b_s$ are

$$\langle n_s + 1 | b_s^\dagger | n_s\rangle = \langle n_s | b_s | n_s + 1\rangle$$
$$= (n_s + 1)^{1/2} \tag{2.17}$$

The eigenstate $|n_s\rangle$, normalized to unity is therefore

$$|n_s\rangle = \frac{1}{\sqrt{n_s!}}(b_s^\dagger)^{n_s} |0\rangle \tag{2.18}$$

where the ground state wave function $|0\rangle$ is defined by

$$b_s |0\rangle = 0 \tag{2.19}$$

The eigenvalues of the sub-Hamiltonian $H_s$ given by equation (2.16) are found from (2.16) to be

$$H_s |n_s\rangle = (n_s + \tfrac{1}{2})\hbar\omega_s |n_s\rangle \tag{2.20}$$

The eigenstate $|n_s\rangle$ of the sub-Hamiltonian $H_s$ is the wave function of an harmonic oscillator of frequency $\omega_s$ when it has been raised to its $n_s$th excited state, in which it has an energy of $(n_s + \tfrac{1}{2})\hbar\omega_s$. The quantum of energy $\hbar\omega_s$ by which the energy of the $s$th oscillator comprising the Hamiltonian $H_0$ given by equation (2.14) is increased or decreased as we go from one excited state to the next higher or next lower state is called a *phonon*. From equations (2.17) we see that the operator $b_s^\dagger$ applied to an eigenstate $|n_s\rangle$, which describes the $s$th oscillator in an excited state in which it contains $n_s$

phonons, yields an eigenstate $|n_s + 1\rangle$ which contains $n_s + 1$ phonons. The operator $b_s^\dagger$ is therefore called a phonon creation operator. Similarly, the operator $b_s$ applied to an eigenstate containing $n_s$ phonons yields an eigenstate containing $n_s - 1$ phonons and is called a phonon destruction operator. Finally, we find from equations (2.13) and (2.14) that the equation of motion of the operator $b_s^\dagger$ is

$$i\hbar \dot{b}_s^\dagger = [b_s^\dagger, H_0] = -\hbar\omega_s b_s^\dagger \qquad (2.21)$$

that is, the phonon creation operator $b_s^\dagger$ oscillates with a frequency $\omega_s$ which is just the excitation energy (divided by $\hbar$) of the elementary excitation which it creates. We will use this result several times in what follows.

We have not assumed in the preceding discussion that the crystal is perfectly periodic. The results obtained therefore apply equally well to a molecule or to a crystal containing an impurity atom. Although in much of what follows we will assume that we are dealing with a perfect (periodic) crystal, this condition was relaxed in the present discussion to bring out the point that the existence of elementary excitations does not rest on the assumption of lattice periodicity. To be sure, the results we have obtained are considerably simplified if the crystal is periodic. But the possibility of the quantization of vibrational energy, of the existence of phonons, ultimately has its origin in the fact that in the harmonic approximation the kinetic and potential energies of the atoms in a crystal are positive-definite quadratic forms, and hence can be simultaneously diagonalized. The same result will be found to hold in one form or another for all of the elementary excitation considered in this lecture.

If a crystal has perfect periodicity, the atomic masses must be independent of the unit cell index $l$, and the atomic force constants $\Phi_{\alpha\beta}(l\kappa; l'\kappa')$ can depend on the cell indices $l$ and $l'$ only through their difference. The expression for the displacement amplitude $U_\alpha(l\kappa)$ in such a crystal now must have the form

$$U_\alpha(l\kappa) = \frac{e_\alpha(\kappa)}{\sqrt{M_\kappa}} e^{i\mathbf{k}\cdot\mathbf{x}(l) - i\omega t} \qquad (2.22)$$

where $\mathbf{x}(l)$ is the position vector of the origin of the $l$th unit cell This form is dictated by the fact that inasmuch as the Hamiltonian $H_0$ commutes with the operation of dispalcing the crystal rigidly through a lattice translation vector, the displacement $U_\alpha(l\kappa)$ must be simultaneously an eigenfunction of both $H_0$ and of the translation operator. If the cyclic boundary condition is adopted for the displacement amplitudes, the wavevector $\mathbf{k}$ in equation (2.22) can take on only certain discrete values, which are uniformly distributed throughout the first Brillouin zone of the crystal with a density $\Omega/(2\pi)^3$, where $\Omega$ is the volume of the crystal.

The eigenvalue equation determining the relation between the frequency $\omega$ and the wavevector $\mathbf{k}$ is found by substitution of equation (2.22) into (2.4) to be

$$\sum_{\kappa'\beta} D_{\alpha\beta}(\kappa\kappa'\,|\,\mathbf{k})e_\beta(\kappa') = \omega^2 e_\alpha(\kappa) \qquad (2.23)$$

where the matrix $\mathbf{D}(\mathbf{k})$ is a $3\tau \times 3\tau$ Hermitian matrix whose elements are given by

$$D_{\alpha\beta}(\kappa\kappa'\,|\,\mathbf{k}) = \frac{1}{(M_\kappa M_{\kappa'})^{1/2}} \sum_{l'} \Phi_{\alpha\beta}(l\kappa;\,l'\kappa')e^{-i\mathbf{k}\cdot(\mathbf{x}(l)-\mathbf{x}(l'))} \qquad (2.24)$$

For each value of $\mathbf{k}$ equation (2.23) has $3\tau$ solutions which we label by an index $j(=1, 2, \ldots, 3\tau)$. If we make explicit the dependence of the solutions of equation (2.23) on the index $j$ and on the wavevector $\mathbf{k}$, we can rewrite it as

$$\sum_{\kappa'\beta} D_{\alpha\beta}(\kappa\kappa'\,|\,\mathbf{k})e_\beta(\kappa'\,|\,\mathbf{k}j) = \omega_j^2(\mathbf{k})e_\alpha(\kappa\,|\,\mathbf{k}j) \qquad (2.23a)$$

The fact that the eigenfrequency equation for a periodic crystal is a matrix equation of dimension $3\tau$ rather than $3\tau N$ is the chief simplification that lattice periodicity introduces into the dynamical theory of crystals. The eigenvectors $\{e_\alpha(\kappa\,|\,\mathbf{k}j)\}$ describe the polarizations of the $3\tau$ normal modes of vibration associated with a given wavevector $\mathbf{k}$. They can also be made orthonormal and complete:

$$\sum_{\kappa\alpha} e_\alpha^*(\kappa\,|\,\mathbf{k}j)e_\alpha(\kappa\,|\,\mathbf{k}j') = \delta_{jj'}$$
$$\sum_j e_\alpha^*(\kappa\,|\,\mathbf{k}j)e_\beta(\kappa'\,|\,\mathbf{k}j) = \delta_{\kappa\kappa'}\delta_{\alpha\beta} \qquad (2.24)$$

The normal coordinate transformation which is the analog of equation (2.10) takes the form

$$U_\alpha(l\kappa) = \left(\frac{\hbar}{2NM_\kappa}\right)^{1/2} \sum_{\mathbf{k}j} \frac{e_\alpha(\kappa\,|\,\mathbf{k}j)}{[\omega_j(\mathbf{k})]^{1/2}} e^{i\mathbf{k}\cdot\mathbf{x}(l)}(b_{\mathbf{k}j} + b_{\mathbf{k}j}^\dagger) \qquad (2.25a)$$

$$P_\alpha(l\kappa) = \frac{1}{i}\left(\frac{\hbar M_\kappa}{2N}\right)^{1/2} \sum_{\mathbf{k}j} [\omega_j(\mathbf{k})]^{1/2}e_\alpha(\kappa\,|\,\mathbf{k}j)e^{i\mathbf{k}\cdot\mathbf{x}(l)}(b_{\mathbf{k}j} - b_{\mathbf{k}j}^\dagger) \qquad (2.25b)$$

where we have adopted the phase convention

$$e_\alpha(\kappa\,|\,-\mathbf{k}j) = e_\alpha^*(\kappa\,|\,\mathbf{k}j) \qquad (2.26)$$

The commutation rules for the operators $b_{\mathbf{k}j}^\dagger$ and $b_{\mathbf{k}j}$ are

$$[b_{\mathbf{k}j}, b^{\dagger}_{\mathbf{k}'j'}] = \Delta(\mathbf{k} - \mathbf{k}')\delta_{jj'}$$
$$[b_{\mathbf{k}j}, b_{\mathbf{k}'j'}] = [b^{\dagger}_{\mathbf{k}j}, b^{\dagger}_{\mathbf{k}'j'}] = 0 \tag{2.27}$$

where $\Delta(\mathbf{k})$ equals unity if $\mathbf{k}$ vanishes (modulo $2\pi$ times a reciprocal lattice vector) and is zero otherwise, and the transformed Hamiltonian is

$$H_0 = \sum_{\mathbf{k}j} \hbar\omega_j(\mathbf{k})\left[b^{\dagger}_{\mathbf{k}j}, b_{\mathbf{k}j} + \frac{1}{2}\right] \tag{2.28}$$

There is one feature of the solutions of equation (2.8) which deserves comment. The right-hand side of (2.4) gives the force on atom $(l\kappa)$ when it and all of the other atoms in the crystal are displaced arbitrarily from their equilibrium positions. This force must vanish if all the atoms are given the same displacement, because this set of displacements merely describes a rigid body displacement of the crystal. Consequently, the atomic force constants $\{\Phi_{\alpha\beta}(l\kappa; l'\kappa')\}$ must satisfy the condition:

$$\sum_{l'\kappa'} \Phi_{\alpha\beta}(l\kappa; l'\kappa') = 0 = \sum_{l\kappa} \Phi_{\alpha\beta}(l\kappa; l'\kappa') \tag{2.29}$$

Combining equations (2.7), (2.8), and (2.29) we see that there are three solutions of the eigenvalue equation (2.8) which have zero frequency, viz.,

$$B^{(s)}_{\alpha}(l\kappa) = \left(\frac{M_{l\kappa}}{M_T}\right)^{1/2} d^{(s)}_{\alpha} \tag{2.30}$$

where $\mathbf{d}^{(1)}, \mathbf{d}^{(2)}, \mathbf{d}^{(3)}$ are any three mutually perpendicular unit vectors, and $M_T$ is the total mass of the crystal. These solutions obviously describe a rigid body displacement of the entire crystal. The same result is found for the normal-mode frequencies of a perfect, periodic crystal, in which case it takes the form

$$\omega_j(\mathbf{0}) - 0 \quad j - 1, 2, 3 \tag{2.31}$$

where $j = 1, 2, 3$ are the labels of the three modes of lowest frequency for each value of $\mathbf{k}$. In this case a little more effort yields the result that the frequency given by equation (2.31) is not that of an isolated mode but is part of a continuum, in the sense that

$$\lim_{\mathbf{k}\to 0} \omega_j(\mathbf{k}) = c_j\left(\frac{\mathbf{k}}{k}\right)k \quad j = 1, 2, 3 \tag{2.32}$$

Similarly, the eigenvectors $e_{\alpha}(\kappa\,|\,\mathbf{k}j)/\sqrt{M_{\kappa}}$ for $j = 1, 2, 3$ in the limit as $\mathbf{k} \to 0$, become independent of $\mathbf{k}$ and describe atomic motions which closely resemble a uniform translation of the crystal.

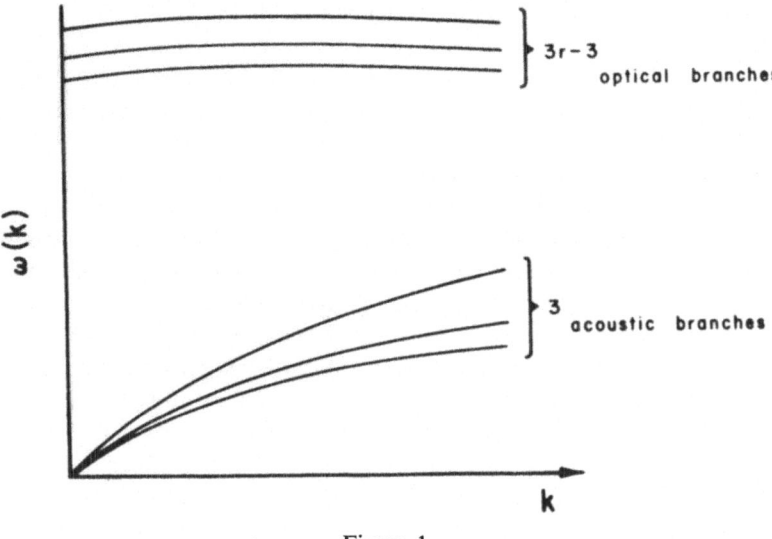

Figure 1

The branches of the multivalued function $\omega(\mathbf{k})$ labeled by $j = 1, 2, 3$ are called *accoustic*, because it is the long-wavelength limits of these branches which are obtained when the discrete crystal is assimilated into an elastic continuum and the equations of motion of the latter are solved.

The remaining $3\tau - 3$ branches of the function $\omega(\mathbf{k})$ are characterized by the fact that for these branches $\omega_j(\mathbf{k})$ does not vanish as $\mathbf{k} \to 0$, but has a nonzero limit, and by the fact that the corresponding eigenvectors $\{e_\alpha(\kappa \,|\, kj)\}$ describe atomic motions in which the sublattices comprising the crystal vibrate rigidly against each other. These branches are called *optical* branches because if the atoms comprising the crystal possess charges, the atomic motions in these modes impart a macroscopic, fluctuating dipole moment to the crystal, which can interact with an incident electromagnetic wave, viz., light. The relation between the frequency of a normal mode and its wavevector, called a *dispersion curve*, therefore takes the form shown schematically in Fig. 1.

Finally, it should be pointed out that the assumption of $H_0$ as the unperturbed Hamiltonian does not mean that we are neglecting interactions between the atoms comprising the crystal. Atomic interactions are taken into account in $H_0$ by the fact that the site index $(l\kappa)$ is not required always to be equal to $(l'\kappa')$. We could rewrite $H_0$ in the form

$$H_0 = \sum_{l\kappa\alpha} \left\{ \frac{P_\alpha^2(l\kappa)}{2M_\kappa} + \tfrac{1}{2}\Phi_{\alpha\alpha}(l\kappa;\, l\kappa)U_\alpha^2(l\kappa) \right\}$$
$$+ \tfrac{1}{2}\sum_{l\kappa\alpha}\sum_{l'\kappa'\beta}{}' \Phi_{\alpha\beta}(l\kappa;\, l'\kappa')U_\alpha(l\kappa)U_\beta(l'\kappa') \tag{2.33}$$

where the prime on the second sum means that the diagonal terms are to be omitted. In this form, the first sum is clearly the Hamiltonian of noninteracting atoms, each of which can be regarded as a three-dimensional Einstein oscillator with frequencies of vibration along the coordinate axes of $\omega^2 = \Phi_{\alpha\alpha}(l\kappa; l\kappa)/M_\kappa$ [for the atom $(l\kappa)$]. The second term on the right-hand side of (2.4) describes the interactions between these independent oscillators. It is an interesting exercise to show that by treating the interaction terms on the right-hand side of equation (2.4) as a perturbation on the Hamiltonian of the independent, Einstein oscillators described by the first term, one obtains all the results that follow from a direct diagonalization of $H_0$ as given by (2.3). Such a demonstration has been given by Mattuck[10] by the use of many-body perturbation theory, but we will not discuss his work further here.

We have here an example where a diagonalization of a Hamiltonian has been carried out by simply omitting the terms which represent the interactions between the constituent particles. However, this is clearly an unsatisfactory simplification of the original Hamiltonian in that the omitted terms have a profound effect on the vibration frequencies obtained from the first line on the right-hand side of (2.4), spreading them from a few discrete values into an essentially continuous band which extends from zero to some finite upper limit.

The Hamiltonian $H_0$ which describes the atomic vibrations in the harmonic approximation, therefore already describes an interacting many-particle system. It is the elementary excitations, the phonons, which make their appearance following a principal axis transformation, which are the noninteracting entities in the harmonic approximation, and which are coupled by the anharmonic terms in the crystal potential energy.

## III. SPIN WAVES

Of all of the various types of elementary excitations in solids, magnons are perhaps the ones which resemble phonons most closely. As is also the case for atomic displacements, the classical and quantum mechanical equations of motion for the spin vectors at the atomic sites of a magnetic crystal are the same. In addition, as we will see below, the matrix elements of the spin operators in a particular representation closely resemble the matrix elements of the phonon creation and destruction operators given by equations (2.16) and (2.17). By exploiting this resemblance we can effect the approximate diagonalization of the Hamiltonian of the spin system and obtain the elementary excitations of the system, which are called magnons. The starting point for our discussion of spin waves is the Heisenberg Hamiltonian for a system of spins localized on the lattice sites of a primitive lattice

$$\mathscr{H} = -Hg\beta \sum_{l} S_{lz} - \tfrac{1}{2}\sum_{ll'}{}' 2J(ll')\mathbf{S}_l\cdot\mathbf{S}_{l'} \tag{3.1}$$

In this expression $\mathbf{S}_l$ is the spin vector of atom $l$ measured in units of $\hbar$. It is assumed that every atom has the same spin quantum number $S$. The first term in the Hamiltonian (3.1) is the Zeeman energy of the spins in a magnetic field of magnitude $H$ which is assumed to be directed along the z-axis. In this term $\beta$ is the Bohr magneton, $e\hbar/2mc$, and $g$ is the spectroscopic splitting factor, which is approximately equal to 2. The second term is the exchange energy. The coefficient $J(ll')$ is the exchange integral between the atom $l$ and $l'$, and the prime on the sum means that the terms with $l = l'$ are to be omitted. We assume that the crystal is perfectly periodic, so that $J(ll')$ depends on $l$ and $l'$ only through their difference.

The z-component of an atomic spin $\mathbf{S}$ has eigenvalues $m$, where $m$ is an integer which takes the values $-S, -S+1, \ldots, S-1, S$. In their fundamental paper on spin waves Holstein and Primakoff [11] introduced the spin deviation quantum number $n = S - m$, which measures the deviation of $S_z$ from its maximum value $S$. The allowed values of $n$ are $0, 1, 2, \ldots, 2S$, and they are the eigenvalues of the spin deviation operator $\hat{n} = S - S_z$. With the use of the commutation relations obeyed by $S_x, S_y, S_z$, namely,

$$[S_x, S_y] = iS_z \qquad \text{and permutations} \tag{3.2}$$

it is found that in a representation in which $\hat{n}$ is diagonal the nonzero matrix elements of $S_x$, $S_y$, and $S_z$ are given by

$$\langle n | S_x | n+1 \rangle = \langle n+1 | S_x | n \rangle^* = \tfrac{1}{2}(n+1)^{1/2}(2S-n)^{1/2}$$
$$\langle n | S_y | n+1 \rangle = \langle n+1 | S_y | n \rangle^* = -\tfrac{1}{2}i(n+1)^{1/2}(2S-n)^{1/2}$$
$$\langle n | S_z | n \rangle = S - n \tag{3.3}$$

The matrices of the components of $\mathbf{S}_l$ have just these forms with respect to $n_l$, but are diagonal with respect to the $n_{l'}$ of all the other atoms of the crystal.

The matrix elements given by equation (3.3) have a vague similarity to the matrix elements of the phonon creation and destruction operators $b^\dagger$ and $b$ given by equations (2.16) and (2.17). We can make the similarity a good deal closer by introducing the new operators $a$ and $a^\dagger$ by

$$a = (S_x + iS_y)\,|\,(2S)^{1/2}$$
$$a^+ = (S_x - iS_y)\,|\,(2S)^{1/2} \tag{3.4}$$

The nonzero matrix elements of these operators are

$$\langle n\,|\,a\,|\,n+1\rangle = \langle n+1\,|\,a^+\,|\,n\rangle = (n+1)^{1/2}\left(1-\frac{n}{2S}\right)^{1/2} \qquad (3.5a)$$

together with

$$\langle n\,|\,\hat{n}\,|\,n\rangle = n \qquad (3.5b)$$

As is also true of the operators $b$ and $b^+$, the only nonzero matrix elements of the operators $a$ and $a^+$ connect states differing only by one quantum number, and the values of the matrix elements are nearly the same. Similarly, the operator $\hat{n}$ is analogous to the operator $b^+b$. However, there are several important differences between the phonon operators and the operators $a$, $a^+$, and $\hat{n}$. The first is the obvious and trivial one that whereas the latter operators are associated with the individual lattice sites of the crystal, the phonon operators appearing in equations (2.16) and (2.17) are associated with a given vibration mode. They are therefore transforms of each other. Second, while the phonon matrices (2.16) and (2.17) are of infinite dimension, the matrices (3.5) are only $(2S + 1)$ dimensional. That is to say, the matrix elements (3.5) corresponding to $n > 2S$ are identically zero. Finally, the factor $(1 - n/2S)^{1/2}$ in the spin matrix is replaced by unity in the phonon matrix elements. The matrix elements (3.5a) and (3.5b) are exactly the same as the matrix elements (2.17) for $n = 0$ and for all values of $S$. They are approximately equal for other values of $n$, provided that $n$ is small compared to $2S$. Thus the operators $a$ and $a^+$ cannot be identified with the operators $b$ and $b^+$, respectively. In fact the relations between these two sets of operators was shown by Holstein and Primakoff to be

$$a = \left(1 - \frac{b^+b}{2S}\right)^{1/2} b$$

$$a^+ = b^+\left(1 - \frac{b^+b}{2S}\right)^{1/2}$$

$$\hat{n} = b^+b \qquad (3.6)$$

provided that we never apply the operators on the right-hand sides of these expressions to eigenstates $|n\rangle$ which contain more than $2S$ spin deviations. In a sense we can do somewhat better in approximating the spin operators $a$ and $a^+$ by combinations of the phonon operators $b$ and $b^+$ than we have done in writing equations (3.6) and (11a). In writing the latter equations we have ensured that the matrix elements of $a$, $a^+$, and $\hat{n}$ are correct when evaluated between states for which $n \leq 2S$, but we have not imposed the condition that the matrix elements vanish identically between states for which $n > 2S$. If for the time being we restrict ourselves to the case that $S = \frac{1}{2}$, we see that the simple approximations

$$a = b \qquad a^+ = b^+ \qquad \hat{n} = b^+b \qquad (3.7)$$

yield the correct matrix elements between states for which $n = 0$ and $n = 1$. The matrix elements between the states for which $n = 0, 1, 2$ will be given correctly if we go to the second approximation and define

$$a = b - b^+bb \qquad a^+ = b^+ - b^+b^+b \qquad \hat{n} = b^+b - b^+b^+bb \qquad (3.8)$$

Presumably one can continue this procedure to yield approximations for $a$, $a^+$, and $\hat{n}$ which yield the correct matrix elements between states for which $n = 0, 1, 2, \ldots, N$ for any given $N$. However, it is not known whether this approximation procedure converges in the limit as $N \rightarrow \infty$. Inasmuch as such approximation schemes lie outside the conventional spin wave approximation, we will not pursue them further here.

The conventional spin wave approximation consists of (i) making the approximations (3.7), or, what is squivalent, the approximations

$$S_x = \left(\frac{S}{2}\right)^{1/2}(b + b^+)$$

$$S_y = \frac{1}{i}\left(\frac{S}{2}\right)^{1/2}(b - b^+)$$

$$S_z = S - b^+b \qquad (3.9)$$

From the transformation inverse to that given by equation (3.9),

$$b = \left(\frac{1}{2S}\right)^{1/2}(S_x + iS_y) \qquad b^+ = \left(\frac{1}{2S}\right)^{1/2}(S_x - iS_y) \qquad (3.10)$$

and the commutation rules (3.2), we find that the operators $b$ and $b^+$ satisfy the commutation rules

$$[b, b^+] = 1 - \frac{n}{S}; \qquad [b, b] = [b^+, b^+] = 0 \qquad (3.11)$$

which are just the phonon commutation rules when $n \ll 2S$, as they should be; (ii) ignoring the restriction to states for which $n \ll 2S$; and (iii) neglecting all terms in the Hamiltonian which are of higher order in the phonon operators than quadratic. With these transformations and approximations the Hamiltonian (3.1) becomes

$$\mathcal{H}_0 = -Hg\beta \sum_i (S - b_i^+ b_i)$$

$$-\tfrac{1}{2}\sum_{ij}' 2J(ij)\left\{\frac{S}{2}(b_i + b_i^+)(b_j + b_j^+)\right.$$

$$\left. -\frac{S}{2}(b_i - b_i^+)(b_j - b_j^+) + S^2 - Sb_i^+ b_i - Sb_j^+ b_j\right\} \qquad (3.12)$$

$$= E_0 - Hg\beta \sum_l b_l^+ b_l - 2S \sum_{ll'} \mathcal{J}_{ll'} b_l^+ b_{l'}$$

where

$$E_0 = -NHg\beta S - S^2 \sum_{ll'}' J(ll') \qquad (3.13)$$

and

$$\mathcal{J}_{ll'} = J(ll') \qquad\qquad l \neq l'$$

$$= -\sum_{l'}' J(ll') \qquad l = l' \qquad (3.14)$$

The terms omitted in going from the Hamiltonian (3.1) to the Hamiltonian (3.12) are anharmonic, i.e., they are trilinear and quadrilinear in the operators $b_l$ and $b_l^+$. The energy $E_0$ is the energy of the completely saturated state in which all the spins are parallel to the external field. The approximations made in writing equation (3.12) can be justified on the assumption that we are working in a temperature range (i.e., at low temperatures) in which few spin deviations have been excited, so that the expectation value $\langle b_l^+ b_l \rangle$ of the number of spin deviations on the atom $l$ is small compared with $2S$. Then the replacement of $(1 - b^+ b/2S)^{1/2}$ by 1 is reasonable because the expectation value of $(1 - b^+ b/2S)^{1/2}$ is approximately equal to $(1 - \langle b^+ b \rangle/2S)^{1/2} \cong 1$. At low temperatures only the low-lying excited states of the system contribute significantly to its thermodynamic functions because the excited states of high energy are damped out rapidly by the statistical weighting factor, which is proportional to $\exp(-E/k_B T)$, where $E$ is the energy of the state, $k_B$ is Boltzmann's constant, and $T$ is the absolute temperature. Thus we make little error in the computation of low-temperature thermodynamic properties of ferromagnets by dropping the restriction $n \leq 2S$ when carrying them out, since the contributions to these functions from states with $n \gtrsim 2S$ is negligible. The neglect of terms in the Hamiltonian which are of higher order in the operators $b_l$ and $b_l^+$ than quadratic is also justified when $\langle b_l^+ b_l \rangle$ is small compared with $2S$. Terms of the type $b_l^+ b_l b_{l'}^+ b_{l'}$, have an expectation value $\langle b_l^+ b_l \rangle \langle b_{l'}^+ b_{l'} \rangle$ (in the absence of correlation between spin deviations situated on different atoms) which is smaller than that of $2S b_l^+ b_l$ by a factor $\langle b_{l'}^+ b_{l'} \rangle/2S \ll 1$. Finally terms like $(2S)^{1/2} b_l^+ b_l b_{l'}^+$, which cause the system to make transitions between states of different total spin, are rarer than terms of the type $2Sb_i^+ b_{i'}^+$, which have the same character, by a factor $\langle b_i^+ b_i \rangle / (2S)^{1/2} \ll 1$.

To complete the diagonalization of the Hamiltonian $H_0$ we carry out the normal coordinate transformations

$$b_l = \frac{1}{(N)^{1/2}} \sum_{\mathbf{k}} e^{-i\mathbf{k}\cdot\mathbf{x}(l)} b_{\mathbf{k}} \tag{3.15a}$$

$$b_l^+ = \frac{1}{(N)^{1/2}} \sum_{\mathbf{k}} e^{i\mathbf{k}\cdot\mathbf{x}(l)} b_{\mathbf{k}}^+ \tag{3.15b}$$

The $N$ allowed values of the wavevector $\mathbf{k}$ are determined by the cyclic boundary condition, and they are uniformly distributed throughout the first Brillouin zone of the crystal with a density $\Omega/(2\pi)^3$, where $\Omega$ is the periodicity volume.

The commutation rules (3.11) satisfied by the operators $\{b_l\}$ and $\{b_l^+\}$ impose the following commutation rules on the operators $\{b_{\mathbf{k}}\}$ and $\{b_{\mathbf{k}}^+\}$:

$$[b_{\mathbf{k}}, b_{\mathbf{k}'}^+] = \Delta(\mathbf{k} - \mathbf{k}')$$
$$[b_{\mathbf{k}}, b_{\mathbf{k}'}] = [b_{\mathbf{k}}^+, b_{\mathbf{k}'}^+] = 0 \tag{3.16}$$

so that the latter are just the analogs of the phonon creation and destruction operators for a perfect periodic crystal given by equation (2.25). In terms of the new operators $\mathcal{H}_0$ takes the form

$$\mathcal{H}_0 = E_0 + \sum_{\mathbf{k}} \hbar\Omega_{\mathbf{k}} b_{\mathbf{k}}^+ b_{\mathbf{k}} \tag{3.17}$$

where

$$\hbar\Omega_{\mathbf{k}} = Hg\beta - 2S \sum_l \mathcal{J}(\bar{l}) e^{i\mathbf{k}\cdot\mathbf{x}(l)} \tag{3.18}$$

and $\bar{l} = l - l'$.

Thus we have succeeded in approximating the Hamiltonian of a system of interacting spins by a Hamiltonian which is the sum of the Hamiltonians of independent harmonic oscillators. The consecutive energy levels of the oscillator labeled by the wavevector $\mathbf{k}$ are separated by the quantum of energy $\hbar\Omega_{\mathbf{k}}$ given by equation (3.18), and this quantum is called a *magnon*. If we invoke the assumption that $\mathcal{G}(\bar{l})$ is nonzero only if $\bar{l}$ connects two nearest-neighbor atoms, we obtain for $\hbar\Omega_{\mathbf{k}}$ the expression

$$\hbar\Omega_{\mathbf{k}} = Hg\beta + 2SJ[z - \sum_l \cos \mathbf{k}\cdot\mathbf{x}(\bar{l})] \tag{3.19}$$

where $z$ is the number of nearest neighbors of any atom, and we have used the fact that every atom is at a center of inversion symmetry in primitive crystals.

In the limit as $\mathbf{k} \longrightarrow 0$, $\hbar\Omega_k$ has the expansion

$$\hbar\Omega_k = Hg\beta + 2SJ[z - \sum_l (1 - \tfrac{1}{2}(\mathbf{k}\cdot\mathbf{l})^2 + \cdots)]$$

$$= Hg\beta + SJ \sum_l (\mathbf{k}\cdot\mathbf{l})^2 + O(\mathbf{k}^4) \qquad (3.20)$$

Thus, unlike the dispersion relation for the frequencies of acoustic phonons, which says that these frequencies vary linearly with $\mathbf{k}$ in the limit as $\mathbf{k} \longrightarrow 0$, the frequencies of magnons are quadratic functions of $\mathbf{k}$ in this limit.

One can now proceed to calculate various thermodynamic properties of the ferromagnet described by the Hamiltonian (3.17) as if we were dealing with a gas of noninteracting bosons.

## IV. ELECTRONS AND PLASMONS

The Hamiltonian of $N$ interacting electrons in a uniform, compensating, background of positive charge is given by the familiar expression

$$\mathscr{H} = \sum_i^N \frac{P_i^2}{2m} + \frac{1}{2} \sum_{ij}^{N}{}' \frac{e^2}{|\mathbf{x}_i - \mathbf{x}_j|} \qquad (4.1)$$

The first term is the kinetic energy of the electrons, and $m$ is the electron mass. The second term is the potential energy of interaction of the electrons arising from their mutual Coulomb repulsions; $e$ is the magnitude of the electronic charge, and the prime on the sum means that the terms with $i = j$ are to be omitted. The condition of charge neutrality is most readily expressed by Fourier-transforming the potential energy and omitting the $\mathbf{k} = 0$ Fourier component of the resulting expression

$$\mathscr{H} = \sum_i^N \frac{P_i^2}{2m} + \frac{1}{2}e^2 \sum_{ij}^{N}{}' \frac{1}{\Omega} \sum_{\mathbf{k}} e^{i\mathbf{k}\cdot(\mathbf{x}_i - \mathbf{x}_j)}\frac{4\pi}{k^2}$$

$$= \sum_i^N \frac{P_i^2}{2m} + \frac{1}{2} \sum_{\mathbf{k}} \frac{4\pi e^2}{\Omega k^2}\{\sum_{ij} e^{i\mathbf{k}\cdot\mathbf{x}_i}e^{-i\mathbf{k}\cdot\mathbf{k}_j} - N\}$$

$$= \sum_i^N \frac{P_i^2}{2m} + \sum_{\mathbf{k}}{}' \frac{2\pi e^2}{k^2}\{\rho_{\mathbf{k}}^+\rho_{\mathbf{k}} - N\} \qquad (4.2)$$

where the normalization volume $\Omega$ has been chosen to be unity for convenience, and

$$\rho_{\mathbf{k}} = \sum_i e^{-i\mathbf{k}\cdot\mathbf{x}_i} \qquad (4.3)$$

is the kth Fourier coefficient of the departure of the electron density from its average value $\rho_0 = N/\Omega = N$.

You are all probably familiar with the results that if we neglect the Coulomb repulsion among the electrons the eigenfunctions of the resulting Hamiltonian

$$\mathscr{H}_0 = \sum_i^N \frac{P_i^2}{2m} \tag{4.4}$$

are

$$\psi_k(\mathbf{x}) = \frac{1}{\Omega^{1/2}} e^{i\mathbf{k}\cdot\mathbf{x}} \tag{4.5}$$

and the corresponding energy eigenvalues are

$$\epsilon(\mathbf{k}) = \frac{\hbar^2\kappa^2}{2m} \tag{4.6}$$

If the cyclic boundary condition has been imposed on the wavefunction $\psi_k(\mathbf{x})$, then the allowed values of $\mathbf{k}$ are

$$k_x = \frac{2\pi n_x}{L_x} \quad k_y = \frac{2\pi n_y}{L_y} \quad k_z = \frac{2\pi n_z}{L_z} \quad n_x, n_y, n_z = 0, \pm 1, \pm 2, \cdots \tag{4.7}$$

where $L_x, L_y, L_z$ are the lengths of the edges of the rectangular parallelopiped of volume $L_x L_y L_z = \Omega$ which is the periodicity and normalization volume for the electron gas.

Each triplet of integers $n_x, n_y, n_z$ specifies an energy eigenstate of the noninteracting electron gas. The Pauli exclusion principle tells us that no more than two electrons, one with spin up and the other with spin down, can occupy each of these states. The ground state of the free-electron gas therefore is obtained by placing two electrons in each of the $N/2$ lowest energy levels given by equation (4.6). In momentum space the occupied states comprise a sphere (the Fermi sphere), whose radius $k_F$ is determined by the condition that it contain the $N/2$ wave vectors closest to the origin:

$$N = \sum_{\kappa < \kappa_F} 2 = \frac{\Omega}{(2\pi)^3} 2 \left( \frac{4\pi}{3} k_F^3 \right) \tag{4.8}$$

so that

$$k_F = (3\pi^2 N)^{1/3} \tag{4.9}$$

The excited states of this system are states in which one or more of the electrons has been removed from the Fermi sphere, leaving a hole behind. The energy required to create an excitation in which an electron of momentum $\hbar\mathbf{p}$ is excited out of the Fermi sea into a state with momentum $\hbar(\mathbf{p}+\mathbf{k})$ is

$$\hbar\omega(\mathbf{k}) = \frac{\hbar^2(\mathbf{p}+\mathbf{k})^2}{2m} - \frac{\hbar^2 p^2}{2m} = \hbar^2\left(\frac{\mathbf{k}\cdot\mathbf{p}}{m} + \frac{k^2}{2m}\right) \tag{4.10}$$

We can say alternatively that this is the energy required to create an electron of momentum $\mathbf{p}+\mathbf{k}$ and a hole of momentum $\mathbf{p}$. Because of the Pauli exclusion principle we must have that $p \le k_F$ and that $|\mathbf{p}+\mathbf{k}| \ge k_F$. The energy spectrum of these electron-hole pairs is continuous and lies between the two curves given by

$$\hbar\omega_{min} = 0 \qquad\qquad k \le 2k_F$$
$$= \hbar^2\left(\frac{k^2}{2m} - \frac{kk_F}{m}\right) \qquad k \ge 2k_F \tag{4.11}$$

and

$$\hbar\omega_{max} = \hbar^2\left(\frac{k^2}{2m} + \frac{kk_F}{m}\right) \qquad k \ge 0 \tag{4.12}$$

as shown in Fig. 2 For a given value of $\mathbf{k}$ these two limiting excitation energies corresponding to the processes depicted in Fig. 3.

The preceding simple picture obtains only as long as we neglect the Coulomb interactions between the electrons. To find what changes are introduced into this picture by the Coulomb interactions it is convenient to rewrite the Hamiltonian for the electron gas in the representation of second quantization as

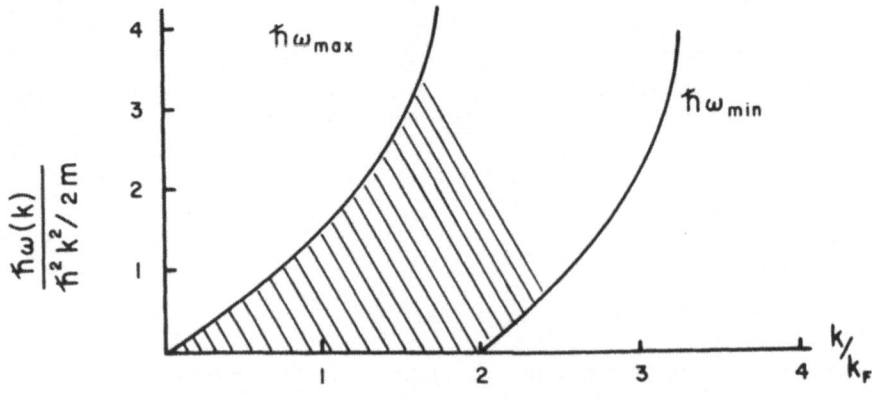

Figure 2

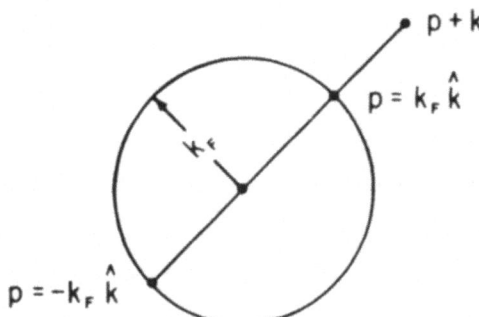

Figure 3

$$\mathcal{H} = \sum_r \epsilon(\mathbf{p}) C_\mathbf{p}^+ C_\mathbf{p} + \sum_{\mathbf{pqk}}{}' \frac{V_\mathbf{k}}{2} C_{\mathbf{p+k}}^+ C_{\mathbf{q-k}}^+ C_\mathbf{q} C_\mathbf{p} \tag{4.13}$$

where

$$\epsilon(\mathbf{p}) = \frac{\hbar^2 p^2}{2m} \tag{4.14a}$$

and

$$V_\mathbf{k} = \frac{4\pi e^2}{\mathbf{k}^2} \tag{4.14b}$$

$C_\mathbf{p}^+$ and $C_\mathbf{p}$ are electron creation and destruction operators, respectively. This expression for the Hamiltonian follows directly from that given by equation (4.2) when the second quantized form for the kth Fourier coefficient of the density fluctuation.

$$\rho_\mathbf{k}^+ = \sum_\mathbf{p} C_{\mathbf{p+k}} C_\mathbf{p} = \rho_{-\mathbf{k}} \tag{4.15}$$

is employed (see, for example, Ref. 5). The ground state $|0\rangle$ of our system is that corresponding to a filled Fermi sphere of radius $k_F$, at the absolute zero of temperature. Therefore, according to the Pauli exclusion principle we must have that

$$C_\mathbf{p}^+ |0\rangle = 0 \qquad p < k_F \tag{4.16a}$$

$$C_\mathbf{p} |0\rangle = 0 \qquad p > k_F \tag{4.16b}$$

These operators obey the anticommutation rules

$$[C_\mathbf{p}, C_{\mathbf{p'}}]_+ = [C_\mathbf{p}^+, C_{\mathbf{p'}}^+]_+ = 0$$

$$[C_\mathbf{p}, C_{\mathbf{p'}}^+]_+ = \delta_{\mathbf{pp'}} \tag{4.17}$$

The expectation value in the ground state of the number operator $C_p^+ C_p$ is

$$n_p = \langle 0 | C_p^+ C_p | 0 \rangle = \begin{cases} 1 \text{ for } p < k_F \\ 0 \text{ for } p > k_F \end{cases} \tag{4.18}$$

The operator which creates an electron of momentum $\mathbf{p} + \mathbf{k}$ and a hole of momentum $\mathbf{p}$ is

$$d_{\mathbf{k}}^+(\mathbf{p}) = C_{\mathbf{p}+\mathbf{k}}^+ C_{\mathbf{p}}$$
$$|\mathbf{p} + \mathbf{k}| > k_F \qquad p < k_F \tag{4.19}$$

The equation of motion of this operator is

$$i\hbar \frac{d}{dt} d_{\mathbf{k}}^+(\mathbf{p}) = [d_{\mathbf{k}}^+(\mathbf{p}), H]$$
$$= [\epsilon(\mathbf{p}) - \epsilon(\mathbf{p} + \mathbf{k})] d_{\mathbf{k}}^+(\mathbf{p})$$
$$+ \sum_{\mathbf{l}} \frac{V_{\mathbf{l}}}{2} \{ [C_{\mathbf{p}+\mathbf{k}}^+ C_{\mathbf{p}+\mathbf{l}} - C_{\mathbf{p}+\mathbf{k}-\mathbf{l}}^+ C_{\mathbf{p}}] \rho_{\mathbf{l}}^+$$
$$+ \rho_{\mathbf{l}}^+ [C_{\mathbf{p}+\mathbf{k}}^+ C_{\mathbf{p}+\mathbf{l}} - C_{\mathbf{p}+\mathbf{k}-\mathbf{l}}^+ C_{\mathbf{p}}] \} \tag{4.20}$$

We could continue by writing down the equations of motion for the quadrilinear operators on the right-hand side of equation (4.20), and then for the new operators to which this step gives rise, and so on. In this way we would generate an infinite hierarchy of successively more complicated combinations of electron creation and destruction operators and their equations of motion. However, we truncate this hierarchy already at this stage, by linearizing equation (4.20). We do this by invoking the random phase approximation ([12]): In determining the motion of an electron—hole pair of momentum $\mathbf{k}$, keep only the interaction term associated with the $k$th momentum transfer. Although this is an approximation, it has been shown that it is exact in the limit as the electron density becomes very large ([13]). With this approximation the equation of motion (4.20) becomes

$$i\hbar \dot{d}_{\mathbf{k}}^+(\mathbf{p}) = -[\epsilon(\mathbf{p} + \mathbf{k}) - \epsilon(\mathbf{p})] d_{\mathbf{k}}^+(\mathbf{p})$$
$$+ V_{\mathbf{k}} [n_{\mathbf{p}+\mathbf{k}} - n_{\mathbf{p}}] \sum_{\mathbf{q}} d_{\mathbf{k}}^+(\mathbf{q}) \tag{4.21}$$

where we have approximated $C_{\mathbf{p}+\mathbf{k}}^+ C_{\mathbf{p}+\mathbf{k}}$ and $C_p^+ C_p$ by their expectation values in the unperturbed ground state. Assuming an harmonic time dependence for $d_{\mathbf{k}}^+(\mathbf{p})$,

$$i\hbar \dot{d}_{\mathbf{k}}^+(\mathbf{p}) = -\hbar \omega_{\mathbf{k}} d_{\mathbf{k}}^+(\mathbf{p}) \tag{4.22}$$

(the sign of the right-hand side is determined by the fact that $d_k^+(\mathbf{p})$ is a creation operator), we obtain as the equation determining $\omega_k$

$$
\begin{aligned}
1 &= V_k \sum_{\mathbf{p}} \frac{n_{\mathbf{p}} - n_{\mathbf{p}+\mathbf{k}}}{\hbar\omega_k - [\epsilon(\mathbf{p}+\mathbf{k}) - \epsilon(\mathbf{p})]} \\
&= \frac{4\pi e^2}{k^2} \sum_{\substack{p<k_F \\ |\mathbf{p}+\mathbf{k}|>k_F}} \left\{ \frac{1}{\hbar\omega_k - [\epsilon(\mathbf{p}+\mathbf{k}) - \epsilon(\mathbf{p})]} - \frac{1}{\hbar\omega_k + [\epsilon(\mathbf{p}+\mathbf{k}) - \epsilon(\mathbf{p})]} \right\} \\
&= \frac{4\pi e^2}{k^2} \sum_{p<k_F} \frac{2[\epsilon(\mathbf{p}+\mathbf{k}) - \epsilon(\mathbf{p})]}{\hbar^2\omega_k^2 - [\epsilon(\mathbf{p}+\mathbf{k}) - \epsilon(\mathbf{p})]^2}
\end{aligned}
\tag{4.23}
$$

In writing the last equation we have dropped the restriction that $|\mathbf{p} + \mathbf{k}| > k_F$ since the terms with $|\mathbf{p} + \mathbf{k}| < k_F$ do not contribute to the sum. If we had neglected the interaction terms on the right-hand side of Eq. (4.21) we would have found that $d_k^+(\mathbf{p})$, i.e., the electron–hole pair creation operator oscillates at a frequency

$$
\hbar\omega_k = \epsilon(\mathbf{p} + \mathbf{k}) - \dot{\epsilon}(\mathbf{p})
\tag{4.24}
$$

This is the free-particle result which we have discussed previously and have depicted in Fig. 2.

In the presence of the Coulomb interactions we find that the oscillation frequencies $\{\omega_k\}$ fall into two radically different categories. In the first category are the analogs of the solutions given by equation (4.24); in the present case they are shifted from the latter values by an amount of $O(\Omega^{-1})$. In the second category are solutions which tend to a finite value as $\mathbf{k} \longrightarrow 0$ and which have no counterpart in the noninteracting system.

The frequencies of these modes are easily obtained by expanding the right-hand side of equation (4.23) in inverse powers of $\omega_k^2$. The result is found to be

$$
\omega_k = \omega_{\text{pl}}\left\{ 1 + \frac{3}{10}\frac{k_F^2}{m^2}\frac{k^2}{\omega_{\text{pl}}^2} + \cdots \right\}
\tag{4.25}
$$

where $\omega_{\text{pl}}$ is the so-called electron plasma frequency,

$$
\omega_{\text{pl}}^2 = \frac{4\pi N e^2}{m}
\tag{4.26}
$$

Some insight into the nature of the excitations whose excitation frequencies are given by Eq. (4.25) can be gained in the following manner. If we take the time derivative of both sides of equation (4.21), and use the fact that the kth Fourier component of the departure of the electron density from its average value is given by

$$
\rho_k^+ = \sum_{\mathbf{p}} d_k^+(\mathbf{p})
\tag{4.27}
$$

we find that the equation of motion of $\rho_k$ can be written in the form

$$-\hbar^2\ddot{\rho}_k^+ = -V_k \sum_p [\epsilon(p+k) - \epsilon(p)][n_{p+k} - n_p]\rho_k^+$$
$$+ \sum_p [\epsilon(p+k) - \epsilon(p)]^2 d_k^+(p) \qquad (4.28)$$

The first term on the right-hand side of this equation can be simplified if we note that the first two terms in the expansion of the first equation on the right hand side of equation (4.23) in inverse powers of $\hbar\omega_k$ are

$$1 = V_k \left\{ \frac{1}{\hbar\omega_k} \sum_p (n_p - n_{p+k}) \right.$$
$$+ \frac{1}{(\hbar\omega_k)^2} \sum_p (n_p - n_{p+k})[\epsilon(p+k) - \epsilon(p)] + \cdots \right\}$$

The first term sums to zero, and we obtain the result that approximately

$$(\hbar\omega_k)^2 \cong (\hbar\omega_{pl})^2 = V_k \sum_p (n_p - n_{p+k})[\epsilon(p+k) - \epsilon(p)] \qquad (4.29)$$

Therefore, equation (4.28) can be rewritten as

$$\ddot{\rho}_k^+ + \omega_{pl}^2 \rho_k^+ = -\frac{1}{\hbar^2} \sum_p [\epsilon(p+k) - \epsilon(p)]^2 d_k^+(p) \qquad (4.30)$$

We see from this result that in the limit of small $k$, $\rho_k^+$ oscillates with a frequency $\omega_{pl}$ provided that the term on the right-hand side of this equation can be neglected. Pines [14] has shown that the condition which must be satisfied in order that the right-hand side of equation (4.30) can be neglected is that

$$\frac{k^2 V_F^2}{\omega_{pl}^2} \ll 1 \qquad (4.31)$$

where $V_F$ is the speed of electrons on the surface of the Fermi sphere. For sufficiently small $k$, this condition will be satisfied. We thus find that the frequency $\omega_k = \omega_{pl}\{1 + O(k^2)\}$, which originally appeared as the frequency of a collective oscillation of the electron–hole pairs which comprise the low-lying elementary excitations of the electron gas, is also the frequency of oscillation of the Fourier coefficients of the fluctuations in the electron density. The quantum of energy $\hbar\omega_k$ required to excite the collective oscillation $\rho_k^+$ is called a *plasmon*. That the electron density can sustain oscillations is not surprising.[15] If there occurs a local deficiency in the number of electrons, the electrons will move into the region of excess positive charge to screen it out. However, in doing so they will generally overshoot their mark, so that they are pulled balck into the region, overshoot again, and so on, in such a way that oscillations in the electron density about the state of charge neutrality are set up.

That the frequency $\omega_k$ associated with a plasmon does not go to zero with vanishing $\mathbf{k}$ is a consequence of the long range of the Coulomb potential [16].

The long-wavelength longitudinal vibrations of the electrons in a plasma oscillation have an electric field associated with them which gives rise to a restoring force acting on the electrons and raises their frequency of oscillation in much the same way as the frequencies of the long-wavelength longitudinal optical modes of an ionic crystal are raised above the frequencies of the transverse optical modes by the electric field associated with the former [17].

## V. EXCITONS*

Excitons are the low-lying excited states of pure insulating crystals. They are linear combinations with suitably adjusted phases of the localized electronic excited states of the individual atoms comprising the crystal in much the same way that phonons are linear combinations with suitably adjusted phases of the displacements from equilibrium of the individual atoms in the crystal.

In this section we present a simple introduction to the theory of excitons in the tight binding approximation. We assume a Bravais crystal made up of $N$ atoms, each of which has a single valence electron. We assume moreover that the electron can exist in only two states: a ground state which is denoted by the index (0), and an excited state which is denoted by an index (1). The Hamiltonian for the crystal can be written in the form

$$H = \sum_{l} H(l) + \tfrac{1}{2} \sum_{ll}' V(ll') \tag{5.1}$$

where the indices $l$ and $l$ label the lattice sites of the crystal, $H(l)$ is the Hamiltonian of the $l$th atom in isolation, and $V(ll')$ is the operator of the Coulomb interaction between the charges on the atoms $l$ and $l'$. The prime on the second sum in equation (5.1) excludes the terms with $l = l'$. The eigenfunctions of the Hamiltonian $H(l)$ will be denoted by $\phi_l^{f_l}$ where the index $f_l(=0,1)$ describes the state of excitation of the $l$th atom:

$$H(l)\phi_l^{f_l} = E_l^{f_l}\phi_l^{f_l} \tag{5.2}$$

If we nelgect the overlap between wavefunctions centered on different atoms $l$ and $l'$, the determinantal wavefunction

---

*The discussion in this section is based largely on a paper by Agranovich [18].

$$\Psi_0 = \frac{1}{\sqrt{N!}} \begin{vmatrix} \phi_1^{(0)}(1) & \phi_1^{(0)}(2) & \cdots & \phi_1^{(0)}(N) \\ \phi_2^{(0)}(1) & \phi_2^{(0)}(2) & \cdots & \phi_2^{(0)}(N) \\ \cdots\cdots\cdots\cdots\cdots\cdots \\ \phi_N^{(0)}(1) & \phi_N^{(0)}(2) & \cdots & \phi_N^{(0)}(N) \end{vmatrix} \tag{5.3}$$

is an antisymmetrized normalized wavefunction for the ground state of the crystal. The ground state energy is then found to be given by

$$\epsilon_0 = \langle \Psi \,|\, H \,|\, \Psi \rangle = \sum_l E_l^{(0)} + \tfrac{1}{2}\sum_{ll'}{}' V_{ll'}(00;00) \tag{5.4}$$

Here, and in all that follows, we use the shorthand notation

$$V_{ll'}(f_1 f_{1'}; f_1' f_{1'}') = \int d\tau_1 \int d\tau_2 \{ \phi_l^{f'}(r_1)^* \phi_{l'}^{f''}(r_2)^* V(ll')$$

$$\times \phi_l^{f'}(r_1) \phi_{l'}^{f''*}(r_2) - \phi_l^{f'}(r_1) \phi_{l'}^{f''}(r_2)^* V(ll') \phi_{l'}^{f''}(r_1) \phi_l^{f'}(r_2) \} \tag{5.5}$$

As the excited states of the crystal we consider those states in which one atom in the crystal is in its excited state, but all the remaining atoms are in their ground states. Again neglecting the overlap of wavefunctions centered on different atoms, we can write the wavefunction of the crystal when the $r$th atom is in its excited states as

$$\Psi_r = \frac{1}{\sqrt{N!}} \begin{vmatrix} \phi_1^{(0)}(1) & \phi_1^{(0)}(2) & \cdots & \phi_1^{(0)}(N) \\ \phi_2^{(0)}(1) & \phi_2^{(0)}(2) & \cdots & \phi_2^{(0)}(N) \\ \cdots\cdots\cdots\cdots\cdots\cdots \\ \phi_r^{(1)}(1) & \phi_r^{(1)}(2) & \cdots & \phi_r^{(1)}(N) \\ \cdots\cdots\cdots\cdots\cdots\cdots \\ \phi_N^{(0)}(1) & \phi_N^{(0)}(2) & \cdots & \phi_N^{(0)}(N) \end{vmatrix} \tag{5.6}$$

The eigenfunction $\Psi_r$ is not an exact eigenstate of the Hamiltonian given by equation (5.1). For if we evaluate the matrix element $\langle \Psi_r \,|\, H \,|\, \Psi_s \rangle$ we find that it has a nondiagonal contribution as well a diagonal contribution:

$$\langle \Psi_r \,|\, H \,|\, \Psi_s \rangle = \delta_{rs} \{ \sum_m E_m^{(0)} + (E_r^{(1)} - E_r^{(0)})$$

$$+ \tfrac{1}{2} \sum_{m,n}{}' V_{mn}(00;00)(1 - \delta_{mr})(1 - \delta_{nr}) + \sum_m V_{rm}$$

$$(10;10)(1 - \delta_{mr}) \} + (1 - \delta_{rs}) V_{rs}(10;10) \tag{5.7}$$

The determination of the excitation energies of the crystal thus requires the diagonalization of the matrix $\langle \Psi_r \,|\, H \,|\, \Psi_s \rangle$, and this diagonalization also provides us with the correct linear combinations of the wavefunctions $\{\Psi_r\}$

which give the true wavefunctions for the excited states of the type we are considering.

However, rather than diagonalizing the matrix $\langle \Psi_r \,|\, H \,|\, \Psi_s \rangle$ directly, we proceed in a somewhat different manner, one which is in keeping with the approach taken in the preceding sections. We try to rewrite the crystal Hamiltonian in the form of the sum of independent harmonic oscillator Hamiltonians, whose frequencies are the excitation frequencies of the low-lying excited states of the crystal. For this purpose we introduce creation and destruction operators $b_{l0}^+$, $b_{l0}$ and $b_{l1}^+$, $b_{l1}$ which are associated with the atom at the site $l$. The operator $b_{l0}$, for example, removes an atom from its ground electronic state at the site $l$, while the operator $b_{l1}^+$ creates an atom in its excited state at this site. If we take as the ground or vacuum state of the crystal the state described by equation (5.3), in which every atom is in its ground state, and if we now denote this state by $|0\rangle$

$$\Psi_0 \equiv |0\rangle \tag{5.8}$$

then the state described by the wave function $\Psi_r$ can be written

$$\Psi_r = b_{r1}^+ b_{r0} \,|0\rangle \tag{5.9}$$

It is clear physically that these creation and destruction operators must satisfy the conditions

$$b_{rf_r}^+ b_{rf_r} \,|0\rangle = \delta_{f_r,0} \,|0\rangle$$
$$b_{rf_r} b_{rf_r}^+ \,|0\rangle = \delta_{f_r,1} \,|0\rangle \tag{5.10}$$

Adding these two equations, we obtain the commutation rule

$$b_{rf_r}^+ b_{rf_r} + b_{rf_r} b_{rf_r}^+ = 1 \tag{5.11}$$

However, since the operators $b_{rf_r}$, $b_{rf_r}^+$ and $b_{rf'_r}$, $b_{rf'_r}^+$ act on different variables for $f_r \neq f'_r$ the following commutation rules hold:

$$b_{rf_r} b_{rf'_r} - b_{rf'_r} b_{rf_r} = 0$$
$$b_{rf_r}^+ b_{rf'_r}^+ - b_{rf'_r}^+ b_{rf_r}^+ = 0$$
$$b_{rf_r} b_{rf'_r}^+ - b_{rf'_r}^+ b_{rf_r} = 0 \tag{5.12}$$

Operators with different $r$ commute in all combinations. The operators which we have introduced are therefore neither Fermi nor Bose operators, but instead are Pauli operators.

With the aid of the definitions (5.8) and (5.9), and the conmutation rules (5.11) and (5.12), it is straightforward to verify that the matrix elements

$\langle \Psi_0 | H | \Psi_0 \rangle$ and $\langle \Psi_r | H | \Psi_s \rangle$ of the following Hamiltonian agree with the results given by equations (5.4) and (5.7):

$$H = \sum_m \{E_m^0 b_{m0}^+ b_{m0} + E_m^{(1)} b_{m1}^+ b_{m1}\} + \tfrac{1}{2} \sum_{mn}' V_{mn}(00;00) b_{m0}^+ b_{m0} b_{n0}^+ b_{n0}$$

$$+ \tfrac{1}{2} \sum_{mn}' V_{mn}(10;10)[b_{m1}^+ b_{m1} b_{n0}^+ b_{n0} + b_{n0}^+ b_{n0} b_{m1}^+ b_{m1}]$$

$$+ \tfrac{1}{2} \sum_{mn}' V_{mn}(10;10)[b_{m1}^+ b_{m0} b_{n0}^+ b_{n1} + b_{n0}^+ b_{n1} b_{m1}^+ b_{m0}] \tag{5.13}$$

We can simplify this Hamiltonian by introducing the new operators

$$B_{r1}^+ = b_{r1}^+ b_{r0} \qquad B_{r1} = b_{r0}^+ b_{r1} \tag{5.14}$$

The commutation rules for these operators can be obtained with the aid of equations (5.11) and (5.12). We have from the latter equations that

$$B_{r1} B_{r1}^+ = b_{r0}^+ b_{r1} b_{r1}^+ b_{r0} = b_{r0}^+ b_{r0}[1 - b_{r1}^+ b_{r1}] \equiv N_{r0}(1 - N_{r1}) \tag{5.15}$$

$$B_{r1}^+ B_{r1} = b_{r1}^+ b_{r0} b_{r0}^+ b_{r1} = b_{r0} b_{r0}^+ b_{r1}^+ b_{r1} \equiv (1 - N_{r0}) N_{r1} \tag{5.16}$$

where we have defined the operator

$$N_{rf} = b_{rf}^+ b_{rf} \tag{5.17}$$

With the aid of the commutation rules (5.11) and (5.12) it is straightforward to establish the results that

$$(N_{r0} + N_{r1}) |0\rangle = |0\rangle$$
$$(N_{r0} + N_{r1}) |\Psi_s\rangle = |\Psi_s\rangle$$

so that within the manifold of the wavefunction with which we work we have the result that

$$N_{r0} + N_{r1} = 1 \tag{5.18}$$

It is also straightforward to establish the results that

$$N_{r1} N_{r0} |0\rangle = 0$$
$$N_{r1} N_{r0} |\Psi_s\rangle = 0$$

so that within the space spanned by the functions $|0\rangle$ and $\{|\Psi_s\rangle\}$ we can set

$$N_{r1} N_{r0} = 0 \tag{5.19}$$

Combining equations (5.18) and (5.19) with equations (5.15) and (5.16) we obtain the commutation rules for the $B$ operators, namely,

$$B_{r1}B_{r1}^+ = N_{r0} = 1 - N_{r1} \tag{5.20a}$$

$$B_{r1}^+ B_{r1} = N_{r1} \tag{5.20b}$$

so that

$$B_{r1}B_{r1}^+ - B_{r1}^+ B_{r1} = 1 - 2N_{r1}$$

However in the states which we are considering, the mean value of $N_{r1}$ is of $O(1/N)$:

$$\frac{1}{N}\sum_s \langle \Psi_s | N_{r1} | \Psi_s \rangle = \frac{1}{N}\sum_s \delta_{rs}\langle \Psi_s | \Psi_s \rangle = \frac{1}{N}$$

Therefore we have to good approximation that

$$B_{r1}^+ B_{r1}^+ - B_{r1}^+ B_{r1} = 1 \tag{5.21}$$

For $r \neq r'$, the operators $B_{r1}$, $B_{r1}^+$ and $B_{r'1}$, $B_{r'1}^+$ commute with one another. Therefore, the operators $B_{r1}$ and $B_{r1}^+$ are operators which obey boson or phonon commutation rules.

We now rewrite the Hamiltonian (5.13) in terms of these new operators:

$$H = \sum_m \{E_m^{(0)} N_{m0} + E_m^{(1)} N_{m1}\} + \tfrac{1}{2} {\sum_{mn}}' V_{mn}(00;00)N_{m0}N_{n0}$$
$$+ {\sum_{mn}}' V_{mn}(10;10)N_{m1}N_{n0} + {\sum_{mn}}' V_{mn}(10;01)B_{m1}^+ B_{n1} \tag{5.22}$$

Using equations (5.18) and (5.20b) and neglecting all terms quadratic in the operator $N_{r1}$ we obtain for $H$ the result

$$H = \epsilon_0 + \sum_m \{[E_m^{(1)} - E_m^{(0)}] + {\sum_n}' V_{mn}(10;10)$$
$$- {\sum_n}' V_{mn}(00;00)\}B_{m1}^+ B_{m1} + {\sum_{mn}}' V_{nm}(01;10)B_{m1}^+ B_{n1} \tag{5.23}$$

The neglect of terms quadratic in $N_{r1}$ is a reflection of the assumption that the atom $r$ has only a single excited electron.

To diagonalize this Hamiltonian we first work out the equation of motion of the operator $B_{r1}^+$. This is given by

$$i\hbar \dot{B}_{r1}^+ = [B_{r1}^+, H] = -\{(E_r^{(1)} - E_r^{(0)}) + {\sum_n}' V_{rn}(10;10)$$
$$- {\sum_n}' V_{rn}(00;00)]\}B_{r1}^+ - {\sum_m}' V_{rm}(01;10)B_{m1}^+ \tag{5.24}$$

If we assume an harmonic time dpendence for $B_{r1}^+$,

$$i\hbar \dot{B}_{r1}^+ = -\hbar\omega B_{r1}^+ \tag{5.25}$$

we can rewrite equation (5.24) as

$$\omega B_{r1}^{+} = \sum_{m} M_{rm} B_{m1}^{+} \tag{5.26}$$

The elements of the matrix $M_{rm}$ are given explicitly by

$$
\begin{aligned}
\hbar M_{rm} &= \delta_{rm}\{(E_r^{(1)} - E_r^{(0)}) + {\sum_{n}}' [V_{rn}(10;10) \\
&\quad - V_{rn}(00;00)]\} + (1 - \delta_{rm})V_{rm}(01;10) \\
&\equiv \delta_{rm}\Delta_r + (1 - \delta_{rm})V_{rm}(01;10) \\
&= \hbar M_{mr}^{*} \tag{5.27}
\end{aligned}
$$

As the matrix $\mathbf{M}$ is an $N \times N$ Hermitian matrix, it possesses $N$ real eigen-values $\{\omega_s\}$ and $N$ corresponding eigenvectors $\{\mathbf{V}^{(s)}\}$, which we label by an index $s(= 1, 2, 3, \ldots, N)$:

$$\sum_{m} M_{rm} V_m^{(s)} = \omega_s V_r^{(s)} \tag{5.28}$$

The eigenvectors $\{\mathbf{V}^{(s)}\}$ can be constructed to be orthonormal and complete:

$$\sum_{r} V_r^{(s)*} V_r^{(s')} = \delta_{ss'}$$

$$\sum_{s} V_r^{(s)*} V_{r'}^{(s)} = \delta_{rr'} \tag{5.29}$$

We accordingly define new operators $\{B_s^{+}\}$ and $\{B_s\}$ by the relations

$$B_{r1}^{+} = \sum_{s} V_r^{(s)} B_s^{+}$$

$$B_{r1} = \sum_{s} V_r^{(s)*} B_s \tag{5.30}$$

In terms of these new operators the Hamiltonian (5.23) takes the form

$$H = \epsilon_0 + \sum_{s} \hbar \omega_s B_s^{+} B_s \tag{5.31}$$

We have thus re-expressed that part of the Hamiltonian (5.1) which de-scribes the low-lying excited states of our crystal in the form of the Hamil-tonian of an assembly of noninteracting harmonic oscillators. The quantum of energy required to excited one of these oscillators from a given state to the next higher state,

$$\hbar \omega_s = \hbar \sum_{rm} V_r^{(s)*} M_{rm} V_m^{(s)} \tag{5.32}$$

is called an *exciton*. Just as in the case of the elementary excitations of the electron gas, the frequency $\omega_s$ in the present case is the frequency of oscillation of a properly phased sum of the electron–hole pair creation operators $B_{r1}^+ = b_{r1}^+ b_{r0}$,

$$B_s^+ = \sum_r V_r^{(s)*} B_{r1}^+ \tag{5.33}$$

In the case when the crystal is perfectly periodic, the eigenvector $V_r^{(s)}$ becomes the eigenvector $V_r(\mathbf{k})$ which is given by

$$V_r(\mathbf{k}) = \frac{1}{\sqrt{N}} e^{i\mathbf{k}\cdot\mathbf{x}(r)} \tag{5.34}$$

where the allowed values of $\mathbf{k}$ are specified by the periodic boundary condition. The frequency $\omega_s$ goes into the frequency $\omega(\mathbf{k})$, which is given by

$$
\begin{aligned}
\hbar\omega(\mathbf{k}) &= \frac{1}{N} \sum_{ll'} e^{-i\mathbf{x}\cdot[\mathbf{x}(l)-\mathbf{x}(l')]} \{\delta_{ll'}\Delta + (1 - \delta_{ll'})V_{l-l'}(01\,;10)\} \\
&= \Delta + \sum_{l(\neq 0)} e^{-i\mathbf{k}\cdot\mathbf{x}(l)} V_l(01\,;10)
\end{aligned}
\tag{5.35}
$$

In obtaining equation (5.35) we have used the fact that in a perfect, periodic, crystal $\Delta$ is independent of the site index $l$, and that the matrix element $V_{ll'}(01\,;10)$ depends on $l$ and $l'$ only through their difference $l - l' = \bar{l}$. The first term on the right-hand side of equation (5.35) gives the energy of excitation of any atom in the crystal, modified by the energy of interaction between the electron and the hole created in the excitation process. The second term on the right-hand side of this equation describes the transfer of the excitation from one lattice site to another, in the form of a traveling wave.

## REFERENCES

1. P.W. Anderson, *Concepts in Solids*, W.A Benjamin, Inc. New York, 1963, p. 99.
2. P.W. Anderson, *Concepts in Solids*, W.A Benjamin, Inc. New York, 1963, p. 102.
3. T.H.K. Barron, *Proceedings International Conference on Lattice Dynamics*, Copenhagen, 1963, Pergamon Press Inc., New York, 1965, p. 247.
4. V. Ambegaokar, in *Astrophysics and the Many-Body Problem*, W.A. Benjamin, Inc., New York, 1964, p. 369.
5. D. Pines, *Elementary Excitations in Solids*, W.A. Benjamin, Inc., New York, 1963.
6. J. Bernoulli, *Petrop. Comm.* 3: 13 (1728).
7. J.L. Lagrange, *Mechanique Analytique*, Gauthier-Villars, Paris, 1888.
8. P. Debye, *Ann. Phys.* 39: 789 (1912).
9. P.A.M. Dirac, *The Principles of Quantum Mechanics*, The Clarendon Press, Oxford, 1947, p. 136.
10. R.D. Mattuck *Ann. Phys.* (*N. Y.*) 27: 216 (1964).

11.   T. Holstein and H. Primakoff, *Phys. Rev.* **58**: 1098 (1940).

11a.  J. Van Kranendonk and J.H. Van Vleck, *Rev. Mod. Phys.* **30**: 1 (1958).

12.   D. Pines and D. Bohm, *Phys. Rev.* **85**: 338 (1952).

13.   K. Sawaka, K.A. Brueckner, N. Fukuda, and R. Brout, *Phys. Rev.* **108**: 507 (1957).

14.   Reference 5, p. 100.

15.   Reference 5, p. 98.

16.   R.V. Lange, *Phys. Rev. Letters* **14**: 3 (1965).

17.   M. Born and K. Huang, *Dynamical Theory of Crystal Lattices*, Oxford, The Clarendon Press, 1954, p. 86.

18.   V.M. Agranovich, *Soviet Physics—JETP* **37**: 307 (1960).

# Theoretical and Experimental Aspects of Localized and Pseudolocalized Phonons

## A. A. Maradudin*

*Department of Physics*
*University of California*
*Irvine, California*

I think it is fair to say that it is well known that the introduction of defects into a crystal perturbs the atomic vibrations and, consequently, also any physical property of the crystal in which the vibrations play a central role.

The effects of point defects on the atomic vibrations of crystals can be of at least two different kinds. First of all they can introduce exceptional modes into the vibration spectrum of the crystal. For example, if the impurity atom is a good deal lighter than the atom it replaces in the crystal, or if it couples much more strongly to the surrounding host crystal than the atom it replaces, we can obtain certain exceptional vibration modes called localized modes. These are characterized by the fact that their frequencies lie above the maximum frequency of the unperturbed host crystal and that the displacement amplitudes of the atoms, when the crystal is vibrating in these modes, decay faster than exponentially with increasing distance from the impurity site. If the crystal has a gap in its frequency spectrum between the acoustic and optical branches, as is the case, for example, in NaI, KI, KBr, and LiCl, then it is possible to have localized vibration modes whose frequencies lie in the gap. These modes are also localized in the sense that the amplitudes of the atoms when the crystal vibrates in such a mode decay faster than exponentially with distance from the impurity site. However, to distinguish between the localized modes whose frequencies are in the gap and localized modes whose frequencies are in the gap and localized modes whose frequencies lie above the maximum frequency of the unperturbed crystal, I refer to the former as gap modes. However, I emphasize that there is no fundamental distinction between a localized mode and a gap mode.

*This research was supported in part by the Air Force Office of Scientific Research, Office of Aerospace Research U.S. Air Force, under AFOSR Grant No. 1080–66.

A third kind of exceptional mode introduced into the crystal by the presence of impurities are the so-called resonance modes. A very crude picture of a resonance mode might be obtained in the following way. We know that in the very low frequency vibration modes of a crystal, i.e., in the low-frequency acoustic branches, all the atoms are moving in phase, in parallel with each other, independently of their masses. This means that if we have a very heavy impurity atom or an impurity atom which is coupled much more weakly to the surrounding host lattice than the atom it replaces, it will be vibrating in phase with its, let us say, lighter neighbors. However, as the frequency of the vibrations increases, the heavy impurity cannot keep up with its lighter neighbors and begins to lag behind them until a point is reached where the heavy impurity finds itself vibrating 180° out of phase with its light neighbors. The frequency at which this switchover in the character of the vibrations take place (from the case in which the heavy impurity vibrates in phase with its neighbors to the one in which it is moving out of phase with its neighbors) is called the frequency of the resonance mode. It is, then, a localized optical type of vibration in the crystal and is characterized by the fact that the mean-squared amplitude of the impurity as a function of frequency undergoes a large increase at the frequency of the resonance mode. However, unlike the localized modes, which are associated with tightly bound or light impurities in the crystal, resonance modes are not true normal modes of the perturbed crystal. They form a kind of collective mode which is the superposition of all the unperturbed modes of the crystal. A resonance mode is localized in frequency but not in space, while a true localized mode is localized both in frequency and in space.

In addition to introducing such exceptional vibration modes into the vibration spectrum of the crystal, impurity atoms can also be used as probes to study the dynamical properties of the unperturbed host crystal, e.g., they can give information about the frequency spectrum of the unperturbed host crystal. This comes about basically because of the destruction of the translational symmetry or periodicity of the crystal which results when the impurity is introduced.

At the present time, localized modes, gap modes, and resonance modes have been studied experimentally by a wide variety of methods: specific heat measurements, tunneling between superconductors, the Mössbauer effect, neutron scattering, infrared absorption, Raman spectroscopy, and the temperature dependence of the relaxation time for spin flips when we have paramagnetic impurities in the crystal. I think, however, at the present time, the method which has yielded the most detailed results about the dynamical properties of imperfect crystals is infrared absorption. A second technique which is being used more and more today for the same purpose is Raman spectroscopy. In my talk this morning, I should therefore like to discuss primarily the type of informations that has been obtained concerning the dynamical properties of imperfect crystals by infrared-absorption and Raman-scattering experiments.

In infrared-absorption experiments, light falling in the crystal is absorbed with no net change in the electronic state of the crystal but with the excitation or de-excitation of one or more phonons. Now, in order for the absorption process to take place, there are several rules which have to be satisfied. Whether the crystal is perfect or imperfect, energy has to be conserved in the process, i.e., the energy, or frequency, of the incident light has to equal the net frequency change of the phonons participating in the absorption process. In a perfect crystal, the translational periodicity of the crystal, i.e., the fact that the crystal is invariant against a displacement through a translation vector of the crystal, gives rise to a second conservation law which has the character of the law of conservation of linear momentum; that is, the wavevector of the incident light has to equal the net wavevector of the phonons participating in the absorption process.

Ordinarily, as has been pointed out by several authors, it is a good approximation to regard the wavevector of the incident light as being equal to zero, so that the second selection rule can be called the $\mathbf{k} = 0$ selection rule, namely, that the sum of the wavevectors of the phonons participating in the absorption processes must equal zero. There is a third selection rule, having its origin in the symmetry of the crystal, which states that only those vibration modes which transform according to the polar vector irreducible representation of the point group of the space group of the crystal will couple to the incident light. Applied to a perfect crystal of, let's say, the rocksalt structure these various restrictions put together imply that the one-phonon absorption spectrum as a function of the frequency of the incident light will consist of $\delta$-function peaks centered at the frequencies of the optical vibration modes of infinite wavelength. However, because light has a transverse character, there is an additional selection rule: namely, that light is coupled directly to the transverse optical mode of infinite wavelength. Such a simple absorption spectrum, consisting of a $\delta$-function line, clearly provides only little information about the dynamical properties of the crystal. Of course, in fact, no crystal is perfectly harmonic, and the absorption spectrum of even a perfect crystal that is free from defects will display an absorption spectrum possessing a structure rather more complex than that of a $\delta$-function peak centered at the frequency of the transverse optical modes. The $\delta$-function peak is broadened into a Lorentzian peak with additional structure such as sidebands. However, even in the harmonic approximation, the introduction of impurities into the crystal can modify the observed absorption spectrum quite significantly, and there are three kinds of modifications that I want to discuss. I shall limit myself to considering the one-phonon spectrum only, i.e., the spectrum that is associated with the excitation or the de-excitation of one quantum of vibrational energy by the incident light. For such processes to occur, the crystal must have a first-order dipole moment, i.e., if we expand the dipole moment of the crystal in powers of the displacements of the atoms from their equilibrium positions, then there have to be nonvanishing terms which are linear in the atomic displacements.

Now, if we are dealing with ionic crystals, such as NaCl, or with more complicated structures such as CsCl or $CaF_2$, the ions which comprise the crystals are charged. Therefore, even in the absence of defects, this kind of crystal has a first-order dipole moment which can couple to the incident light. If we then introduce into the crystal an impurity which is also charged, for example, a $H^-$ or $D^-$ impurity which replaces a negative ion in the crystal, then we do not destroy the first-order dipole moment of the crystal. Further, because the impurity itself is vibrating in a localized vibration mode it has its own dipole moment which can also couple directly to the incident radiation. In this case it is found that the absorption spectrum of the crystal, in the harmonic approximation, is now proportional to the frequency spectrum of the perfect host crystal weighted by the frequency dependence of the mean-squared vibration amplitude of the impurity atom. This is, in fact, a result which has a somewhat general character. In the case of ionic crystals the situation is complicated by the fact that even in the absence of impurities the crystal displays a fundamental (reststrahl) absorption which is also present in the absorption spectrum of the perturbed crystal. The additional, impurity induced, structure will reflect the frequency spectrum of the host crystal and this structure will be added to the fundamental absorption. If there are resonance or localized modes which have the appropriate symmetry properties, i.e., which transform as the polar vector irreducible representation of the point symmetry group of the crystal, they will be infrared-active and will show up as additional peaks in the absorption spectrum.

I will now consider a second kind of impurity: charged impurities in a homopolar crystal, e.g., diamond, germanium, and silicon. Such crystals are composed of neutral atoms. Therefore, if we expand the dipole moment of the crystal in powers of the atomic displacements, the term linear in the displacements vanishes, because these terms have the form of atomic displacements multiplied by an effective charge which is zero. In such crystals, when they are perfect, the absorption is due to two-phonon processes, i.e., to processes in which the light excites or de-excites two phonons. However, if we introduce into diamond, silicon, or germanium an impurity from a neighboring column of the periodic table, i.e., one with a different valence, the impurity atom now acquires a charge in the host lattice. Because the impurities are assumed to be randomly distributed in the crystal, they therefore give rise to a random first-order dipole moment which can couple to the incident light ([1]). In such a case, one induces a one-phonon absorption spectrum that vanishes in the absence of impurities. Again, the absorption spectrum is essentially the frequency spectrum of the perfect host crystal, weighted by the frequency dependence of the mean-square amplitude of the impurity atom.

The third kind of impurity that gives rise to a nontrivial, nonzero, one-phonon absorption in crystals is a nominally neutral impurity atom in the lattice of a homopolar crystal, e.g., carbon as an impurity in silicon ([2]). In such a case the nominally neutral carbon atom acquires an effective charge and the surrounding silicon atoms acquire a compensating opposite charge, so that the whole crystal remains neutral. The mechanims by which the carbon impurity acquires an effective charge has its origin in charge transfer, and in the fact that the polarizabilities of the carbon and of silicon atoms are different. In a very crude way this result can be seen as follows: if the atomic volume of the silicon atom is smaller than the atomic volume of the impurity

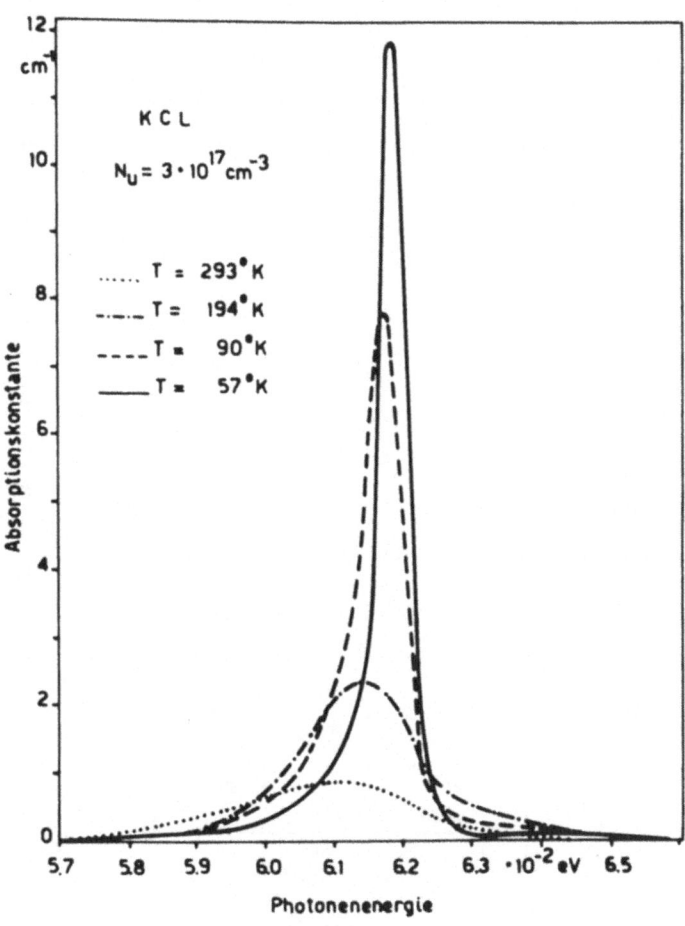

Figure 1

introduced into the crystal, so that the electronic wavefunction of the impurity extends over its nearest neighbors in the crystal, then within the atomic volume occupied by the atom replaced by the impurity there is a deficiency of electrons compared to the situation which existed before the introduction of the impurity. So the impurity acquires a slight positive charge, and the excess electrons in the vicinity of the nearest neighbors contribute to a weak negative effective charge on these atoms. Therefore, it is possible to induce a nonzero one-phonon absorption in silicon containing carbon impurities ([3]) or solid argon containing xenon impurities ([4]) by this mechanism, and again the frequency dependence of the absorption spectrum is the reflection of the frequency spectrum of the perfect host crystal weighted by the frequency dependence of the mean-squared amplitude of the impurity. So much for the qualitative theoretical aspect of the kinds of absorption which can be induced by different kinds of crystals. What I would like to do now is to go on to describe how these different phenomena manifest themselves experimentally.

In Fig. 1 is shown one of the first experimental observations of the

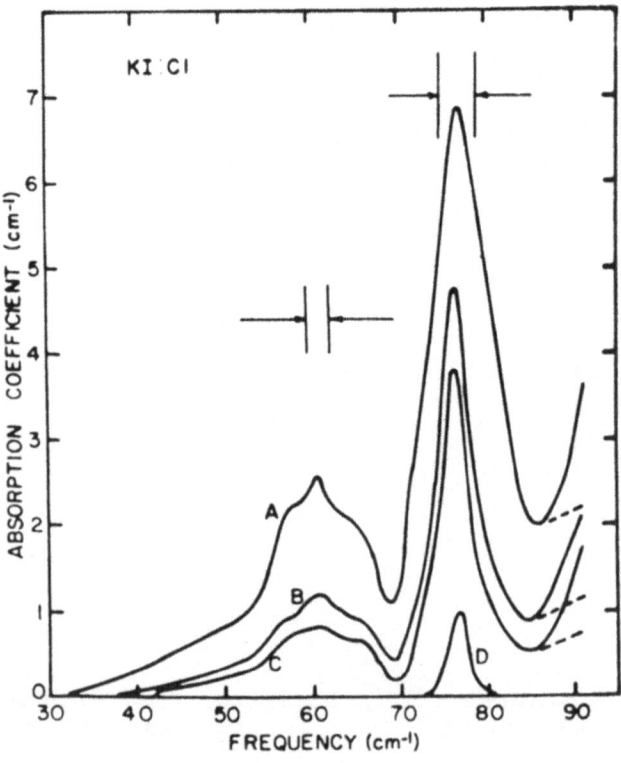

Figure 2

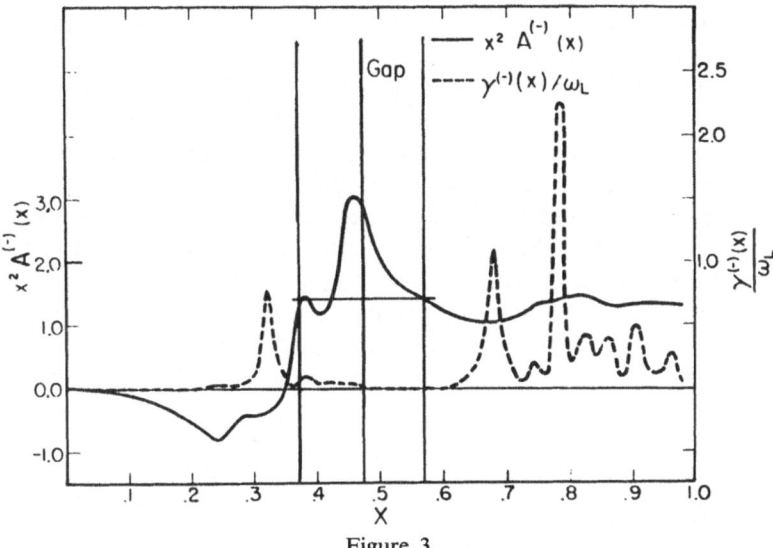

Figure 3

existence of localized modes. This is a result of Schaefer ([5]) showing the absorption spectrum of H$^-$ in KCl. The concentration of H$^-$ ions is of the order of $10^{18}$/cm$^3$. At the low temperatures at which the experiment was carried out, the experimental results show a well-defined sharp, narrow peak which has been identified as the absorption associated with the H$^-$ localized mode.

I remarked earlier that if the frequency spectrum of the host crystal has a gap between the acoustic and optical branches, we can have localized vibration modes whose frequencies fall into this gap. An example of this situation is shown in Fig. 2. The host crystal is KI, which has a gap in the frequency spectrum which runs from about 70 to about 96 cm$^{-1}$. This figure shows the absorption spectrum associated with Cl$^-$ impurities substituting for the I$^-$ ions. These are the results of Sievers ([6]). The different curves shown correspond to different concentrations of impurities. We see, in any case, a well-defined peak in the frequency region of the gap in the frequency spectrum of the perfect host crystal. This is attributed to absorption by localized modes due to the Cl$^-$ impurities whose frequencies lie in the gap. If one stretches one's imagination, one can see in the frequency region just below 63 cm$^{-1}$ three poorly resolved peaks at frequencies which lie in the frequency range between the maximum frequency of the transverse and longitudinal acoustic modes for the crystal. If the interpretation of these peaks is correct, then they are presumably associated with resonance modes whose frequencies lie in this range, and this is depicted theoretically in Fig. 3. This is a result of calculations by Sievers, Jaswal, and Maradudin ([6]).

I don't want to go into the details of how curves of this sort are obtained, but I just want to point out that the intersections of the solid curve with the horizontal line drawn in the figure give the frequencies of localized or resonance modes in the crystal when Cl⁻ ions replace I⁻ ions. The region of the gap lies between the second and third vertical lines and the maximum frequency of the TA mode is given by the left-hand vertical line.

It is seen that there is an intersection in the gap which gives the frequency of the gap modes, but the agreement between the experimental (77 cm⁻¹) and theoretical (77.6 cm⁻¹) values of its frequency, while gratifying, is probably fortuitous. This is because the calculations have been done assuming that Cl⁻ differs from I⁻ only in its mass. At the same time, we do see three additional intersections which could, in principle, give the frequencies of three resonance modes. In fact, the situation is more complicated than that because one cannot guarantee that every intersection will give a resonance mode. There are some other conditions to be satisfied before this identification can be made, and in addition the frequencies given by these intersections are not in good agreement with the experimental frequencies, except that they all lie between the maximum frequency of the TA branches and the maximum frequency of the LA branch.

In Fig. 4 I have again displayed a result of work by Sievers (⁷), which is the first experimental demonstration of the existence of a resonance mode by infrared-absorption methods. The system is KI containing Ag⁺ impurities

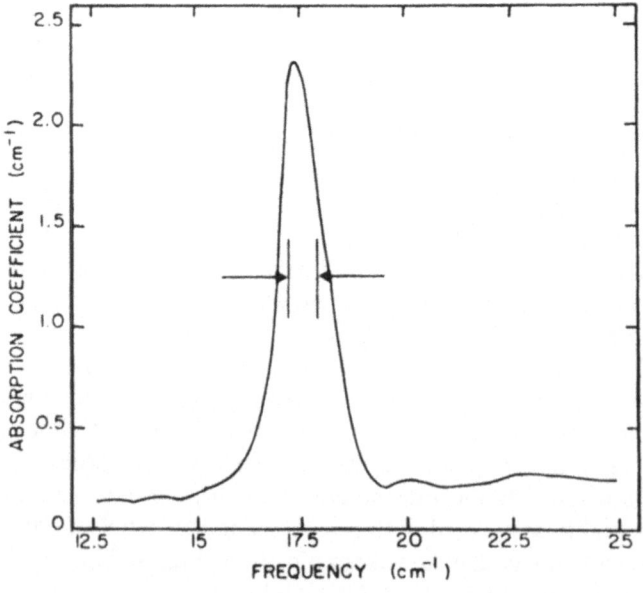

Figure 4

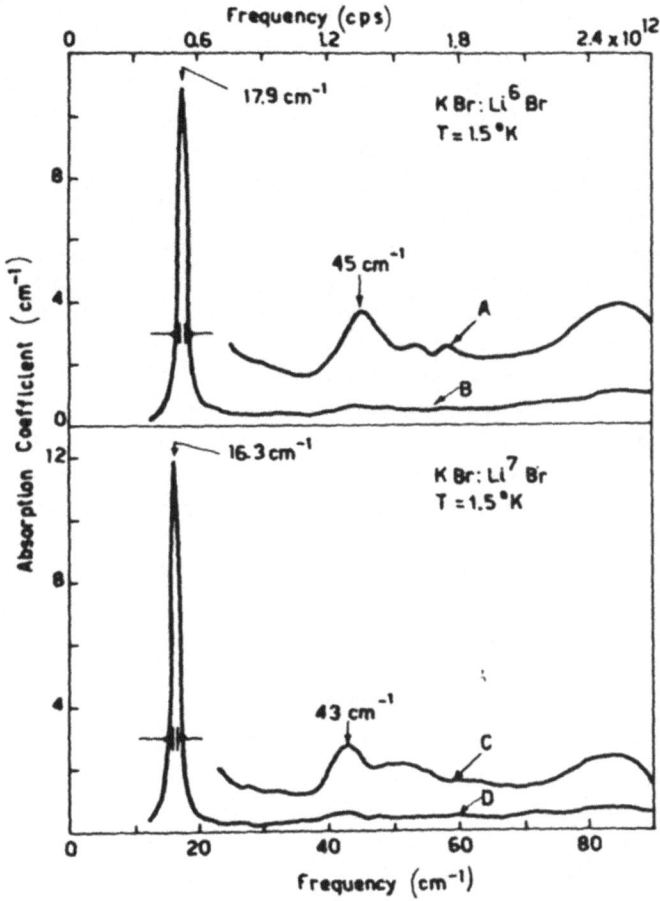

Figure 5

in the $K^+$ sublattices. The $Ag^+$ is heavier than $K^+$ and gives rise to a low-frequency resonance mode centered at about $17.5 \text{ cm}^{-1}$.

Having shown the existence of resonance modes, we now proceed to study some of their properties. One of these properties is depicted in Fig. 5, and is the isotope shift in the frequency of the resonance mode when, in this particular case, $Li^7$ is substituted for $Li^6$ in KBr [8] With the heavier $Li^7$ impurity, the resonance-mode frequency is $16.3 \text{ cm}^{-1}$ while, for the $Li^6$ impurity it is $17.9 \text{ cm}^{-1}$. Now, it is to be expected that an isotope shift will occur. The fact that one observes the resonance modes in infrared absorption in the first place implies that the impurity ion is moving rather than sitting still, as could be the case for certain kinds of resonance modes. Therefore the change in the mass of the impurity should be felt in the vibration frequency.

If the resonance mode could be approximated simply by an independent particle vibrating in an harmonic potential well, we know then that the ratio of the frequency of the Li[7] resonance mode to that of the Li[6] resonance mode would be just the ratio of the square root of the mass of Li[6] to that of Li[7]. However, the shift observed in this case is somewhat larger than would be predicted on the basis of this simple relation, and this suggests that the force constants with which the impurity couples to its nearest neighbors play an important role in determining these frequencies. It is the fact that the nearest neighbors are also vibrating that gives rise to a somewhat softer well than it would have been otherwise. The peaks observed at 45 and 43 cm$^{-1}$ are attributed to third harmonics of the resonance mode. The second harmonics are forbidden by symmetry when the point group at the impurity site is $O_h$, as it is in this case, but the third harmonic is allowed.

In addition to this kind of work, Prof. Sievers also looked into the temperature dependence of the area under a resonance mode absorption peak associated with an Ag$^+$ impurity in KI and found an anomalous temperature dependence ([9]). His results are shown in Fig. 6. If we have a Lorentzian peak, we expect that with an increase in temperature, the peak itself decreases in amplitude and the width increases but in such a manner that the area under the curve is preserved. However, in the present case, as is perhaps most

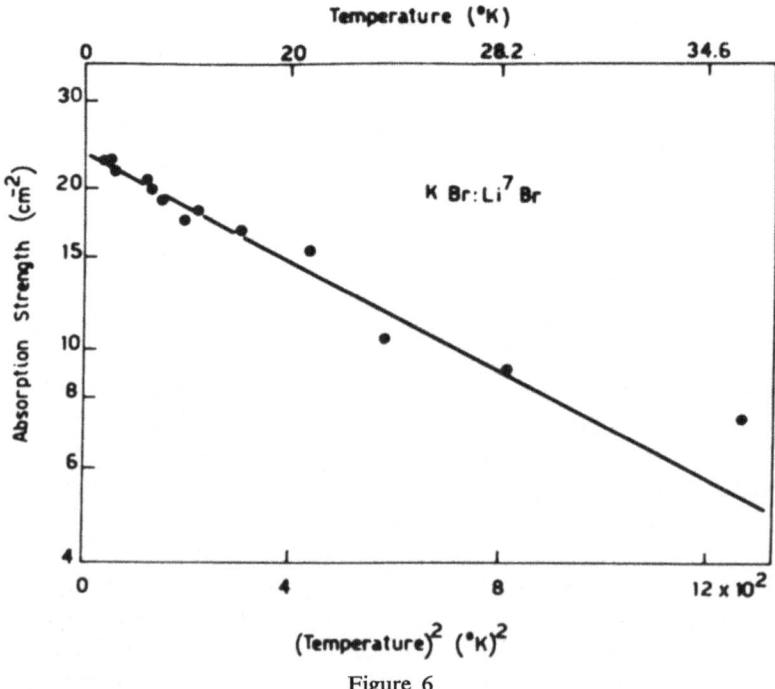

Figure 6

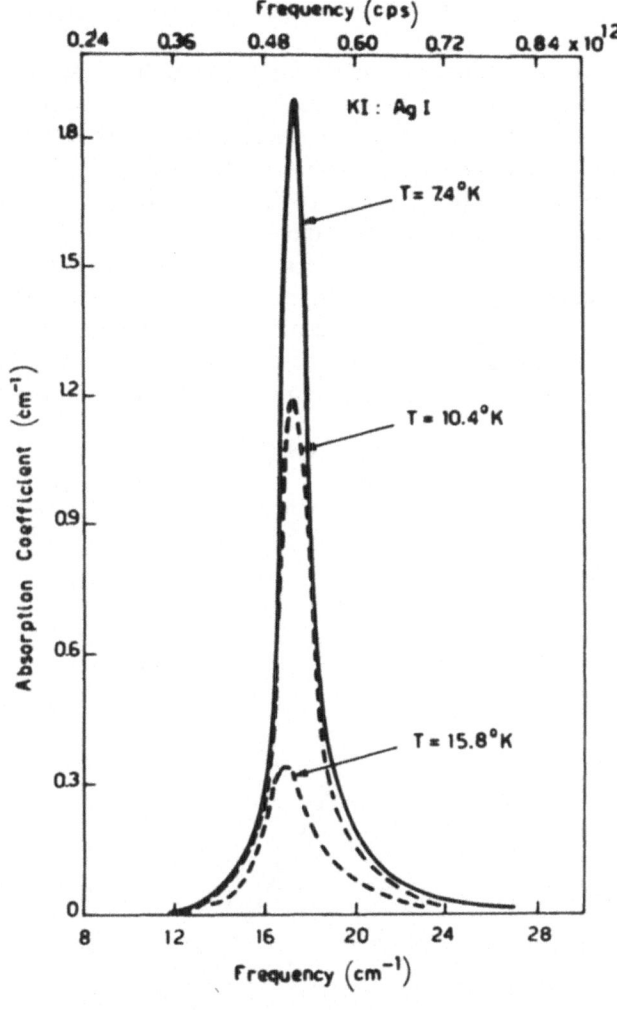

Figure 7

clearly demonstrated in Fig. 7, the area under the curve decreases in a dramatic fashion. In fact, it decreases with increasing temperature in this very low
temperature range in a manner which strongly suggests that the absorption
coefficient as a function of temperature has a factor similar to a Debye–
Waller factor multiplying the line shape of the absorption by the resonance
mode. Sievers and Takeno (⁹) have given a theoretical discussion of how such
a Debye–Waller factor might arise in the expression for the absorption spectrum. It is based on an adiabatic approximation in which the resonance
mode is coupled weakly to the lattice modes through anharmonic terms in

the potential energy and this, as in the Mössbauer effect, gives rise to a Debye Waller factor.

The results of a similar experiment for Li[7] in KBr are shown in Fig. 8, and again the decrease in the logarithm of the absorption strength is proportional to $T^2$ at low temperatures, which again implies that the absorption coefficient is multiplied by a factor similar to the Debye–Waller factor, because at low temperatures tha exponent in the Debye–Waller factor is proportional to the square of the absolute temperature.

This behavior, however, does not seem to be confined only to resonance modes but is apparently displayed by absorption associated with certain localized modes as well. In Fig. 9 the results for the temperature dependence of the area under the absorption curve for localized modes due to $H^-$ and $D^-$ $U$-centers in KCl are presented ([10]). Again we see that the area under the

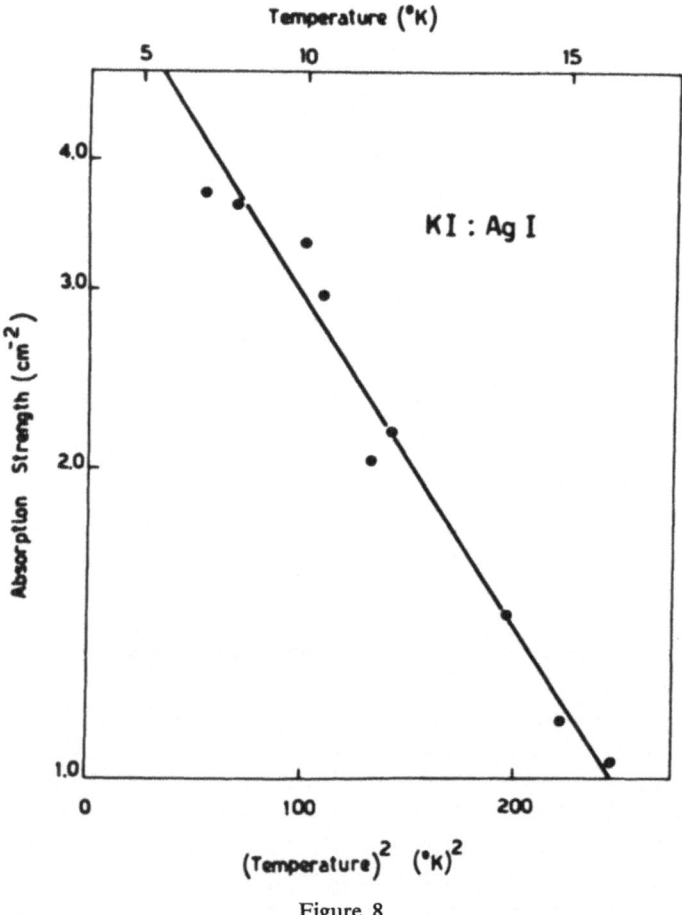

Figure 8

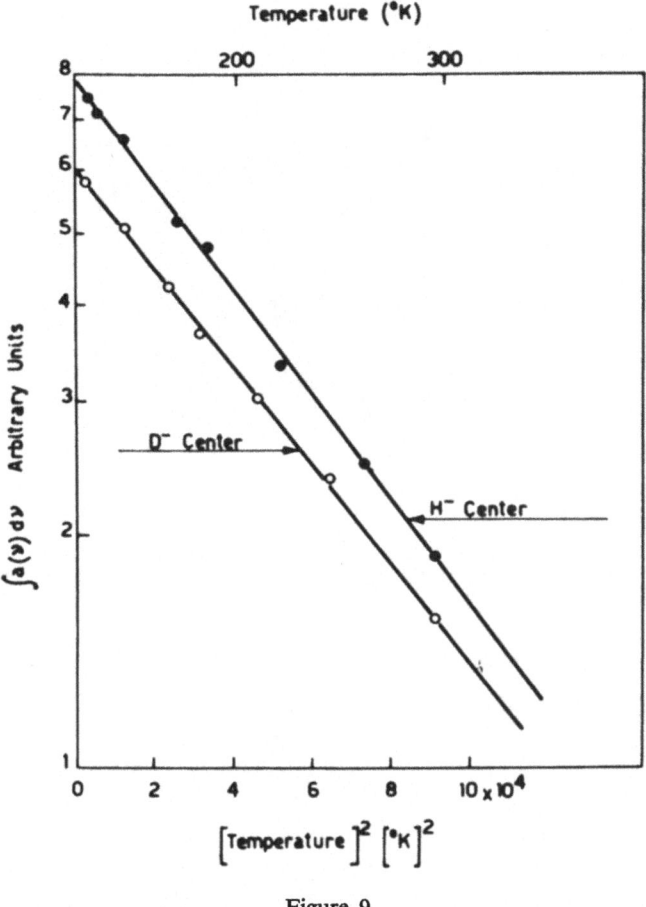

Figure 9

curve decreases with increasing temperature in an exponential fashion, with the exponent proportional to $T^2$. This is somewhat interesting because a result of this sort was also reported in the paper by Schaefer [5] in 1960, namely, that the area under the localized mode peak decreases with increasing temperature. However, about two to three years ago, Mirlin and Reshina [12] re-examined the problem using the same host crystals and impurities as used in the work of Schaefer, and they found in contrast with Schaefer's results that the area under the curve remains constant. This conclusion was subsequently confirmed by work of Fritz [13]. However, more recent work by Mitra and Singh [14] again gives the result that the area under the curve decreases with increasing temperature, and I think that the situation calls for some clarification. [15].

So far I have been discussing charged impurities in alkalihalide crystals.

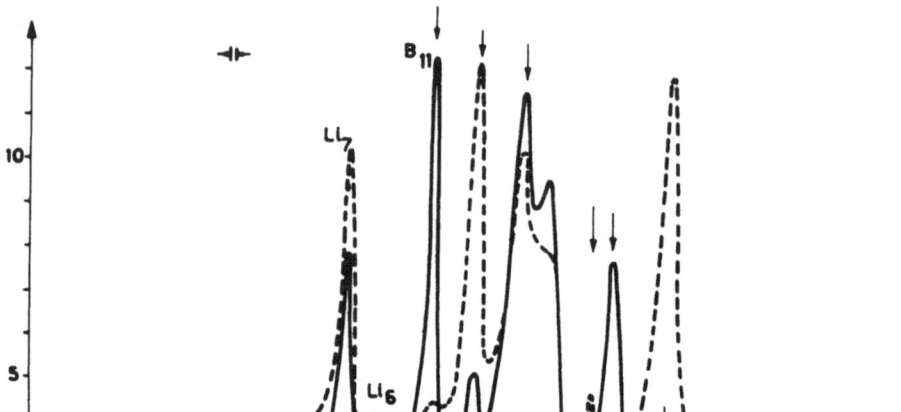

Figure 10

In Fig. 10 are shown the results, due to Balkanski and Nazarewicz ([16]), for the one-phonon absorption spectrum of silicon containing boron impurities compensated by lithium. The boron impurities are substitutional impurities whereas the lithium impurities are interstitial. However, both of these impurities are situated in different columns of the periodic table from that of silicon, so that they become charged impurities when they are present in a silicon lattice and give rise to a nonzero first order dipole moment. The measured spectrum is complicated but interesting. First of all, I should point out that the maximum frequency (the Raman frequency) of silicon is 518 $cm^{-1}$. All the peaks above that frequency are then associated with localized modes. The two peaks at 534 and 522 $cm^{-1}$ are associated with localized modes due to $Li^6$ and $Li^7$, respectively. The relative areas under these two peaks are an indication of the relative abundance of $Li^6$ and $Li^7$ in naturally occurring lithium. The solid curve corresponds to the case that both $B^{10}$ and $B^{11}$ are present in the crystal, and the dashed curve corresponds to the case when only $B^{10}$ is present. The two peaks at 644 $cm^{-1}$ and 620 $cm^{-1}$ are the localized modes associated with $B^{10}$ for the higher-frequency peak and $B^{11}$ for the lower-frequency peak. However, associated with each of these two peaks are two satellite peaks. For the $B^{11}$ peak, the satellite peaks are at 564 and at 653 $cm^{-1}$, and for the $B^{10}$ peak the satellite peaks are at 584 and 681 $cm^{-1}$. The origin of these two satellite peaks is the following: the site symmetry of the boron impurity (since it is a substitutional impurity) is $T_d$. This means that the localized modes associated with a boron impurity site are triply degenerate. However, the lithium impurities which compensate the sample occupy interstitial sites, and they like to be close to the boron impurities. The proximity of the lithium impurities to the boron impurities reduces the site symmetry of the boron from $T_d$ to $C_{3v}$ and leads to a splitting of the triple degeneracy of the boron localized modes into a singlet and a doublet. The singlets are low-frequency peaks, the doublets are the high-frequency peaks. The singlet and doublet for $B^{11}$ are at 564 and at 653 cm$^{-1}$, respectively. The singlet and doublet for $B^{10}$ are at 584 and at 681 cm$^{-1}$, respectively.

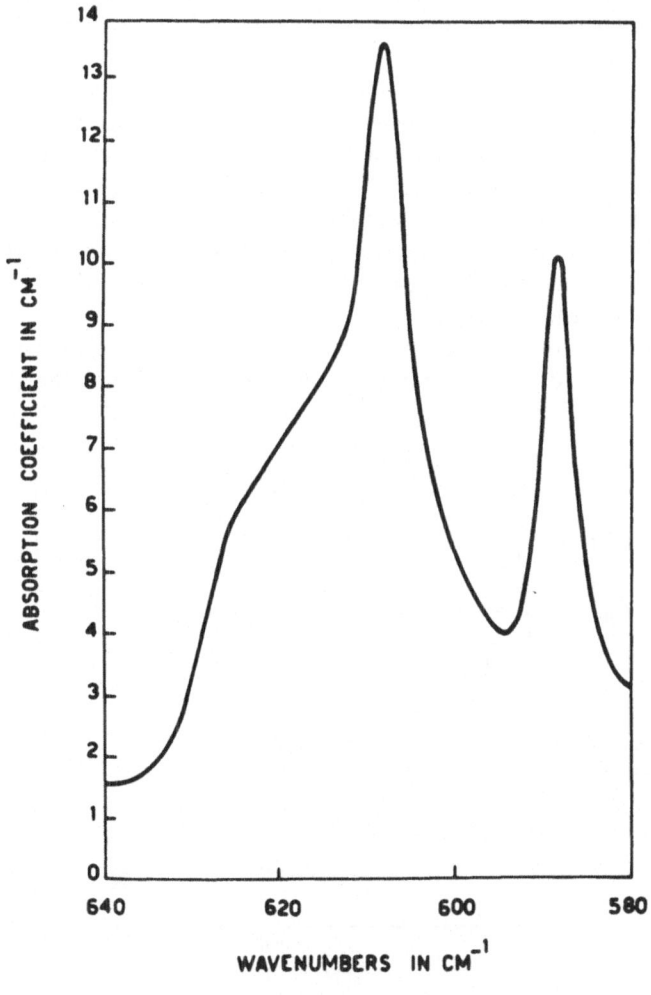

Figure 11

The integrated areas under these two curves are in fairly good agreement with the ratio two-to-one which would be expected for a singlet and a doublet.

Recently, Mirlin and Reshina ([17]) have also carried out similar experiments on ionic crystals, for example, KCl containing $H^-$ impurities into which NaCl has been introduced. Again the same kind of effect is observed, namely, the triple degeneracy of the localized modes associated with the $H^-$ impurities is split by the fact that the $H^-$ impurities now can have as one of its nearest neighbors a sodium impurity, which lowers the site symmetry of the $H^-$ site. The situation theoretically speaking is somewhat simpler in the latter case because the $H^-$ and the $Na^+$ impurities in KCl occupy substitutional sites whereas in the Si(Li + B) system, the lithium impurities occupy interstitial sites.

Results of Newman and Willis ([3]) for the absorption spectrum of silicon containing $C^{12}$ and $C^{13}$ are shown in Fig. 11. The heavier $C^{13}$ atoms give rise to a localized mode of frequency somewhat lower than that due to the $C^{12}$ impurities. The absorption mechanism here is that suggested by Lax and Burstein ([2]) to explain certain observations in diamond. The normally neutral carbon atom acquires an effective charge when introduced into the silicon host crystal, and the effective charge on this ion is apparently very close to the electronic charge.

A different kind of defect from those considered until now can be introduced into crystals by radiation damage. In Fig. 12 we display the absorption

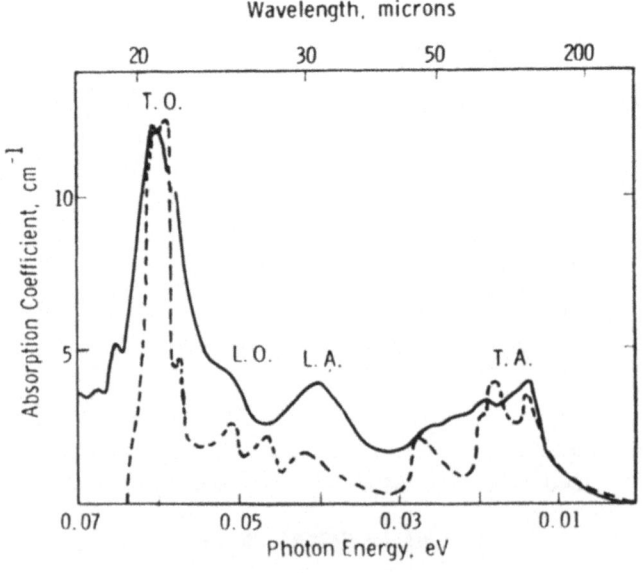

Figure 12

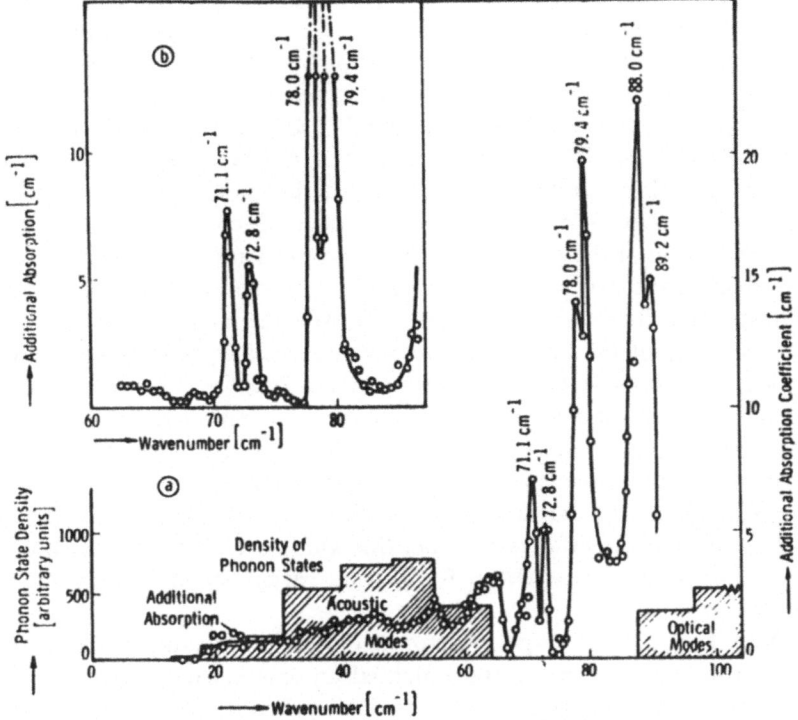

Figure 13

spectrum of silicon which has been irradiated by neutrons. This is the work of Angress, Smith, and Renk ([18]). The solid curve gives the experimental absorption spectrum. The dashed curve is a theoretical estimate of the frequency spectrum of perfect silicon computed by Bilz on the basis of phonon dispersion curves obtained by Dolling ([19]) by neutron spectroscopy methods. We see that there exists a crude resemblance between the absorption curve and the frequency spectrum of the perfect host crystal. The difficulty with infrared absorption induced by neutron irradiation is that one is never sure what kinds of defects are introduced by the irradiation.

Apart from radiation damage, I have so far talked only about monatomic impurities, in crystals. In Fig. 13 is shown the result of Renk ([20]) for the infrared absorption spectrum of KI containing $NO_2^-$ ions as substitutional impurities in the $I^-$ sublattice. As we have already mentioned, KI has a gap in the frequency spectrum between the acoustic and optical branches, and the experimental results show strong absorption at frequencies which lie in the gap. The richness of the absorption spectrum in the frequency region of the gap had been attributed by Sievers and Lytle ([21]), who had studied the same system earlier, to the low-symmetry ($C_{2v}$) of the nitrite ion. The site symmetry of the $I^-$ ion which is replaced by the $NO_2^-$ ion is $O_h$ and the lower symmetry of the substitutional impurity can lead to a splitting of the triply degenerate impurity modes which would result if the impurity also had $O_h$ symmetry. An interesting feature of Renk's results is the double structure of each of these peaks. Recently Sievers et al. ([22]) have re-examined the spectrum and found that in fact the doublet structure is due to the presence of a small concentration of nitrate ions in the crystal, which was introduced during the introduction of the nitrite impurities into the sample. In fact, in this case the weaker of the two peaks in each doublet is associated with the presence of nitrate impurities and has nothing to do with the nitrite impurities.

The study of the absorption spectrum associated with polyatomic impurities in crystals is particularly interesting because, as was shown, for example, by Wagner ([23]) some five years ago, if one of the intramolecular frequencies of the molecule falls in the frequency region allowed to the normal modes of the unperturbed crystals then one can expect to see a strong resonance effect with corresponding structure in the absorption spectrum. But I think that at the present time, this kind of experiment has not been carried out in any great detail for a wide variety of impurities and host crystals.

Now I should like to go over to the discussion of some of the effects of the anharmonicity of the crystal potential on dynamical properties of point defects in crystals. These effects are generally of three types. First, in an anharmonic crystal the localized mode is no longer an exact eigenstate of the Hamiltonian. In fact, it is coupled through the anharmonicity of the crystal potential to the modes of the continuum and as a result of this, the frequency of the localized modes undergoes a complex. shift of the type discussed by Prof. Ludwig in his lectures. The real part of the shift of the frequency of the localized mode due to anharmonic coupling to the band modes reflects a change in the actual numerical value of the frequency, while the imaginary part is related through the uncertainty principle to the lifetime of the localized mode. It is in fact the length of time over which an harmonic localized mode introduced into an anharmonic crystal would retain a recognizable identity as an harmonic phonon before it decays into a much more complicated combination of the anharmonic eigenstates. Depicted in Fig. 14 are results of Mirlin and Reshina ([24]) for the localized mode peak associated with $H^-$ impurities in $KBr$. The lower peak is due to heavier $D^-$ impurities, the higher to the lighter $H^-$ impurities. The frequency of the localized mode associated with the $D^-$ impurities is less than twice the maximum frequency of the unperturbed host crystal. The frequency of the $H^-$ localized mode is more than twice the maximum frequency of the unperturbed crystal. This gives rise to two different effects. The $D^-$ localized mode can decay by a cubic

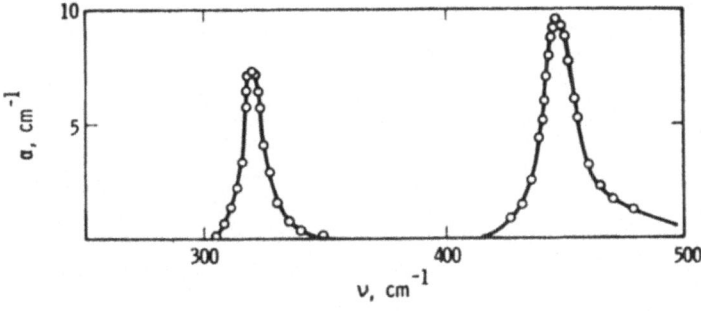

Figure 14

anharmonic process into two band modes, i.e., the conservation of energy allows decay processes in which the localized mode decays into two band modes. The H⁻ localized mode cannot decay into two band modes by cubic anharmonic processes because its frequency is more than twice the maximum frequency of the unperturbed host crystal, so that the lowest-order mechanism by which it can decay is by quartic anharmonic processes, i.e., it can decay into three band modes. However, as has been pointed out by Sennett ($^{25}$), there is a second quartic anharmonic mechanism which gives a contribution to the lifetime of a localized mode. In this mechanism a localized mode is scattered by a second localized mode, yielding two band modes. In fact, it turns out that for the hydride impurity the second mechanism is the dominant one for determining the width. It is perhaps surprising that the width of the higher-frequency hydride localized mode is greater than the width of the lower-frequency D mode simply because the width of the D⁻ mode comes both from the cubic and quartic anharmonic processes whereas the width of the higher-frequency H⁻ localized mode is determined essentially by quartic anharmonic processes. But the reason for this is twofold; one is the fact that the scattering of the localized mode by a second localized mode into two band modes has a greater volume in the phase space of the band mode frequencies assigned to it; and second, the mean-squared amplitude of the lighter H⁻ impurities is greater than the mean-squared amplitude of the heavier D⁻ impurities, and this is also reflected in the width. The temperature dependence of the width varies essentially as $T^2$ for the H⁻ localized mode, for this is characteristic of quartic anharmonic processes, and it is between $T$ and $T^2$ for the D⁻ localized mode, because this is characteristic of the temperature dependence of the decay by cubic and quartic anharmonic processes.

A second effect of lattice anharmonicity is to allow overtones of the localized mode to be observed in the absorption spectrum of a crystal possessing the appropriate symmetry at the impurity site. In Fig. 15 are the

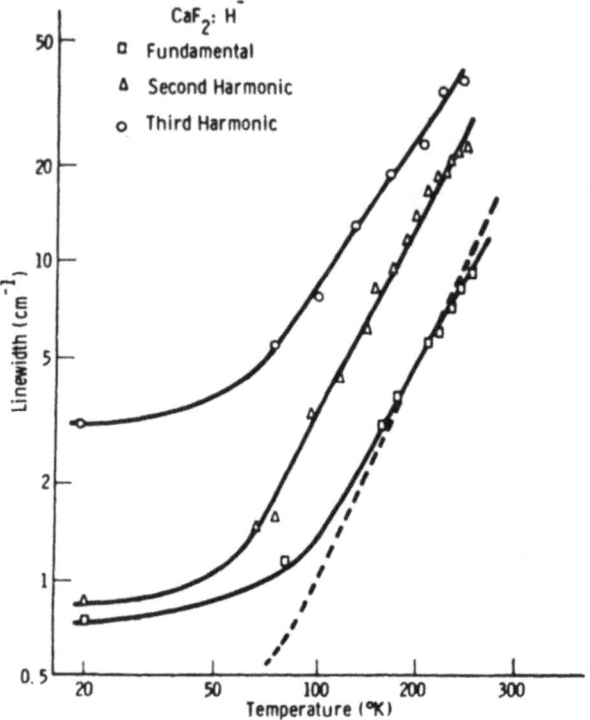

Figure 15

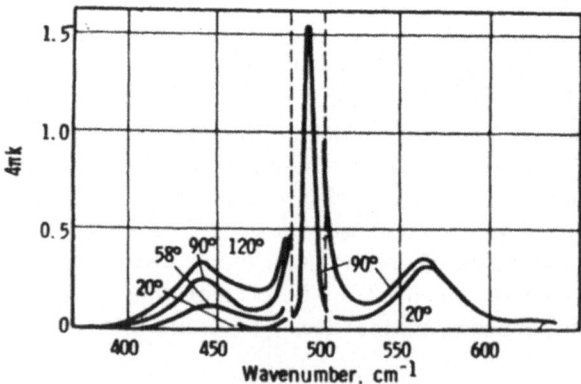

Figure 16

experimental results of Hayes, et al. ([26]) for the widths of the localized mode peaks associated with $H^-$ impurities in $CaF_2$. In the simplest approximation. if the localized mode is very highly localized in space, as is the case when the impurity is very light, the impurity atom can be approximated as a particle vibrating in an anharmonic potential well. For a three-dimensioned oscillator in an harmonic potential well, the ground state is nondegenerate, the first excited state is triply degenerate, the second excited state is sixfold degenerate, and so on; if we introduce anharmonic terms into the potential of the well, it is found that these degeneracies are partially split. More than that, the ground state and the various excited states now possess a certain admixture of the wavefunctions of the excited states. The usual harmonic oscillator selection rules which state that the dipole moment operator can have nonzero matrix elements only between states which differ by only one quantum number are now relaxed, and it becomes possible to have nonzero matrix elements of the dipole moment operator between states which differ by more than one quantum number. These matrix elements of course are proportional to the anharmonic coefficients in the potential energy. Transitions can now occur between the ground state and the second harmonic of the localized mode, and between the ground state and the third harmonics.

Figure 15 shows the temperature dependence of the width of the fundamental, second, and third harmonics of the $H^-$ ion in $CaF_2$. I should remark that in the case of $CaF_2$ one second harmonic is allowed by symmetry and it is observed, while two third harmonics are allowed by symmetry and both are observed. The temperature dependence of the widths is like $T^2$ at high temperatures, as would be expected from the fact that the widths of these modes are governed by the quartic anharmonic terms in the crystal potential energy.

Another effect of the anharmonic terms in the crystal potential energy is the occurence of satellites or sidebands to the localized mode peaks. In Fig. 16 are shown the results of Fritz for KCl containing $H^-$ impurities ([27]). The scale is reduced by a factor of 10 in the center of the figure. One can see sidebands on either of the peaks whose intensity varies rapidly with temperature. Experimental results of Timusk ([28]) for the absorption spectrum of $H^-$ in KBr are shown in Fig. 17. The figure shows the higher frequency sideband. The frequency spectrum of KBr has a gap between the acoustic and optical branches which is reflected in the structure of the sidebands. This fact lends support to the idea that the sideband is a reflection of the frequency spectrum of the unperturbed KBr crystal. (The upper curve is the same as the lower curve but multiplied by a factor of 8).

The theory of the sideband structure has been worked out by Elliott and Sennett ([25]), Timusk and Klein, ([28]) by Nguyen Xuan Xinh ([29]), and by Page and Dick ([30]). There are three possible mechanisms which can give

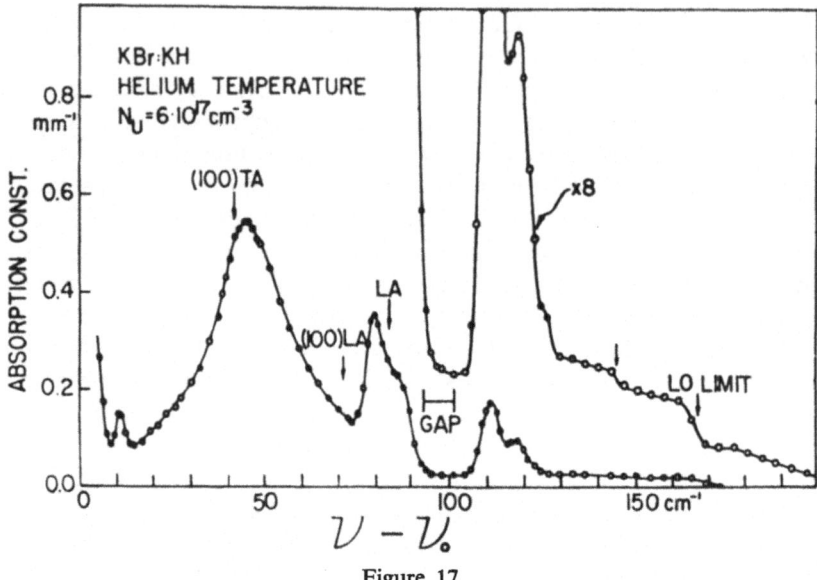

Figure 17

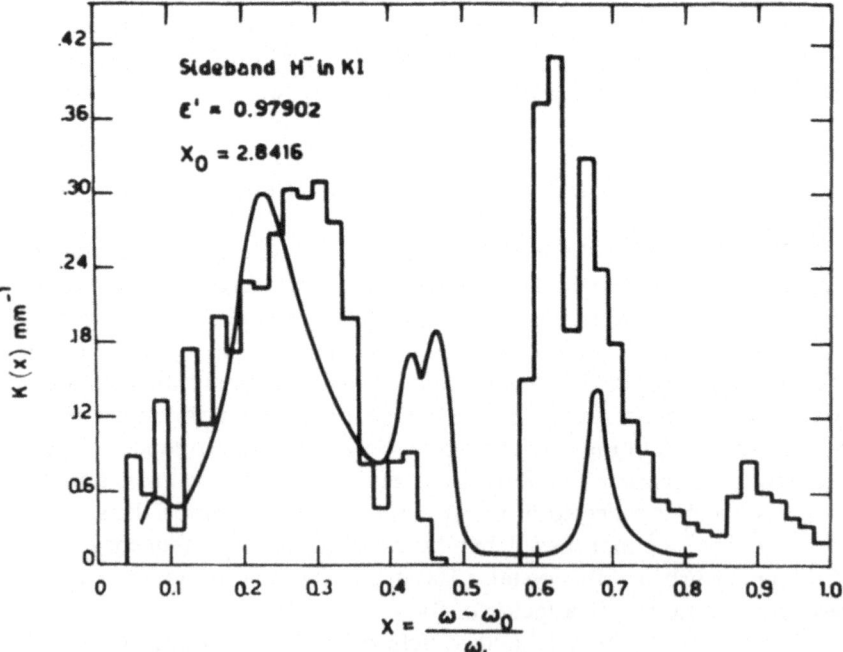

Figure 18

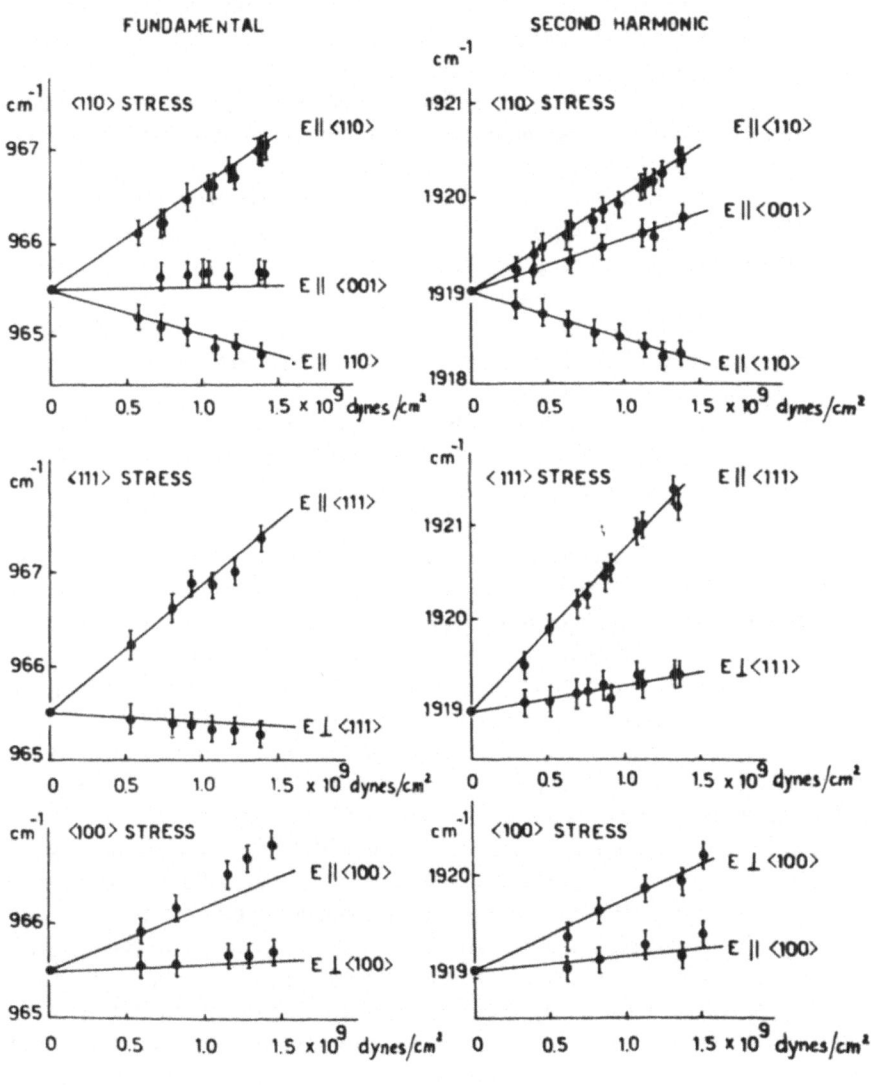

Figure 19

rise to the sideband structure. One is cubic anharmonicity, by which the localized mode decays into a second localized mode and one band mode. A second mechanism is the second-order dipole moment, i.e., those terms in the expansion of the crystal dipole moment in powers of the atomic displacements which are quadratic in these displacements. Here again it is possible for a localized mode to decay into a second localized mode and a band mode. The third principal mechanism contributing to the formation of sidebands is the constructive interference between the first two processes. All of these mechanisms can give rise to the sideband structure. However, it now appears form the theoretical calculations that the cubic anharmonic mechanism is the most important. The theoretical results of Nguyen Xuan Xinh ($^{29}$) for the $H^-$ sideband in KI are shown in Fig. 18. The fact that the quantitative agreement is only fair probably is a reflection of the fact that it is not easy to estimate the magnitude of the coefficients in the second-order dipole moment and the cubic anharmonic force constants. The gap in the frequency spectrum of the host crystal also appears in the frequency dependence of the sideband absorption.

Another effect, a so-called morphic effect, which can be used for studying localized modes in crystal is shown in Fig. 19. This figure shows the results obtained by Hayes, Mac Donald, and Elliott ($^{31}$) in an experiment in which an uniaxial stress has been applied to a crystal of $CaF_2$ containing $H^-$ impurities. The triple degeneracy of the localized modes associated with the $H^-$ impurities is split by the application of such an uniaxial stress. The symmetry of the crystal is reduced and, depending on the direction of the application of the stress with respect to the crystal axes, one finds different splittings, either into a singlet and a doublet or a complete splitting into three singlets. Both the fundamental and the harmonics of the localized modes will have their frequencies split by the application of the stress. The results for the fundamental and for the second harmonic are shown in Fig. 19. The stresses used in these measurements are of the order of $10^9$ dynes/cm$^2$.

In Fig. 20 is shown a theoretical curve which gives the frequencies of resonant and localized modes associated with $C^{12}$ and germanium impurities in silicon, assuming that these impurities are simple mass defects. The intersection of the hyperbolas with the curve gives the frequencies of the resonant and localized modes. The disagreement between the calculated and the measured frequencies for the localized mode due to $C^{12}$ in silicon is only of the order of 4%. In the case of alkali halides, the agreement is much poorer if an $H^-$ ion is approximated by an isotopic impurity (the theoretical frequencies are $\sim 50\%$ too high), and the changes in force constants associated with the impurity has to be taken into account. According to this figure there are resonant modes and several potential resonant modes associated with the heavy germanium impurity in silicon and with the light carbon impurity.

This figure is shown to preface the next topic which I wish to discuss. I would like now to say something about the Raman effect in crystals containing impurities. The Raman effect is the inelastic scattering of light by crystals. Qualitatively the theory has some points of similarity with the theory of the absorption of light by lattice vibrations. There are similar selection rules. Since energy has to be conserved, the frequency of the incident light must equal the net frequency change of the phonons that scatter the light. The scattering mechanism here is the fluctuations of the electronic polarizability of the crystal due to the atomic vibrations. In perfect crystals there is also a $\mathbf{k} = 0$ selection rule if we assume that the wavevector of the incident light is essentially zero. There is also a group theoretical selection rule which says that only those modes which transform according to the second-rank tensor irreducible representations of the point group of the crystal will scatter light. Again, for perfect crystals, these selection rules lead to rather uninteresting results for the one-phonon scattering processes. The cross section for the scattering of light by one-phonon processes in crystals

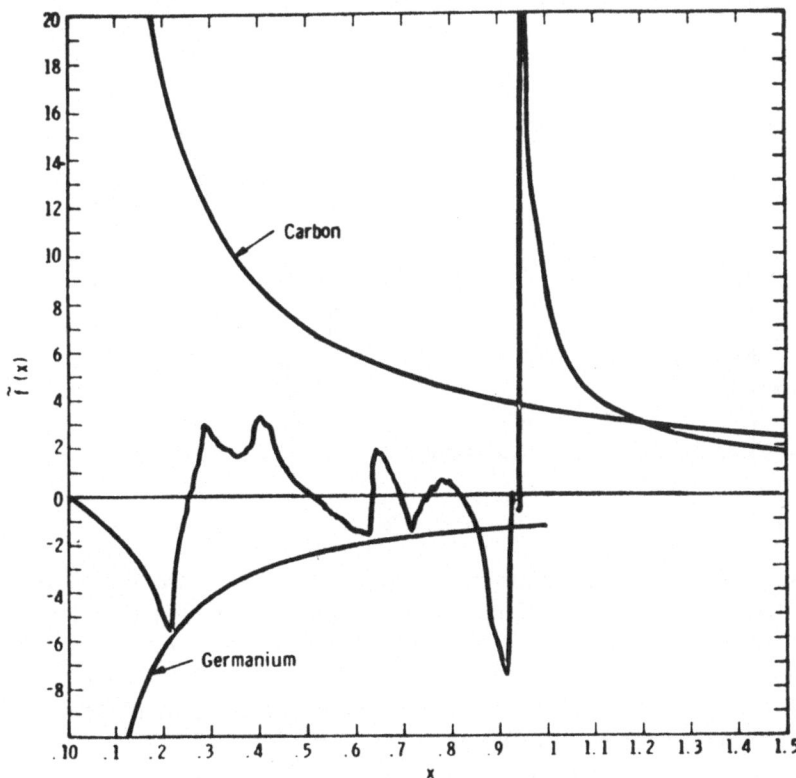

Figure 20

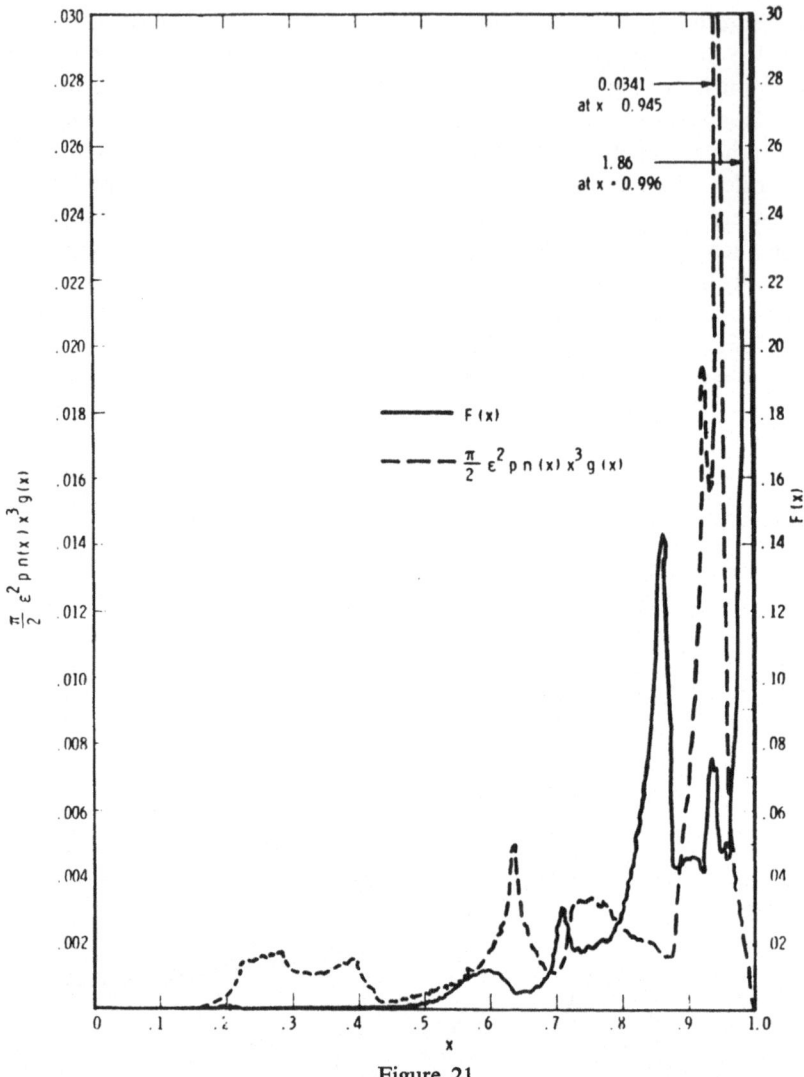

Figure 21

consists only of $\delta$-function peaks at frequencies which are shifted up or down from the frequency of the incident light by the frequencies of the optical modes of infinite wavelength. In the case of diamond, for example, one can see a high-frequency line (anti-Stokes) and a low-frequency line (Stokes). This does not give much interesting information about the dynamical properties of the perfect crystal.

However, just as in the case of infrared absorption, if we introduce defects into the crystal, two different kinds of things will happen. For example, in the case of crystals of the diamond structure which, in the absence of defects, possess a first-order Raman spectrum, the presence of defects can smear out this line spectrum into a continuous spectrum which is nonzero over the frequency range allowed the vibrations of the perfect crystal and which again, to some extent, reflects the frequency spectrum of the perfect host crystal. Second, in crystals such as those of the rocksalt structure, which in the absence of impurities possess no first-order Raman spectrum, one can induce a nonzero continuous first-order Raman spectrum by breaking down the translational symmetry of the crystal by the introduction of defects.

In Fig. 21 is shown a result of a calculation by Nguyen Xuan Xinh

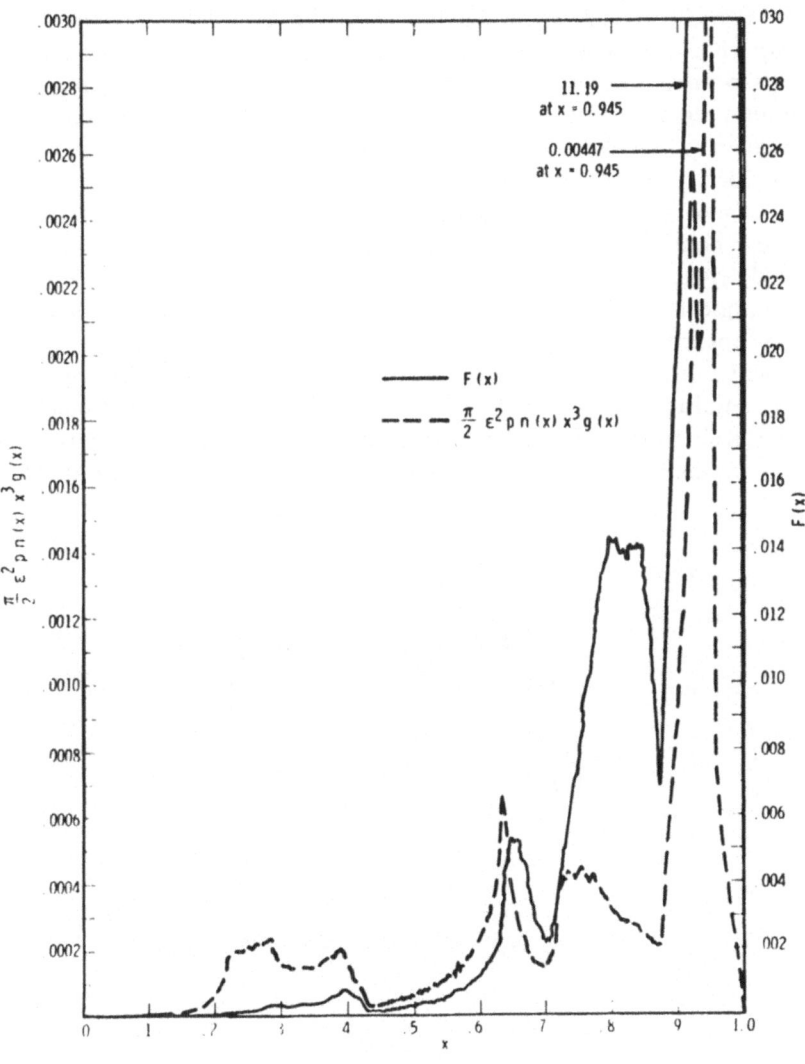

Figure 22

([32]) for the first-order Raman spectrum to be expected in silicon into which heavy germanium impurities are introduced. From Fig. 20, one saw that there are resonance modes at $x = 0.85$ and at $x = 0.95$, and what I call near-resonance modes at the frequencies $x = 0.22$, 0.63, and 0.71, so that the Raman spectrum one expects to obtain does not reflect the frequency spectrum of the crystal, which is represented in this figure weighted by an appropriate temperature-dependent coefficient to make the comparison more meaningful. In general, if one has resonance modes then these distort the scattering spectrum sufficiently so that any details of the frequency spectrum are lost.

In Fig. 22 is shown the Raman spectrum of silicon containing light $C^{12}$ impurities ([32]). There should be a peak at high frequencies, which is off-scale, associated with the localized modes, which are Raman-active. In the continuum region, apart from some distortions, the Raman spectrum follows the frequency spectrum, at least as far as the singularities are concerned. In Fig. 23 is displayed a theoretical curve due to Nguyen Xuan Xinh ([32]) of the Raman spectrum to be expected from $H^-$ impurities in KCl. It shows a nonzero continuous Raman spectrum in the presence of impurities which is completely absent in the absence of impurities.

I should like to point out that a Raman spectrum is somewhat richer than an absorption spectrum because each of the different modes of symmetry which are Raman-active gives rise to its own characteristic spectrum. Experimental results of Stekhanov and Eliashberg ([33]) for the Raman spectrum of KCl containing $Li^+$ impurities are shown in Fig. 24. Most of the structure in the spectrum is noise. The upper curve was measured at 300° K,

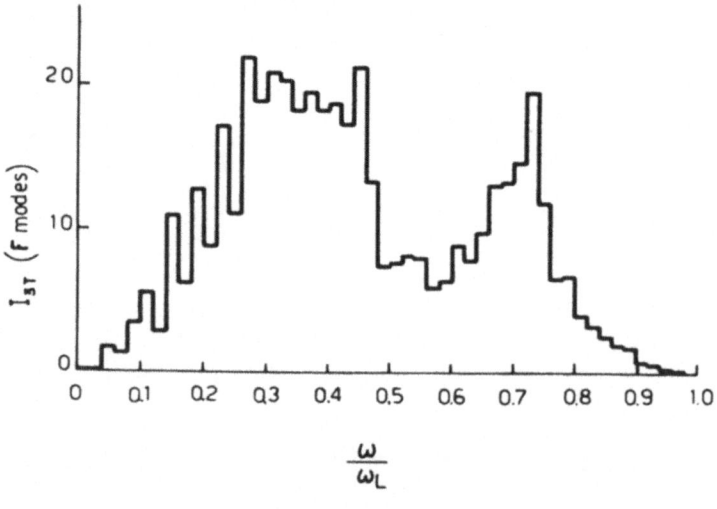

Figure 23

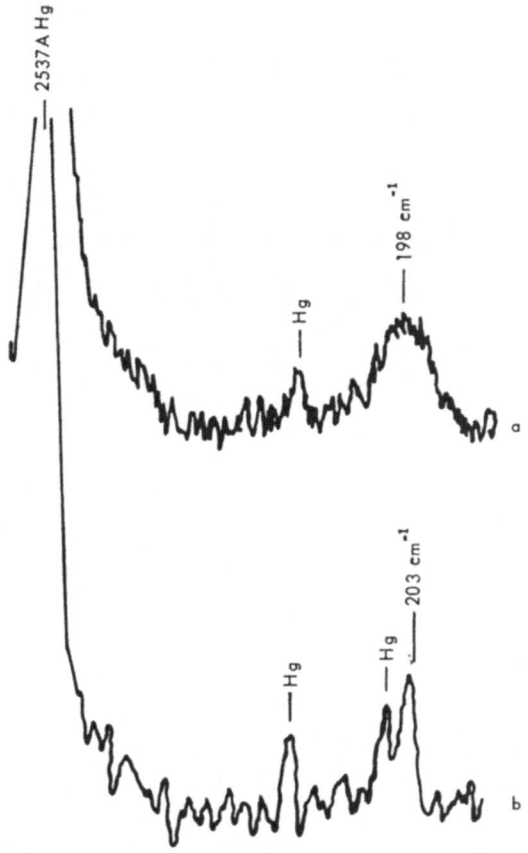

Figure 24

and the lower curve at 77° K. The salient features of the spectrum are the two peaks at 130 and 200 cm⁻¹ which are better resolved at low temperatures.

In the report of the work of Stekhanov and Eliashberg, no indication has been given of the polarization of the light and the directions of the incident and scattered beams with respect to the axes of the crystal, so that it is very difficult to try to compare theory with experiment.

Several theoretical Raman spectra ($^{32}$) due to a positive impurity ion in KCl with different values of the central force constant are shown in Fig. 25. The top curve is that associated with the fully symmetric $A_{1g}$ modes due to the impurity. The two other curves represent the contributions to the Raman spectrum from the doubly degenerate $E_g$ and the triply degenerate $F_{2g}$ modes. Although there is a peak present at about 60% of the maximum frequency which would correspond to the 130 cm⁻¹ peak observed, there is

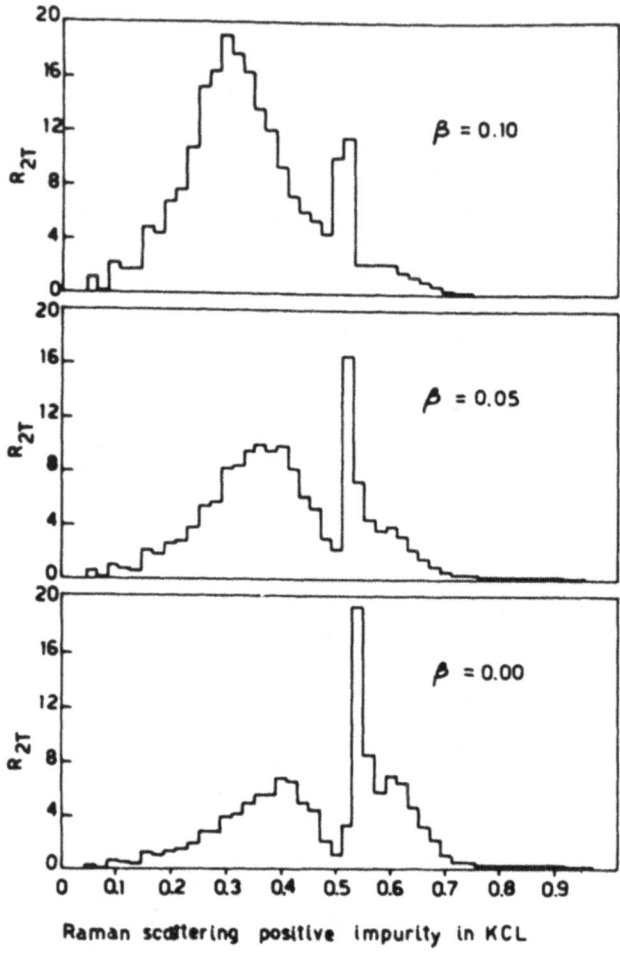

Figure 25

no indication of a second peak at 90% of the maximum frequency of the crystal which would correspond to the 200 cm⁻¹ line observed by Stekhanov and Eliashberg, so that a theoretical understanding of the experimental results of Stekhanov and Eliashberg is still lacking.

## ACKNOWLEDGMENTS

I would like to thank Dr. Nguyen Xuan Xinh for his help in the preparation of the manuscript of this lecture.

# REFERENCES

1. P.G. Dawber and R.J. Elliott, *Proc. Phys. Soc.* **81**: 453 (1963).
2. M. Lax, and E. Burstein, *Phys. Rev.* **97**: 309 (1955).
3. R.C. Newman and J.B. Willis, *J. Phys. Chem. Solids* **26**: 373 (1965).
4. G.O. Jones and J.H. Woodfine, *Proceedings IX International Conference on Low Temperature Physics*, Columbus, Ohio, 1964, Plenum Press, New York, 1965, p. 1089.
5. G. Schaefer, *J. Phys. Chem. Solids* **12**: 233 (1960).
6. A.J. Sievers, A.A. Maradudin, and S.S. Jaswal, *Phys. Rev.* **138**: A272 (1965).
7. A.J. Sievers, *Phys. Rev. Letters* **13**: 320 (1964).
8. A.J. Sievers and S. Takeno, *Phys. Rev.* **140**: A1030 (1965).
9. A.J. Sievers and S. Takeno, *Phys. Rev. Letters* **15**: 1020 (1965).
10. This plot is due to A.J. Sievers and S. Takeno, Ref. 9, and is based on experimental results of Price and Wilkinson, Ref. 11.
11. W.C. Price and G.R. Wilkinson, Final Technical Report No. 2 (December, 1960) on U.S. Army Contract DA-91-591-EUC-1308 OI-4201-60(R & D 260), U.S. Army through the European Research Office.
12. D.N. Mirlin and I.I. Reshina, *Fiz. Tver. Tela* **6**: 945 (1964); *Soviet Physics-Solid State* **6**: 728 (1964).
13. B. Fritz, U. Gross, and D. Bäuerle, *Phys. Stat. Sol.* **11**: 231 (1965).
14. S.S. Mitra and R.S. Singh, *Phys. Rev. Letters* **16**: 694 (1966).
15. I.P. Ipatova, A.V. Subashiev, and A.A. Maradudin, *Proceedings International Conf. on Localized Excitations in Solids*, Edited by R.F. Wallis, Plenum Press, New York, 1968, p. 93.
16. M. Balkanski and W. Nazarewicz, *J. Phys. Chem. Solids* **25**: 437 (1964).
17. D.N. Mirlin and I.I. Reshina, *Fiz Tver. Tela* **8**: 152 (1966); *Soviet Physics-Solid State* **8**: 116 (1966).
18. J.F. Angress, S.P. Smith, and K.F. Renk, *Proceedings International Conference on Lattice Dynamics*, Copenhagen 1963, Pergamon Press, Inc. N.Y. 1965, p. 467.
19. C. Dolling, *Inelastic Scattering of Neutrons in Solids and Liquids*, Vol. II, (International Atomic Energy Agency, Vienna, 1963), p. 37.
20. K.F. Renk, *Phys. Letters* **14**: 281 (1965).
21. A.J. Sievers and C.D. Lytle, *Phys. Letters* **14**: 271 (1965).
22. C.D. Lytle, M.S. Thesis, Cornell University, September 1965. Available as Report #390 of the Materials Science Center, Cornell University.
23. M. Wagner, *Phys. Rev.* **131**: 2520 (1963); **133**: A750 (1964).
24. D.N. Mirlin and I.I. Reshina, *Fiz. Tver. Tela* **6**: 3078 (1964); *Soviet Physics-Solid State* **6**: 2454 (1964).
25. C.T. Sennett, Thesis Oxford University (1964), unpublished.
26. W. Hayes, H.F. MacDonald, G. Jones, C.T. Sennett, and R.J. Elliott, *Proc. Roy. Soc.* **289**: 1 (1965).
27. B. Fritz, in *Lattice Dynamics*, edited by R. F. Wallis, Pergamon Press, New York, 1965, p. 485.
28. T. Timusk and M.V. Klein, *Phys. Rev.* **141**: 664 (1964).
29. Nguyen Xuan Xinh, *Phys. Rev.* **163**: 896 (1967).
30. J.B. Page and B.G. Dick, *Phys. Rev.* **163**: 910 (1967).
31. W. Hayes, H.F. MacDonald, and R. J. Elliott, *Phys. Rev. Letters* **15**: 961 (1965).
32. Nguyen Xuan Xinh, Westinghouse Research Laboratories Scientific paper 65-9F5-442-P8 (1965).
33. A.I. Stekhanov and M.B. Eliashberg, *Fiz. Tver. Tela* **5**, 2985 (1963); *Soviet Physics-Solid State* **5**: 2185 (1964).

# Localized and Pseudolocalized Excitations in Solids, as a Scattering Problem*

G. F. Nardelli

*Institute of Physics, University of Milan*
*and*
*Gruppo Nazionale Struttura della Materia del Consiglio Nazionale Recerche*
*Milan, Italy*

## I. INTRODUCTION

In a perfect solid the invariance properties of the crystal Hamiltonian against the operations of the translational space group make it possible to attribute a wave character to its elementary excitations. Under the action of local perturbations which destroy the translationsl invariance, the elementary excitations lose such character however and become, in principle, more similar to the excitations of a large macromolecule. Nevertheless, if the range of the perturbation is extremely small as compared to the crystal size, it is still meaningful to retain the wave description and represent the action of the perturbation in the framework of the scattering formalism. The aim of this lecture is to present from a unifying point of view how the theory approaches the description of the elementary excitations in the presence of local perturbations in two simple, but physically significant, examples: phonons and ferromagnetic magnons.

The Hamiltonians are as follows:

$$\mathscr{H} = \sum_x \frac{1}{2M(x)} \mathbf{p}^2(x) + \frac{1}{2}\sum_x\sum_{x'} \mathbf{u}(x)\cdot\mathbf{\Phi}(x, x')\cdot\mathbf{u}(x') \qquad (1)$$

for phonons ([1]), and

---

*This research has been sponsored in part by EOAR under Grant N. 65–05 with the European Office of Aerospace Research-U.S. Air Force.

$$\mathcal{H} = -\sum_x \sum_{x'} J(x, x') S(x) \cdot S(x') - 2\mu_0 H_0 \sum_x S_z(x) \tag{2}$$

for magnons ([2]).

In the phonon Hamiltonian, $p(x)$ is the momentum operator and $u(x)$ the displacement operator of the atom at site $x$; $\Phi$ is the fore-constant tensor matrix and $M$ the mass matrix.

In the magnon Hamiltonian, $S(x)$ is the spin operator, in dimensionless units, for the atom at site $x$, $J(x, x')$ is the exchange integral ($J > 0$), $\mu_0 = (g/2)\mu_B$ is the magnetic moment ($g$ gyromagnetic factor, $\mu_B$ Bohr magneton) and $H_0$ denotes the intensity of an externally applied magnetic field. It is usually assumed that $J(x, x')$ has nonvanishing values only when $x$ and $x'$ are nearest neighbors. In both expressions (1) and (2) $x$ denotes the discrete variable $x_{l\kappa} = x_l + x_\kappa$, where $x_l$ is the Bravais vector and $x_\kappa$ is the site vector in the primitive cell; $p(x)$, $u(x)$, and $S(x)$ are then recognized to be field operators in the discrete field of the variable $x$.

Before entering the theory of local perturbations, it seems useful to analyze the structure of these two Hamiltonians in order to reduce them to a common form; this form is the basic Hamiltonian for both phonons and magnons. The advantage of doing this is twofold: on one hand, the theory of local perturbations can be developed formally in the same way without making any reference to the particular system we are concerned with; on the other hand, the similarities between the properties of the imperfect phonon and magnon fields can be appreciated in their full extent.

The common form we have mentioned for our Hamiltonians is

$$\mathcal{H} = E_0 + \sum_\lambda \hbar\omega_\lambda b_\lambda^\dagger b_\lambda \tag{3}$$

where $E_0$ is the ground-state energy, $b_\lambda$, $b_\lambda^\dagger$ are Bose operators satisfying the well-known commutation relations

$$[b_\lambda, b_{\lambda'}^\dagger] = \delta_{\lambda, \lambda'} \tag{4}$$

and $\hbar\omega_\lambda$ is the single-excitation energy; $\lambda$ labels the degree of freedom of the system, and $\delta_{\lambda, \lambda'}$ is the Kronecker symbol. Hereafter we assume that the system consists of a large but finite crystal.

Consider first the phonon Hamiltonian and put

$$u(x) = (1/M(x))^{1/2} Q(x) \quad \text{and} \quad p(x) = M^{1/2}(x) P(x) \tag{5}$$

Hamiltonian (1) reads

$$\mathcal{H} = \tfrac{1}{2} \sum_x P^2(x) + \tfrac{1}{2} \sum_x \sum_{x'} Q(x) \cdot L(x, x') \cdot Q(x') \tag{6}$$

where

$$L(x, x') = M^{-1/2}(x)\Phi(x, x')M^{-1/2}(x') \tag{7}$$

is what is called the dynamical matrix L.

Thus, the principal-axis transformation of the matrix L, once introduced the phonon variables $b_\lambda$ and $b_\lambda^\dagger$ related to the Q's and P's by

$$Q(x) = \left(\frac{\hbar}{2}\right)^{1/2} \sum_\lambda \omega_\lambda^{-1/2} \varphi_\lambda(x)\{b_\lambda + b_{-\lambda}^\dagger\} \tag{8a}$$

$$P(x) = -i\left(\frac{\hbar}{2}\right)^{1/2} \sum_\lambda \omega_\lambda^{1/2} \varphi_\lambda(x)\{b_\lambda - b_{-\lambda}^\dagger\} \tag{8b}$$

brings Hamiltonian (6) into the diagonal form (3).

The phonon variables $b_\lambda$ and $b_\lambda^\dagger$ operate on the quantum states of harmonic oscillators in the occupation-number representation, and are recognized to be the destruction and creation operators, respectively. The $c$ numbers $\varphi_\lambda(x)$ which appear in (8a) and (8b) are the coefficients of the principal-axis transformation; they are nothing else than the representatives of the eigenvectors of the dynamical matrix, i.e., the normal modes of the crystal lattice. In matrix notation the eigenvector equation reads

$$L|\varphi_\lambda) = \omega_\lambda^2|\varphi_\lambda) \tag{9}$$

with the normalization condition

$$(\varphi_\lambda|\varphi_{\lambda'}) = \delta_{\lambda,\lambda'} \tag{10}$$

As usual, $(\ldots|\ldots)$ denotes a scalar product.

The circular frequency $\omega_\lambda$ which appears in (3), (8a), and (8b) is the positive square root of the $\lambda$th eigenvalue of the dynamical matrix. In writing expressions (8a) and (8b) we have labeled by $-\lambda$ the complex conjugate of $\varphi_\lambda(x)$; since L is a real and symmetric matrix, $\varphi_\lambda(x)$ and $\varphi_{-\lambda}(x) \equiv \varphi_\lambda^*(x)$ are two linearly independent solutions of (9) for the same eigenvalue $\omega_\lambda^2$ except for the case in which $\varphi_\lambda(x)$ is real.

Consider now the magnon Hamiltonian. It can be written as

$$\mathcal{H} = -\sum_x\sum_{x'} J(x, x')\{S_z(x)S_z(x') + S_-(x')S_+(x)\} - 2\mu_0 H_0 \sum_x S_z(x) \tag{11}$$

where, as usual, $S_+$ and $S_-$ are related to the $x$ and $y$ components of S by

$$S_+(x) = S_x(x) + iS_y(x)$$
$$S_-(x) = S_x(x) - iS_y(x) \tag{12}$$

and $S_z(x)$ is the $z$ component of the spin operator at the lattice site x. $S_+(x)$, $S_-(x)$, and $S_z(x)$ can be expressed in terms of Bose operators $b(x)$ and $b^\dagger(x)$ satisfying the commutation relations

$$[b(\mathbf{x}), b^\dagger(\mathbf{x}')] = \delta_{\mathbf{x},\mathbf{x}'} \tag{13}$$

by means of the Holstein–Primakoff transformation [3]

$$S_+(\mathbf{x}) = (2S)^{1/2}\left(1 - \frac{b^\dagger(\mathbf{x})b(\mathbf{x})}{2S}\right)^{1/2} b(\mathbf{x}) \simeq (2S)^{1/2}b(\mathbf{x})$$

$$S_-(\mathbf{x}) = (2S)^{1/2}b^\dagger(\mathbf{x})\left(1 - \frac{b^\dagger(\mathbf{x})b(\mathbf{x})}{2S}\right)^{1/2} \simeq (2S)^{1/2}b^\dagger(\mathbf{x})$$

$$S_z(x) = S - b^\dagger(\mathbf{x})b(\mathbf{x}) \tag{14}$$

By retaining terms up to the second power in the Bose operators, the magnon Hamiltonian reads

$$\mathscr{H} = E_0 + \hbar \sum_{\mathbf{x}}\sum_{\mathbf{x}'}(\mathbf{x}\,|\,\Omega\,|\,\mathbf{x}')b^\dagger(\mathbf{x})b(\mathbf{x}') \tag{15}$$

where

$$E_0 = -S^2 \sum_{\mathbf{x}}\sum_{\mathbf{x}'} J(\mathbf{x}, \mathbf{x}') - 2N\mu_0 S \tag{16}$$

is the ground-state energy, and $\Omega$ is the matrix whose elements are given by

$$(\mathbf{x}\,|\,\Omega\,|\,\mathbf{x}') = \left(\frac{2}{\hbar}\right)\{S\sum_{\mathbf{x}''} J(\mathbf{x}, \mathbf{x}'') + \mu_0 H_0\}\delta_{\mathbf{x},\mathbf{x}'} - \left(\frac{2}{\hbar}\right)SJ(\mathbf{x}, \mathbf{x}') \tag{17}$$

For convenience we have considered here a system of $N$ equivalent spins. As for the previous case, the principal-axis transformation of the matrix $\Omega$ brings hamiltonian (15) into the diagonal form (3). The magnon variables $b_\lambda$ and $b_\lambda^\dagger$ are seen to be

$$b_\lambda \equiv \sum_{\mathbf{x}} \varphi_\lambda(\mathbf{x})b(\mathbf{x})$$

$$b_\lambda^\dagger \equiv \sum_{\mathbf{x}} \varphi_\lambda^*(\mathbf{x})b^\dagger(\mathbf{x}) \tag{18}$$

where, as before, $\varphi_\lambda(\mathbf{x})$ are the coefficients of the principal-axis transformation, i.e., the representative of the $\lambda$th eigenvector of the matrix $\Omega$; the magnon frequency $\omega_\lambda$ is the $\lambda$th eigenvalue of this matrix. It is worthwhile to note that the diagonal form (3) just represents an approximate expression (the so-called free-magnon approximation) for the Hamiltonian of an exchange-interacting system of spins. Indeed, in deducing (3) from (2) we have neglected the magnon–magnon interaction which reads

$$\mathscr{H}_{\text{int}} = \mathscr{H}_4 + \cdots$$

where

$$\mathscr{H}_4 = O(b^{\dagger 2} b^2)$$

At the same time, the diagonal form (3) represents the true Hamiltonian for a system of harmonically-interacting particles. The free-magnon approximation for a system of exchange interacting spins is then recognized to be the counterpart of the harmonic approximation for a system of interacting particles.

We may conclude that, by a convenient choice of the origin on the energy scale, the diagonal bilinear form in the Bose operators $b_\lambda$ and $b_\lambda^\dagger$

$$\mathscr{H} = \sum_\lambda \hbar \omega_\lambda b_\lambda^\dagger b_\lambda \tag{19}$$

can be considered as the basic Hamiltonian for both phonons and magnons. Furthermore, in the derivation of (19) form (1) or (2) no assumption has been made about the perfection (or imperfection) of the crystal lattice. In the present approach the possible local perturbations affect the matrix $\mathsf{L}$ (phonons) or the matrix $\Omega$ (magnons) and, consequently, both eigenvectors $\varphi_\lambda$ and eigenfrequencies $\omega_\lambda$; in spite of this, they leave formally unchanged the basic Harmiltonian (19). Note that we can define a matrix $\Omega$ also for phonons; we have

$$\Omega \equiv +\mathsf{L}^{1/2} \tag{20}$$

## II. SINGLE-EXCITATION STATE

Starting from the expression (19) it is possible to carry out the formal theory of localized and pseudolocalized excitations, either for a system of harmonically interacting particles or a system of exchange interacting spins, without making any explicit reference to the particular system we are concerned with. Let $|0\rangle$ denote the ground state of our system, i.e., the vacuum state for the elementary excitations. Since the set of Hilbert rectors $b_\lambda^\dagger |0\rangle$ constitutes an orthogonal and complete set in the one-excitation subspace, the quantum state $|\Psi_t\rangle$ which corresponds to a single excitation at time $t$ can be written as

$$|\Psi_t\rangle = \sum_\lambda \psi(\lambda; t) b_\lambda^\dagger |0\rangle \tag{21}$$

where $\psi(\lambda; t)$ are $c$ numbers.

The normalization condition

$$\langle \Psi_t | \Psi_t \rangle = 1 \tag{22}$$

implies that

$$\sum_\lambda \psi^*(\lambda; t)\psi(\lambda; t) = 1 \tag{23}$$

The equation of motion (Schroedinger equation) reads:

$$i\hbar\frac{\partial}{\partial t}|\Psi_t\rangle = \mathcal{H}|\Psi_t\rangle \tag{24}$$

From the Bose commutation relations (4) it then follows

$$i\frac{\partial}{\partial t}\psi(\lambda; t) = \omega_\lambda\psi(\lambda; t) \tag{25}$$

that is,

$$\psi(\lambda; t) = \psi(\lambda)e^{-i\omega_\lambda t} \tag{26}$$

where $\psi(\lambda)$ are free parameters which have to satisfy (23), and are at our disposal for specifying the structure of the one-excitation state.

In order to find the connection between the parameters $\psi(\lambda)$ entering the definition of the one-excitation state and the eigenvectors $|\varphi_\lambda\rangle$ of the matrix $\mathbf{L}$ or $\mathbf{\Omega}$, we introduce the linear vector space $\{|\lambda\rangle\}$ by the defining equations

$$\psi(\lambda) \equiv (\lambda|\psi)$$
$$(\lambda|\lambda') = \delta_{\lambda,\lambda'} \tag{27}$$

A one-to-one correspondence exists between the time-dependent vector $|\psi_t\rangle = \sum_\lambda |\lambda\rangle\psi(\lambda; t)$ in this linear vector space and the vector $|\Psi_t\rangle$ in the one-excitation Hilbert space; then $\psi(\lambda; t) \equiv (\lambda|\psi_t)$ can be considered to be the representative of the one-excitation state in the former space.

Let us now define a new basis $\{|x\rangle\}$ in this linear vector space by the equations

$$|\lambda) = \sum_x |x)(x|\lambda)$$
$$= \sum_x |x)\varphi_\lambda(x)$$
$$(x|x') = \delta_{x,x'} \tag{28}$$

where $\varphi_\lambda(x)$ are the coefficients of our principal-axis transformation, and we have identified $(x|\lambda)$ with $\varphi_\lambda(x)$.

In (28) $x$ labels the lattice coordinates as well as the Cartesian components of $\boldsymbol{\varphi}_\lambda(\mathbf{x})$. The inverse equation

$$|x) = \sum_\lambda |\lambda)\varphi_\lambda^*(x) \tag{29}$$

follows from the completeness relation

$$\sum_\lambda \varphi_\lambda(x)\varphi_\lambda^*(x') = \delta_{x,x'} \tag{30}$$

which holds among the coefficients of our principal-axis transformation. Thus,

$$|\psi_t) = \sum_x |x)(x|\psi_t)$$
$$= \sum_x |x)\psi(x;t) \tag{31}$$

and $\psi(x;t)$ can be considered as the Schroedinger representative of the one-excitation state, i.e., the representative in the $x$-space. The relation between the $x$-space and the $\lambda$-space representatives is easily seen to be

$$\psi(\lambda;t) = \sum_x \varphi_\lambda^*(x)\psi(x;t) \tag{32}$$

By taking into account (32) and the completeness relation (30), the evolution equation for the Schroedinger representative follows immediately from equation (25). It reads

$$i\frac{\partial}{\partial t}\psi(x;t) = \sum_{x'} (x|\Omega|x')\psi(x';t) \tag{33}$$

where we have put

$$\Omega = \sum_\lambda |\lambda)\omega_\lambda(\lambda| \tag{34}$$

$\Omega$ is the matrix we have defined in the previous section; in (33) $\Omega$ appears in the $x$-space representation, i.e.,

$$(x|\Omega|x') = \sum_\lambda \varphi_\lambda(x)\omega_\lambda \varphi_\lambda^*(x') \tag{35}$$

A second-order equation holds also; it is seen to be

$$\frac{\partial^2}{\partial t^2}\psi(x;t) = -\sum_{x'} (x|\Omega^2|x')\psi(x';t) \tag{36}$$

and is particularly useful in dealing with phonon problems, as $\Omega^2$ is recognized to be nothing but the dynamical matrix $L$. For magnons it is more useful to work with the first-order equation (33). Indeed the matrix $\Omega$ which is involved in this equation is in such a case the matrix which enters into the definition of the magnon Hamiltonian (15).

We are now in the position to give, on quantum-mechanical ground, a physical interpretation of the coefficients $\varphi_\lambda(x)$ of our principal-axis transformation. In order that the single-excitation state involve a well-defined frequency, say $\omega_{\lambda'}$, it turns out from (25) and (26) that we have to choose the particular solution

$$\psi(\lambda) = \delta(\lambda, \lambda') \equiv \begin{cases} 1, \text{ if } \lambda = \lambda' \\ 0, \text{ if } \lambda \neq \lambda' \end{cases} \tag{37}$$

and this solution is seen from (26) and (32) to correspond to choosing

$$\psi(x; t) = \varphi_\lambda(x)e^{-i\omega_{\lambda'}t} \tag{38}$$

for the Schroedinger representative. From the quantum-mechanical point of view the coefficients $\varphi_\lambda(x)$ of our principal-axis transformation are then recognized to be the Schroedinger representative (the time-oscillating phase factor apart) of the single-excitation state of frequency $\omega_\lambda$. In other words, $|\varphi_\lambda(x)|^2$ gives the probability of finding the elementary excitation at the lattice site $x_{l\kappa}$ along the Cartesian axis specified by $x$ [$\varphi_\lambda(x)$ is usually a vectorial quantity]. [Compare with the physical meaning of $\varphi_\lambda(x)$ in classical mechanics.]

## III. WAVE-FIELD OPERATOR

In the preceding section we have considered the quantum state for a single excitation in the Schroedinger picture. We proceed now with the description of the one-excitation state in the Heisenberg picture.

The wavefield operator is

$$\psi_{op}^\dagger(x; t) \equiv \sum_\lambda \varphi_\lambda^*(x)b_\lambda^\dagger(t) \tag{39}$$

where, as before, $\varphi_\lambda(x)$ are the coefficients of the principal-axis transformation of either $\Omega$ or $L$. The equation of motion reads

$$i\hbar\dot{\psi}_{op}^\dagger(x; t) = [\psi_{op}^\dagger(x; t), \mathcal{H}] \tag{40}$$

and the commutator at the right-hand member can be easily evaluated by keeping in mind the expression of the commutator of Bose operators $b_\lambda^\dagger$ with our basic Hamiltonian, i.e.,

$$[b_\lambda^\dagger(t), \mathcal{H}] \equiv i\hbar\dot{b}_\lambda^\dagger(t) = -\hbar\omega_\lambda b_\lambda^\dagger(t) \tag{41}$$

The Green function for the wavefield operator is

$$G(x, x'; t) = i\langle 0 | \vartheta_t\{\psi_{op}(x; t)\psi_{op}^\dagger(x'; 0)\} | 0 \rangle \tag{42}$$

where $|0\rangle$ denotes vacuum state, and $\vartheta_t$ is the time-ordering operator. The

Fourier transform of (42) can be evaluated in a straightforward manner, and reads

$$
\begin{aligned}
G(x, x'; \omega) &= \int_{-\infty}^{+\infty} dt \, e^{i\omega t} G(x, x'; t) \\
&= i \sum_\lambda \sum_\lambda \varphi_\lambda(x) \varphi_\lambda^*(x') \int_0^\infty dt \, e^{i\omega t} \langle 0 \, | \, b_\lambda(t) b_\lambda^\dagger(0) \, | \, 0 \rangle \\
&= 2\pi i \sum_\lambda \varphi_\lambda(x) \delta^{(+)}(\omega_\lambda - \omega) \varphi_\lambda^*(x') \\
&= \sum_\lambda \varphi_\lambda(x) \frac{1}{\omega_\lambda - \omega - i\eta} \varphi_\lambda^*(x') \\
&= \left( x \left| \frac{1}{\Omega - \omega - i\eta} \right| x' \right) = \left( x \left| \frac{1}{L^{1/2} - \omega - i\eta} \right| x' \right) \quad (43)
\end{aligned}
$$

Use is made of the integral representation of the $\delta^{(+)}$ function, i.e.,

$$
\delta^{(+)}(\omega) = \left( \frac{1}{2\pi} \right) \int_0^\infty dt \, e^{-i\omega t} = \left( \frac{1}{2\pi i} \right) \frac{1}{\omega - i\eta} \quad (44)
$$

where $\eta$ denotes a positive infinitesimal quantity. The last two expressions give the relation between the Fourier transform of the Green function for the Green function for the wavefield operator and the matrix $\Omega$, or L which enters the definition of our physical system.

Consider the matrix $(\Omega - \mathfrak{z})^{-1} = (L^{1/2} - \mathfrak{z})^{-1}$ in the complex-frequency plane $\mathfrak{z} = \omega + i\eta$. $G(x, x'; \omega)$ is seen to be the limit of the $x$-space representation of this matrix at the point $\omega$ of the real axis, coming from the upper half plane. In phonon problems is more useful to work in the complex-squared-frequency plane $z = \omega^2 + i\epsilon$, since the dynamical matrix deals with the squares of the excitation frequencies. The Green function $\mathscr{G}(x, x'; \omega^2)$ which is appropriate for this plane can be introduced by considering the displacement-field Green function

$$
G_{uu}(x, x'; t) \equiv \frac{i}{\hbar} \langle 0 \, | \, \vartheta_t \{ u_{op}(x, t) u_{op}(x', 0) \} \, | \, 0 \rangle \quad (45)
$$

where

$$
u_{op}(x, t) = \left[ \frac{\hbar}{2M(x)} \right]^{1/2} \sum_\lambda \omega_\lambda^{-1/2} \varphi_\lambda(x) \{ b_\lambda(t) + b_{-\lambda}^\dagger(t) \} \quad (46)
$$

is the displacement-field operator (compare with expressions (5) and (8)). Following the same procedure as in the derivation of (43) yields

$$
\begin{aligned}
G_{uu}(x, x'; \omega) &= \left( x \, | \, M^{-1/2} \frac{1}{L - \omega^2 - i\epsilon} M^{-1/2} \, | \, x' \right) \\
&\equiv [M(x) M(x')]^{-1/2} \mathscr{G}(x, x'; \omega^2) \quad (47)
\end{aligned}
$$

The last equality defines $\mathscr{G}(x, x'; \omega^2)$.

It would be easy to show that $\mathscr{G}(x, x'; \omega^2)$ is nothing but the Green function for the classical equation of motion of the displacements $\mathbf{u}(x)$ (see Ludwig's lectures). We emphasize that, as a function of $\mathfrak{z} = \omega + i\eta$, $(\mathsf{L} - \mathfrak{z}^2)^{-1}$ displays on the positive-frequency axis the same singularities as $(\mathsf{L}^{1/2} - \mathfrak{z})^{-1}$, so that $\mathscr{G}(x, x'; \omega^2)$ can be used just as well as $G(x, x'; \omega)$ in looking for the excitation frequencies of our system.


## IV. LOCALIZED AND PSEUDOLOCALIZED EXCITATIONS: T—MATRIX APPROACH

We consider now the effect of a local perturbation. The matrix $\Omega = \mathsf{L}^{1/2}$ splits as

$$\Omega = \Omega_0 + \delta\Omega \tag{48}$$

where $\Omega_0$ refers to the perfect lattice and is here considered to specify the unperturbed system.

We denote by $\phi_q(x) = (x \,|\, \phi_q)$ the Schroedinger representative of the unperturbed one-excitation state of frequency $\omega_q$. It satisfies the stationary equation

$$\Omega_0 \,|\, \phi_q) = \omega_q \,|\, \phi_q) \tag{49}$$

together with the normalization condition

$$(\phi_q \,|\, \phi_{q'}) = \delta_{q,q'} \tag{50}$$

$\phi_q(x)$ has the well-known plane-wave structure

$$\phi_q(\mathbf{x}) = \frac{1}{N^{1/2}} \epsilon_q(x_\kappa) e^{i\mathbf{q} \cdot \mathbf{x}_{l\kappa}} \tag{51}$$

where label $q \equiv (\mathbf{q}, s)$ corresponds to a wavevector $\mathbf{q}$ inside the first Brillouin zone (B.z.) and eventually to a branch index $s$; $\epsilon_q(x_\kappa)$ are the polarization vectors and $N$ is the number of primitive cells comprising the lattice.

The $T$-matrix can be introduced starting from the stationary equation $(\Omega_0 + \delta\Omega) \,|\, \psi) = \omega \,|\, \psi)$, which can be written as

$$|\psi) = |\phi_g) + \frac{1}{\omega_g \pm i\eta - \Omega_0} \delta\Omega \,|\, \psi) \tag{52}$$

Equation (52) can be solved formally, and we obtain

$$|\psi_g^{(\pm)}\rangle = \left[\mathbf{1} - \frac{1}{\omega_g \pm i\eta - \Omega_0}\delta\Omega\right]^{-1}|\phi_g\rangle \tag{53a}$$

$$= |\phi_g\rangle + \frac{1}{\omega_g \pm i\eta - \Omega_0}\mathsf{T}\,|\phi_g\rangle \tag{53b}$$

$$= |\phi_g\rangle + |\sigma_g^{(\pm)}\rangle \tag{53c}$$

where $|\phi_g\rangle$ denotes a symmetrized lattice wave, i.e., a system of plane waves which transforms according to the irreducible representations of the point group of the perturbation $\delta\Omega$. Label $g$ splits as $g \equiv (\mathbf{q}, s, \Gamma)$, where, for cubic groups, $\mathbf{q}$ is a wavevector inside $(\frac{1}{48})$-th of the B.z., and $\Gamma$ labels the irreducible representations of the point group. The second equally defines our $T$-matrix; it turns out that $\mathsf{T}$ satisfies the equation

$$\mathsf{T} = \delta\Omega + \delta\Omega\frac{1}{\mathfrak{z} - \Omega_0}\mathsf{T} \tag{54}$$

which is particularly useful for the practical evaluation of $\mathsf{T}$. An equivalent definition of $\mathsf{T}$ in the $g$-space is the following:

$$T_{g'g} = [\phi_{g'}\,|\,\delta\Omega\,|\,\psi_g^{(+)}] = (\psi_g^{(-)}\,|\,\delta\Omega\,|\,\phi_g) \tag{55}$$

It appears from (53b) that the matrix $\mathsf{T}$ enables us to find the stationary solutions of the perturbed equation of motion (33), once the unperturbed problem has been solved completely.

Note that our stationary solutions, as given by (53), do not constitute, in general, an orthogonal nor a normalized set. Furthermore, we have to decide about the sign in front of the infinitesimal quantity $\eta$ which appears in the unperturbed Green-function matrix. The question can be answered by considering the Fourier transform of the time-dependent state vector; it is an easy matter to verify ([5]) that, in the adiabatic hypothesis, the small positive or negative imaginary addition to the frequency selects, automatically, outgoing or incoming diffused waves $|\sigma_t\rangle$.

Hereafter we are interested in solutions of the $|\psi^{(+)}\rangle$ type, so we drop the superscript $(+)$, if unnecessary, and we choose the sign $+$ in front of the infinitesimal imaginary frequency $i\eta$.

The expressions in (53) apply equally well to phonons and magnons. However, in phonon problems the matrix which enters the dynamical description of the system most naturally is the dynamical matrix $\mathsf{L}$ and not its positive square root $\Omega = +\mathsf{L}^{1/2}$; then, for practical purposes, the use of (53b) and (54) is made difficult by the evaluation of $\Omega_0$ and $\delta\Omega$. It is therefore useful to develop a $T$—matrix formalism starting from the second-order evolution equation (36). We cannot apply the general results ([5]) of the scattering theory straightforwardly, since equation (36) involves the second-order time derivatives. We overcome this formal difficulty by the same procedure one uses in giving the relativistic extension of the Klein–Gordon equation ([6]). We start from

$$\frac{\partial^2}{\partial t^2}|\psi_t) = -\mathsf{L}\,|\psi_t) \tag{56}$$

and we put

$$i\frac{\partial}{\partial t}|\psi_t) = \Omega\,|\chi_t) \tag{57}$$

Thus the evolution equation reads

$$i\frac{\partial}{\partial t}|\psi_t) = \Omega\,|\chi_t) \tag{58a}$$

$$i\frac{\partial}{\partial t}|\chi_t) = \Omega\,|\psi_t) \tag{58b}$$

Consider the two-component vector

$$\hat{\varphi}(x, t) \equiv \begin{pmatrix} \psi(x, t) \\ \chi(x, t) \end{pmatrix} \tag{59a}$$

In our linear vector space we will write it as

$$|\hat{\varphi}_t) \equiv |\begin{pmatrix} \psi_t \\ \chi_t \end{pmatrix}) \tag{59b}$$

The evolution equations (58) can be written as

$$i\frac{\partial}{\partial t}|\hat{\varphi}_t) = \mathscr{L}\,|\hat{\varphi}_t) \tag{60}$$

where we have denoted by $\mathscr{L}$ the $2N \times 2N$ matrix

$$\mathscr{L} \equiv \begin{pmatrix} 0 & \Omega \\ \Omega & 0 \end{pmatrix} \tag{61}$$

$N$ denotes now the number of degrees of freedom of our system. In the $2 \times 2$ subspace the matrix $\mathscr{L}$ can be diagnoalized; we obtain

$$\mathscr{L}_d = \begin{pmatrix} \Omega & 0 \\ 0 & -\Omega \end{pmatrix} \tag{62}$$

and the eigenvectors can be divided into two groups, i.e.,

$$|\hat{\varphi}_{(+)}\rangle = \begin{pmatrix} 1 \\ 0 \end{pmatrix} |\psi_{(+)}\rangle \tag{63a}$$

and

$$|\hat{\varphi}_{(-)}\rangle = \begin{pmatrix} 0 \\ 1 \end{pmatrix} |\psi_{(-)}\rangle \tag{63b}$$

according to whether their frequency falls in the negative or positive frequency part of the spectrum of the matrix $\mathscr{L}$. Indeed, the stationary solutions of (60) have to satisfy the equations

$$\Omega |\psi_{(+)}\rangle = \omega_{(+)} |\psi_{(+)}\rangle \tag{64a}$$

and

$$-\Omega |\psi_{(-)}\rangle = \omega_{(-)} |\psi_{(-)}\rangle \tag{64b}$$

for positive $\omega_{(+)} = \omega_\lambda$ and negative $\omega_{(-)} = -\omega_\lambda$ frequencies, respectively. The positive-frequency solution $|\psi_{(+)}\rangle$ is seen to satisfy the same equation as the first-order equation (33). Note that it is always possible to choose $|\psi_{(+)}\rangle = |\psi_\lambda\rangle = |\psi_{(-)}\rangle$.

Now we split $\Omega$ into the unperturbed part $\Omega_0$ and perturbation $\delta\Omega$. Correspondingly we have

$$\mathscr{L} = \mathscr{L}_0 + \delta\mathscr{L} \tag{65a}$$

where

$$\mathscr{L}_0 \equiv \begin{pmatrix} 0 & \Omega_0 \\ \Omega_0 & 0 \end{pmatrix} \tag{65b}$$

$$\delta\mathscr{L} \equiv \begin{pmatrix} 0 & \delta\Omega \\ \delta\Omega & 0 \end{pmatrix} \tag{65c}$$

The matrix $L$ splits analogously into $L_0$ and $\delta L$, where

$$L_0 = \Omega_0^2$$

and

$$\delta L = \Omega_0 \delta\Omega + \delta\Omega \Omega_0 + \delta\Omega^2 \tag{66}$$

In view of (64a) it turns out that the positive-frequency stationary solution of (60) has to satisfy the equation

$$|\psi\rangle = |\phi_g\rangle + \frac{1}{\omega^2 + i\epsilon - L_0}\delta L\,|\psi\rangle \tag{67}$$

which is equivalent to (52) in terms of $L_0$ and $\delta L$. In writing (67) we have dropped the subscript $(+)$ on $\psi$.

Starting from (67) a new matrix $\mathcal{T}$ can be introduced, exactly in the same way as for $T$. Indeed (67) can be written as

$$|\psi_g^{(\pm)}\rangle = |\phi_g\rangle + \frac{1}{\omega_g^2 \pm i\epsilon - L_0}\mathcal{T}\,|\phi_g\rangle \tag{68a}$$

$$= |\phi_g\rangle + |\sigma_g^{(\pm)}\rangle \tag{68b}$$

where, in the $g$-space representation, $\mathcal{T}$ is defined by

$$\mathcal{T}_{g'g} \equiv (\phi_{g'}\,|\,\delta L\,|\,\psi_g^{(+)}) = (\psi_{g'}^{(-)}\,|\,\delta L\,|\,\phi_g) \tag{69}$$

and satisfies the equation

$$\mathcal{T} = \delta L + \delta L \frac{1}{z - L_0}\mathcal{T} \tag{70}$$

where $z = \omega^2 + i\epsilon$ is the complex squared frequency. Equations (68) and (69) or (70) solve our problem for phonons in terms of $L_0$ and $\delta L$.

In order to clarify the physical meaning of our matrix $T$ or $\mathcal{T}$ it seems useful to go back to the evolution equation (33) and to summarize briefly the main results of the standard scattering theory. We use Lippman–Schwinger notations ([5]). The interaction representation of our state vector $|\psi_t\rangle$ is defined by

$$|\psi_t^{(I)}\rangle \equiv \exp\,[i\Omega_0 t]\,|\psi_t\rangle \tag{71}$$

and the equation

$$|\psi_{+\infty}^{(I)}\rangle = S\,|\psi_{-\infty}^{(I)}\rangle \tag{72}$$

defines the collision operator, which generates the final state $|\psi_{+\infty}^{(I)}\rangle$ from an arbitrary initial state $|\psi_{-\infty}^{(I)}\rangle$. It is well known that $S$ is unitary, so we can write it as

$$S = \frac{1 - \frac{1}{2}iK}{1 + \frac{1}{2}iK} \tag{73}$$

where $K$, the reaction operator, is Hermitian.

The operator $T$ is defined by

$$T \equiv S - 1 \tag{74}$$

and the unitary property of $S$ implies that

$$T^\dagger T = -(T + T^\dagger) \tag{75}$$

where a dagger means Hermitain conjugate.

It turns out that the operator $T$ can be written as

$$T = -\frac{iK}{1 + \frac{1}{2}iK} \tag{76}$$

The principal result of the scattering theory is the following

$$T_{q'q} = -2\pi i\delta(\omega_{q'} - \omega_q)\mathbf{T}_{q'q} \tag{77}$$

where $T_{q'q}$ is the $q$-space representation of the operator $T$, as defined by (74), and $\mathbf{T}$, the association matrix, is the matrix defined by (54) and (55). Note that in scattering problems only states $|\phi_q\rangle$ and $|\phi_{q'}\rangle$ of equal frequency are involved. The matrix elements $\mathbf{T}_{q'q}$ which enter into equation (77) then correspond to the diagonal part of $\mathsf{T}$ in the energy representation. Furthermore, it is more significant to work in the $q$-space than in the $g$-space; to emphasize this we use labels $q$, $q'$ instead of $g$, $g'$.

On writing

$$K_{q'q} = 2\pi\delta(\omega_{q'} - \omega_q)\mathbf{K}_{q'q} \tag{78}$$

the relationship between the matrices $\mathbf{T}$ and $\mathbf{K}$ is obtained from (76); it reads

$$\mathbf{T}_{q'q} + i\pi \sum_{q''} \mathbf{K}_{q'q''}\delta(\omega_{q''} - \omega)\mathbf{T}_{q''q} = \mathbf{K}_{q'q} \tag{79}$$

where $\omega$ is the common frequency state $q$ and $q'$.

Keeping in mind the equation (54), the matrix $\mathbf{K}$ can be written as

$$\left[ I - \delta\Omega\frac{P}{\omega - \Omega_0} \right]^{-1}\delta\Omega \tag{80}$$

Where we have to consider only matrix elements for states of equal frequency $\omega_q = \omega_{q'} = \omega$. In deducing (80) from (54) and (79) we have used the well-known statement

$$\frac{1}{\omega \pm i\eta - \Omega_0} = \frac{P}{\omega - \Omega_0} \mp i\pi\delta(\Omega_0 - \omega) \tag{81}$$

where $P$ denotes principal part.

Consider the eigenvalue equation

$$K\delta(\Omega_0 - \omega)|f_A\rangle = \mathbf{K}_A|f_A\rangle \tag{82}$$

K is an Hermitian matrix and the eigenvectors $|f_A\rangle$ will then constitute an orthogonal set. If the eigenvectors are normalized according to

$$(f_A|\,\delta(\Omega_0 - \omega)\,|f_B) = \delta_{A,B} \tag{83}$$

the matrix K can be written as

$$\mathbf{K} = \sum_A |f_A)\mathbf{K}_A(f_A| \tag{84}$$

Since the eigenvalues $\mathbf{K}_A$ are real, they can be conveniently expressed by

$$\mathbf{K}_A = -\frac{1}{\pi}\tan\delta_A \tag{85}$$

where the real angles $\delta_A$ are nothing but the phase shift of the scattering process. Indeed, equation (79) for T is satisfied by putting

$$\mathbf{T} = \sum_A |f_A)\mathbf{T}_A(f_A| \tag{86}$$

with

$$\mathbf{T}_A = \frac{\mathbf{K}_A}{1 + i\pi\mathbf{K}_A} \tag{87a}$$

$$= -\frac{1}{\pi}\sin\delta_A \exp[i\delta_A] \tag{87b}$$

On account of (85) and (87b), we call $\mathbf{K}\delta(\Omega_0 - \omega)$ the phase-shift operator. From (53), (83), (86), and (87) it appears that the eigenvectors $|f_A\rangle$ and eigenvalues $\mathbf{K}_A$ of $\mathbf{K}\delta(\Omega_0 - \omega)$ provide us with all the elements for the evaluation of the diffused part $|\sigma_q\rangle$ of the representative of our elementary excitation. We can now proceed with the physical interpretation.

Since the matrix T connects only states of equal frequency, the elastically diffused part $|\sigma_q\rangle$ has the following expression:

$$|\sigma_q) = \sum_A i\sin\delta_A e^{i\delta_A}\delta(\Omega_0 - \omega)|f_A)(f_A|\phi_q) \tag{88}$$

Due to the normalization (50) for the unperturbed states $|\phi_q\rangle$, the resulting expression for the transition probability

$$W_{q'q} = 4\pi^2\,|(\phi_{q'}|\sigma_q)|^2 \tag{89}$$

can be interpreted as the fraction of time that an elementary excitation, switched at $t = -\infty$ in the state $|\phi_q\rangle$, spends in the diffused part $|\sigma_q\rangle$ as a state $|\phi_{q'}\rangle$ with $\omega_{q'} = \omega_q(q' \neq q)$, and the total probability

$$\sum_{q'} W_{q'q} = 4\pi^2 \sum_{q'} |(\phi_{q'}|\sigma_q)|^2$$
$$= 4\pi^2(\sigma_q|\sigma_q) \tag{90}$$

as the fraction of time that the elementary excitation spends in the elastically diffused part $|\sigma_q)$.

The transition probability per unit time turns out to be

$$w_{q'q} = 2\pi\delta(\omega_{q'} - \omega_q)|T_{q'q}|^2 \tag{91a}$$

$$= \frac{2}{\pi}\delta(\omega_{q'} - \omega_q)|\sum_A \sin\delta_A e^{i\delta_A} f^*_{q'A} f_{qA}|^2 \tag{91b}$$

The result (91a) represents a familiar expression in the conventional phase-shift analysis of the scattering of a particle by a central field of force.

The total probability per unit time for transitions from a particular state $|\phi_q)$, i.e.,

$$\sum_{q'} w_{q'q} = \frac{2}{\pi}\sum_A \sin^2\delta_A |f_{qA}|^2 \tag{92}$$

has then to be interpreted as the probability per unit time that an elementary excitation traveling through the crystal with wavevector $\mathbf{q}$ is elastically diffused by the defect responsible for the local perturbation $\delta\Omega$.

In the derivation of (92) from (91a) use has been made of the normalization condition (83). We can also write

$$\sum_{q'} w_{q'q} = -2\,\mathrm{Im}(T_{qq}) \tag{93}$$

once equations (75), (77), and (91a) are taken into account. Resonance scattering is said to occur when $|T_A|$ exhibits the maximum value $|T_A| = 1/\pi$, i.e., when $\delta_A = \pm\pi/2$. In such a case the scattering amplitude (93) exhibits, indeed, a maximum.

In phonon problems it is easier to work with equation (67), so that, in order to go back to the scattering formalism it is useful to know the matrix elements $T_{q'q}$ in terms of the correspondent matrix elements of $\mathscr{T}$. The connection between $T_{q'q}$ and $\mathscr{T}_{q'q}$ come out from (55), (66), and (69), if we keep in mind that $|\phi_q)$ and $|\psi_q)$ are stationary eigenstates of $\Omega_0$ and $\Omega_0 + \delta\Omega$, respectively; it turns out that

$$\mathbf{T}_{q'q} = \frac{1}{2\omega}\mathscr{T}_{q'q} \tag{94}$$

where $\omega$ is the common frequency state $q'$ and $q$. Equation (94) further clarifies the equivalence between matrices $\mathbf{T}$ and $\mathscr{T}$. In particular, it appears that the phase shifts, as deduced from **T** or $\mathcal{T}$, are exactly the same. (See also Section V.)

We emphasize that only matrix elements of **T** or $\mathcal{T}$ which correspond to states of equal frequency are involved in the scattering problem, while the stationary solutions of equations (33) or (36), i.e., the eigensolutions of the matrix $\Omega_0 + \delta\Omega$ (or $L_0 + \delta L$) involve all the matrix elements of **T** (or $\mathcal{T}$), with no restriction about the frequency of the states $|\phi_{q'}\rangle$ and $|\phi_q\rangle$. In the $x$-space representation our stationary solutions, as given by expressions (53) or (68), are in the form of the symmetrized lattice wave $\phi_g(x)$ plus a diffused wave $\sigma_g(x)$. *The elementary excitations have in this case wavelike structure.*

So far, we have developed our theory on the assumption that, as the volume of the crystal tends to infinity, the unperturbed operator $\Omega_0$ or $L_0$ has practically a continuous spectrum. What happens in solids is that the unperturbed operator has, indeed, a continuous spectrum, but this spectrum covers only a limited interval of the frequency (or squared-frequency) axis. It turns out that, besides the solutions we have considered above, which fall at frequencies inside the continuous spectrum, the eigenvector equation

$$(\Omega_0 + \delta\Omega)|\psi\rangle = \omega|\psi\rangle \qquad (95a)$$

or

$$(L_0 + \delta L)|\psi\rangle = \omega^2|\psi\rangle \qquad (95b)$$

may have discrete solutions for frequencies outside the spectrum of the unperturbed operator. Notice that, in such a case, the normalization constant for $\psi(x)$ does not depend on the volume. We denote by $\omega_l$ the frequencies for these solutions. It is quite clear that the equation which is appropriate for such a case is the homogeneous part of the equation (52) or equation (67), i.e.,

$$|\psi\rangle = \frac{1}{\omega_l - \Omega_0}\delta\Omega|\psi\rangle \qquad (96a)$$

or

$$|\psi\rangle = \frac{1}{\omega_l^2 - L_0}\delta L|\psi\rangle \qquad (96b)$$

The eigenvalue equation reads

$$\left\|1 - \frac{1}{\omega_l - \Omega_0}\delta\Omega\right\| = 0 \qquad (97a)$$

or

$$\left\|1 - \frac{1}{\omega_l^2 - L_0}\delta L\right\| = 0 \qquad (97b)$$

It is well known ([7]) that the $x$-space representatives of the eigensolutions of equations (96) exhibit an exponential decay, as one goes far from the local perturbation. This is only to say that *the elementary excitations which correspond to the roots of equations. (97) are localized excitations*

It is instructive to find the connection between the $T$-matrix and the Green-function matrix. With no restriction on the matrix elements of T, the following relationship exists between the $T$-matrix and the perturbed Green-function matrix

$$\frac{1}{z - (\Omega_0 + \delta\Omega)}\delta\Omega = \frac{1}{z - \Omega_0}\mathsf{T} \tag{98a}$$

or

$$\frac{1}{z - (\mathsf{L}_0 + \delta\mathsf{L})}\delta\mathsf{L} = \frac{1}{z - \mathsf{L}_0}\mathscr{T} \tag{98b}$$

As the excitation frequencies of our system are the poles of the perturbed Green function, it turns out that the $T$-matrix has to exhibit poles at the frequencies $\omega_l$ of the localized excitation. Then, the frequencies of localized excitations can be found by looking at the poles that T (or $\mathscr{T}$) may exhibit on the real axis of the complex-frequency (or squared complex-frequency) plane. In scattering language this is only to say that a stationary state has to occur when the scattering amplitude (93) becomes infinite. We can now go back to the wavelike excitations and give a physical picture of resonance scattering.

We have shown that resonance scattering occurs when the phase shift becomes equal to $\pm\pi/2$. For this value of the phase shift, the matrix **K** becomes singular and the matrix elements of Im **T**, in the representation in which the eigenvectors of equation (82) [say $|A)$] are normalized according to

$$(A \,|\, B) = \delta_{A,B} \tag{99}$$

exhibit, as a function of $\omega$, a Lorentzian peak at the resonance frequency $\omega_R$. Since a narrow Lorentzian peak approaches a $\delta$-function, while Im **T** exhibits $\delta$-type singularities at the frequencies $\omega_l$ of localized excitations (as could be shown easily ([8])), at the resonance frequency $\omega_R$ *it is quite natural to treat the elementary excitations as pseudolocalized excitations.*

The resonance condition reads

$$\left\| \mathsf{I} - \frac{P}{\omega_R - \Omega_0}\delta\Omega \right\| = 0 \tag{100a}$$

or

$$\left\| \mathbf{1} - \frac{P}{\omega_R^2 - \mathbf{L}_0} \delta\mathbf{L} \right\| = 0 \tag{100b}$$

It looks quite similar to the condition (97) for localized excitations, so it is reasonable to assume that, in the neighborhood of the perturbation, the representative of a pseudolocalized excitation has to satisfy the equation

$$|\psi^{(P)}\rangle = \frac{P}{\omega_R - \Omega_0} \delta\Omega \, |\psi^{(P)}\rangle \tag{101a}$$

or

$$|\psi^{(P)}\rangle = \frac{P}{\omega_R^2 - \mathbf{L}_0} \delta\mathbf{L} \, |\psi^{(P)}\rangle \tag{101b}$$

The central point is, however, the normalization of $|\psi^{(P)}\rangle$. Except for the case in which the pseudolocalized excitation is embedded in the continuum (in this case $\psi^{(P)}(x)$ behaves as $1/|x|^2$, as we go far from the perturbation) the normalization constant of $|\psi^{(P)}\rangle$ still depends on the volume of the crystal, so that we cannot treat $|\psi^{(P)}\rangle$ as a localized excitation in a strict sense i.e., in the framework of a stationary approach). Nevertheless, in a time-dependent approach it turns out that near the resonance frequency, the envelope of the diffused waves actually behaves as a localized excitation, at least for a time interval of the order of the inverse of the resonance width. We donot enter this problem in detail and we refer the reader to Wagner's paper [9].

We report now, as a final consideration, the relationship between the perturbed Green function and the matrix $\mathsf{T}$ when a random distribution of local perturbations at finite concentration $p$ is considered in the lattice. Let $\mathsf{T}^{(n)}$ denote the $T$-matrix for the total system $(n = pN)$ of local perturbations; it accounts for the single-defect scattering as well as for the multiple scattering in which two, three, etc. scattering centers are involved at the time. The contribution of such processes to $\mathsf{T}^{(n)}$ can be analyzed by diagram technique; following Langer [10], a horizontal line with label $q$ reqresents the unperturbed Green function

$$G_q(\zeta) = \frac{1}{\zeta - \omega_q} \tag{102}$$

in wavevector representation. We further write $\delta\Omega^{(n)} = \sum_i \delta\Omega_{y_i}$. The interactions are denoted by dashed lines which start at dots representing the scattering centers (say $y_i$) where the interactions occur and connect to the continuous lines in the order in which they occur in the perturbation expansion of $\mathsf{T}^{(n)}$ in powers of $\delta\Omega_{y_i}$. Typical graphs entering $\mathsf{T}^{(n)}$ are the following

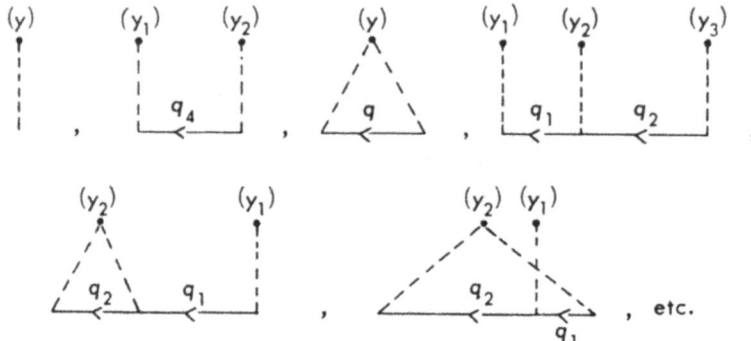

If scattering centers are distributed in the crystal according to a random distribution, what we are interested in knowing is the statistical average of the perturbed Green function over the distribution of the scattering centers. This is, indeed, the average perturbed propagator of our elementary excitation. If we call irreducible a graph which can not be separated into two disconnected parts by breaking a single continuous line, it could be shown easily ([8]) that the statistical average makes the parturbed Green function become

$$\left\langle \frac{1}{\mathfrak{z} - (\Omega_0 + \delta\Omega^{(n)})} \right\rangle_{\mathrm{av}} = \frac{1}{\mathfrak{z} - \Omega_0 - \langle T^{(n)}_{\mathrm{irr}} \rangle_{\mathrm{av}}} \tag{104a}$$

or

$$\left\langle \frac{1}{z - (L_0 + \delta L^{(n)})} \right\rangle_{\mathrm{av}} = \frac{1}{z - L_0 - \langle \mathscr{T}^{(n)}_{\mathrm{irr}} \rangle_{\mathrm{av}}} \tag{104b}$$

This is an exact result. In the statistical average the label $y_i$, which gives the position in the lattice of the $i$th scattering center, plays the role of a random variable. $T^{(n)}_{\mathrm{irr}}$, or $\mathscr{T}^{(n)}_{\mathrm{irr}}$, is defined as the sum of all the irreducible graphs entering the perturbation expansion of $T^{(n)}$ or $\mathscr{T}^{(n)}$. Furthermore, it is found that the statistical average makes $\langle T^{(n)}_{\mathrm{irr}} \rangle_{\mathrm{av}}$, or $\langle \mathscr{T}^{(n)}_{\mathrm{irr}} \rangle_{\mathrm{av}}$, become diagonal with respect to wavevector indices. We define then a matrix $T_0$, or $\mathscr{T}_0$, by

$$\langle T^{(1)} \rangle_{\mathrm{av}} \equiv p T_0 \tag{105}$$

or

$$\langle \mathscr{G}^{(1)} \rangle_{\mathrm{av}} \equiv p \mathscr{T}_0$$

where $T^{(1)}$ (or $\mathscr{T}^{(1)}$) is the $T$-matrix for a single scattering center situated at the random variable $y$, and $p$ is the fractional concentration of scattering centers. It is easy to see that $T_0$ (or $\mathscr{T}_0$) is a matrix which is diagonal with respect to wavevectors and has matrix elements which do not depend on the volume of the lattice. In order to put expressions (104) in a useful form for practical purposes, we introduce some simplifying assumptions. The simplest assumption is

$$\langle \mathsf{T}^{(n)}_{irr} \rangle_{av} \simeq p\mathsf{T}_0 \tag{106a}$$

or

$$\langle \mathscr{T}^{(n)}_{irr} \rangle_{av} \simeq p\mathscr{T}_0 \tag{106b}$$

which corresponds to saying that the sum of all the irreducible graphs which involve a single scattering center (this is just the term proportional to the first power of $p$) is the leading term in the expansion of

$$\langle \mathsf{T}^{(n)}_{irr} \rangle_{av} \qquad \text{or} \qquad \langle \mathscr{T}^{(n)}_{irr} \rangle_{av}$$

in powers of $p$.

Then, equations (104a) and (104b) read

$$\left\langle \frac{1}{\mathfrak{z} - (\Omega_0 + \delta\Omega^{(n)})} \right\rangle_{av} \simeq \frac{1}{\mathfrak{z} - \Omega_0 - p\mathsf{T}_0} \tag{107a}$$

and

$$\left\langle \frac{1}{z - (\mathsf{L}_0 + \delta\mathsf{L}^{(n)})} \right\rangle_{av} \simeq \frac{1}{z - \mathsf{L}_0 - p\mathscr{T}_0} \tag{107b}$$

Note that if two or more branches occur in the spectrum of $\Omega_0$ or $\mathsf{L}_0$, the matrix $\mathsf{T}_0$ (or $\mathscr{T}_0$) is not diagonal with respect to branch indices $s$ and $s'$.

$$p\mathsf{T}_{oq'q} = p\mathsf{T}_{oq}\delta_{q',q}$$

which is still a matrix with respect to branch indices, is recognized as playing the role of irreducible self-energy (in frequency units) of the elementary excitation [4] traveling in a lattice containing a random distribution of defects.

## V. POINT-GROUP ANALYSIS OF THE T-MATRIX

We start by considering the symmetry properties of the perturbation $\delta\Omega$ or $\delta\mathsf{L}$. In a large class of problems of physical interest, the $x$-space representation of $\delta\Omega$ or $\delta\mathsf{L}$ has nonvanishing matrix elements only in a small region of the lattice, so that $\delta\Omega$ or $\delta\mathsf{L}$ covers a small portion of the total linear vector space we have introduced in Section II. Let $\Gamma$ label the irreducible representations (irr rep) of the point group which belongs to $\delta\Omega$ or $\delta\mathsf{L}$, and let $(x \,|\, \Gamma, jr)$ denote the $x$-representative of the symmetry vector which transforms according to the $\Gamma$ irr rep; $j$ labels the symmetry vectors which belong to a same irr rep $\Gamma$, while $r$ is a repetition index. Index $r$ labels the so-called oriented symmetry vectors. It is needed when $\Gamma$ occurs more than a single time in the subspace of the perturbation. We may assume that $|\Gamma, jr)$ are orthogonal and normalized to unity, so that the identity

$$(x \,|\, \delta\Omega \,|\, x') \equiv \sum_{\Gamma} \sum_{j} \sum_{rr'} (x \,|\, \Gamma, jr) \delta\Omega_{rr'}^{(\Gamma)} (r'j, \Gamma \,|\, x')$$

or

$$(x \,|\, \delta L \,|\, x') \equiv \sum_{\Gamma} \sum_{j} \sum_{rr'} (x \,|\, \Gamma, jr) \delta L_{rr'}^{(\Gamma)} (r'j, \Gamma \,|\, x') \tag{108}$$

give the definition of the $\Gamma$ component $\delta\Omega^{(\Gamma)}$ or $\delta L^{(\Gamma)}$ of our perturbation. The symmetry vectors can be easily found by applying the projection operator [11]

$$P_{\Gamma} \equiv \frac{d_{\Gamma}}{g} \sum_{R} \chi_{\Gamma}^{*}(R) O_{R} \tag{109}$$

to a generic vector $|v)$ in tje subspace of the perturbation, i.e.,

$$|\Gamma, jr) = P_{\Gamma} \,|\, v) \tag{110}$$

A suitable choice of few vectors $|v)$ then enables us to specify indices $j$ and $r$.

    $R$ labels the elements of the point group, i.e., the symmetry operations; $\chi_{\Gamma}(R)$ is the character of the $\Gamma$th irr rep of the operator $O_R$, $g$ is the total number of elements in the group, and $d_{\Gamma}$ the dimension of the irreducible representation.

    It appears from (54) or (70) that the matrix $T$ or $\mathscr{T}$ covers the same subspace as the perturbation, and remains unchanged under the application of the symmetry operations. Then, the set of symmetry vectors $|\Gamma, jr)$ enable us to evaluate $T$ or $\mathscr{T}$ in its reduced form $T_{rr'}^{(\Gamma)}$, or $\mathscr{T}_{rr'}^{(\Gamma)}$. Note that from Schur's lemma only indices $r, r'$ are relevant. The inspection of the formal solution of equation (54) or (70) tells us that the $\Gamma$th component of $T$ or $\mathscr{T}$ can be written in the form

$$T_{rr'}^{(\Gamma)} = \frac{N_{rr'}^{(\Gamma)}(\mathfrak{z})}{D^{(\Gamma)}(\mathfrak{z})} \tag{111a}$$

or

$$\mathscr{T}_{rr'}^{(\Gamma)} = \frac{\mathscr{N}_{rr'}^{(\Gamma)}(z)}{\mathscr{D}^{(\Gamma)}(z)} \tag{111b}$$

where $D^{(\Gamma)}(\mathfrak{z})$, or $\mathscr{D}^{(\Gamma)}(z)$, is what is called the "resonance denominator," and $N^{(\Gamma)}(\mathfrak{z})$, or $\mathscr{N}^{(\Gamma)}(z)$, is a matrix having the same dimension as the number of times, say $n_\Gamma$, that the $\Gamma$th irr rep appears in the perturbation. The resonance denominator turns out to be

$$D^{(\Gamma)}(\mathfrak{z}) = \left\| 1 - P_\Gamma \delta\Omega \frac{1}{\mathfrak{z} - \Omega_0} \right\| \tag{112a}$$

or

$$\mathscr{D}^{(\Gamma)}(z) = \left\| 1 - P_\Gamma \delta L \frac{1}{z - L_0} \right\| \tag{112b}$$

By comparing (112a) with the eigenvalue equation of the phase-shift operator for the $\Gamma$th irr rep, which reads

$$\left\| \left( 1 - P_\Gamma \, \delta\Omega \frac{P}{\omega - \Omega_0} \right)^{-1} P_\Gamma \, \delta\Omega \, \delta(\Omega_0 - \omega) + \frac{1}{\pi} \tan \delta_{\Gamma r} \right\| = 0 \tag{113}$$

it is easy to see that, in the limit $\mathfrak{z} = \omega + i0^+$, the resonance denominator $D^{(\Gamma)}(\omega + i0^+)$ can be written as

$$D^{(\Gamma)}(\omega + i0^+) = \left\| \left( 1 - P_\Gamma \, \delta\Omega \, \frac{P}{\omega - \Omega_0} \right)^2 \right.$$
$$\left. + \, \pi^2 (P_\Gamma \, \delta\Omega \, \delta(\Omega_0 - \omega))^2 \right\|^{1/2} \exp\left( -id_\Gamma \sum_r \delta_{r r} \right) \tag{114a}$$

The correspondent expression for $\mathscr{D}^{(\Gamma)}(\omega^2 + i0^+)$ is

$$\mathscr{D}^{(\Gamma)}(\omega^2 + i0^+) = \left\| \left( 1 - P_\Gamma \, \delta L \, \frac{P}{\omega^2 - L_0} \right)^2 \right.$$
$$\left. + \, \pi^2 (P_\Gamma \, \delta L \, \delta(L_0 - \omega^2))^2 \right\|^{1/2} \exp\left( -id_\Gamma \sum_r \delta_{r r} \right) \tag{114b}$$

Note that both $D^{(\Gamma)}$ and $\mathscr{D}^{(\Gamma)}$ have the same argument $-id_\Gamma \sum_r \delta_{r r}$, and this argument, the factor $-i$ apart, is the sum of the phase shifts which belong to the $\Gamma$th irr reps. This comes from equation (94). However, it can be also seen in the following way: In principle, the argument of $\mathscr{D}^{(\Gamma)}$, say $-id_\Gamma \sum_r \pi\xi_{r r}$, has to satisfy the equation

$$\left\| \left( 1 - P_\Gamma \, \delta L \, \frac{P}{\omega^2 - L_0} \right)^{-1} P_\Gamma \, \delta L \, \delta(L_0 - \omega^2) + \frac{1}{\pi} \tan \pi\xi_{r r} \right\| = 0 \tag{115}$$

By keeping in mind the Lifschitz derivation of the "trace formula" ([12]), $\xi_{r r}$ is recognized to be the fractional shift of squared frequencies, i.e.,

$$\zeta_{\Gamma r} = \frac{(\delta\omega^2)_{\Gamma r}}{\omega_{n+1}^2 - \omega_n^2} = \frac{(\delta\omega^2)_{\Gamma r}}{\Delta\omega^2} \tag{116}$$

$\omega_n$ and $\omega_{n+1}$ are two subsequent frequencies in the discrete spectrum of a large, but finite, crystal, and $\delta\omega^2$ is the shift that $\omega_n^2$ suffers under the action of the perturbation.

However, both $(\delta\omega^2)_{\Gamma r}$ and $\Delta\omega^2 \equiv (\omega_{n+1}^2 - \omega_n^2)$ are to be considered infinitesimal quantities, so that

$$\frac{(\delta\omega^2)_{\Gamma r}}{\Delta\omega^2} = \frac{(\delta\omega)_{\Gamma r}}{\Delta\omega} = \frac{1}{\pi}\delta_{\Gamma r} \tag{117}$$

The above statement is then justified.

Expressions (116) and (117) supply the information that, essentially, the phase shift is nothing but the fractional shift that an unperturbed frequency suffers under the action of the perturbation. This physical meaning of $\delta_{\Gamma r}$ suggests how to evaluate the contribution, coming from the continuum of states, to the change of any extensive property of the system under the action of a local perturbation.

Let

$$F = \sum_\lambda f(\omega_\lambda) = F_0 + \delta F \tag{118}$$

be our extensive property. If the unperturbed frequency spectrum extends from $\omega = 0$ to a maximum $\omega_M$, the change $\delta F$ is then given by

$$\delta F = \frac{1}{\pi}\sum_\Gamma d_\Gamma \sum_r \int_0^{\omega_M} d\omega f'(\omega)\delta_{\Gamma r}(\omega) + \sum_l \{f(\omega_l) - f(\omega_M)\} \tag{119}$$

where the former term concerns the continuum, while the latter the possible localized excitations. Equation (119) shows the role played by the phase shifts in the thermodynamics of imperfect crystals.

We may conclude this lecture with some remarks about the lifetime of a pseudolocalized excitation. For convenience, we assume that the resonance falls in an irr rep $\Gamma$ which is singly contained in the perturbation $\delta\Omega$. The matrix $T^{(\Gamma)}$ reads

$$T^{(\Gamma)} = \frac{\delta\Omega^{(\Gamma)}}{1 - \bar{p}_\Gamma(\omega)\delta\Omega^{(\Gamma)} + i\pi p_\Gamma(\omega)\delta\Omega^{(\Gamma)}}|\Gamma)(\Gamma| \tag{120}$$

where

$$p_\Gamma(\omega) \equiv (\Gamma|\delta(\Omega_0 - \omega)|\Gamma) \tag{121}$$

is the projected density of states, and

$$\tilde{\rho}_\Gamma(\omega) \equiv \int d\omega' \rho_\Gamma(\omega') \frac{P}{\omega - \omega'} \tag{122}$$

is the Hilbert transform of $\rho_\Gamma(\omega)$.

In the neighborhood of a resonance, $T^{(\Gamma)}$ can be written as

$$T^{(\Gamma)} \simeq -\frac{1}{|\tilde{\rho}'_\Gamma(\omega_R)|} \left\{ \frac{\pm(\omega_R - \omega)}{(\omega_R - \omega)^2 + (\gamma_0/2)^2} + i \frac{(\gamma_0/2)}{(\omega_R - \omega)^2 + (\gamma_0/2)^2} \right\} \tag{123}$$

where

$$\gamma_0(\omega_R) = \frac{2\pi \rho_\Gamma(\omega_R)}{|\tilde{\rho}'_\Gamma(\omega_R)|} = 2 \frac{D_2^{(\Gamma)}(\omega_R)}{|D_1^{(\Gamma)'}(\omega_R)|} \equiv \tau_R^{-1} \tag{124}$$

is the width of the resonance, i.e., the inverse of the lifetime $\tau_R$ of the pseudolocalized excitation. A prime on $\tilde{\rho}_\Gamma$ or $D_1^{(\Gamma)}$ means first-order derivative with respect to $\omega_R$, i.e.,

$$\tilde{\rho}'_\Gamma(\omega_R) \equiv \left( \frac{\partial}{\partial \omega} \tilde{\rho}_\Gamma(\omega) \right)_{\omega = \omega_R}$$

In writing the second equality of (124) we have split the resonance denominator as $D^{(\Gamma)} = D_1^{(\Gamma)} + iD_2^{(\Gamma)}$. It could be easily shown [8] that this expression for $\tau_R^{-1}$ retains its validity also when $\Gamma$ occurs more than one time in $\delta\Omega$, provided that $\rho_\Gamma(\omega_R)$ refers to the resonant symmetry vector.

When a random distribution of scattering centers is present in the lattice, expression (124) becomes [8]

$$\tau_R^{-1} = \gamma_0(\omega_R) + p\gamma_1(\omega_R) \tag{125}$$

where $\gamma_0$, the intrinsic width, is given by (124), and

$$\gamma_1(\omega_R) \simeq \frac{2(N_1' D_2 - N_2 D')}{[(\omega_\Gamma - \omega_R) D_1'^2]} \tag{126}$$

$\omega_\Gamma$ is the center of zone frequency of the unperturbed excitation which transforms according to the same irr rep (say $\Gamma$) in which falls the resonance. At the usual concentration of defects the width of the resonance is dominated by the intrinsic width $\gamma_0$. Of course, a further broadening comes from the anharmonic interaction. This may be of particular importance for low-acoustic resonances.

## REFERENCES

1. M. Born and K. Huang, in: *Dynamical Theory of Crystal Lattices*, Oxford, 1954.
2. C. Kittel, *Quantum Theory of Solids*, John Wiley Inc. 1963.

3. T. Holstein and H. Primakoff, *Pys. Rev.* **58**: 1098 (1940).
4. See for instance. D.J. Thouless in the *Quantum Mechanics of Many-Body Systems*, Academic Press, 1961.
5. B.A. Lippmann and J. Schwinger, *Phys. Rev.* **79**: 569 (1959).
6. H. Feshbach and F. Villars, *Phys. Rev.* **30**: 24, (1958).
7. A. Maradudin in: *Phonons and Phonon Interactions*, Benjamin, 1964.
8. G. Benedek and G.F. Nardelli, *Phys. Rev.* **155**: 1004 (1967), see also "Calculation of the Properties of Vacancies and Interstitials" N.B.S. publ. 287. Nov. 1966; see also A.A. Maradudin in: *Astrophysics and Many-Body Problem*, Benjamin, 1963.
9. M. Wagner, *Phys. Rev.* **136: B 562** (1964); L. Fonda, *Annals of Physics* **22**: 123 (1963); L. Fonda and G.C. Ghirardi, *Annals of Physics* **26**: 240 (1964).
10. J. Langer, *J. Math. Phys.* **2**: 584 (1961).
11. M. Hamermesh in: *Group Theory*, Addison-Wesley, 1962.
12. M. Lifschitz, *Nuovo Cimento Suppl. to Vol. III, serie X* **4**: 716 (1956).

# Phonons and Their Interactions

### W. Ludwig

*Institute of Theoretical Physics*
*University of Giessen*
*Germany*

---

## I. GENERAL CONSIDERATIONS

In the first lecture we will sketch the mathematical tools for dealing with temperature-dependent Green functions and correlation functions of an ideal crystal. For simplicity we confine ourselves to infinite Bravais-lattices with periodic boundary conditions.

The Hamiltonian of the crystal is

$$\mathscr{H} = \sum_{m,n} \frac{(p_i^m)^2}{2M} + \phi$$

with

$$\phi = \frac{1}{2} \sum_{\substack{m,n \\ i,j}} \phi_{i,j}^{m,n} u_i^m u_j^n + \frac{1}{3!} \sum_{\substack{m,n,p \\ i,j,k}} \phi_{i,j,k}^{m,n,p} u_i^m u_j^n u_k^p + \cdots$$

where $u_i^m$ is the displacement of the atom $m$ in the direction $i$ and $p_i^m$ is associated the momentum. The coupling constants $\phi_{i,j}^{m,n}$ have the following properties:

i) $\phi_{i,j}^{m,n} = \phi_{j,i}^{n,m}$     Actio = reactio

ii) $\sum_{n} \phi_{i,j}^{m,n} = 0$     $\begin{cases} \text{Conservation of momentum} \\ \text{Translation invariance in space} \end{cases}$

iii) $\phi_{i,\ j}^{m+h,\ n+h} = \phi_{i,j}^{m,n}$     $\begin{cases} \text{Conservation of quasi-momentum} \\ \text{Translation invariance of displacement} \\ \text{pattern in the lattice} \end{cases}$

Analogous relations hold for the higher-order coupling constants. The equations of motion are

$$\ddot{M u}_i^m = -\sum_j \phi_{ij}^{mn} u_j^n - \frac{1}{2} \sum_{jk}^{np} \phi_{ijk}^{mnp} u_j^n u_k^p - \cdots$$

In the harmonic approximation, where only the first term of $\phi$ is taken into account, this equation has the stationary solutions

$$\varphi_i^m(s) = \frac{1}{\sqrt{NM}} e_i(\mathbf{q}, \sigma) e^{i[\mathbf{q}R^m - \omega(s)t]} = \phi_i^m(s) e^{-\omega_s t} \qquad s = (\mathbf{q}, \sigma)$$

The $\mathbf{q}$'s are reciprocal lattice vectors (in the first Brillonin zone)

$$\mathbf{q} = \frac{2\pi}{N^{1/3}} B\nu$$

with

$$B'A = 1, \quad \mathbf{R}^m = Am, \quad B' \text{ is the transpose of } B$$

and $\nu$ is a vector with integer components. $e_i(\mathbf{q}, \sigma)$ is a polarization vector, chosen so that

$$e_i(\mathbf{q}, \sigma) = -e_i^*(-\mathbf{q}, \sigma) \qquad \text{(Convention!)}$$

The frequencies $\omega(\mathbf{q}, \sigma)$ are determined by the secular equation

$$t_{ij} e_j = \omega^2 e_i \qquad \text{with} \qquad t_{ij} = \sum_h \phi_{ij}^{oh} e^{i\mathbf{q}R^h}$$

An arbitrary displacement $u_i^m$ can be written as

$$u_i^m = \left(\frac{h}{2NM}\right)^{1/2} \sum_{\mathbf{q}\sigma} \frac{1}{\sqrt{\omega(\mathbf{q}, \sigma)}} [b_{\mathbf{q}\sigma}(t) - b_{-\mathbf{q}\sigma}^+(t)] e_i(\mathbf{q}, \sigma) e^{i\mathbf{q}R^m}$$

with time-dependent amplitudes $b_{\mathbf{q}\sigma}(t)$ and $b_{\mathbf{q}\sigma}^+(t)$. In the quantum mechanical treatment $b_{\mathbf{q}\sigma}(t)$ and $b_{\mathbf{q}\sigma}^+(t)$ are creation and annihilation operators of a phonon $(\mathbf{q}, \sigma)$. (For the details see the lecture of A. A. Maradudin.)

An arbitrary state vector can be built up from the vacuum state $|0\rangle$

$$|\cdots n_{\mathbf{q}\sigma} \cdots\rangle = \prod_{\mathbf{q}, \sigma} \frac{1}{\sqrt{n_{\mathbf{q}\sigma}!}} (b_{\mathbf{q}\sigma}^+)^{n_{\mathbf{q}\sigma}} |0\rangle = |N\rangle$$

The diagonal elements of $b_s^+ b_{s'}$ are

$$\langle \ldots n_s \ldots | b_s^+ b_s | \ldots n_s \ldots \rangle = n_s \delta_{s,s'} \qquad s = (\mathbf{q}, \sigma)$$

and $b_s^+ b_s$ is called the phonon number operator.

Later on we will use instead of $b_s$ and $b_s^+$ the following operators

$$B_{q,\sigma} = b_{q,\sigma} - b^+_{-q,\sigma}, \; B^+_{q,\sigma} = b^+_{q,\sigma} - b_{-q,\sigma} = -B_{-q,\sigma}$$

for these is

$$\langle N | B^+_s B_{s'} | N \rangle = \langle N | B_{-s'} B^+_{-s} | N \rangle = -\langle N | B_{-s'} B_s | N \rangle = -\langle N | B^+_s B^+_{-s'} | N \rangle$$

$$= n_s \delta_{ss'} + (n_{-s} + 1)\delta_{ss'} \qquad -s = (-q, \sigma)$$

With $b^+b$ operators the harmonic part $\mathcal{H}_0$ of the Hamiltonian can be written

$$\mathcal{H}_0 = \sum_s \hbar\omega_s(b^+_s b_s + \tfrac{1}{2})$$

with the eigen-values

$$E^{(0)} = \sum_s \hbar\omega_s(n_s + \tfrac{1}{2})$$

For a crystal of temperature $T$ the mean occupation number is given by

$$\bar{n}_s = \frac{1}{Z_0} \text{Trace} \{b^+_s b_s e^{-\beta\mathcal{H}_0}\} = \langle b^+_s b_s \rangle_\beta = \frac{1}{e^{\beta\hbar\omega_s} - 1}$$

where $\beta = 1/kT$ and $Z_0$ is the partition-function

$$Z_0 = \text{Trace}\{e^{-\beta\mathcal{H}_0}\}$$

The mean energy is

$$\bar{E} = \frac{1}{Z_0} \text{Trace} \{\mathcal{H} \, e^{-\beta\mathcal{H}_0}\} = \sum_s \hbar\omega_s(\bar{n}_s + \tfrac{1}{2})$$

The Green function for $T = 0$ can be defined by

$$G^{mn}_{ij}(t, t') = \frac{i}{\hbar}\langle 0 | T_D u^m_i(t) u^n_j(t') | 0 \rangle$$

$$= \frac{i}{\hbar}\langle 0 | u^m_i(t) u^n_j(t') | 0 \rangle \theta(t - t') + \frac{i}{\hbar}\langle 0 | u^n_j(t') u^m_i(t) | 0 \rangle \theta(t' - t)$$

with the step function $\theta(t) = \begin{cases} 1 \text{ for } t > 0 \\ 0 \text{ for } t < 0 \end{cases}$

In the harmonic case one obtains for $G$

$$G^{mn}_{ij}(t - t') = \frac{i}{2NM} \sum_s \frac{1}{\omega_s} e_i(s) e^*_j(s) e^{iq(R^m - R^n) - i\omega_s|t - t'|}$$

One can easily verify, that this is a solution of the equation

$$M\ddot{G}^{mn}_{ij} + \sum_{pk} \phi^{mp}_{ik} G^{pn}_{kj} = \delta_{m,n}\delta_{i,j}\delta(t - t')$$

Therefore $G$ can be interpreted as the response of the crystal to a unit force at the atom $m$ in direction $i$ at time $t'$.

The Fourier transform of $G$ with respect to time is

$$G_{ij}^{mn}(\omega) = \lim_{\epsilon \to 0^+} \int_{-\infty}^{\infty} G_{ij}^{mn}(t) e^{i\omega t - \epsilon |t|} \, dt = \frac{1}{NM} \sum_s \frac{e_i(s) e_j^*(s)}{\omega_s^2 - \omega^2 - i\epsilon} e^{-iq(R^m - R^n)}$$

Analogously the Fourier transform of $G$ with respect to $m$ and $n$ is

$$G(s, s', t) = \sum_{\substack{mn \\ ij}} \phi_i^{*m}(s) G_{ij}^{mn}(t) \phi_j^n(s') = \frac{i}{2M\omega_s} e^{-i\omega_s |t|} \delta_{s,s'}$$

The "double" Fourier transform $G(s, s', \omega)$ is then

$$G(s, s', \omega) = \frac{1}{M(\omega_s^2 - \omega^2 - i\epsilon)} \delta_{ss'}$$

$G(s, s', \omega)$ has a pole on the real axis at the point $\omega = \pm\omega_s$, the frequency of the phonon $s = (q, \sigma)$. The lifetime is infinite, because $\omega_s$ has no imaginary part. ($G(s, s', \omega)$ is called a phonon propagator.)

For finite temperature a generalized Green function can be defined:

$$G_{ij}^{mn}(t - t', \beta) = \frac{1}{Z} \frac{i}{\hbar} \sum_{\ldots n_s \ldots} e^{-\beta E \ldots n_s \ldots} \langle \cdots n_s \cdots | T_D u_i^m(t) u_j^n(t') | \cdots n_s \cdots \rangle$$

$$= \frac{i}{\hbar} \langle T_D u_i^m(t) u_j^n(t') \rangle_\beta$$

Analogously we obtain for the Fourier transform with respect to $m$ and $n$:

$$G(s, s', t, t') = \langle T_D B_s(t) B_{s'}^+(t') \rangle_\beta$$

As in the case of $T = 0$, $G(s, s', t, t')$ depends only on the difference $t - t'$. In the harmonic approximation one obtains the result

$$G(s, s', t, t') = \frac{i}{\hbar} \delta_{ss'} \{(\bar{n}_s + 1) e^{-i\omega_s |t - t'|} + \bar{n}_s e^{i\omega_s |t - t'|}\}$$

The "double" Fourier transform is

$$G(s, s', \omega) = \frac{2\omega_s}{\hbar} \delta_{ss'} \left\{ P \frac{1}{\omega_s^2 - \omega^2} + i\pi \coth \frac{\beta\hbar\omega_s}{2} \delta(\omega_s^2 - \omega^2) \right\}$$

$$= G' + iG''$$

The following dispersion relations are valid:

$$G'(s, s', \omega^2) = \frac{1}{\pi} \mathscr{P} \int_0^\infty dy^2 \frac{G''(s, s', y^2)}{y^2 - \omega^2} \tanh\left(\frac{\beta\hbar\sqrt{y^2}}{2}\right)$$

or

$$G'(s, s', \omega) = \frac{1}{\pi} \mathscr{P} \int_{-\infty}^\infty dx \frac{G''(s, s', x)}{x - \omega} \tanh\frac{\beta\hbar x}{2} \qquad G''(x) = G''(-x)$$

With the linear combinations

$$G_\pm(s, s', \omega) = G' \pm iG'' \tanh\left(\frac{\beta\hbar\omega}{2}\right) = \frac{2\omega_s}{\hbar} \frac{1}{\omega_s^2 - (\omega \pm i\epsilon)^2} \delta_{ss'}$$

one obtains the usual dispersion relations

$$\mathrm{Re}\, G_\pm(\omega) = \pm\frac{1}{\pi} \mathscr{P} \int_{-\infty}^{+\infty} \frac{\mathrm{Im}\, G_\pm(x)}{x - \omega} dx$$

We will now calculate for a general hamiltonian $\mathscr{H}$ the correlation function

$$\tilde{\rho}(t) = \frac{i}{\hbar} \langle B_s(t) B_s^+(0) \rangle_\beta$$

$$\widehat{B(t)} \quad \hat{A} \quad \text{are used as abbreviations}$$

$$\tilde{\rho}(t) = \frac{i}{\hbar Z} \sum_N e^{-\beta E_N} \langle N | e^{i\mathscr{H}t/\hbar} B e^{-i\mathscr{H}t/\hbar} A | N \rangle$$

$$= \frac{i}{\hbar Z} \sum_{NN'} e^{-\beta E_N} e^{iE_N t/\hbar} \langle N | B | N' \rangle e^{-iE_{N'}t/\hbar} \langle N' | A | N \rangle$$

The Fourier transform of $\tilde{\rho}(t)$ is

$$\rho(\omega) = \frac{\hbar}{Z} \sum_{NN'} e^{-\beta E_N} \langle N | B | N' \rangle \langle N' | A | N \rangle \delta(E_{N'} - E_N - \hbar\omega)$$

We now introduce Green functions with imaginary time

$$f_{BA}(\tau) = \langle T_D e^{\tau\mathscr{H}/\hbar} B e^{-\tau\mathscr{H}/\hbar} A \rangle_\beta \equiv \langle T_D B(\tau) A \rangle_\beta$$

with

$$\langle T_D B(\tau_1) A(\tau_2) \rangle = \begin{cases} \langle B(\tau_1) A(\tau_2) \rangle & \text{for } \tau_1 > \tau_2 \\ \langle A(\tau_2) B(\tau_1) \rangle & \text{for } \tau_2 > \tau_1 \end{cases}$$

The following relations are valid:

i) $f_{BA}(\tau + \hbar\beta) = f_{BA}(\tau) \qquad -\beta\hbar < \tau < 0$

ii) $f_{BA}(\tau) = f_{AB}(-\tau)$

iii) $f_{BA}(+0) = f_{BA}(-0) = \langle [B, A]_- \rangle_\beta$

After periodic continuation, $f_{BA}(\tau)$ may be expanded into a Fourier series

$$f_{BA}(\tau) = \sum_{\nu=-\infty}^{+\infty} a_\nu e^{i\tau\omega_\nu} \qquad \omega_\nu = \frac{2\pi\nu}{\beta}$$

$$a_\nu = \frac{1}{\hbar\beta} \int_0^{\beta\hbar} f_{BA}(\tau) e^{-i\tau\omega_\nu} \, d\tau$$

$$= \frac{1}{\beta Z} \sum_{NN'} e^{-\beta E_N} \langle N|B|N'\rangle \langle N'|A|N\rangle \frac{e^{\beta(E_N - E_{N'})} - 1}{E_N - E_{N'} - i\hbar\omega_\nu}$$

We continue the $\nu$-dependence into the complex $z$-plane putting $i\omega_\nu = z$. Apart from a cut on the real axis $a(z)$ is a regular function with the property

$$\lim_{R\to\infty} a(x \pm iR) = 0$$

Then

$$a(x \pm i\omega) = \frac{\hbar}{\beta Z} \sum_{NN'} e^{-\beta E_N} \langle N|B|N'\rangle \langle N'|A|N\rangle (e^{\beta(E_N - E_{N'})} - 1)$$

$$\times \left\{ \mathscr{P}\left(\frac{1}{E_N - E_{N'} - \hbar x}\right) \pm i\pi\delta(E_N - E_{N'} - \hbar x)\right\}$$

One can easily verify that

$$\rho(\omega) = \frac{\hbar\beta}{e^{-\hbar\omega\beta} - 1} \lim_{\epsilon\to 0} \frac{a(-\omega + i\epsilon) - a(-\omega - i\epsilon)}{2\pi i}$$

## II. ANHARMONIC GREEN FUNCTIONS

The anharmonic Hamiltonian can be written as

$$\mathscr{H} = \mathscr{H}_0 + \phi^{(3)} + \phi^{(4)} + \cdots = \mathscr{H}_0 + W$$

where $\mathscr{H}_0$ is the harmonic Hamiltonian,

$$\phi^{(3)} = \frac{1}{3!} \sum_{\substack{mnp \\ ijk}} \phi_{ijk}^{mnp} u_i^m u_j^n u_k^p \quad \text{etc.}$$

Using the transformation of the $u_i^m$ to the phonon operators $B_s$, we have[1]

$$\phi^{(\nu)} = \frac{1}{\nu!} \sum_{s_1 \cdots s_\nu}' \phi_{s_1 \cdots s_\nu} B_{s_1}(t) \cdots B_{s_\nu}(t)$$

---

[1] Terms with $\omega_s = 0$, center of mass motion have to be dropped.

$$\phi_{s_1 \cdots s_\nu} = \left(\frac{\hbar}{2NM}\right)^{1/2} \frac{1}{(\omega_{s_1} \cdots \omega_{s_\nu})^{1/2}} \sum_{\substack{mn\cdots \\ ij\cdots}} \phi_{ij\cdots}^{mn\cdots} e_i(s_1) \cdots e_i(s_\nu)$$

$$\times \exp\{i\mathbf{q}_1 \cdot \mathbf{R}^m + \cdots + i\mathbf{q}_\nu \cdot \mathbf{R}^h\}$$

Because of the translational invariance

$$\phi_i^{m+h, n+h, \cdots}_{\cdots} = \phi_{ij\cdots}^{mn\cdots}$$

the $\phi_{s_1 \cdots s_\nu}$ contain the selection factor

$$\Delta^P(\mathbf{q}_1 + \cdots + \mathbf{q}_\nu)$$

which is periodic in reciprocal space ($\mathbf{q}$- space). We want to calculate

$$f_{BA}(\tau) = \langle T_D B(\tau) A \rangle_\beta = \frac{1}{Z} \,\text{Trace}\, e^{-\beta\mathscr{H}} T_D e^{\tau\mathscr{H}/\hbar} B e^{-\tau\mathscr{H}/\hbar} A$$

Because

$$[\mathscr{H}_0, W] \neq 0$$

in expanding $f_{BA}$ with respect to $W$, we have to use the perturbation expansion

$$f_{BA}(\tau) = \frac{\langle T_D \tilde{B}(\tau) \tilde{A} \sum\limits_{n=0}^{\infty} \frac{(-1)^n}{n!} \int_0^\beta \cdots \int d\tau_1 \cdots d\tau_n \tilde{W}(\tau_1) \cdots \tilde{W}(\tau_n) \rangle_{\beta, 0}}{\langle T_D \sum\limits_{n=0}^{\infty} \frac{(-1)^n}{n!} \int_0^\beta \cdots \int d\tau_1 \cdots d\tau_n \tilde{W}(\tau_1) \cdots \tilde{W}(\tau_n) \rangle_{\beta, 0}}$$

where

$$\tilde{B}(\tau), \tilde{W}(\tau) = e^{\tau\mathscr{H}_0/\hbar} B, W e^{-\tau\mathscr{H}_0/\hbar}$$

and the average is now with the harmonic Hamiltonian $\mathscr{H}_0$. In our case

$$\tilde{B} = \tilde{B}_s; \qquad \tilde{A} = \tilde{B}_{s'}^+; \qquad f_{BA}(\tau) = f_{ss'}(\tau)$$

## Lowest-Order Calculation

This contains only the terms $n = 0$ in the sums for $f_{ss'}$. Simple straightforward calculations give

$$f_{ss'}^0(\tau) = \langle T_D \tilde{B}_s(\tau) \tilde{B}_{s'}^+ \rangle_{\beta, 0} = \langle T_D \tilde{B}_s^+(\tau) \tilde{B}_{s'} \rangle_{\beta, 0}$$
$$= \langle T_D \tilde{B}_s(\tau) \tilde{B}_{-s'} \rangle_{\beta, 0} = \langle T_D \tilde{B}_s^+(\tau) \tilde{B}_{-s'}^+ \rangle_{\beta, 0}$$

$$f^0_{ss'}(\tau) = \delta_{ss'}\{\bar{n}_s e^{|\tau|\omega_s} + (\bar{n}_s + 1)e^{-|\tau|\omega_s}\}$$
$$= \delta_{ss'}g_s(\tau)$$

Note   that

$$g_s(\tau) = g_s(-\tau)$$

**Fourier Transformation**

$$g_s(\tau) = \sum_{\nu=-\infty}^{\infty} a^s_\nu e^{i\tau\omega_\nu}$$

$$\omega_\nu = \frac{2\pi\nu}{\beta}$$

gives

$$a^s_\nu = a^s_{-\nu} = \frac{2\omega_s}{\hbar\beta} \cdot \frac{1}{\omega_s^2 + \omega_\nu^2}$$

$$a^s(z) = \frac{2\omega_s}{\hbar\beta} \cdot \frac{1}{\omega_s^2 - z^2} \qquad (z = i\omega_\nu)$$

This gives immediately

$$p_s(\omega) = \frac{2\omega_s}{e^{-\rho\hbar\omega} - 1} \, \text{sgn}\,(\omega)\cdot\delta(\omega_s^2 - \omega^2)$$

$$\langle B_s(t)B^\dagger_{s'}\rangle = \frac{\hbar}{i}\int_{-\infty}^{\infty} p_s(\omega)e^{-i\omega t}d\omega\cdot\delta_{ss'} = \delta_{ss'}\frac{\hbar}{i}\{(\bar{n}_s + 1)e^{-i\omega_s t} + \bar{n}_s e^{i\omega_s t}\}$$

As an example for the anharmonic contribution, we calculate the contribution of $\phi^{(3)}$ in lowest order. The first, linear, contribution of $\phi^{(3)}$ vanishes, because it contains an average over an odd number of $B_s$. The second-order term gives, in the numerator

$$\frac{1}{2}\cdot\frac{1}{36}\sum_{s_1\cdots s_6}\phi_{s_1 s_2 s_3}\cdot\phi_{s_4 s_5 s_6}\cdot T_D\iint_0^\beta d\tau_1 d\tau_2$$

$$\langle\tilde{B}_s(\tau)\tilde{B}^\dagger_{s'}\tilde{B}_{s_1}(\tau_1)\tilde{B}_{s_2}(\tau_1)\tilde{B}_{s_3}(\tau_1)\tilde{B}_{s_4}(\tau_2)\tilde{B}_{s_5}(\tau_2)\tilde{B}_{s_6}(\tau_2)\rangle_{\beta,0}$$

Only those terms contribute, where at least pairwise indices are equal. We neglect terms which have more than two indices equal, because they are often less by factors $1/N$, $1/N^2$, etc. Thus we consider only terms like

$$s = s', \quad s_1 = -s_2, \quad s_3 = -s_4, \quad s_5 = -s_6, s \neq s_1 \neq s_3 \neq s_4, \text{ etc.}$$

Then the averages for the four different products

$$\tilde{B}_s \tilde{B}_{s'}^+, \quad \tilde{B}_{s_1} \tilde{B}_{s_1}^+, \quad \tilde{B}_{s_3} \tilde{B}_{s_3}^+, \quad \tilde{B}_{s_5} \tilde{B}_{s_5}^+$$

can be done independently.[2]

We consider different cases:

i) $s = s'$, $s_1 = -s_2$. Then there is a factor $\Delta^P(\mathbf{q}_1 - \mathbf{q}_2 + \mathbf{q}_3) = \Delta_P(\mathbf{q}_3)$ meaning $\mathbf{q}_3 = 0$ and this term vanishes, in Bravais-crystals.

ii) $s = s', s_1 = -s_4, s_2 = -s_5, s_3 = -s_6$. This gives a contribution; every pair of averages like $\langle \tilde{B}_s(\tau)\tilde{B}_s^+(0)\rangle$ gives a contribution $g_s(\tau)$ etc. We represent this graphically by phonon lines. The lines are connected by the interaction factor $\phi_{s_1 s_2 s_3}$, which we represent by vertices, here with three "corners." Because the $g_s(\tau)$ is not connected with an interaction, the term is graphically

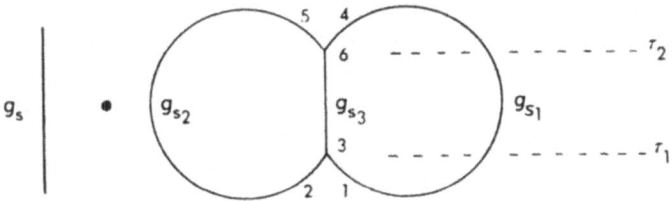

Such a diagram is called disconnected.

iii) $s = -s_1, s' = s_2, s_3 = -s_4, s_5 = -s_6$. This contribution, shown in the diagram, contains

$$\Delta^P(-\mathbf{q}_3 + \mathbf{q}_5 - \mathbf{q}_5) = \Delta^P(\mathbf{q}_3)$$

and vanishes. It is a connected diagram.

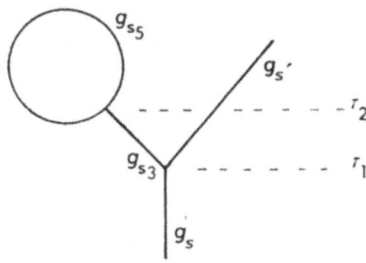

---

[2] This is essentially Wick's theorem, but with the above assumptions it does not to be used explicitly.

iv) $s = -s_3, s' = s_6, s_2 = -s_5, s_1 = -s_4$. This gives a contribution

$$\phi_{s_1 s_2 - s}\phi_{-s_1 - s_2 s'} \cdot g_s(\tau - \tau_1)g_{s'}(\tau_2)g_{s_1}(\tau_1 - \tau_2)g_{s_2}(\tau_1 - \tau_2)$$

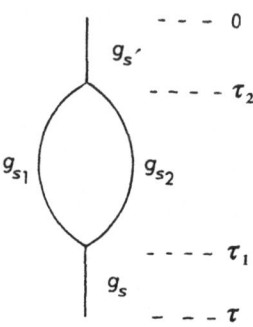

which has to be integrated over $\tau_1$ and $\tau_2$ from zero to $\hbar\beta$. It is a connected diagram. The evaluation of the integrals can be done simply if the $g_s$ are represented by their Fourier-transforms $a_\nu^s$. Then we obtain

$$\sum_\nu a_\nu^s \beta^3 \sum_{s_1 s_2}\phi_{s_1 s_2 - s}\phi_{s' - s_1 - s_2}\sum_{\nu_1} a_{\nu_1}^{s_1} a_{\nu - \nu_1}^{s_2} a_\nu^{s'} e^{i\tau\omega_\nu}$$

which gives the contribution to $f_{ss'}(\tau)$. Because of the involved factors

$$\Delta^P(\mathbf{q}_1 + \mathbf{q}_2 - \mathbf{q})\Delta^P(\mathbf{q}_1 + \mathbf{q}_2 - \mathbf{q}')$$

it is $\mathbf{q} = \mathbf{q}'$, and $s$ and $s'$ are distinguished only by the polarization indices $\sigma, \sigma'$. Other combinations of the $s_\nu$ give the same contribution, so that there is a numerical factor (36, e.g.).

With these rules we can construct all the contributions. $\phi^{(4)}$-interactions have vertices like $\times$, etc. So we have disconnected diagrams

and connected ones

The latter ones can be distinguished in the following way: The first four cannot be divided into two parts by cutting a single interior phonon line, whereas the last two can be cut. The first sort is called irreducible, the second sort reducible.

Two things can be done. The numerator of $f_{ss'}$ contains certain disconnected diagrams. Separating out these disconnected diagrams in factors, it can be shown, that the factors just cancel out the diagrams of the denominator, not discussed up till now. This is the linked cluster theorem. We can obtain $f_{ss'}$ just by considering only the connected diagrams of the numerator.

Second, if the sum over all irreducible diagrams, without the incoming and outgoing phonon lines, is denoted by ,

It can be seen immediately, that the sum over all connected diagrams, say , is which is the graphical description of an integral equation. Denoting the Fourier transformed by $\sum_{\nu}^{ss'}$, and the Fourier transformed complete sum by $F_{\nu}^{ss'}$ we have the equation

$$F_{\nu}^{ss'} = a_{\nu}^{s}\delta_{ss'} + a_{\nu}^{s}\delta_{ss'''} \sum_{\nu}^{s'''s''}F_{\nu}^{s''s'}$$

or

$$\{\delta_{ss''} - a_{\nu}^{s} \sum_{\nu}^{ss''}\}F_{\nu}^{s''s'} = a_{\nu}^{s}\delta_{ss'}$$

The above calculated contribution to $\sum_{\nu}^{ss'}$ is, e.g., (apart from factors)

$$\sum_{\nu(3)}^{ss'} \sim \beta^{3} \sum_{s_{1}s_{2}} \phi_{-ss_{1}s_{2}}\phi_{s'-s_{1}-s_{2}} \sum_{\nu_{1}} a_{\nu_{1}}^{s_{1}}a_{\nu-\nu_{1}}^{s_{2}}$$

Because of the selection factors $\Delta^{p}$, as already mentioned, $\sum_{\nu}^{ss'}$ is diagonal with respect to $\mathbf{q}, \mathbf{q}'$, but not ot $\sigma, \sigma'$. The determination of $F_{\nu}^{s's'}$ needs the solution of a $3 - \lambda$ demensional equation, $\lambda$ being the number of atoms in a unit-cell. This has been discussed by Cowley. For a first insight we nelgect the nondiagonal elements in $\sum_{\nu}^{ss'}$ and write $\sum_{\nu}^{ss''} = \sum_{\nu}^{s} \delta_{ss''}$ and obtain the simple solution

$$F_{\nu}^{ss'} = \frac{\delta_{ss'}}{1/a_{\nu}^{s} - \sum_{\nu}^{s}}$$

$$= \delta_{ss'}\frac{2\omega_{s}/\beta}{\omega_{s}^{2} + \omega_{\nu}^{2} - 2(\omega_{s}/\beta) \sum_{\nu}^{s}}$$

In $\sum_{\nu}^{s}$ we include the above-calculated third-order contribution and the lowest-order fourth-order contribution which can be calculated along the same lines giving

$$\sum_{\nu(4)}^{ss'} \sim -\beta \sum_{s_{1}} \phi_{-ss'-s_{1}s_{1}} \sum_{\nu_{1}} a_{\nu_{1}}^{s_{1}}$$

From the continuation of the Fourier transforms $F_{\nu}^{ss'}$ into the complex $z$-plane, $F_{\nu}^{ss'}(z)$ with $i\omega_{\nu} \rightarrow z$, we obtain the spectral representation $\rho_{s}(\omega)$ by forming $F^{ss'}(-\omega + i\epsilon)$. There enters $\sum_{\nu}^{s}(-\omega + is)$ which we divide into its real and imaginary parts:

$$\lim_{\epsilon \to 0+} \sum_{\nu}^{s}(-\omega + is) = -\beta\hbar\{\Delta_{s}(\omega) \pm i\Gamma_{s}(\omega)\}$$

The sums

$$\sum_{\nu_{1}} a_{\nu_{1}}^{s_{1}} \text{ and } \sum_{\nu_{1}} a_{\nu_{1}}^{s_{1}}a_{\nu-\nu_{1}}^{s_{2}}$$

can be performed in a straightforward manner (see Maradudin *et al.*). Finally we obtain with

$$\Delta_s(\omega) = \frac{1}{\hbar} \sum_{s_1} \phi_{-ss-s_1s_1} \frac{\epsilon_{s_1}}{\omega_{s_1}} - \frac{4}{3\hbar^3} \sum_{s_1s_2} |\phi_{-ss_1s_2}|^2 \Delta^P(\mathbf{q} - \mathbf{q}_1 - \mathbf{q}_2) \cdot$$

$$\cdot \mathscr{P} \left\{ \left[ \frac{\epsilon_1}{\omega_1} + \frac{\epsilon_2}{\omega_2} \right] \left[ \frac{1}{\omega + \omega_1 + \omega_2} - \frac{1}{\omega - \omega_1 - \omega_2} \right] \right.$$

$$\left. + \left[ \frac{\epsilon_1}{\omega_1} - \frac{\epsilon_2}{\omega_2} \right] \left[ \frac{1}{\omega - \omega_1 + \omega_2} - \frac{1}{\omega + \omega_1 - \omega_2} \right] \right\}$$

$$\Gamma_s(\omega) = \frac{\pi}{2\hbar^3} \sum_{s_1s_2} |\phi_{-ss_1s_2}|^2 \Delta^P(\mathbf{q} - \mathbf{q}_1 - \mathbf{q}_2) \cdot$$

$$\cdot \left\{ \left( \frac{\epsilon_1}{\omega_1} + \frac{\epsilon_2}{\omega_2} \right) \left[ \delta(\omega + \omega_1 + \omega_2) - \delta(\omega - \omega_1 - \omega_2) \right] \right.$$

$$\left. + \left( \frac{\epsilon_1}{\omega_1} - \frac{\epsilon_2}{\omega_2} \right) \left[ \delta(\omega - \omega_1 - \omega_2) - \delta(\omega + \omega_1 + \omega_2) \right] \right\}$$

for the spectral representation

$$\rho_s \sim \left\{ \frac{\Gamma_s(\omega)}{[\omega + \omega_s + \Delta_s(\omega)]^2 + \Gamma_s^2(\omega)} + \frac{\Gamma_s(\omega)}{[\omega - \omega_s - \Delta_s(\omega)]^2 + \Gamma_s^2(\omega)} \right\}$$

By Fourier transformation we obtain the displacement-correlation function from this. It clearly shows the meaning of $\Delta_s(\omega)$ and $\Gamma_s(\omega)$. $\Delta_s(\omega)$ describes the frequency shift by anharmonic processes, which in our cases depends on $\phi^{(4)}$ linearly and on $\phi^{(3)}$ quadratically. $1/\Gamma_s(\omega)$ is the lifetime of the phonon $s$, and depends on $\phi^{(3)}$ only in the above approximation, thus $\Gamma_s$ is the line width of phonon-involved scattering processes. The detailed calculation of $\Delta$ and $\Gamma$ needs a model (or a knowledge) of the anharmonic force-constants. Even then the calculation is difficult and can be done with a reasonable amount of work only in the classical high-temperature limit, where $\Delta$ and $\Gamma$ in the above approximation are proportional to the temperature. Numerical values have been gotten mainly by Maradudin and coworkers (see references). Typical values for lead are, e.g.,

$$\frac{\Delta}{\omega_D} \simeq \frac{-0.006T}{\theta_D}$$

$$\frac{\Gamma}{\omega_D} \simeq 10^{-3} \text{ to } 10^{-1} \frac{T}{\theta_D}$$

for $q$-values being about $1/2a$ ($\omega_D$, $\Theta_D$ Debye-frequency, temperature; $a$: lattice constant)

## III. APPLICATIONS

### a. Phonon–Neutron Interaction

In the Fermi approximation the neutron–nucleus interaction is

$$W(\mathbf{r}) = \frac{2\pi\hbar^2}{m} \sum_n a_n \delta(\mathbf{r} - \mathbf{r}^n)$$

$\mathbf{r}$ is the coordinate of the neutron, $m$ is its mass, and $a_n$ is the scattering length of the nucleus $n$ at the point $\mathbf{r}^n = \mathbf{R}^n + \mathbf{u}^n$. The total Hamiltonian is given by

$$\mathscr{H} = \mathscr{H}_{\text{crystal}} + \mathscr{H}_{\text{neutron}} + W(\mathbf{r}); \quad -\frac{\hbar^2}{2m}\frac{\partial^2}{\partial\mathbf{r}^2} = \mathscr{H}_{\text{neutron}}$$

We want to calculate the transition-probability $W_{N,i\to M,f}$ from an initial state $|N\rangle$ of the crystal and $|k_i\rangle = 1/\sqrt{V}\,e^{i\mathbf{k}_i\mathbf{r}}$ of the neutron to a final state $|M\rangle$ of the crystal and $|k_f\rangle$ of the neutron. The cross section is related to the transition probability by

$$\frac{d^2\sigma_{N,i\to M,f}}{d\omega\,d\Omega} = \frac{m^2 V}{\hbar^2}\frac{k_f}{k_i} w_{N,i\to M,f}$$

Here $\hbar\omega = \hbar^2(k_i^2 - k_f^2)/2m$ is the transferred energy and $\Omega$ is the solid angle element in the direction $\mathbf{k}_f$ of the outgoing neutron.

In the Born approximation the transition amplitude is

$$W_{N,i\to M,f} = \frac{2\pi}{\hbar}|\langle M\mathbf{k}_f|W|N k_i\rangle|^2\delta(E_N - E_M + \hbar\omega)\frac{V}{(2\hbar)^3}$$

Performing the integration over the neutron coordinate one obtains for the cross section

$$\frac{d^2\sigma_{N,i\to M,f}}{d\omega\,d\Omega} = \frac{k_f}{k_i}\sum_{n,m} a_n a_m^* \langle N|e^{-i\mathbf{K}\mathbf{r}^m}|M\rangle\langle M|e^{i\mathbf{K}\mathbf{r}^n}|N\rangle\delta\left(\omega + \frac{E_N - E_M}{\hbar}\right)$$

with $\mathbf{K} = \mathbf{k}_i - \mathbf{k}_f$. Summing over all final states $M$ and initial state $N$ (with the appropriate probability $P_N = (1/Z)e^{-\beta E_N}$ of the crystal, we obtain the cross section of the transition $|k_i\rangle \longrightarrow |k_f\rangle$ which can be measured directly.

$$\frac{d^2\sigma}{d\omega\,d\Omega} = \frac{k_f}{k_i}\sum_{n,m} a_n a_m^* S^{nm}(\mathbf{K}, \omega)$$

$$S^{nm}(\mathbf{K}, \omega) = \sum_{N,M} \frac{1}{Z}e^{-\beta E_N}\langle N|e^{-i\mathbf{K}\mathbf{r}^m}|M\rangle\langle M|e^{i\mathbf{K}\mathbf{r}^n}|N\rangle\delta\left(\omega + \frac{E_N - E_M}{\hbar}\right)$$

The Fourier transform $\chi^{mn}(\mathbf{K}, t)$ with respect to $\omega$ and the "double" transform $g^{mn}(\mathbf{r}, t)$ with respect to $\omega$ and $\mathbf{k}$ are given by

$$\chi^{mn}(\mathbf{K}, t) = \langle e^{-i\mathbf{K}\mathbf{r}^m(0)} e^{i\mathbf{K}\mathbf{r}^n(t)} \rangle_\beta$$

$$= e^{i\mathbf{K}(\mathbf{R}^n - \mathbf{R}^m)} \langle e^{-i\mathbf{K}\mathbf{u}^m(0)} e^{i\mathbf{K}\mathbf{u}^n(t)} \rangle_\beta$$

$$g^{mn}(\mathbf{r}, t) = \int d\mathbf{r}' \langle \delta(\mathbf{r}' - \mathbf{r}^m(0)) \delta(\mathbf{r}' + \mathbf{r} - \mathbf{r}^n(t)) \rangle_\beta$$

$\mathbf{r}^n(t)$ and $\mathbf{u}^n(t)$ are the time-dependent Heisenberg operators

$$\mathbf{r}^n(t) = e^{i/\hbar \mathcal{H}_{\text{cryst.}} t} \mathbf{r}^n e^{-i/\hbar \mathcal{H}_{\text{cryst.}} t}$$

So we obtain the result that the observable $S^{mn}(\mathbf{K}, \omega)$ is the double Fourier transform of the correlation function $g^{mn}(\mathbf{r}, t)$. If the atoms with different scattering length are randomly distributed, we obtain by averaging over the scattering length

$$\frac{d^2\sigma}{d\omega d\Omega} = \frac{k_f}{k_i} \{ \sum_m \overline{a^2} S^{mm}(\mathbf{K}, \omega) + \sum_{n \neq m} \bar{a}^2 S^{mn}(\mathbf{K}, \omega) \}$$

$$= \frac{k_f}{k_i} \{ (\overline{a^2} - \bar{a}^2) S_s(\mathbf{K}, \omega) + \bar{a}^2 S(\mathbf{K}, \omega) \}$$

$$S_s(\mathbf{K}, \omega) = \sum_m S^{mm}(\mathbf{K}, \omega)$$

$$= N S^{mm}(\mathbf{K}, \omega)$$

$$S(\mathbf{K}, \omega) = \sum_{n, m} S^{mn}(\mathbf{K}, \omega)$$

$$= N \sum_{m-n} S^{m-n}(\mathbf{K}, \omega)$$

The first term is called the incoherent cross section, and the second term the coherent one. For calculating $\chi^{mn}$ or $S^{mn}$ we need the following theorem:

$$\langle e^A e^B \rangle = e^{-W_A - W_B} \exp\{ \langle AB \rangle - \langle A \rangle \langle B \rangle$$
$$+ \tfrac{1}{2} \langle A^2 B \rangle - \tfrac{1}{2} \langle A^2 \rangle \langle B \rangle - \langle A \rangle \langle AB \rangle + \langle A \rangle^2 \langle B \rangle$$
$$+ \tfrac{1}{2} \langle AB^2 \rangle - \tfrac{1}{2} \langle A \rangle \langle B^2 \rangle - \langle AB \rangle \langle B \rangle + \langle A \rangle \langle B \rangle^2$$
$$+ \text{fourth-order terms} \ldots \}$$

$$-W_A = \langle A \rangle + \tfrac{1}{2}(\langle A^2 \rangle - \langle A \rangle^2) + \tfrac{1}{6}(\langle A^3 \rangle - 3\langle A^2 \rangle \langle A \rangle + 2\langle A \rangle^3) + \cdots$$

In the Bravais lattices considered here $\langle A \rangle = \langle \mathbf{u}^m \rangle$ vanishes.

$$\langle e^A e^B \rangle_\beta = e^{-W_A - W_B} \exp\{ \langle AB \rangle + \tfrac{1}{2} \langle A^2 B \rangle + \tfrac{1}{2} \langle AB^2 \rangle + \cdots \}$$

$$-W_A = \tfrac{1}{2} \langle A^2 \rangle + \tfrac{1}{6} \langle A^3 \rangle + \cdots$$

Taking only correlations of second order into account and moreover expanding $e^{\langle AB \rangle}$ linearly, we obtain for $\chi^{mn}$ putting $A = -i\mathbf{k}u^m$ and $B = i\mathbf{k}u^n(t)$

$$\chi^{mn}(\mathbf{K}, t) = e^{i\mathbf{K}(\mathbf{R}^n - \mathbf{R}^m)}e^{-2W}\{1 + K_i K_j\langle u_i^m(0)u_j^n(t)\rangle_\beta\}$$

$$W = W_A = W_B = \tfrac{1}{2}K_i K_j\langle u_i^n u_j^m\rangle \qquad \text{(Debye-Waller-factor)}$$

The first term of $\chi^{mn}$ is the zero phonon contribution, and the second the one phonon contribution. Tranforming back, one obtains

$$S^{mn} = S_0^{mn} + S_1^{mn}$$

$$S_0^{mn}(\mathbf{K}, \omega) = e^{-2W}e^{i\mathbf{K}(\mathbf{R}^m - \mathbf{R}^n)}\delta(\omega)$$

$$S_1^{mn}(\mathbf{K}, \omega) = e^{-2W}e^{i\mathbf{K}(\mathbf{R}^m - \mathbf{R}^n)}\frac{1}{2\pi}\sum_{i,j}K_i K_j\int dt\, e^{-i\omega t}\langle u_i^m(0)u_j^n(t)\rangle_\beta$$

In the harmonic approximation one obtains for $S_1^{mn}$

$$S_1^{mn}(\mathbf{K}, \omega) = e^{-2W}e^{i\mathbf{K}(\mathbf{R}^m - \mathbf{R}^n)}\frac{\hbar^2}{2NM}\sum_s\frac{(\mathbf{K}\cdot\mathbf{e}(s))(\mathbf{K}\mathbf{e}^*(s))}{\omega_s}e^{i\mathbf{q}(\mathbf{R}^n - \mathbf{R}^m)}$$

$$\times \{\bar{n}_s\delta(\omega + \omega_s) + (\bar{n}_s + 1)\delta(\omega - \omega_s)\}$$

With anharmonic contributions it is

$$S_1^{mn}(\mathbf{K}, \omega) = \frac{\hbar}{2\pi NM}e^{-2W}\frac{1}{1 - e^{-\hbar\omega\beta}}\sum_s e^{i(\mathbf{K}-\mathbf{q})(\mathbf{R}^m - \mathbf{R}^n)}\frac{|\mathbf{K}\cdot\mathbf{e}(s)|^2}{\omega_s}$$

$$\times \left\{\frac{\Gamma_s(\omega)}{[\omega + \omega_s + \Delta_s(\omega)]^2 + \Gamma_s^2(\omega)} + \frac{\Gamma_s(\omega)}{[\omega - \omega_s - \Delta_s(\omega)]^2 + \Gamma_s^2(\omega)}\right\},$$

Reforming the sum over $m$ and $n$, one obtains the coherent and incoherent contributions to the cross section.

$$\bar{p}_s(\omega) = \frac{\Gamma_s(\omega)}{[\omega + \omega_s + \Delta_s(\omega)]^2 + \Gamma_s^2(\omega)} + \frac{\Gamma_s(\omega)}{[\omega - \omega_s - \Delta_s(\omega)]^2 + \Gamma_s^2(\omega)}$$

$$S_0^{inc} = e^{-2W}\delta(\omega)$$

$$S_1^{inc} = \frac{\hbar e^{-2W}}{2\pi NM}\sum_s\frac{|\mathbf{K}\cdot\mathbf{e}(s)|^2}{\omega_s}\frac{1}{1 - e^{-\hbar\omega\beta}}\bar{p}_s(\omega)$$

$$S_0^{coh} = Ne^{-2W}\Delta^P(\mathbf{K})\delta(\omega)$$

$$S_1^{coh} = \frac{\hbar e^{-2W}}{2\pi NM}\sum_s\frac{|\mathbf{K}\cdot\mathbf{e}(s)|^2}{\omega_s}\Delta^P(\mathbf{K} - \mathbf{q})\frac{1}{1 - e^{-\hbar\omega\beta}}\bar{p}_s(\omega)$$

$$\frac{1}{1 - e^{-\hbar\omega\beta}}\bar{p}_s(\omega) \Rightarrow 2\pi\{\bar{n}_s\delta(\omega + \omega_s) + (\bar{n}_s + 1)\delta(\omega - \omega_s)\}$$

In this formula several approximations are involved in the anharmonic approximation. Higher-order phonon processes have been neglected. Further such averages as $\langle A^2 B \rangle$ in the formula above do not vanish in anharmonic theory. Rather they contribute to the same order of magnitude as $\Delta_s$ and $\Gamma_s$; they mean an asymmetry of the phonon emission or absorption line.

We will mention further, that the correlation function enters several other scattering cross-sections in the same way as for neutron scattering, for example, the X-ray scattering at crystals and the electron scattering at crystals, as well as the Mössbauer–nuclear-resonance absorption or emission cross sections.

### b. Optical Absorption

We now treat the interaction of electromagnetic radiation with the crystal. The total Hamiltonian is

$$\mathscr{H}_0 = \mathscr{H}_{\text{cryst.}} + \mathscr{H}_{\text{el.m.}} + W$$

$W$ being the interaction energy.

$$W = -\mathbf{M} \cdot \mathbf{E} e^{\pm i\omega t + \epsilon t} \qquad -\mathbf{M} \cdot \mathbf{E} = D, \ \epsilon > 0, \ -\infty \leq t \leq 0$$

The dipole moment $\mathbf{M}$ of the crystal depends on the displacements $\overset{m}{\mathbf{u}}{}^{\mu}$ of the atoms. ($\mu$ is the atom number in the unit cell)

$$\mathbf{M} = \sum_{m, \mu} Q^{\mu} \overset{m}{\mathbf{u}}{}^{\mu} + \cdots$$

Inserting the representation of $\overset{m}{\mathbf{u}}{}^{\mu}$ by the $B_s$ operators, one obtains

$$M_i = \sum_{\mu} \sqrt{\frac{\hbar N}{2M_{\mu}}} \, Q^{\mu} \sum_{\sigma}' \frac{1}{\sqrt{\omega_{\sigma}}} e_i^{\mu}(0, \sigma) B_{0, \sigma}(t) = \sum_{\sigma} M_i(0, \sigma) B_{0, \sigma}(t) + \cdots$$

The average of an observable represented by an operator $A$ is

$$\langle A \rangle = \text{Trace} \, \{ \rho(t) A \}$$

where $\rho(t)$ is the density matrix corresponding to the complete Hamiltonian $\mathscr{H} = \mathscr{H}_0 + W$. In our case, for $t \to -\infty$

$$\rho_0 = \frac{1}{Z_0} e^{-\beta \mathscr{H}_0}$$

The equation of motion of $\rho(t)$ is:

$$\dot{\rho} = \frac{i}{\hbar}[\rho, \mathcal{H}_0 + W]$$

Putting $\rho = \rho_0 + \rho_1$ and linearizing in the perturbation, we have the solution

$$\rho_1(t) = \frac{i}{\hbar} e^{-i\mathcal{H}_0 t/\hbar} \int_{-\infty}^{t} e^{i\mathcal{H}_0 t'/\hbar}[\rho_0, W] e^{-i\mathcal{H}_0 t'/\hbar} \, dt' \, e^{i\mathcal{H}_0 t/\hbar}$$

and

$$\langle A(t)\rangle_\beta = \langle A\rangle_{\beta,\mathcal{H}_0} - \frac{i}{\hbar} \int_{-\infty}^{t} \langle [\tilde{A}(t), \tilde{W}(t')]_- \rangle_{\beta,\mathcal{H}_0} dt'$$

with

$$\tilde{A}(t) = e^{i\mathcal{H}_0 t/\hbar} A e^{-i\mathcal{H}_0 t/\hbar}$$

By introducing a step function $\theta(t - t')$ into the integrand, the upper limit can be replaced by $+\infty$.

In our case

$$\tilde{W} = -\sum_i \tilde{M}_i E_i e^{i\omega t + \epsilon t}$$

and the quantity which shall be considered is the polarization $\tilde{A} = \tilde{P}_k = (1/V)\tilde{M}_k$. Therefore

$$\langle P_k(t)\rangle_\beta = \langle P_k\rangle_{\beta\mathcal{H}_0} + \sum_i \frac{iE_i}{\hbar V} \int_{-\infty}^{+\infty} \langle \theta(t - t')[\tilde{M}_k(t), \tilde{M}_i(t')]\rangle_{\beta\mathcal{H}_0} e^{i\omega t' + \epsilon t'} dt'$$

If there is no permanent polarization $\langle P_k\rangle_{\beta\mathcal{H}_0} = 0$. Taking the Fourier transform of the equation we have

$$P_k(\omega) = -\frac{i}{\hbar V} \int_{-\infty}^{+\infty} \theta(t)\langle [\tilde{M}_k(t), \tilde{M}_i(0)]\rangle_{\beta\mathcal{H}_0} e^{i\omega t - \epsilon t} dt \cdot E_i$$

where we have used the homogeneity property

$$[\tilde{M}_k(t), \tilde{M}_i(t')] = [\tilde{M}_k(t - t'), \tilde{M}_i(0)]$$

The coefficient between $P_k(\omega)$ and $E_i(\omega)$ is the frequency-dependent susceptibility

$$\chi_{ki}(\omega) = +\frac{i}{\hbar V} \int_{-\infty}^{+\infty} \theta(t)\langle [\tilde{M}_k(t), \tilde{M}_i(0)]\rangle_{\beta\mathcal{H}_0} e^{i\omega t - \epsilon t} dt$$

$$= +\frac{i}{\hbar V} \sum_{\sigma\sigma'} M_k(0, \sigma) M_i(0, \sigma') \int_{-\infty}^{+\infty} \theta(t)\langle [\tilde{B}_{\sigma 0}(t), \tilde{B}_{0\sigma'}(0)]\rangle_{\beta\mathcal{H}_0} e^{i\omega t - \epsilon t} dt$$

With the above representation of $M_i$, the function $\theta(t)\langle[\tilde{M}_k(t), \tilde{M}_i]\rangle_{\beta\mathcal{H}_0}$ is different from zero only for $t > 0$ and is called the retarded Green function. It describes the response of a dielectric system to an external field.

In the harmonic approximation we have

$$\tilde{B}_{0\sigma}(t) = e^{i\mathcal{H}_0 t/\hbar} B_{0\sigma} e^{-i\mathcal{H}_0 t/\hbar}$$

and by straightforward calculation

$$\text{Re } \chi_{ki}(\omega) = \frac{1}{\hbar V} \sum_\sigma M_k(0\sigma) M_i(0\sigma) \mathscr{P} \frac{2\omega_\sigma}{\omega_\sigma^2 - \omega^2}$$

$$\text{Im } \chi_{ki}(\omega) = \frac{2\pi}{\hbar V} \sum_\sigma M_k(0\sigma) M_i(0\sigma) \omega_\sigma \text{ sgn } \omega \delta(\omega_\sigma^2 - \omega^2)$$

The imaginary part describes the absorption, which in the harmonic approximation has the shape of a $\delta$-function. The real part gives the usual dispersion. Anharmonic effects can be represented again by shifts $\Delta_s(\omega)$ and widths $\Gamma_s(\omega)$, and $\Gamma_s(\omega)$ now is the width of the absorption of light by the dispersion ocsillators for optical branches $\sigma$ with $\mathbf{q} = 0$.

In general, one can say that all the responses of the systems to external fields and the scattering cross sections can be described by correlation functions and retarded Green functions. Thus, these having calculated, all measurable quantities can be expressed by them.

## BIBLIOGRAPHY

1. G. Leibfried, *Encyclopeadia Physics VII/I*, Springer Verlag, Berlin, 1955, p. 104.
2. M. Born and K. Huang, *Dynamical Theory of Crystal Lattices*, Oxford Univ. Press, 1954.
3. A.A. Maradudin, E.W. Montroll, and G.H. Weiss, *Solid State Physics, Suppl. III*, 1963.
4. G. Leibfried, W. Ludwig, *Solid State Physics*, **12**: 275 (1961).
5. T. Bak, editor, *Phonons and Phonon Interactions*, Benjamin, New York, 1964.
6. R.A. Cowley, *Adv. Phys.* **12**: 421 (1963).
7. W. Ludwig, *Ergeb. der exakt. Naturw.* **35**: 1–102 (1964); and Springer Tracts in Modern Physics **43**: 1–301 (1967).
8. A.A. Abrikosov, L.P. Gorkov, and I.E. Dzyaloskinski, *Quantum Field theory in Statistical Physics*, Prentice Hall, Englewood Cliffs, 1963.
9. V.L. Bonch-Bruevich and S.V. Tyablikov, *The Green function Method in Statistical Mechanics*, North Holland, Amsterdam, 1962.
10. Ph. Nozières, *The Theory of interacting Fermi-Systems*, Benjamin, New York, 1964.
11. L.P. Kadanoff and G. Baym, *Quantum Statistical Mechanics*, Benjamin, New York, 1962.
12. D.J. Thouless, *The Quantum Mechanics of Many-Body Systems*, Academic Press, New York, 1961.

## Special Papers

13.  D.N. Zubarev, *Sov. Phys. Uspekhi* **3**: 320 (1960).
14.  A.A. Maradudin, A.E. Fein, *Phys. Rev.* **128**: 2589 (1962).
15.  A.A. Maradudin and A. Ambegpakar, *Phys. Rev.* **135**: A1071 (1964).
16.  J.J.J. Kokkedee, *Physica* **28**: 374 (1962).
17.  R. Wehner, *Phys. Stat. Solidi* **15**: 725 (1966).
18.  V.S. Vinogradov, *Sov. Phys. Solid State* **4**: 519 (1963).
19.  V.V. Mitskevich, *Sov. Phys. Solid State* **4**: 2224 (1963).
20.  R.A. Cowley, Princeton lecture January 1966.

# Photon-Phonon Interaction in Semiconductors

## M. Balkanski

*Laboratoire de Physique*
*Ecole Normale Supérieure*
*Paris, France*

## I. INTRODUCTION

The interaction of a radiation field with fundamental lattice vibrations generally results in two effects: absorption or emission of the electromagnetic wave due to the creation or annihilation of lattice vibrations, on the one hand, and scattering of the eletromagnetic wave by the lattice vibrational modes on the other.

Optical absorption due to lattice vibrations results in the infrared region absorption bands attributed either to a first- or second-order electric dipole moment or to anharmonic terms in the potential energy. In ionic crystals, the first-order dipole moment gives rise to a very strong absorption band associated with optical modes of essentially zero $k$-vector. For homopolar crystals, in which the first-order dipole moment vanishes, intrinsic absorption due to lattice vibrations is explained in the basis of higher-order electric dipole moments.

Higher-order optical absorption involves simultaneous interaction of the radiation field with two or more vibrational modes. Raman scattering, instead of being related to the electric dipole moment, as it is in the case of optical abscription, is due to changes in the electrical polarizability of the crystal induced by the lattice vibrational modes. The scattered radiation differs from the incident light by the frequency of a normal mode of lattice vibration. When the difference measures the frequency of an optical vibration mode the process is the Raman scattering, and when it measures the frequency of an acoustical vibration mode the process is Brillouin scattering. The second-order scattering involves two vibrational modes which may be either optical or acoustical or a combination of the two.

Selection rules for photon–phonon interactions can be established for different crystal structures through group-theoretical considerations. The introduction of electrically charged point defects at high density has the effect of inducing a sufficient density of dipole moments to make the photon–phonon interaction at the impurity site observable. A high density of impurities also perturbs the vibrational spectrum of the lattice. Whereas for a simple perfect lattice, the frequency spectrum is split into two bands, accoustic and optical, with a gap in between, the introduction of point defects gives rise to allowed frequency levels in the gap of forbidden frequencies and above the limiting optical frequency.

The presence of defects influences the normal mode frequency in two ways: the single phonon frequencies inside the bands of allowed frequencies are shifted by small amounts and a certain number of these can be displaced toward the forbidden frequency gap. The particular frequencies which experience this displacement are associated with localized modes whose space-dependent factor decreases rapidly with distance from the defects. The optical spectra of the localized modes are shown to be not only specific for the defect masses, but also for its space position through the variation of the restroing force constants. After a brief examination of the theoretical background and selection rules, experimental results will be discussed on the single and multiphonon absorption spectra, direct optical transitions on the crystal points, and localized modes' infrared absorption.

## II. LATTICE VIBRATIONS

### 1. General Theory

We first review the theory of lattice vibrations in a perfect periodic lattice. In the harmonic approximation, the Hamiltonian is

$$\mathcal{H} = \tfrac{1}{2}\sum_{\alpha k l}\frac{p_\alpha^2(^l_k\ t)}{M(^l_k)} + \tfrac{1}{2}\sum_{\substack{\alpha k l \\ kk'}}\sum_{\beta k'l'} u_\alpha(^l_k\ t)\Phi_{\alpha\beta}(^l_k\ ^{l'}_{k'})u_\beta(^{l'}_{k'}\ t) \tag{1}$$

where $u_\alpha(^l_k\ t), p_\alpha(^l_k\ t)$ are the displacement and momentum of an atom at time $t$ in the unit cell specified by $l$; $k$ defines the atom in the cell and $\alpha$ the defines the Cartesian coordinates of the atoms; $M(^l_k)$ is the atomic mass; $\Phi$ is the force constant matrix and depends only on $\mathbf{r}(l) - r(l')$ the distance between cells because of the periodic symmetry. The equations of motion of the crystal obtained from this Hamiltonian is then

$$M(^l_k)\frac{d^2}{dt^2}u_\alpha(^l_k\ t) + \sum_{\beta k'l'}\Phi_{\alpha\beta}(^{l\ l'}_{k\ k'})u_\beta(^{l'}_{k'}) = 0 \tag{2}$$

The normal modes and their frequencies are obtained by diagonalizing the matrix:

$$\Phi_{\alpha\beta}(^{l\,l'}_{k\,k'}) - M(^l_k)\omega^2\delta_{\alpha\beta}\delta(^{l\,l'}_{k\,k'})$$

Because of the periodic symmetry the modes have a wavelike phase variation from cell to cell and specific relative motions in the cell:

$$u_\alpha(^l_k\,t) = M_k^{-1/2}\,v_\alpha(k)e^{i\mathbf{q}\cdot\mathbf{r}(l)-i\omega t} \tag{3}$$

The $v(k)$ are the eigenvectors of the dynamical matrix

$$C_{\alpha\beta}\,|^{\,q}_{k\,k'}| = \sum_{l'}\phi_{\alpha\beta}(^{l\,l'}_{k\,k'})e^{i\mathbf{q}[\mathbf{r}(l)-\mathbf{r}(l')]}(M_kM_{k'})^{-1/2} \tag{4}$$

the characteristic frequencies $\omega(\mathbf{q})$ are solutions of the secular equation $|C - \omega^2\,I| = 0$

We obtain the dispersion relations $\omega_j(\mathbf{q})$ of the crystal where $j$ means the branch and is equal to $3s$; $s$ is the number of atoms per cell. The matrix $C_{\alpha\beta}|^{\,q}_{k\,k'}|$ is a $3s \times 3s$ matrix for instance in the case of the blende structure $s = 2$ and the dynamical matrix contains 6 rows and 6 columns. This matrix is Hermitian and the eigenvalues are real and positive. $\mathbf{q}$ is defined in the first Brillouin zone because of the periodicity of the crystal.

## 2. Application to Ionic Crystals

In a rigid-ion approximation when the ionic charge is centered on the ion the dynamical matrix can be written in two terms:

$$C_{\alpha\beta}(^q_{k\,k'}) = C^c_{\alpha\beta}(^q_{k\,k'}) + C^N_{\alpha\beta}(^q_{k\,k'}) \tag{5}$$

The first term concerns only the Coulomb interaction and is due to long-range forces. The polar modes which are infrared-active arise from these Coulomb forces. The second term is due to short-range forces between next neighbors and gives all the normal modes. For $q \simeq 0$ we obtain the following expression for $C^N$

$$C^N_{\alpha\beta}(^0_{k\,k'}) = \sum_l \Phi_{\alpha\beta}(^{0\,l}_{k\,k'})(M_kM_{k'})^{-1/2} \tag{6}$$

These matrices satisfy some relations due to the symmetry of the lattice.

$$C^N(^0_{k\,k'}) = C^N(^0_{k'\,k}) \tag{7}$$

the $3 \times 3$ matrices $C^N_{\alpha\beta}(^0_{k\,k'})$ are diagonal. We shall apply these results and study the lattice vibrations in two cases:

1. for the blende structure with two ions in the unit cell.
2. For the wrutzite structure with four atoms in the unit cell.

a. *Blende*

From the symmetry properties we deduce the relation

$$C_{xx}(^0_{kk'}) = C_{yy}(^0_{kk'}) = C_{zz}(^0_{kk'}) = \sum_l \frac{\Phi_{xx}(^l_{kk'})}{(M_k M_{k'})^{1/2}} \tag{8}$$

$$C^N_{xx}(1\,2) = C^N_{xx}(2\,1) = -\left(\frac{M_1}{M_2}\right)^{1/2} C^N(1\,1) = -\left(\frac{M_2}{M_1}\right)^{1/2} C^N(22)$$

The secular equation $|C - \omega^2 I|$ gives two distinct roots threefold-degenerate

$$\omega_{ac} = 0$$

$$\omega_{opt} = \left\{ \frac{M_1 + M_2}{M_1 M_2} [-\sum_l \Phi^N_{xx}(^l_{12})] \right\}^{1/2} \tag{9}$$

$\omega_{ac}$ corresponds to the acoustical branches and $\omega_{opt}$ to the optical branches. The displacement vectors corresponding to $\omega_{ac}$ are $\mathbf{u}_1 = \mathbf{u}_2$ with the two sublattices moving in the same direction and these corresponding to $\omega_{opt}$ are $\mathbf{u}(1) = M_2/M_1 \mathbf{u}(2)$ with the two sublattices moving in opposite directions. To the optical modes corresponds an induced dipole moment equal to

$$\mathbf{M} = e^* \mathbf{u}(1) \frac{M_1 + M_2}{M_2}$$

where $e_1 = -e_2 = e^*$.

b. *Wurtzite*

The unit cell contains four atoms two anions and two cations; 12 branches are expected in the dispersions curves. From the symmetry of the crystal we deduce the following relations between the $C$ coefficients.

$$C^N_{xx}(kk') = C^N_{yy}(kk') \neq C^N_{zz}(kk') \tag{10}$$

When we solve the secular equations we obtain six different values of $\omega$. One is threefold-degenerate $\omega_{ac} = 0$ and corresponds to the acoustical branches and from the five other, three are twofold degenerate and correspond to optical transverse branches and three are nondegenerate corresponding to the optical longitudinal branches. There appears an accidental degeneracy for $\omega_{1\perp} = \omega_{1\parallel}$ If a model of central forces between first and second neighbors is used, we obtain the following relative values of

$$\omega_{3\perp} < \omega_{3\parallel} < \omega_{2\parallel} < \omega_{2\perp}$$

for

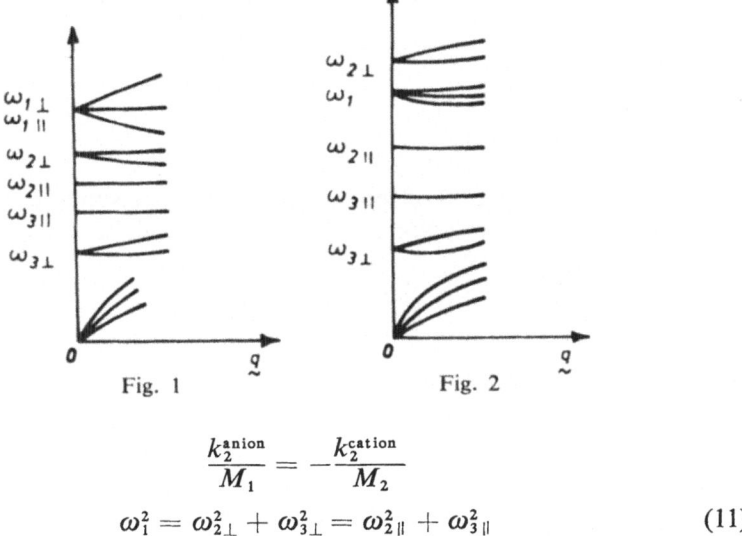

Fig. 1        Fig. 2

$$\frac{k_2^{\text{anion}}}{M_1} = -\frac{k_2^{\text{cation}}}{M_2}$$

$$\omega_1^2 = \omega_{2\perp}^2 + \omega_{3\perp}^2 = \omega_{2\|}^2 + \omega_{3\|}^2 \tag{11}$$

where $k_2$ is a force constant between second neighbors and we obtain the dispersion curves shown in Fig. 1.

For:

$$\frac{k_2^{\text{anion}}}{M_1} = \frac{k_2^{\text{cation}}}{M_2}$$

$$\omega_1^2 = \omega_{2\perp}^2 - \omega_{3\perp}^2 = \omega_{2\|}^2 + \omega_{3\|}^2 - S\omega_{3\perp}^2 \tag{12}$$

and we obtain the dispersion curves shown in Fig. 2.

The acoustical branch corresponds to the following displacement vectors

$$u_\alpha(1) = u_\alpha(2) = u_\alpha(3) = u_\alpha(4)$$

The displacement vectors corresponding to $\omega_1$ are related by the following relations

$$u_\alpha(1) = -\frac{M_2}{M_1}u_\alpha(2) = u_\alpha(3) = -\frac{M_2}{M_1}u_\alpha(4) \tag{13}$$

and are associated with a total dipole moment per unit cell $\mathbf{M} = 2e^*\mathbf{u}_1(M_1 + M_2)/M_2$ which is different from zero. The sublattices with the same kind of atoms (1) and (3) are moving in the same direction and the two other (2) and (4) in opposite direction.

We obtain displacement vectors of the same type for $\omega_2$ and $\omega_3$ but the important point is that the induced dipole moments per unit cell are always equal to zero for those modes which are not infrared-active. Up to now we only considered the short-range forces between first and second neighbors and neglected completely the long-range interaction Coulomb forces. Let us now consider these Coulomb forces.

## 3. Coulomb Forces Effect

We consider the model of the rigid ions where we suppose that the ions are not polarizable. The Coulomb potential energy can be written

$$\Phi = \tfrac{1}{2} \sum_{\substack{ll' \\ kk'}} \frac{e_k e'_{k'}}{|\mathbf{r}(^l_k) - \mathbf{r}(^{l'}_{k'})|} \tag{14}$$

and the coefficients $C^c$ deduced from $\Phi$ are

$$C^c_{\alpha\beta}(^q_{kk}) = -\frac{e_k^2}{M_k} \sum_{l \neq 0} \left[\frac{\partial^2}{\partial_\alpha \partial_\beta}\left\{\frac{1}{|\mathbf{r} - \mathbf{b}_l|}\right\}\right]_{\mathbf{r}=0} e^{i\mathbf{q}\cdot\mathbf{b}_l}$$

$$C^c_{\alpha\beta}(^q_{kk'}) = -\frac{l_k l_{k'}}{(M_k M_{k'})^{1/2}} \sum_l \left[\frac{\partial^2}{\partial_\alpha \partial_\beta}\left\{\frac{1}{|\mathbf{r} - \mathbf{b}_l|}\right\}\right]_{\mathbf{r}=\mathbf{r}_{kk'}} e^{i\mathbf{q}(\mathbf{b}_l - \mathbf{r})} \tag{15}$$

where $\mathbf{b}_e$ is a translation vector of the Bravais lattice and $\mathbf{r}_{kk'}$ a vector joining two atoms $kk'$ in the unit cell. For $\mathbf{q} = 0$ this becomes

$$C^c_{\alpha\beta}(^0_{kk}) = -\frac{4\pi N e_k^2}{M_k}\left(\frac{\delta_{\alpha\beta}}{3} - \frac{q_\alpha q_\beta}{q^2}\right)$$

$$C^c_{\alpha\beta}(^0_{kk'}) = -\frac{4\pi N e_k e_{k'}}{(M_k M_{k'})^{1/2}}\left(\frac{\delta_{\alpha\beta}}{3} - \frac{q_\alpha q_\beta}{q^2}\right) \tag{16}$$

where $N$ is the number of ion pairs per unit volume.

### a. Blende

Let us suppose $\mathbf{q}$ along a certain axis $O_z$; then in the case of the blende structure we obtain two optical frequencies instead of one. The first one $\omega_{T0}$ which is twofold degenerate corresponds to vibrations perpendicular to the propagation vector $\mathbf{q}$ and the other $\omega_{L0}$ is nondegenerate and corresponds to vibrations parallel to the progagation vector:

$$\omega_{T0}^2 = \omega_0^2 - \frac{4\pi N e_1^2}{3\mu} \tag{17}$$

where $1/\mu = 1/M_1 + 1/M_2$. $\omega_0$ is the optical frequency when we neglect the Coulomb forces:

$$\omega_{L0}^2 = \omega_0^2 + \frac{8\pi N e_1^2}{3\mu} = \omega_{T0}^2 + \frac{4\pi N e_1^2}{\mu} \tag{18}$$

b. *Wurtzite*

In the case of wurtzite we have

$$C^c_{\alpha\beta}\binom{0}{kk} = -\frac{Ne^2_k}{2M_k}\left(L^{kk}_\alpha \delta_{\alpha\beta} - \frac{4\pi q_\alpha q_\beta}{q^2}\right) \tag{19a}$$

$$C^c_{\alpha\beta}\binom{0}{kk'} = -\frac{Ne_k e_{k'}}{(M_k M_{k'})^{1/2}}\left(L^{kk'}_\alpha \delta_{\alpha\beta} - \frac{4\pi q_\alpha q_\beta}{q^2}\right) \tag{19b}$$

where $L^{\kappa\kappa'}_\alpha$ are *Lorentz factors in anisotropic crystals* and their numerical values are known. Without Coulomb forces we have obtained the solutions.

$$\omega_{ac} = 0$$
$$\omega_{opt} = \left\{\mu\left[-\sum_l \Phi^N_{xx}\binom{l}{1\,2}\right]\right\}^{1/2}$$

With Coulomb interaction we have to consider two cases: (1) The propagation vector parallel to the optical one

$$q_x = q_y = 0; \qquad q_z = |\mathbf{q}|$$

One obtains $\omega^2_x = \omega^2_y$ which corresponds to a degenerate $\top 0$ mode, and $\omega^2_z$ which corresponds to a $\llcorner 0$ mode

$$\omega^2_{\top 0\perp} = \omega^2_{i\perp} - Ne^2_i \frac{4.09}{\mu}$$

and

$$\omega^2_{\llcorner 0\|} = \omega^2_{i\|} + Ne^2_i \frac{8.07}{\mu} \tag{20}$$

(2) The propagation vector is perpendicular to the optical axis:

$$q_y = q_z = 0; \qquad q_x = |\mathbf{q}|$$

then $\omega^2_x$ is longitudinal $\perp C$

$$\omega^2_{\llcorner 0\perp} = \omega^2_{i\perp} + Ne^2_i \frac{8.37}{\mu}$$

$\omega^2_y$ is transverse $\perp C$

$$\omega^2_{\top 0\perp} = \omega^2_{i\perp} - Ne^2_i \frac{4.09}{\mu} \tag{21}$$

$\omega^2_z$ is transverse to atoms moving along the axis

$$\omega_{T0\parallel}^2 = \omega_{i\parallel}^2 - Ne_i^2\frac{4.39}{\mu}$$

If we consider only the nearest neighbors we have $\omega_{1\perp} = \omega_{1\parallel}$, and the effect of the Coulomb forces is to split the frequencies. When we change the relative orientation of the propagation vector to the crystal axes, we obtain the relations

$$\omega_{L0\parallel} < \omega_{L0\perp}$$

$$\omega_{T0\parallel} < \omega_{T0\perp}$$

## 4. Polar Modes

Let us first consider the polar modes. In those case the two sublattices move in opposite directions and a polarization appears beside the electronic polarization due to the movement of the charged ions in opposite way:

$$\mathbf{P} = Ne^x\mathbf{u} + \frac{\varepsilon_\infty - 1}{4\pi} \mathbf{E}_{\text{int}} \tag{22}$$

where $e^x$ is a certain effective charge and $\mathbf{E}_{\text{int}}$ is the internal field equal to the macroscopic electric field in the crystal. A depolarization field appears only for longitudinal waves and is equal in the case to $4\pi\mathbf{P}_L$, if $\mathbf{P}_L$ is the longitudinal polarization. The effective charge introduced previously is different for different cases:

1. The Callen effective charge: $e_L$

If we consider a longitudinal wave, in that case the longitudinal polarization is given by:

$$\mathbf{P}_L = Ne^x\mathbf{u}_L + \frac{\epsilon_\infty - 1}{4\pi}(-4\pi\mathbf{P}_L)$$

$$\mathbf{P}_L = \frac{Ne^x}{\epsilon_\infty}\mathbf{u}_L = Ne_L\mathbf{u}_L \qquad e_L = \frac{e^x}{\epsilon_\infty} \tag{23}$$

2. The transverse effective charge obtained for a transverse wave and equal to $e_T = e^x$

3. The Szigeti effective charge obtained for spherical sample and equal to

$$e_s = \frac{3e^x}{\epsilon_\infty + 2}$$

The equation of motion for a pair of ions can be written:

$$\mu\frac{d^2\mathbf{u}}{dt^2} = -\mu\omega_0^2\mathbf{u} + e_s\mathbf{E}_{\text{eff}} \tag{24}$$

where $\mu$ is the reduced mass

$$\frac{1}{\mu} = \frac{1}{M_1} - \frac{1}{M_2}$$

and **u** the relative displacement $= u_1 - u_2$. $\omega_0$ is the optical frequency when not considering the Coulomb forces between the ionic charges.

$$\mathbf{E}_{\text{eff}} = \mathbf{E}_{\text{dip}} + \mathbf{E}_{\text{int}}$$

$$\mathbf{E}_{\text{dip}} = \mathbf{P} \times \frac{4\pi}{3}$$

$$\mathbf{E}_{\text{int}} = \mathbf{E}_{\text{ext}} + \mathbf{E}_{\text{dep}}$$

For a transverse wave

$$\mathbf{E}_{\text{dep}}^T = 0 \qquad \mathbf{E}_{\text{eff}} = \frac{4\pi}{3}\mathbf{P}_T + \mathbf{E}_{\text{ext}} \tag{25}$$

For a longitudinal wave

$$\mathbf{E}_{\text{dep}}^L = -4\pi\mathbf{P}_L \qquad \mathbf{E}_{\text{eff}} = -\frac{4\pi}{3}\mathbf{P}_L + \mathbf{E}_{\text{ext}}$$

Finally we obtain the expression

$$\mathbf{E}_{\text{eff}} = \frac{\epsilon_\infty + 2}{3}\mathbf{E}_{\text{int}} + \frac{4\pi Ne_T}{3}\mathbf{u} \tag{26}$$

for transverse and longitudinal modes as well. The equation of motion (24) becomes

$$\frac{d^2\mathbf{u}}{dt^2} + \left(\omega_0^2 - \frac{4\pi Ne_s e_T}{3\mu}\right)\mathbf{u} = \frac{e_T}{\mu}\mathbf{E}_{\text{int}} \tag{27}$$

## III. PHOTON–PHONON INTERACTIONS

### 1. Infrared Dispersion

*a. Blende*

We have studied the vibrations of the crystal and we consider now the interactions of these vibrational waves with electromagnetic waves. The electromagnetic field is described by the Maxwell equations:

$$\text{rot } \mathbf{H} = \frac{1}{c}\frac{\partial \mathbf{D}}{\partial t}$$

$$\text{rot } \mathbf{E}_{int} = -\frac{1}{c}\frac{\partial \mathbf{H}}{\partial t}$$

$$\text{div } \mathbf{D} = 0$$

$$\text{div } \mathbf{H} = 0 \tag{28}$$

and the equations

$$\mathbf{D} = \mathbf{E}_{int} + 4\pi\mathbf{P}$$

$$\mathbf{P} = \mathcal{X}\mathbf{E}_{int} = \frac{\varepsilon - 1}{4\pi}\mathbf{E}_{int}$$

This system of equations is coupled with the two previous equations concerning the vibrations in the crystal:

$$\mathbf{P} = Ne_T\mathbf{u} + \frac{\varepsilon_\infty - 1}{4\pi}\mathbf{E}_{int}$$

$$\frac{d^2\mathbf{u}}{dt^2} + \left(\omega_0^2 - \frac{4\pi Ne_s e_T}{3\mu}\right)\mathbf{u} = \frac{e_T}{\mu}\mathbf{E}_{int} \tag{29}$$

if we use a waveform for **u** this equation becomes

$$\mathbf{u}\left(\omega_0^2 - \omega^2 - \frac{4\pi Ne_s e_T}{3\mu}\right) = \frac{e_T}{\mu}\mathbf{E}_{int} \tag{30}$$

We deduce then

$$\mathbf{u} = \frac{(e_T/\mu)\cdot\mathbf{E}_{int}}{\omega_{T0}^2 - \omega^2} \tag{31}$$

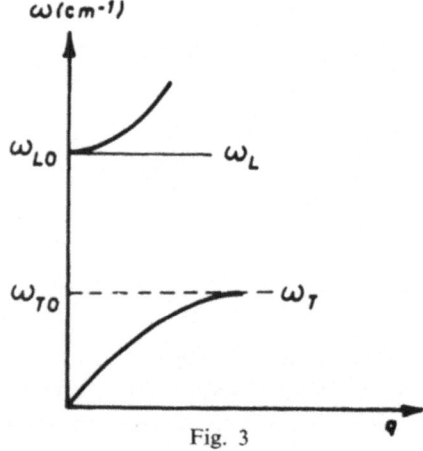

Fig. 3

$$\epsilon = \epsilon_\infty + \frac{4\pi Ne_T^2/\mu}{\omega_{T0}^2 - \omega^2} \tag{32}$$

for a longitudinal wave with $\mathbf{u} \parallel \mathbf{P} \parallel \mathbf{E}_{int}$; $\epsilon = 0$, and we obtain the following dispersion relation shown in Fig. 3:

$$\omega_L^2 = \omega_{T0}^2 + \frac{4\pi Ne_L e_T}{\mu} = \omega_{L0}^2 \tag{33}$$

for a transverse wave where $\mathbf{E}_{int} \perp q$ the equations become

$$\mathbf{D} = \epsilon \mathbf{E}_{int} = \frac{cq}{\omega}\mathbf{H} = \frac{c^2 q^2}{\omega^2}\mathbf{E}_{int} \tag{34}$$

$\mathbf{E}_{int} \neq 0$ and the dispersion relation shown in Fig. 4 is written

$$\frac{c^2 q^2}{\omega^2} = \epsilon = \epsilon_\infty + \frac{4\pi Ne_T^2/\mu}{\omega_{T0}^2 - \omega^2}$$

$$q^2 = \frac{1}{c^2}\epsilon_\infty \omega^2 \frac{\omega^2 - \omega_L^2}{\omega^2 - \omega_T^2} \tag{35}$$

We see that $\omega_T$ is a pole of $\epsilon(\omega)$ and is equal to $\omega_{T0}$, and $\omega_L$ is a root of the equation $\epsilon(\omega) = 0$ and is equal to $\omega_{L0}$. They satisfy the Lyddane–Sachs–Teller relation:

$$\left(\frac{\omega_{L0}}{\omega_{T0}}\right)^2 = \frac{\epsilon_0}{\epsilon_\infty} \tag{36}$$

If we introduce a damping term in the equations of motion

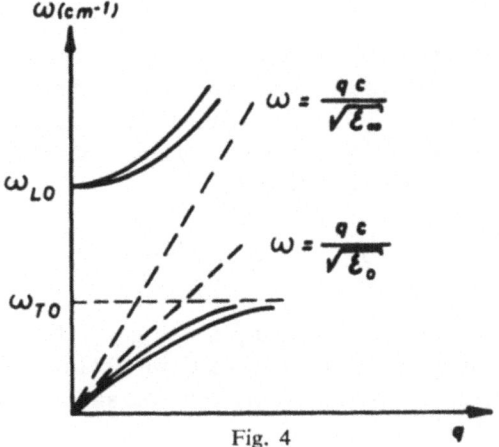

Fig. 4

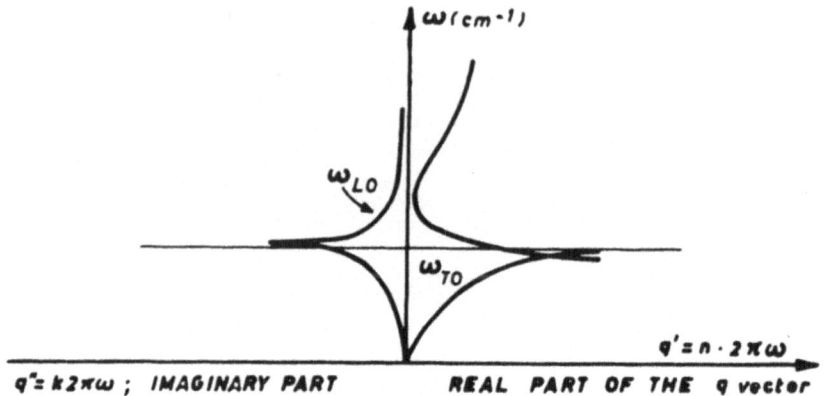

Fig. 5

$$\mu\frac{d^2\mathbf{u}}{dt^2} + \mu\gamma\frac{d\mathbf{u}}{dt} + \mu\omega_{T0}^2\mathbf{u} = e_T\mathbf{E}_{int} \tag{37}$$

then

$$\epsilon(\omega) = (\eta + ik)^2 = \epsilon_\infty + \frac{(\epsilon_0 - \epsilon_\infty)\omega_{T0}^2}{\omega_{T0}^2 - \omega^2 - ij\omega} \tag{38}$$

$$\omega_T = \left(\omega_{T0}^2 - \frac{\gamma^2}{4}\right)^{1/2} - \frac{i\gamma}{2} \neq \omega_{T0}$$

$$\omega_L = \left(\omega_{L0}^2 - \frac{\gamma^2}{4}\right)^{1/2} - \frac{i\gamma}{2} \neq \omega_{L0} \tag{39}$$

The dispersion curves are shown in Fig. 5.

### b. Wurtzite

In the case of wurtzite there is an anisotropy effect. The infrared dispersion curves depend on the propagation direction and we obtain purely transverse and longitudinal waves only for $\mathbf{q}$ parallel or perpendicular to the $C$ axis. There is always a transverse mode whose frequency independent of the direction of propagation and given by (Fig. 6)

$$\epsilon(\omega) = \epsilon_{\infty\perp} + \frac{(\epsilon_{0\perp} - \epsilon_{\infty\perp})}{\omega_{T0\perp}^2 - \omega^2}\,\omega_{T0\perp}^2 \tag{40}$$

We have now to consider separately $\mathbf{q} \parallel C$ and $\mathbf{q} \perp C$ because of the anisotropy effect: (1) For $\mathbf{q}$ parallel to the axis $C$ we obtain a second transverse mode frequency degenerated with the previous one and the longitudinal mode frequency is equal to (Fig. 7)

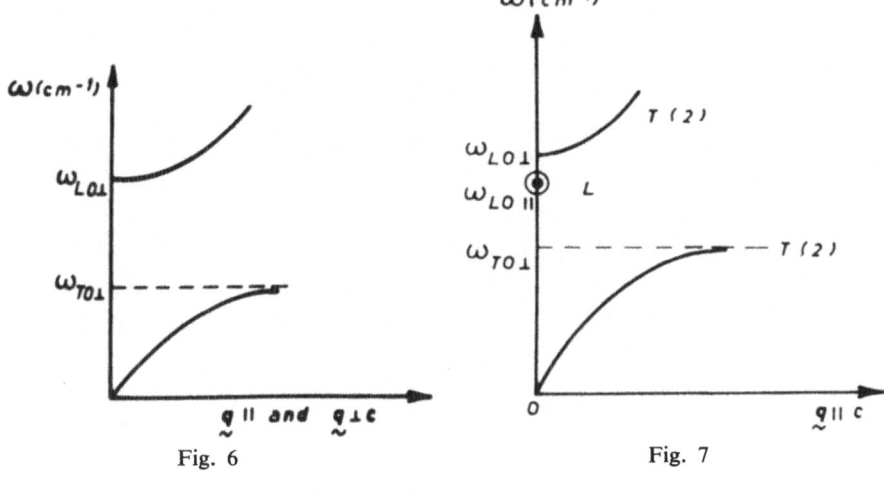

Fig. 6                                          Fig. 7

$$\omega_{L0\,\parallel} = \left(\frac{\epsilon_{0\,\parallel}}{\epsilon_{\infty\,\parallel}}\right)^{1/2} \omega_{T0\,\parallel} \tag{41}$$

(2) For **q** perpendicular to the $C$ axis a second transverse mode frequency appears given by

$$\epsilon(\omega) = \epsilon_{\infty\,\parallel} + \frac{(\epsilon_{0\,\parallel} - \epsilon_{\infty\,\parallel})\omega_{T0\,\parallel}^2}{\omega_{T0\,\parallel}^2 - \omega^2} \tag{42}$$

and the longitudinal mode frequency (Fig. 8.)

$$\omega_{L0\,\perp} = \left(\frac{\epsilon_{0\,\perp}}{\epsilon_{\infty\,\perp}}\right)^{1/2} \omega_{T0L} \tag{43}$$

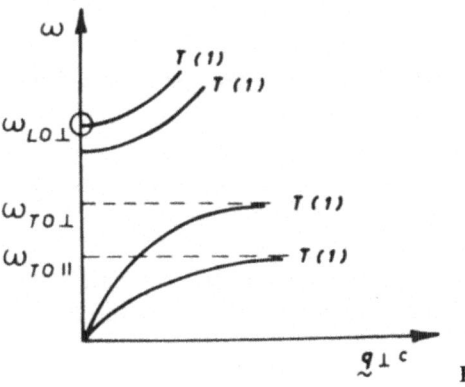

Fig. 8

If $\theta$ is the angle between the propagation vector $\mathbf{q}$ and the axis $C$ then:

$$\epsilon(\omega) = \eta^2(\omega) = \frac{\left(\dfrac{\omega_{T0\parallel}^2 \epsilon_{0\parallel} - \omega^2 \epsilon_{\infty\parallel}}{\omega_{\perp 0\parallel}^2 - \omega^2}\right)\left(\dfrac{\omega_{T0\perp}^2 \epsilon_{0\perp} - \omega^2 \epsilon_{\infty\perp}}{\omega_{T0\perp}^2 - \omega^2}\right)}{\left(\dfrac{\omega_{T0\parallel}^2 \epsilon_{0\parallel} - \omega^2 \epsilon_{\infty\parallel}}{\omega_{T0\parallel}^2 - \omega^2}\right)\cos^2\theta + \left(\dfrac{\omega_{T0\perp}^2 \epsilon_{0\perp} - \omega^2 \epsilon_{\infty\perp}}{\omega_{T0\perp}^2 - \omega^2}\right)\sin^2\theta} \tag{44}$$

We can determine experimentally these frequencies by reflection measurements on thick samples or by transmission measurements on thin films.

1. *Thick crystals.* The axis $C$ is in the surface plane and polarized light is used. The electric field $E$ can be either parallel or perpendicular to $C$:

(a) $\mathbf{E} \parallel C$. We obtained two frequencies

$$\omega_{T0\parallel} = 232 \text{ cm}^{-1}$$

$$\omega_{L0\perp} = 302 \text{ cm}^{-1}$$

(b) $\mathbf{E} \perp C$. We again obtain two frequencies

$$\omega_{T0\perp} = 240 \text{ cm}^{-1}$$

$$\omega_{L0\parallel} = 298 \text{ cm}^{-1}$$

2. *Thin films.* For CdS film on silicon, the $C$ axis is perpendicular to the film and the observed absorption peaks correspond to the frequencies

$$\omega_{T0\perp} = 242 \text{ cm}^{-1}$$

$$\omega_{L0\parallel} = 297 \text{ cm}^{-1}$$

3. *CdS film Evaporated on a Mirror.* The crystallites axes are in mirror plan but with different orientations and two absorption bands are observed by reflection corresponding to $\omega_{L0\perp} = 301.5 \text{ cm}^{-1}$ and $\omega_{T0} = 238.5 \text{ cm}^{-1}$ between $\omega_{T0\perp}$ and $\omega_{T0\parallel}$.

## 2. Symmetry Properties and Selection Rules for Photon–Phonon Interaction

### a. Symmetry Properties

ZnS-blende has an elementary cell with two atoms and the crystal has the symmetries of the group $T_d^2$. There are six normal vibrational modes for the blende. If one considers the symmetry of the crystal[1] looking for the traces of the corresponding matrices, which will be characters of the representation corresponding to the vibrations, one obtains the equation

$$\Gamma = 2\Gamma_{15}$$

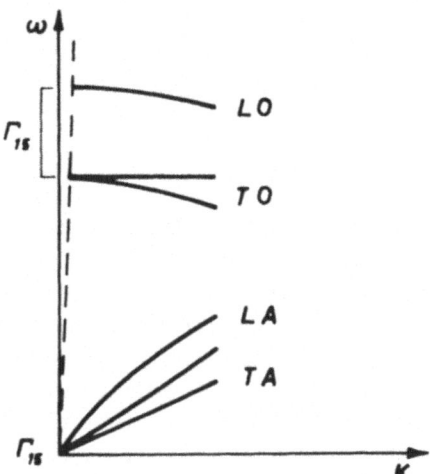

Fig. 9—Dispersion curves of ZnS (blende).

This shows that at the center of the Brillouin zone there are two triply-degenerate energy levels corresponding to the optical and the acoustical modes. For the optical modes the degeneracy is lifted along different symmetric directions as shown in Fig. 9.

ZnS and CdS wurtzite have elementary cells containing four atoms. The hexagonal wurtzite-type crystal has no symmetry center and its point group is $C_{6v}^4$ which corresponds to the plane hexagonal group. The set of vibrational states of the four atoms constituting the elementary cell is a 12-dimensional vectorial space; its matrix trace forms the representation $\Gamma$ of the group. From the character table of wurtzite (ZnS and CdS), one finds that the representation $\Gamma$ can be written

$$\Gamma = 2(\Gamma_5 \oplus \Gamma_6) \oplus 2\Gamma_1 \oplus 2\Gamma_4 \tag{45}$$

where $\oplus$ means direct sum. The twelve possible levels at the center of the Brillouin zone are decomposed in eight levels, four of which are doubly-degenerate. There is also an evident degeneracy, the acoustic branches for $k = 0$ being triply-degenerate.

## b. Selection Rules

1. *Optical Absorption.* (1) An incident radiant beam falling on a solid when interacting with the matter could be either absorbed or scattered. The optical absorption is given by the transition probability proportional to a matrix element which is different from zero when obeying the energy and momentum censervation rules. If the fundamental unperturbed vibrational state is defiened by a wavefunction $\psi_0$ and a possible excited vibrational

state by the wavefunction $\psi_1$ the transition probability under an external perturbation, whose effect is defined by the operator $\theta$, is proportional to the square of the matrix element:

$$\int \psi_0^* \theta \psi_1 \tag{46}$$

The translation is *allowed* when the matrix element does not vanish and *forbidden* when it vanishes. The optical absorption is related to the electric dipole moment, consequently the operator will be $\vec{\nabla}$. The Raman scattering is related to the polarizability tensor $\pi$, therefore, the operator will be $\pi$. The crystal being subject to a group $G$ of symmetry elements, the wavefunction $\psi_0$ belongs to the vector space defining a representation $\Gamma_0$ of the group and $\psi_1$ a representation $\Gamma_\alpha$. If the operator belongs to the representation of the group, one can assume for purely geometrical reasons that a sufficient condition to have the matrix element nonvanishing is to have in the decomposition in direct sum of irreducible representations of the direct product $\Gamma_0 \otimes \Gamma$ an irreducible representation contained in $\Gamma_\alpha$. The optically active modes are those which admit at least one common irreducible representation with $\Gamma_0 \otimes \Gamma$ and the ground state representation $\Gamma_1$ in their decomposition. By virtue of the momentum conservation selection rule

$$\vec{K}_{ph} + \vec{K}_s = 0 \tag{47}$$

The wavevector of the light being $\vec{K}_{ph} \simeq 0$, we are limited to normal modes with approximately $\vec{K}_s \sim 0$.

The $\nabla$ operator related to the optical absorption is transformed like the *xyz* coordinates of a space point by the symmetry operations of the group. The representation of $\vec{\nabla}$ is consequently the one which is composed of the matrices characterizing the operation $\Phi$ in each element $\{\Phi | \mathbf{v}\}$ of the space group. The representation is therefore $\Gamma_{15}$ in the group $T_d^2$ and $\Gamma_1 + \Gamma_5$ in the group $C_{6v}^4$.

The representations of the optical modes for the blende structure are thus identical to those of the $\vec{\nabla}$ operator. The transverse modes in this representation are therefore optically active. The only transverse optical mode in the wurtzite structure having a $\Gamma_1$ or a $\Gamma_5$ representation is the TO$_1$; therefore, we shall have only one absorption band for the wurtzite structure in spite of a quite complex dispersion diagram.

2. *Raman Scattering in Blende and Wurtzite.* In an anisotropic crystal the operator is a symmetrical tensor. In an orthogonal transformation such a symmetrical tensor is trsnsformed like the vectorial space of the six functions

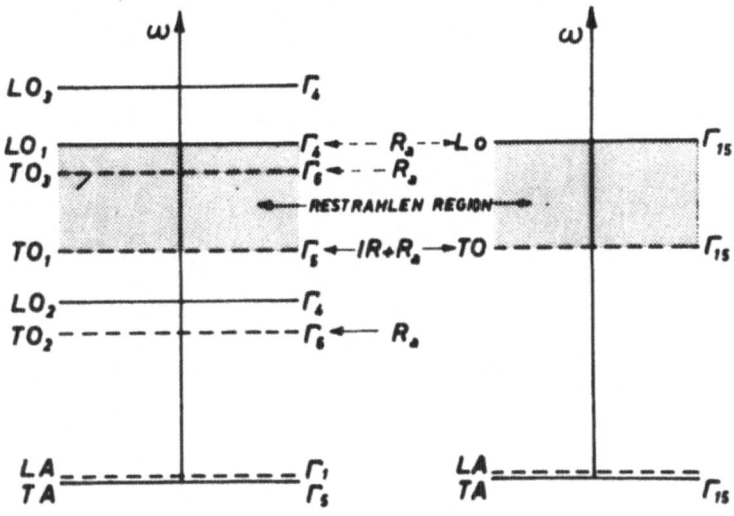

Fig. 10—Dispersion curve and representation $k \simeq 0$

$$x^2 \qquad y^2 \qquad z^2$$

$$xy \qquad xz \qquad yz$$

In order to obtain the representation of the polarizability tensor, we have to find how the six-dimensional vectorial space is transformed by the operations of the group $T_d^2$. The decomposition for the representation of $\Gamma$ the polarizability tensor into irreducible representation gives

$$\Gamma = \Gamma_1 \otimes \Gamma_{15} \oplus \Gamma_{12} \tag{48}$$

Thus the optical modes in the Blende corresponding to the representation $\Gamma_{15}$ are Raman-active.

The transformations of the group $C_{6v}^4$ give the characters allowing the decomposition of $\Gamma$ into irreducible representations

$$\Gamma = 2\Gamma_1 \oplus \Gamma_5 \oplus \Gamma_6 \tag{49}$$

The optical modes corresponding to the representations $\Gamma_1$, $\Gamma_5$, and $\Gamma_6$ will be Raman-active. Figure 10 summarizes the results on selection rules obtained by group-theoretical considerations.

## 3. Kramers–Kronig Analysis

From reflection spectra given in Fig. 11, it is easy to obtain by the Kramers–Kronig inversion the real part and the imaginary part of the dielec-

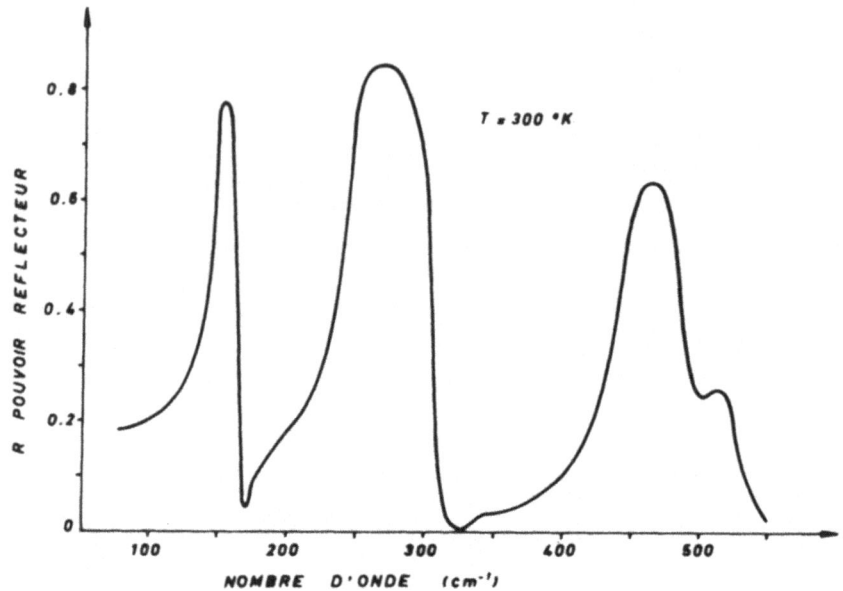

Fig. 11—Reflectivity spectrum of KNiF₃ between 70 and 550 cm⁻¹ at 300°K.

tric constant. If we consider the approximation of the classical dispersion theory the lattice vibrations can be explained as due to noncoupled harmonic oscillators with a damping factor. An oscillator $j$ is characterized by a resonance frequency (transverse optical frequency) $v_j$ an oscillator strength $4\pi pj$, and a damping coefficient $\gamma_j$

$$\epsilon' = n^2 - k^2 = \epsilon_\infty + \sum_j 4\pi p_j \frac{(v_j^2 - v^2)}{(v_j^2 - v^2)^2 + \gamma_j^2 v^2 v_j^2} \tag{50}$$

$$\epsilon'' = 2nk = \sum_j 4\pi p_j \frac{\gamma_j^v v_j}{(v_j^2 - v^2)^2 + \gamma_j^2 v^2 v_j^2} \tag{51}$$

The $v_j$ correspond to the maximum in the curve $\epsilon''(\omega)$ where $\gamma_j = \dfrac{\Delta v_j}{v_j}$ represents the width of the peak corresponding to $v_j$ and $4\pi p_j = \gamma_j \epsilon_N''$ where $\epsilon_N''$ represents the maximum of this $j$th peak.

The longitudinal optical frequencies $v_{Lj}$ are the values of $v$ where $\epsilon' = 0$ going from a negative to a positive value. All the absorption or reflection spectra can be then recalculated using these values for the oscillators and compared with the experimental results. For KNiF₃ the reflection spectrum contains three main peaks associated with three different infrared-active modes (see Fig. 11). $\epsilon'$ and $\epsilon''$ have been calculated (see Fig. 12) and the parameters of the independent corresponding oscillators deduced. The fit between

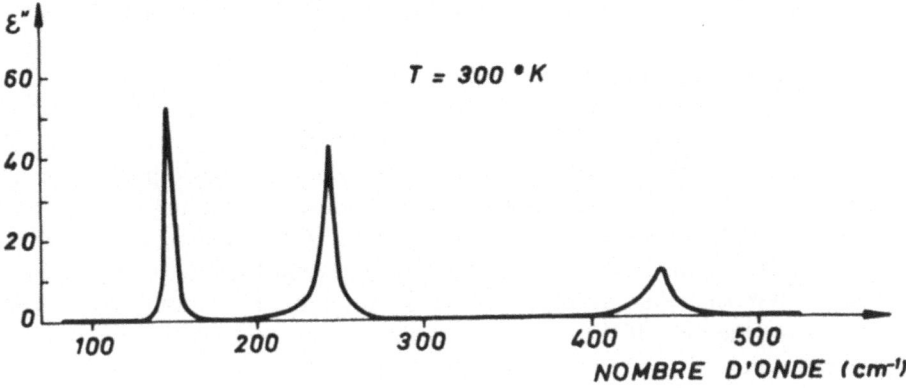

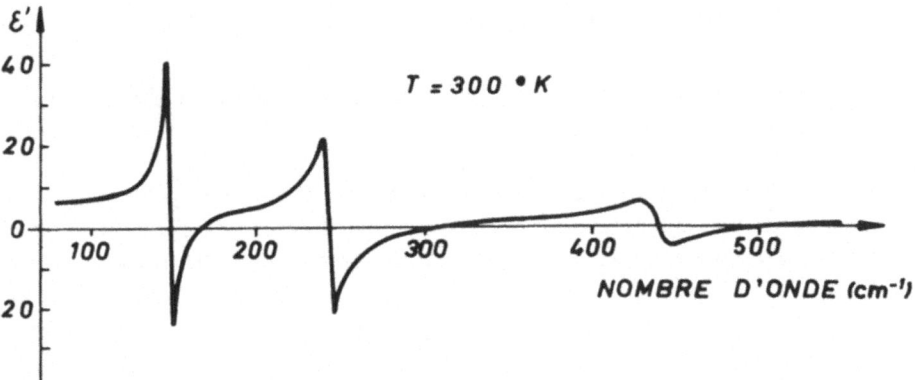

Fig. 12—Imaginary part of dielectric constant in KNiF$_3$ at 300°K; real part of the dielectric constant of KNiF$_3$ at 300°K.

the recalculated spectrum and the previous experimental one is good. This analysis method appears to be useful especially for noncoupled modes.

### 4. Multiphonon Interactions

In addition to single phonon–photon interaction, which corresponds to a very high absorption coefficient when the selection rules are satisfied, a complex absorption spectrum containing series of absorption bands in the infrared region with much lower absorption coefficient has been observed in many cases. This weak absorption is attributed to the interaction of two or more phonons with one photon. The theory of the interaction of the radiation fields with two or more normal vibrational modes has been given in terms of second-order electric dipole moments ([2]) or as due to the presence of anharmonic terms in the potential energy associated with lattice vibrations ([3]) and also as a combination of both.

Recently Wallis and Maradudin ([4]) have discussed the interaction of lattice anharmonicity and higher-order electric dipole moments in the optical absorption of crystals. Selection rules for a nonvanishing dipole moment contribution in the optical absorption are obtained considering the quantity $\vec{a}_0 \vec{M}(\vec{\alpha}, \vec{u})$ determining the probability of optical transitions $\mathbf{a}_0$ as the polarization vector of the radiation and $\vec{M}(\vec{\alpha}, \vec{u})$ the retarded electric dipole moment with $\vec{\alpha}$ being the wavevector of the radiation and $\vec{u}$ a set of displacement vectors for the ions. If the crystal has no dipole moment when the ionic displacements are all zero an expansion in powers of the displacement gives

$$\vec{M}(\vec{\alpha}, \vec{u}) = \sum_{l'k'} e^{2\pi i \vec{\alpha} \cdot \vec{x}(l'k')} M(k') \cdot \vec{u}(l'k')$$
$$+ \frac{1}{2} \sum_{l'k'} \sum_{l''k''} e^{2\pi i \vec{\alpha} \cdot \vec{x}(l'k')} \vec{u}(l'k') M(k'k'', l' - l'') \cdot \vec{u}(l''k'') + \cdots \quad (52)$$

In this expression $\vec{u}(lk) = \vec{u}\binom{l}{k}$ is the displacement vector of the $k$th ion in the $l$th unit cell, $\vec{x}(lk) \equiv \vec{x}\binom{l}{k} = \vec{x}(l) + \vec{x}(k)$ is the equilibrium position vector of the ion $lk$. $M(k)$ is the first-order and $M(kk', l - l')$ is the secondorder electric dipole moment coefficient. Assuming cyclic boundary conditions, one can introduce the normal coordinates $Q(kj)$ for the normal vibrational mode with wavevector $\mathbf{k}$ and branch $j$ through the transformation

$$\vec{u}(lk) = (Nm_k)^{-1/2} \sum_{k,j} \vec{e}\left(\frac{k}{k_j}\right) \vec{Q}(E_j) e^{2\pi i \vec{k} \cdot \vec{x}(l)} \quad (53)$$

Here $N$ is the number of unit cells, $m_k$ the mass of the atom $k$, and $e(k/k_j)$ is an orthogonal eigenvector of the dynamical matrix of the crystal. Using this transformation, one can write the quantity $\mathbf{a}_0 \cdot \mathbf{M}(\alpha, \mathbf{u})$ in the form

$$\vec{a}_0 \cdot \vec{M}(\vec{\alpha}, \vec{u}) = N^{-1/2} \sum_{k'} m_{k'}^{-1/2} \sum_{k'j'} \Delta(\vec{\alpha} + \vec{k}')$$
$$\times e^{2\pi i \vec{\alpha} \cdot \vec{x}(k')} \vec{a}_0 \cdot \vec{M}(k') e\left(\frac{k'}{k'j'}\right) Q(k'j')$$
$$+ \frac{1}{2} \sum_{k'k''} (m_{k'}, m_{k''})^{-1/2} \sum_{l} \sum_{k'j'} \sum_{k''j''} \Delta(\vec{\alpha} + \vec{k}' + \vec{k}'')$$
$$\times e^{2\pi i \vec{\alpha} \cdot \vec{x}(k)} e^{2\pi i k'' \cdot \vec{x}(l)} \vec{a}_0 \cdot \vec{e}\left(\frac{k'}{k'j'}\right) \vec{M}(k'k'', l' - l'')$$
$$\times e\left(\frac{k'}{k''j''}\right) Q(k'j') Q(k''j'') + \cdots \quad (54)$$

$\Delta|\mathbf{y}|$ is unity if $\mathbf{y}$ is zero or a reciprocal lattice vector $\vec{k}$, and is zero otherwise.

From this we can deduce directly the selection rules for momentum conservation. First-order dipole moment will contribute to the optical absorption only if the phonon wavevector $\vec{k}$ is equal and opposite in sign to the photon wavevector

$$-\vec{k} = \vec{\alpha} \tag{55}$$

For the second-order dipole moment contribution we have

$$\vec{k}' + \vec{k}'' = -\vec{\alpha} \tag{56}$$

This means that the wavevector of the radiation must be parallel and equal in magnitude to the resultant of the wavevectors which are to be created or destroyed. In the infrared region, since it is extremely small, we can consider that only phonons with equal-in-magnitude and opposite-in-sign wavevectors will interact with the radiation field in collective processes. The condition for optical transition being that the resulting wavevectors should be zero, we may deduce that only phonons from equivalent points of the Brillouin zone will interact. If the wto interacting phonons have the same energy their contribution results in *overtone* states and if the two phonons have different energies they give *combination* states. As an example for multiphonon transitions, let us consider the infrared absorption spectra of the II-IV compounds CdS and ZnS. Figure 13 shows the absorption spectrum between 15 and 30 $\mu$ for CdS ([5]) hexagonal structure. It is easy to count eight absorption peaks in this curve. These peaks result from the combination of four phonon frequencies. Then all possible combinations are found between the four fundamental frequencies:  $A = 299.5 \pm 0.5\,\text{cm}^{-1}$,  $B = 281.5 \pm 0.5\,\text{cm}^{-1}$,  $C =$

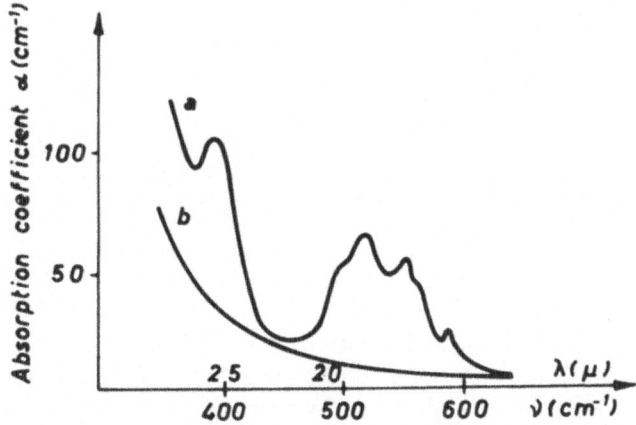

Fig. 13—Absorption spectrum of CdS: (a) calculated, (b) fundamental absorption.

$262.5 \pm 1$ cm$^{-1}$, D $= 201.5 \pm 2$ cm$^{-1}$. Note careful measurements show that the two supplementary peaks correspond to much weaker absorption coefficients.

The temperature dependence of the multiphonon optical transition is characteristic of the nature of the process. The variations of the absorption coefficient due to multiphonon coupling with the radiation field may be deduced from statistical considerations. In a crystal of atoms bound by harmonic forces the vibrational states of the lattice may be represented by those of an assembly of independent oscillators. The creation or annihilation of a phonon corresponds to a variation of the quantum number of an oscillator having the energy $|n_k + \frac{1}{2}| \hbar\omega$ per one unit of energy $\hbar\omega$. Considering the action of creation and annihilation operators $a^+$ and $a$ which express the transition from a state $n_k$ to a state $n_k + 1$ or a state $n_k - 1$, the wave function $\psi(n)$ of the crystal varies as follows

for absorption
$$\begin{cases} a_k^+ \,|\, \psi(n_k) \rangle = (n_k + 1)^{1/2} \,|\, \psi(n_k + 1) \rangle & (57) \\ a_k \,|\, \psi(n_k) \rangle = (n_k)^{1/2} \,|\, \psi(n_k - 1) \rangle & (58) \end{cases}$$

for emission
$$\begin{cases} \langle \psi(n_k - 1) \,|\, a_k \,|\, \psi(n_k) \rangle = (n_k)^{1/2} & (59) \\ \langle \psi(n_k + 1) \,|\, a_k^+ \,|\, \psi(n_k) \rangle = (n_k + 1)^{1/2} & (60) \end{cases}$$

In the case where the coupling of two phonons of wavevectors $+\mathbf{k}$ and $-\mathbf{k}$ is considered, the transition probabilities being proportional to the square of the corresponding matrix elements $(n_k + 1)^{1/2}$, $(n_{-k} + 1)^{1/2}$ and $(n_k)^{1/2}$, $(n_{-k})^{1/2}$, one obtains

$$n_k n_{-k} + 1 + n_k + n_{-k} \qquad (61)$$

for absorption, and

$$n_k n_{-k} \qquad (62)$$

for emission. This yields a total absorption for the two phonons addition process

$$\alpha_{\mathrm{add}} = K(1 + n_k + n_{-k}) \qquad (63)$$

$K$ being a proportionality coefficient.

The subtraction process, where the absorption of a photon corresponds to the annihilation of one phonon of wavevector $\vec{\mathbf{k}}$ and the creation of one phonon of wavevector $-\mathbf{k}$ for example, should also be considered. In this case the absorption yields

$$n_k(n_{-k} + 1) \tag{64}$$

and the emission

$$n_{-k}(n_k + 1) \tag{65}$$

The total absorption for the substraction process will be

$$\alpha_{\text{sub}} = K(n_k - n_{-k}) \tag{66}$$

The system being subject to Bose statistics it is possible to excite at the same time any number of identical phonons in the same quantum state. The total number of particles in the system is determined by the equilibrium conditions. The average number of phonons in a quantum state defined by the quasi-momentum $\mathbf{p}$ and energy $\mathcal{E}$ is given, at thermal equilibrium, by the Planck function

$$\tilde{n}_p = \frac{1}{e^{\mathcal{E}|\mathbf{p}|/kT} - 1} \tag{67}$$

At high temperature, $kT \gg \epsilon$, $\tilde{n}_p = kT/\epsilon$, the phonon density in a given state is proportional to the temperature. For phonon energies $h\nu_1$ and $h\nu_2$ we have

$$n_k = (e^{h\nu_1/kT} - 1)^{-1} \tag{68}$$

and

$$n_{-k} = (e^{h\nu_2/kT} - 1)^{-1} \tag{69}$$

$\nu_1$ and $\nu_2$ are the frequencies of the phonons with wavevectors $\vec{k}$ and $-\vec{k}$, respectively. The absorption coefficient for an addition process thus becomes

$$\alpha_T^{\nu_1 + \nu_2} = k[1 + (e^{h\nu_1/kT} - 1)^{-1} + (e^{h\nu_2/kT} - 1)^{-1}] \tag{70}$$

The variation of $\nu_1$ and $\nu_2$ with temperature, because of the thermal expansion being weak, could be neglected for our purposes.

Systematic [6] investigation shows that this law is pratically followed very closely. Figure 14 shows the theoretical curve and a certain number of a experimental points representing the temperature dependence of the absorption peak at 677 cm$^{-1}$ in ZnS, considered to be due to the coupling of two phonons with frequencies: $\nu_1 = 379$ cm$^{-1}$ and $\nu_2 = 298$ cm$^{-1}$.

In the case of phonon subtraction, the temperature dependence of the absorption coefficient is given by

$$\alpha_T^{\nu_1 - \nu_2} = k[(e^{h\nu_1[kT} - 1)^{-1} - (e^{h\nu_2/kT} - 1)^{-1}] \tag{71}$$

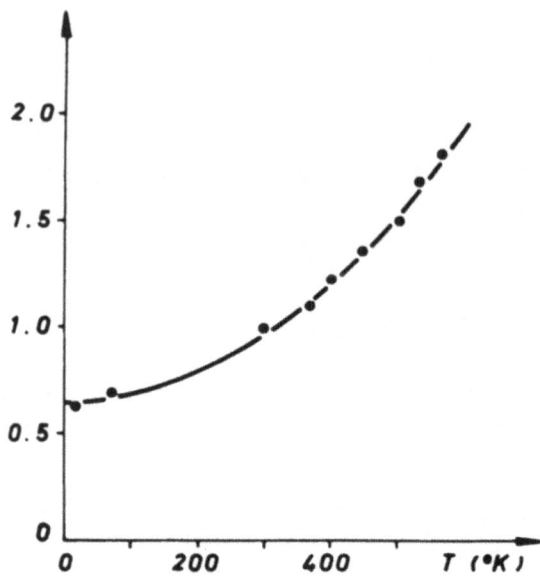

Fig. 14—Temperature dependence of the absorption
coefficient — theoretical × experimental

Figure 15 represents three different cases of photon-phonon interaction:
(1) Single phonon–photon interaction is temperature independent, (2) Two-

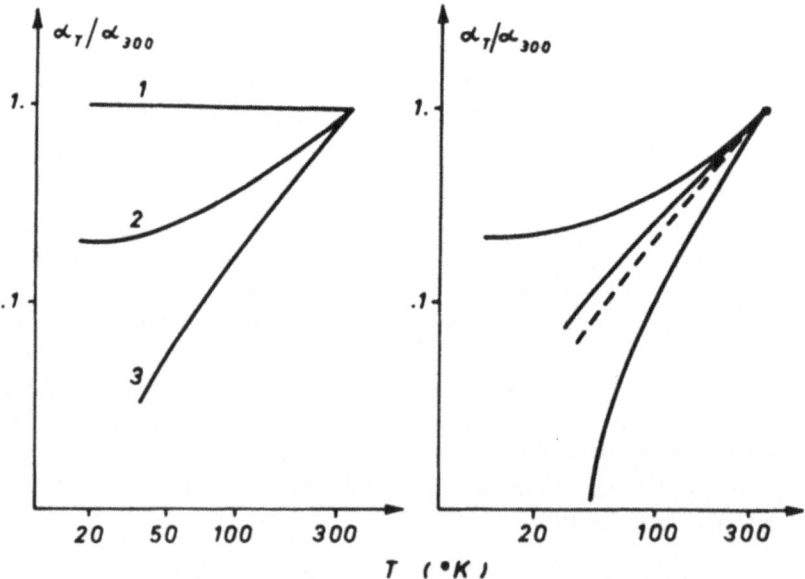

Fig. 15—Three different cases of photon–phonon interaction.

phonon addition $\alpha_T^{\nu_1+\nu_2}$ is given by equation 70, (3) Two-phonons subtraction, $\alpha_T^{\nu_1+\nu_2}$ is given by equation 71.

The absorption spectrum in the near-infrared region has been attributed (²) to second-order processes, and in the case of Si the first experimental results have been obtained by Johnson (⁷). All observed peaks in this region have been attributed to additive combinations of two phonons from the edge of the Brillouin zone as to have

$$\vec{k}_1 + \vec{k}_2 \cong 0 \tag{72}$$

Symmetry consideration on the selection rules for multiphonon transitions lead to the list of allowed and forbidden phonon combinations from different critical points in the zone. The phonon operator $\mathbf{V}$ having the symmetry of the representation $\Gamma_{15}^-$, the only optically active vibrational modes will correspond to the representation $*k_j^{(m)}$, such as to have

$$[*\vec{k}_j^{(m)} *\vec{k}^{1\,(m)} \,|\, \Gamma_{15}^-] \neq 0 \tag{73}$$

A restricted but sufficient multiplication table of the group representations $G_X/T_X$, $G_L/T_L$, $G_W/T_W$ is given by Balkanski and Nusimovici (⁸). The results are that at the point $K$ the allowed phonon combinations are the following

$$L + L, L + TO, TA + L, TA + TO$$

and the combination of the branches $TO + TO$ and $TA + TA$ are forbidden.

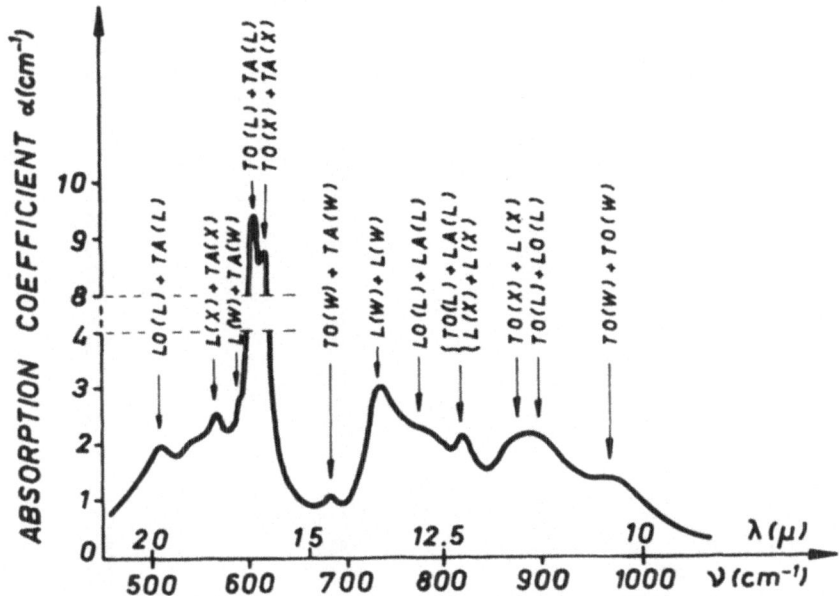

Fig. 16—Absorption spectrum in the region of two-phonon transitions.

At the point $L$, the allowed combinations are from the following branches:

$$TO + TA, LO + TO, LA + TA, LO + LA$$

and the forbidden ones:

$$LO + LA, LA + LA, LO + TA, LA + TA, TO + TO, TA + TA$$

At the point $W$ are allowed

$$L + L, L + TO, L + TA, TA + TA, TO + TO, TA + TO$$

Figure. 16 gives the absorption spectrum in the region of two-phonon transitions.

In Table I the combination frequencies obtained from optical spectra are compared with the values obtained from neutron scattering data. The observed combination modes at different critical points are given in Table II. The results given in Table II compared with the theorical results on selection rules show a satisfactory agreement between theory and experi-

## TABLE I

### Frequencies of Absorption Bands Due to Coupled Phonons

| Combination | Critical points | Our result, cm$^{-1}$ | According to Dolling, cm$^{-1}$ |
|---|---|---|---|
| $TO + TO$ | $L$ | — | 980 |
|  | $W$ | 965 | 961 |
|  | $X$ | 940 | 928 |
| $TO + LO$ | $L$ | 902 | 911 |
|  | $W$ | 856 | 853 |
|  | $X$ | 880 | 875 |
| $TO + LA$ | $L$ | 823 | 869 |
|  | $W$ | 856 | 843 |
|  | $X$ | 880 | 875 |
| $LO + TA$ | $L$ | 780 | 800 |
|  | $W$ | 743 | 734 |
|  | $X$ | 823 | 822 |
| $TO + TA$ | $L$ | 608.5 | 604 |
|  | $W$ | 685 | 700 |
|  | $X$ | 620 | 614 |
| $LO + TA$ | $L$ | 515 | 535 |
|  | $W$ | 585 | 591 |
|  | $X$ | 563 | 561 |

## TABLE II

### Coupling of Experimentally Observed Phonons

|  |  |  |
|---|---|---|
| At point X (1, 0, 0) | $L + L$ | mingled with $TO(L) + LA(L)$ |
|  | $L + TO$ |  |
|  | $TA + L$ |  |
|  | $TO + TA$ |  |
| At point L (1, 1, 1) | $TO + TA$ |  |
|  | $TO + LO$ |  |
|  | $TO + LO$ | mingled with $L(X) + L(X)$ |
|  | $LO + LA$ |  |
| $L +$ |  |  |
| At point $W(1, \frac{1}{2}, 0)$ | $L + L$ | $TO + TO$ |
|  | $L + TA$ | $TO + TA$ |
|  | Weak peak |  |

ment. Similar analyses of the multiphonon spectra in homopolar crystals have been carried recently by Biltz, Geick, and Renk ([9]) and also by Loudon are Johnson ([10]).

An experimental comparison between additive and subtractive processes can be found in the far-infrared phonon spectra of CdS ([11]) given in Fig. 17.

Some of the weaker peaks observed in the infrared absorption spectra

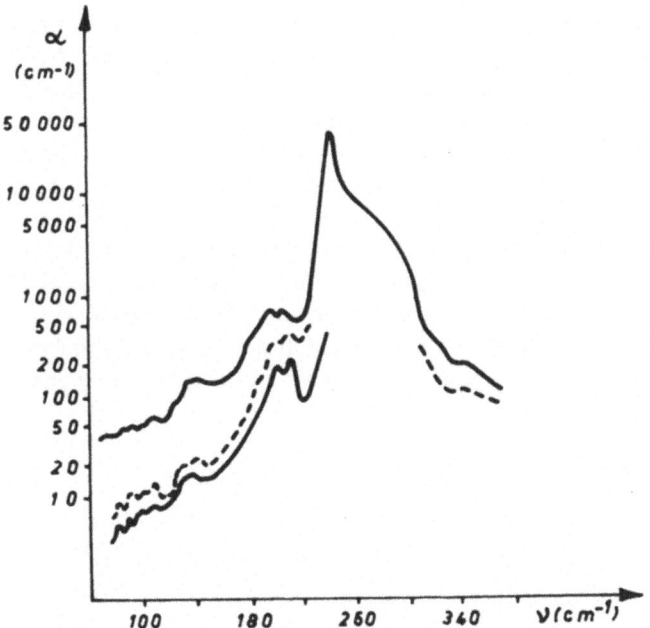

Fig. 17—Absorption spectra for CdS in the far infrared.

are interpreted in terms of three phonon combinations. In order to satisfy the selection rule

$$\vec{k}_1 + \vec{k}_2 + \vec{k}_3 = 0 \qquad (74)$$

one has to consider each combination of two normal modes at the edge of the Brillouin zone and one mode at the center of the zone with $k = 0$.

## IV. LOCALIZED VIBRATIONS

### 1. Localized Surface Modes of Vibrations

The occurrence of a surface or a boundary means that a certain number of interatomic forces are set equal to zero. The qualitative consequence is a lowering of the "stiffness" of the lattice and therefore a lowering of some or all of the normal-mode frequencies. Another consequence of this situation is a shift of some of the normal-mode frequencies out of the allowed frequency range. These frequencies are now within the forbidden frequency gap and the corresponding mades are called localized surface modes. Their amplitude is maximum at the surface and decreases exponentially with increasing distance from the surface.

Theoretical investigations of surface vibrational waves were first carried out by Lord Rayleigh ([12]) who studied an isotropic elastic continuum with a planar surface free. For this case, surface waves called Rayleigh waves exist and propagate with a velocity less then that of bulk transverse waves and a fortiori less than that of bulk longitudinal waves. The frequencies of Rayleigh waves lie in the acoustical branch of the phonon spectrum. More recent studies on surface elastic waves have been developed by Gazis, Herman and Wallis ([13]).

Wallis ([14]) has investigated the effect of free ends on an alternating diatomic lattice with nearest neighbor Hooke's law interactions and has shown that in this case surface-localized modes of vibration exist with frequencies in the forbidden gap between the optical and acoustical branches. Following Wallis, let us rewrite the equations of motion for a diatomic chain including those of the end atoms.

$$m\ddot{u}_1 = \gamma(u_2 - u_1) \qquad (75)$$

$$M\ddot{u}_{2j} = \gamma(u_{2j+1} + u_{2j-1} - 2u_{2j}) \qquad 1 \leq j \leq N-1 \qquad (76)$$

$$m\ddot{u}_{2j+1} = \gamma(u_{2j} + u_{2j+2} - 2u_{2j+1}) \qquad 2 \leq j \leq N \qquad (77)$$

$$M\ddot{u}_{2N} = \gamma(u_{2N-1} - u_{2N}) \qquad (78)$$

One can see that for all normal modes, except one, the amplitudes involve sine and cosine functions of $j$. These modes are therefore wavelike in character and have frequencies lying either in the acoustical or optical branches. The remaining normal mode is a surface-localized mode and this frequency is given by

$$\omega = \sqrt{\gamma \frac{M + m}{Mm}} \tag{79}$$

The amplitude of the surface mode can be expressed in the form

$$A_{2j+1} = U(-1)^{j+1}\left(\frac{m}{M}\right)^{j+1} \tag{80}$$

$$A_{2j} = U(-1)^{j}\left(\frac{m}{M}\right)^{j} \tag{81}$$

where $U$ is a constant. The exponential decrease of the amplitudes away from the end is illustrated in Figure 18.

A general criterion can be given for the existence of surface modes: the mass of the light atoms must be less than the total mass of heavy atoms. If surface modes exist, their number is equal to the number of ends of the lattice which have light atoms. When one end atom is light and the other heavy, the criterion is satisfied for any mass ratio less than unity. When both atoms are light, the criterion is satisfied only if the mass ratio is less than the number of heavy atoms divided by the number of light atoms.

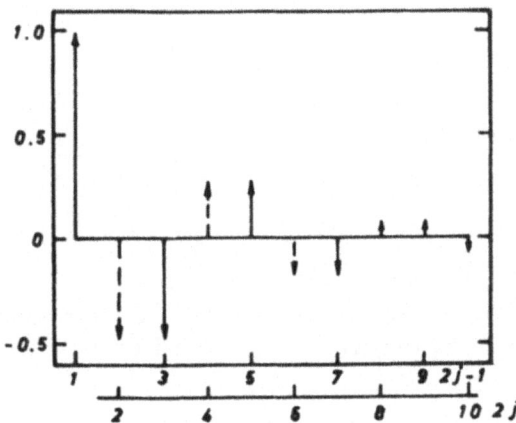

Fig. 18—Maximum atomic displacement in arbitrary units as a function of position in the lattice for the surface modes when $m/M = \frac{1}{2}$ and $N = 5$. The solid arrows refer to mass $m$ and the dashed arrows to mass $M$ [from R.F. Willis, *Phys. Rev.* **105**: 540 (1957).]

When both end atoms are heavy, the criterion is not satisfied for any mass ratio.

Wallis ([15]) has also generalized the treatment for three-dimensional diatomic crystals. In ionic crystals were atoms of different masses have opposite electric charges, a net electric dipole moment localized near the surface may be set due to one excited surface mode. Such a mode should interact with the electromagnetic radiation and give an absorption in the infrared for frequencies lying below the transverse optical mode frequency. Because of the very high absorption intensity of the transverse optical mode to compare with the surface mode, the experiment should be done on a thin film deposited on a metallic mirror, and one should use multiple reflections to increase the absorption due to the surface modes.

If a foreign atom is absorbed on the surface, localized modes should occur provided the mass of the absorbed atom is small enough or the bonding of the surface is sufficiently strong. An analogous situation has been studied ([16]) in the case of a substitutional impurity in bulk material, where it is found that the appearance of localized modes depends on the mass of the impurity and on the force constants. In the case of impurities on the surface, the situation can be illustrated by the semi-infinite monatomic layer chain with nearest-neighbor interactions, the absorbed impurity being treated as an atom of different mass at the free end. The equations of motion may be written then ([17])

$$M\ddot{u}_0 = \gamma'(u_1 - u_0) \tag{82}$$

$$M\ddot{u}_1 = \gamma(u_2 - u_1) + \gamma'(u_0 - u_1) \tag{83}$$

$$M\ddot{u}_n = \gamma(u_{n+1} + u_{n-1} - 2u_n) \qquad n \geq 2 \tag{84}$$

The displacements for the surface modes can be found as solutions of this equation in the form

$$u_0 = U_0 e^{i\omega t} \tag{85}$$

$$u_n = U(-1)^n e^{-qn} e^{i\omega t} \qquad n \geq 1 \tag{86}$$

Where $U_0$ and $U$ are constant amplitudes.

In order for the displacements to be solutions of the equations of motion, it is necessary that the frequency $\omega$ be related to the attenuation constant $q$ by

$$M\omega^2 = 2\gamma(1 + \cosh q) \tag{87}$$

and $q$ be a solution of the equation

$$[\gamma' - \gamma(1 + e^q)][\gamma' - 2\frac{M'}{M}\gamma(1 + \cosh q)] - \gamma'^2 = 0 \tag{88}$$

The value of $q$ must be real and positive; this condition requires that the ratios $\gamma'/\gamma$ and $M'/M$ satisfy the inequality

$$\frac{\gamma'}{\gamma} > 4\frac{M'/M}{2[(M'/M) + 1]} \tag{89}$$

It seems that infrared absorption has been observed ([18]) by surface-localized modes associated with hydrogen absorbed on platinum. Two different absorption peaks have been detected and attributed to two different types of boundaries of the hydrogen on the platinum surfaces. Infrared absorption due to localized modes not only allows us to specify the chemical nature of the absorbed species but also gives an indication of the force constants and therefore provides significant insight into the details of the absorption processes. This is also a powerful means for the investigation of surface states.

The theory of surface impurity states can be developed further, including a specific force model for the impurity state. Kaplan ([19]) and Hori and Asahi ([20]) have studied theoretically the case of surface impurities in a two-dimensional lattice. This investigation confirms that surface localized modes may arise in the forbidden frequency gap on either side of the allowed frequency bands for normal lattice vibrations. Furthermore, Kaplan showed that the tendency of the pure semi-infinite solid to exhibit Rayleigh waves is suppressed for short wavelengths by the surface impurities if the surface atom mass is sufficiently small.

Studies of lattice dynamics, and particularly those studies including surface effects, should become very important tools in the physics of thin films because many of the phenomena, such as infrared absorption and scattering, X-ray and electron scattering, and conduction of heat and electricity occurring in solid thin films will be affected by the existence of surface modes. In all the cases of interactions, phonon–phonon, phonon–photon, or phonon electron in thin films, one would have to take into consideration the surface modes of vibrations, especially when the films become very thin.

## 2. The Effect of Defects on Lattice Vibrations

We have considered in the preceding section the vibrations of a perfect crystal lattice. The presence of defects and imperfections such as vacancies, impurities, and dislocations have an effect in two ways on the normal mode frequency: the individual frequency levels inside the branch of allowed frequencies are shifted by small amounts, and a small number of frequencies which normally lie near the band edges can emerge out of the allowed bands into the gap of forbidden frequencies. The shift of these particular frequencies may be considerable and is associated with the normal mode of the impurity atom vibrations whose space-dependent factors decrease rapidly with distance from the defect.

The models of crystal lattices which we consider can be described in terms of a periodic array of masses coupled to each other by a system of springs. The changes in the normal-mode frequencies of such a dynamical system due to a partial change in the masses and the springs has been studied long ago by Routh ([21]) and Rayleigh ([22]). Further studies of the effects of defects on lattice vibrations have been developed by Lifshitz ([23]), Montroll ([24]), and Elliott ([25]). We shall try to show with a very simple example the salient features of the results from these investigations generally referred to as "Rayleigh's theorems."

The introduction of an impurity of mass $M$ in a linear chain of atoms having the mass $m$, changes the kinetic energy of the normal system into $T = \frac{1}{2}m \sum \dot{u}_j^2$:

$$T = \frac{1}{2}m \sum \dot{u}_j^2 + \frac{1}{2}(M - m)\dot{u}_0^2 \tag{90}$$

The potential energy is

$$V = \frac{1}{2}f \sum (u_j - u_{j+1})^2 \tag{91}$$

With the substitution

$$\xi_k = (e^{ik \cdot r_j/N^{1/2}})u_j \tag{92}$$

we obtain

$$T = \frac{1}{2}m \sum \dot{\xi}_k \dot{\xi}_{-k} + \frac{M - m}{2}[\sum \dot{\xi}_k]^2 \tag{93}$$

and

$$V = \frac{1}{2}m \sum \omega_k^2 \xi_k \xi_{-k} \tag{94}$$

The equation of motion of the altered system yields

$$m\ddot{\xi}_k + \frac{M - m}{N} \sum \ddot{\xi}_n = -m\omega^2 \xi_k \tag{95}$$

$$\omega^2 \left[ m\xi_k + \frac{M - m}{N} \sum \xi_n \right] = m\omega_k^2 \xi_k \tag{96}$$

wherefrom we obtain

$$\frac{1}{N} \sum \xi_n = \xi \tag{97}$$

and

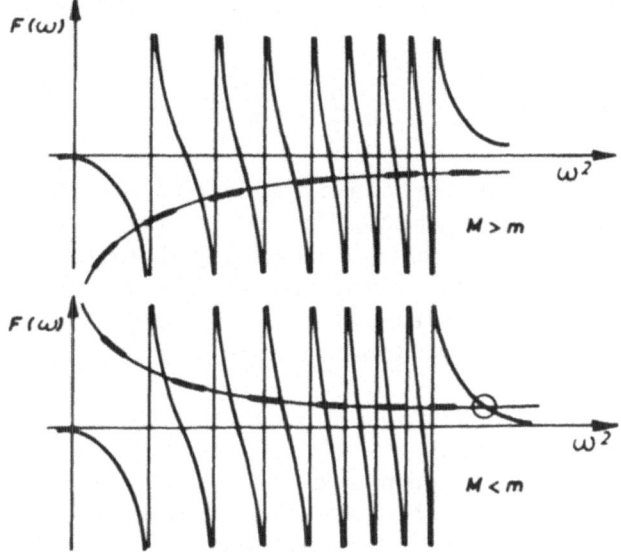

Fig. 19—Dispersion relation in a perturbed lattice: (a) $M > m$, (b) $M < m$.

$$\xi_n = -\frac{M - m}{m} \frac{\zeta \omega^2}{\omega^2 - \omega_k^2} \tag{98}$$

The characteristic equation for the determination of normal-mode frequencies is thus

$$F(\omega^2) \equiv \frac{1}{N} \sum_{k \neq 0} \frac{1}{\omega^2 - \omega_k^2} = -\frac{m}{M - m} \frac{1}{\omega^2} \tag{99}$$

Generally the changes in masses and force constants with which we are concerned do not cause the lattice to be unstable to small oscillations, we therefore assume that $\omega^2$ is never negative.

Plots of $F(\omega^2)$ vs. $\omega^2$ are shown in Fig. 19. The dotted curves represent $F_2(\omega^2) = 1/(M - m)\omega^2$, when $M > m$ all the frequencies are increased; when $M < m$ all the frequencies are lowered, but not more than the distance to the next unperturbed frequency. The band mode could then suffer only a slight shift in frequency due to the presence of impurities. When $M < m$ one frequency is removed from the dense collection of normal mode frequencies. This is possible only for the extreme frequencies of the dense set; the isolated frequency is that of the localized mode of vibration. More elaborate calculations using the Green's function confirm this result.

Considering isotopic defects and using the notation $\epsilon = (m - M)/m$,

Montroll and Potts ([24]) show that the equation whose solutions are the perturbed normal-mode frequencies reduces to

$$\epsilon \tan \frac{\theta}{2} = \tan \frac{N\theta}{2} \tag{100}$$

In absence of defects, the unperturbed normal-mode frequencies $\omega = \omega_L \sin \hbar k/N$ are doubly degenerate. The isotopic defect being the center of the lattice, this degeneracy is split in that only the symmetric $(u_n = u_{-n})$ modes of vibrations are affected by the perturbation. This is a consequence of the fact that the antisymmetric modes $(u_n = -u_{-n})$ have an $n$-mode at the position of the defect, and the lattice as a result does not notice the presence of the defect.

It is found that for a heavy isotope defect $(\epsilon < 0)$, there is a one-to-one correspondence between the perturbed and unperturbed frequencies, and that the former are slightly depressed relative to the latter. In the case of a light isotope $(0 < \epsilon < 1)$, the perturbed frequencies are all raised relative to the unperturbed frequencies, but there is one solution less than the number of unperturbed solutions. This solution corresponds to a complex value of $\theta$, $\theta = \pi + iz$ and the equation determining $z$ becomes

$$\varepsilon = \coth \frac{z}{2} = \tanh \frac{Nz}{2} \xrightarrow[N \to \infty]{} 1 \tag{101}$$

The solution of this equation is

$$z = \log \frac{1 + \epsilon}{1 - \epsilon} \tag{102}$$

and leads to a value of the frequency

$$\omega = \frac{\omega_L}{(1 - \epsilon^2)^{1/2}} \tag{103}$$

This frequency lies above the band of allowed frequencies and, since it corresponds to a complex value of the wavevector $\mathbf{k}$, is associated with a localized mode of vibration in which the amplitudes fall off exponentially with increasing distance from the defect.

As $M \to m$ $(\epsilon \to 0)$, the impurity frequency returns to the top of the band of allowed frequencies, while as $M \to 0$ $(\epsilon \to 1)$, the impurity frequency becomes infinite, corresponding to the loss of one degree of freedom in the lattice. The increase or decrease of a single force constant $f$ hes the same effect as a reduction or an enhancement respectively of the single mass. The greater the difference between host lattice mass and impurity mass, the greater the separation of the localized mode from the band modes.

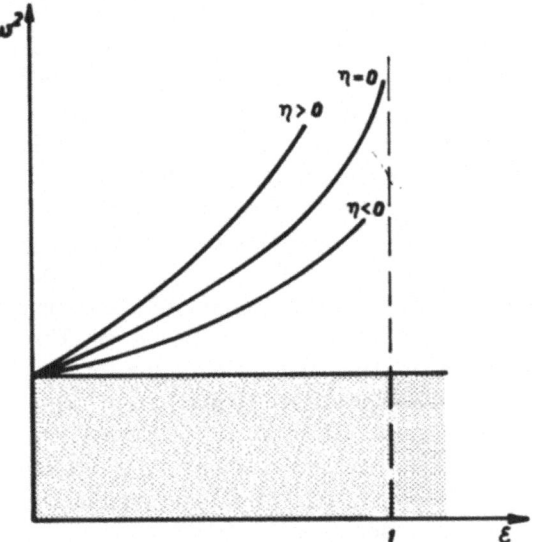

Fig. 20—Schematic representation of the local mode frequencies as a function of $\epsilon = (m - M)/m$ and $\eta = (f - f')/f$

Two defects at large separation give rise to degenerate impurity frequencies corresponding to the impurity frequencies of isolated defects. As the two defects are brought together this frequency is split. These considerations are summarized in Fig. 20.

## 3. Optical Absorption in Perturbed Lattices

In a homopolar crystal with a density of charged impurities of the order of $10^{19}$ or $10^{20}$ cm$^{-3}$, normal modes become optically active due to the polarization in the immediate vicinity of the defects. The band modes, solutions of the perturbed equation of motion for all practical purposes, may be considered to have the same frequencies as the normal vibration modes, the spectral resolution being not better than $N^{-\alpha}$; $N$ in the total number of masses in the system and $\alpha$ depends on the dimensionality of the lattice.

The singularities in the dispersion curve corresponding to the critical points remain practically unchanged in the perturbed lattice and hence become observable by means of optical transitions. This possibility has been experimentally explored in the case of silicon [26] and diamond [27]. The defects are created by means of fast-neutron irradiation. Lark-Horowitz [28] has suggested that the lattice defects created in silicon by neutron bombardment act as charged centers. The vacancies are acceptors, negatively charged, when ionized, and the interstitials are donors, positively charged, when ionized. There is, in addition, an electrical compensation in the whole crystal which prevents the rise of the absorption coefficient due to free carrier optical absorption in the frequency region of lattice vibrations.

The optical absorption resulting from the possibility to activate the ionized impurities depends on the real amplitude of the defect vibration, which is a function of the lattice atoms' amplitude, of the frequency of the mode, the defect mass, and of the force constant relating the defect to the neighboring atoms. Dawber and Elliott ([29]) have calculated the mean-square amplitude of the band modes assuming isotopic defects without charges in the force constants for a simple cubic lattice. The amplitude of the mode at the origin is $O(1/N)$ which is the value for a nonlocalized mode. This demonstration that the perturbed frequencies are indistinguishable from the unperturbed ones could be confirmed experimentally. A plot of the mean-square amplitude of the band mode at the origin normalized to $1/SN$, where $N$ is the number of unit cells and $S$ the number of atoms per cell, is given in Fig. 21 for two different impurity masses.

The absorption coefficient for the band modes calculated by Dawber and Elliott ([30])

$$\alpha(\omega) = \frac{2\pi^2 De^2 \Lambda}{3nc} \sum_{\alpha} |\Lambda_\alpha(f, 0)|^2 F(\omega) \tag{104}$$

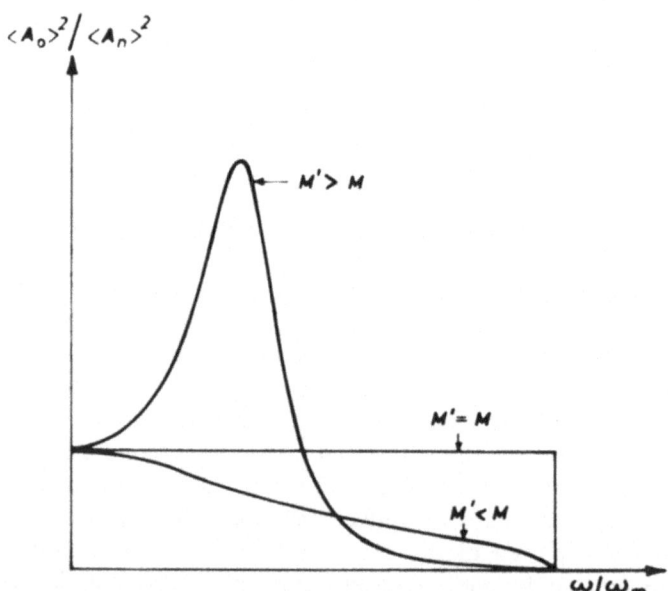

Fig. 21—The mean-square amplitude of the band modes at the origin normalized to $1/SN$

depends on the number of defects per unit volume $D$, on the density of modes per unit frequency range per unit volume $F(\omega)$, and a function of $\epsilon = (m - M)/m$; $n$ is the refractive index; $\Lambda$ is the local field correction which in terms of the refractive index is given by $\Lambda(n^2 + 2)^2/9$.

A sum rule has been also deduced where the integrated absorption coefficient could be directly compared to the density of defects:

$$\int_0^\infty \alpha(\omega) \, d\omega = \frac{2\pi^2 D e^2 \Lambda}{ncM} \tag{105}$$

According to the value of $\epsilon$, one can distinguish the following cases:

(a) $\epsilon < 0$, $M > m$, *Heavy Impurities.* The absorption spectrum contains the normal vibrational modes' spectrum. There is a resonance in the $A_d/A_n = f(\omega)$ curve. This corresponds to a broad maximum at $\omega < \omega_m$ ($\omega_m$ is the maximum frequency of the lattice) in the curve of density of states. Physically it is known that the width of the absorption band of a defect is proportional to the density of lattice modes at the frequency at which the absorption takes place. A heavy impurity which has a resonant frequency at $\omega < \omega_m$ has its vibrations easily propagating through the lattice by the normal modes numerous at this frequency. This leads to only a modification in the shape of the density-of-states curve.

(b) $\epsilon > 0$, $M < m$, *Light Impurities.* The light defect has a frequency in a part of the spectrum where theoretically in the unperturbed lattice no susceptible exists normal mode to take up its energy. The resonance will therefore be sharp and the space attenuation of the vibrational amplitude will be particularly rapid with distance from the defect. The absorption spectrum is a good representation of the density of states, but in this case the largest part of the energy is concentrated in the localized modes at a frequency higher than $\omega_m$. The remaining energy is spread over the whole phonon spectrum and is practically not observable.

(c) $\epsilon \geq 0$, $m - M$, *Very Small.* For an impurity, such as to have $|\epsilon|$ very small and the force constants unchanged, it will be possible to observe optically the singularities in the dispersion curve practically unchanged by the presence of defects. In this case, the normal mode becomes optically active in the perturbed lattice and observable at the critical points in the Brillouin zone.

## 4. Infrared Absorption at the Critical Points and Selection Rules from Group-Theoretical Considerations

The electric dipole moment transforms by the operations of the symmetry group as a vector. If we neglect the representation of the incident phonon, the representation of the group corresponding to the electric dipole moment operator is $\Gamma_{15}^-$. The transition from a state characterized by a wavefunction $|1\rangle$ belonging to the representation $\gamma_1$ of the space group to a state characterized by $|2\rangle$ belonging to $\gamma_2$ is possible if the decomposition of the direct sum of the direct product $\gamma_1 \otimes \gamma_2$ contains $\Gamma_{15}^-$.

It has been demonstrated that it is possible to establish correspondence relations between the space group $G$, leaving the crystal invariant, and the translational group $T$, and the subgroups $G_{\mathbf{k}}$ and $T_{\mathbf{k}}$, giving the possibility to use only the representation of the group $G_{\mathbf{k}}/T_{\mathbf{k}}$. The representations of the group factor $G_x/T_x$ have been established ([31]) for the different critical points in the Brillouin zone.

One has then to find out the corresponding representations of the normal vibration modes at these points. For crystals like silicon with two atoms in the unit cell, we have six branches in the dispersion curve. The translation from one atom to another is a $(1/4, 1/4, 1/4)$. The longitudinal vibrations of two atoms are hence in quadrature at the points $X$ and $W$, the energies of two vibrations in advance quadrature and retard quadrature are the same, $LO$ and $LA$ are degenerate at $X$ and $W$. It is also possible to demonstrate that $TO_1$ and $TO_2$ are degenerate at the points $T, X, L,$ and $W$; this is also true for $TA_1$ and $TA_2$. At the center of the zone we have two sets of triply degenerate branches, optical with the representation $T_{25}^-$, and acoustical with $\Gamma_{25}$. The representation of the normal modes at the critical points $X, L,$ and $W$ are given in Table III.

The maximum density of modes is at these critical points, and, if activated by the presence of electric dipole moments, the optical transitions will show absorption bands corresponding to the critical point frequencies.

Neutron-irradiated silicon has been studied in the infrared region corresponding to single-phonon frequencies ([26]). The absorption peak attributed by Fan and Ramdas ([32]) to one-phonon transitions has been investigated with

**TABLE III**

**Representation of Different Normal Modes at Critical Points**

|   |                  |             |               |
|---|------------------|-------------|---------------|
| $X$ | $L = LO = LA$  | $X(1)$      |               |
|     | $TA$           | $X(3)$      |               |
|     | $TO$           | $X(4)$      |               |
| $L$ | $LA$           | $L(2^-)$    |               |
|     | $TA$           | $L(3^+)$    |               |
|     | $LO$           | $L(1^+)$    |               |
|     | $TO$           | $L(3^-)$    |               |
| $W$ | $TO$           |             |               |
|     | $LO = LA$      | $W(m)$      | $m = 1, 2, \ldots$ |
|     | $TA$           |             |               |

## TABLE IV

### The Frequencies of Vibration Modes in Irradiated Silicon at Different Critical Points

| Branches | Critical points | Wave number cm$^{-1}$ | |
| | | Our results | Dolling's results |
|---|---|---|---|
| TO | $\Gamma$ | 517 ± 2 | 518 ± 3 |
| | L | 493 ± 2 | 490 ± 10 |
| | W | 478 ± 2 | 481 ± 13 |
| | X | 449 ± 3 | 464 ± 10 |
| TA | L | 113 ± 2 | 114 ± 2 |
| | X | 155 ± 5 | 150 ± 2 |
| | W $\Big\}$ $\Sigma$ | 212 ± 3 | 219 ± 5 |

high resolution and the spectrum extended further in the infrared. The results of these studies, summarized in Fig. 22, reveal indications for structures corresponding to the singularities in the dispersion curve (³³). The attributions of the absorption peaks corresponding to the different critical points has been done in comparison with the neutron-scattering results obtained by Dolling (³⁴). Table IV gives a comparison of the results obtained from optical spectra and those from neutron scattering data.

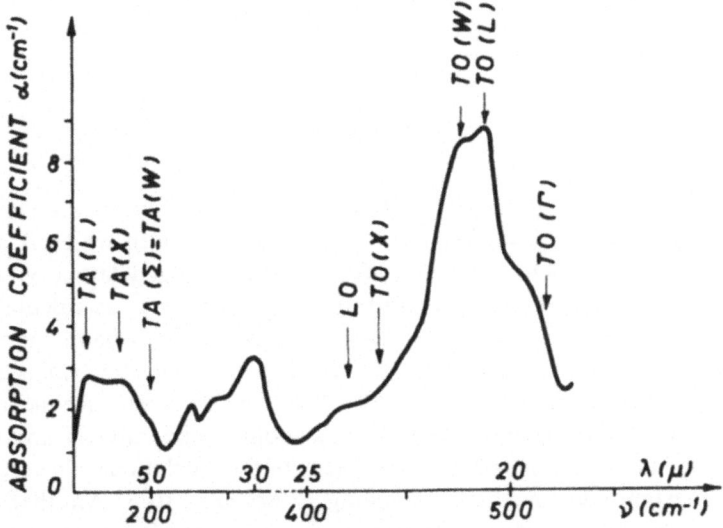

Fig. 22—Absorption spectrum of Si due to single phonon–photon interaction (band modes at critical points).

We do not have optical data for the point $\Sigma$. A good qualitative approximation could be to admit that $\omega(W) = \omega(\Sigma)$ based on the fact that different
force models give values for the frequency for $W$ and $\Sigma$ close to each other.

## 5. Localized Modes

Outside of the frequency band corresponding to the normal modes, the
optical absorption of semiconductors containing a high impurity density
exhibits sharp peaks due to local modes. The total absorption in these sharp
local peaks is of the order of all the band modes added together. The absorption coefficient is related to the squared matrix element of

$$\sum_{\alpha, l} e_\alpha(l) u_\alpha(l) \tag{106}$$

$e_\alpha(l)$ being the charge and $u_\alpha(l)$ the displacement of the atom in the $l$ cell.
The frequency-dependent absorption for the local modes, as calculated by
Dawber and Eliott ([30]) is

$$\alpha(\omega) = \frac{2\pi^2 e^2 N \Lambda}{uc} |A(0)|^2 g(\omega) \tag{107}$$

where $A(0)$ is the relative amplitude of the defect atom and $\Lambda$ is a local field
correction. The $N$ atoms per unit volume are randomly distributed in the
lattice. The randomness and anharmonicity are supposed to lead to a finite
energy width of the localized modes with a normalized shape function of
frequency

$$\int_0^\infty g(\omega)\, d\omega = 1 \tag{108}$$

The maximum value of the absorption depends not only on the density of
defects and the mass deviation but also on the line width which is an estimate of the rate at which an excitation of the localized lattice vibrations
around a defect will return to equilibrium. Klemens ([35]) has calculated the
lifetime of the local mode limited by its decay into two band modes due
to anharmonic forces. The product of lifetimes are frequency. $\tau\omega$, is a measure of the relative broadening $\Delta\omega/\omega$ of the defect mode frequency and
$\Delta\omega$ describes the broadening due to anharmonic interaction of any optical
spectral line involving the vibrational energy of the local mode. The estimated
order of magnitude for this broadening is $\Delta\omega/\omega \simeq 10^{-2}$. Another contribution to the line broadening originates from the random distribution for
the isotopes in the lattice. The relative width due to this effect is

$$\frac{\Delta\omega}{\omega} = \frac{1}{\omega}\frac{d\omega}{dM}\Delta M \tag{109}$$

where $\Delta M$ is the root-mean square deviation in the mass

$$\langle M \rangle^2 = \sum_i c_i M_i^2 - [\Sigma c_i M_i]^2 \tag{110}$$

$c_i$ and $M_i$ are the concentration and mass of the $i$th isotope. The estimated effect in the case of silicon is

$$\frac{\Delta\omega}{\omega} = 0.1\left(\frac{\Delta M}{M}\right) \simeq 1.4 \times 10^{-3} \tag{111}$$

The interaction of the impurities among themselves leads also to an additional broadening. For a small concentration $c$ of defect mass $M$ the extra randomness in the effective host mass is

$$\frac{\Delta M}{M} = c^{1/2}\left(1 - \frac{M'}{M}\right) \tag{112}$$

The first experimental studies in the case of semiconductors were carried out on silicon doped simultaneously with B and Li in order to achieve charge compensation and avoid absorption on the free carriers ([36]). Almost simultaneously analogous work was reported by Smith and Angress ([37]).

## 6. Localized Absorption Spectrum in Boron-Doped Lithium-Compensated Silicon

The absorption spectrum (see Fig. 23) has been obtained from transmission measurements at room temperature between 500 and 700 cm$^{-1}$ for impurities in natural abundance. This more complicated spectrum containing eight lines is due to complex defects. In fact two lines at 620 and 644 cm$^{-1}$ can be associated with the substitutional impurities $B^{11}$ and $B^{10}$ alone by comparison with other experiments with boron-phosphorus or with boron-antimony doped silicon. From the six other lines, a doublet at 564–653 is associated with $B^{11}$ and a doublet at 584–681 with $B^{10}$. Two singlets at 522 and 534 cm$^{-1}$ are associated respectively with Li$^6$ and Li$^7$. In fact the lithium may occupy an interstitial site with the symmetry $T_\alpha$ and the boron a substitutional site with the same symmetry. If we suppose that these two impurities are vibrating independently, the localized mode spectrum contains no more than two threefold degenerate lines. In fact a pairing occurs and the local site symmetry is reduced to $C_{3v}$ and the threefold degenerate lines are each split into two lines, one singlet and one doublet.

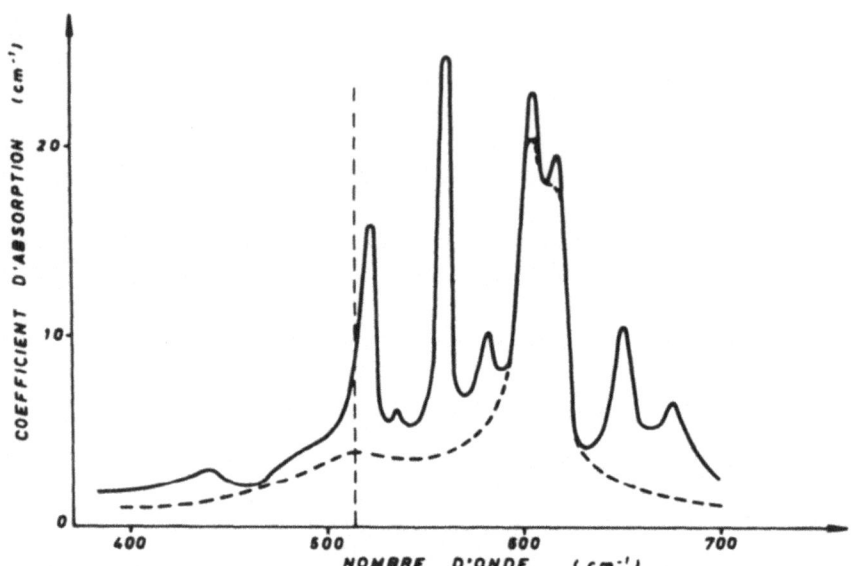

Fig. 23—Absorption spectrum of Si pure (broken line) and of Si doped boron and compensated Li (solid line). [M. Balkanski, W. Nazarewics, *J. Phys. Chem. Solids* **26**: 437 (1964).]

The relative intensities of these two lines are almost in the ratio 1 to 2: The doublet with the high frequency and the singlet with the low frequency. Calculations have been recently made using a tetrahedral defect containing the pair boron–lithium ([38]). We suppose only a local change of the dynamical matrix of the lattice $\Phi$ by introducing a finite perturbation matrix with the defect symmetry $C_{3v}$. The equation giving the local mode frequencies is written $|1 - gc| = 0$ where $C$ is the perturbation matrix and $g = (-M\omega^2 + \Phi)^{-1}$ the Greens function matrix; $g$ has the symmetry $T_\alpha$ and $gc$ the symmetry $C_{3v}$ The secular equation can be reduced by using symmetry coordinates in a certain number of independent equations associated with the irreducible representations of the symmetry group present in the mechanical representation. We obtain two equations

$$|1 - g_\alpha C_\alpha| = 0$$

where $\alpha$ corresponds to the $\Gamma_1$ and $\Gamma_{12}$ representations. The comparison between the calculations and the experiments shows that the force constant between the interstitial and the first neighbor is fairly large (1/2 of the force constants in Silicon) and that the force constant between the boron and the first neighbor are reduced in a large amount when we introduce the interstitial. In that example the absorption occurs due to the response of the positive and negative charges, centred on the two impurities. But only certain modes corresponding to the representation $\Gamma_1$ and $\Gamma_{12}$ are infrared active.

# REFERENCES

1. M. Balkanski, M. Nusimovici, and R. Le Toullec, *J. Physique* **25**: 305 (1964).
2. M. Lax and E. Burstein, *Phys. Rev.* **97**: 39 (1955).
3. M. Born and M. Blakman, *Z. Physik* **82**: 551 (1933).
4. R. F. Wallis and A.A. Maradudin, *Proceedings of the Conference of the Physics of Semiconductors*, Exeter, (1962).
5. M. Balkanski and J.M. Besson, *J. Appl. Phys.* **32**: 2292 (1961).
6. M. Morse, "Functional Topology and Abstract Variational Theory" *Memorial Sciences Mathematiques Gauthier-Villars*, Paris (1938).
7. F. Johnson, *Proc. Phys. Soc. London* **73**: 265 (1959).
8. M. Balkanski and Nusimovici, *Phys. Stat. Solidi* **5**: (1964).
9. M. Biltz, R. Geick, and K.F. Renk, *Proceedings International Conference Lattice Dynamics*, Copenhagen, 1963.
10. R. Loudon and F.A. Johnson *Congrès Intern. sur la Physique des Semiconducteurs*, Paris, 1964.
11. M. Balkanski, J.M. Besson, and R. Le Toullec, *Proceedings International Conference on Physics of Semiconductors*, Paris, 1964.
12. Lcrd Rayleigh *Proc. London Math. Soc.* **17**: 4 (1887).
13. D.C. Gazis, R. Herman, and R.F. Wallis *Phys. Rev.* **119**: 553 (1960).
14. R.F. Wallis *Phys. Rev.* **105**: 540 (1957).
15. R.F. Wallis *Phys. Rev.* **116**: 302 (1959).
16. E.W. Montroll and P.B. Potts *Phys. Rev.* **100**: 525 (1955); *Phys. Rev.* **102**: 72 (1956).
17. R.F. Wallis, *Surface Science* **2**: 146 (1964).
18. W.A. Plistein and R.P. Eischens, *Z. Physik Chem.* **24**: 11 (1960).
19. H. Kaplan *Phys. Rev.* **125**: 1271 (1962).
20. J. Hori and T. Asahi *Prog. Theor. Physics* **31**: 49 (1964).
21. E.J. Routh, *Dynamics of a System of Rigid Bodies*, Reprint Dover N.Y., 1955.
22. Lord Rayleigh, *Theory of Sound Vol. I*, reprint Dover N.Y., 1945.
23. I. M. Lifshitz, *Nuovo Cimento* **3**: 716 (1956) (Suppl.)
24. E.W. Montroll and R.B. Potts, *Phys. Rev.* **100**: 525 (1955).
25. R.J. Elliott *Phil. Mag.* **1**: 298 (1956).
26. M. Balkanski and W. Nazarewics, *J. Phys. Soc. Japan* **18**: 37 (1963).
27. S.D. Smith and R. J. Hardy, *Phyl. Mag.* **5**: 1311 (1960).
28. M. James and K. Lark Horowitz, *Z. Phys. Chem.* **198**: 107 (1951).
29. P. G. Dawber and R. J. Elliott, *Proc. Roy. Soc.* **A273**: 222 (1963).
30. P. G. Dawber and R. J. Elliott, *Proc. Roy. Soc.* **81**: 453 (1963).
31. J. L. Birman, *Phys. Rev.* **127**: 1013 (1962).
32. H. Y. Fan and A. K. Ramdas, *J. Appl. Phys.* **30**: 1127 (1959).
33. M. Balkanski and M. Nusirovici, *Phys. Stat. Solidi* **5**: (1964).
34. G. Dolliap, *Symposium on Inelastic Scattering on Neutrons in Solids and Liquids*, Chalk River, Canada, 1962.
35. P. G. Klemens, *Phys. Rev.* **122**: 443 (1961).
36. M. Balkanski and W. Nazarewicz, *J. Phys. Chem. Solids* **25**: 437 (1964).
37. S. D. Smith and J. F. Angress, *Phys. Rev. Letters* **6**: 131 (1963).

# Determination of Phonon and Magnon Dispersion Curves by Neutron Spectroscopy

## M. F. Collins

*A.E.R.E., Harwell, Berks*
*England, United Kingdom*

## I. THE SCATTERING PROCESSES AND METHODS OF NEUTRON SPECTROMETRY

In this Section I want to talk mainly about neutron spectrometry itself while in the other two lectures I shall discuss what has been learned about phonons and magnons by the application of the technique. To begin it is necessary to discuss what it is that the neutron sees in various scatterers. There are two mechanisms for the scattering of neutrons in solids. The first is scattering of neutrons by the nuclei of the atoms and this tells us about the positions and motions of these nuclei in the solid. The second is an inter-action between the magnetic moment of the neutron and any unpaired magnetic moment that the electrons of the scatterer may have. This gives information about the direction in which the spins are pointing and about their rotations.

Let us consider scattering by phonons first. It can be shown that, in Born approximation for the scattering, the cross section per unit solid angle $\Omega$ of scattered neutron and per unit interval in the energy of the scattered neutron $E_1$, that

$$\frac{d^2\sigma}{d\Omega dE_1} = \frac{k_1}{k_0} \sum_{ll'} e^{i\mathbf{k}(\mathbf{R}_{l'} - \mathbf{R}_l)} \sum_{if} p_i b_l b_{l'} \langle i | e^{-i\mathbf{k}\cdot\mathbf{u}_l} | f \rangle$$
$$\times \langle f | e^{i\mathbf{k}\cdot\mathbf{u}_{l'}} | i \rangle \delta(E_i - E_f + \hbar\omega)$$

where subscripts 0 and 1 refer to the incident and scattered neutrons and $i$ and $f$ refer to initial and final states of the crystal, $p_i$ is the probability of the $i$th initial state occurring. The atoms $l$ and $l'$ have lattice sites $\mathbf{R}_l$ and $\mathbf{R}_{l'}$

and displacements $\mathbf{u}_l$ and $\mathbf{u}_{l'}$ from these positions due to lattice vibrations. Their scattering lengths are $b_l$ and $b_{l'}$, respectively. Also

$$\mathbf{k} = \mathbf{k}_0 - \mathbf{k}_1$$

$$\hbar\omega = E_0 - E_1 = \frac{\hbar^2}{2m}(k_0^2 - k_1^2)$$

with $m$ the mass of the neutron.

The displacements $\mathbf{u}_l$ are expanded in terms of the harmonic phonon modes of the crystal using the techniques described in Professor Maradudin's lectures and those of Dr. Ludwig. For a Bravais lattice this gives

$$\mathbf{u}_l = \sum_{qj} \left(\frac{\hbar}{2NM\omega_{qj}}\right)^{1/2} [a_{qj}e^{i\mathbf{q}\cdot\mathbf{R}_l}\mathbf{V}_{qj} + a_{qj}^*e^{-i\mathbf{q}\cdot\mathbf{R}_l}\mathbf{V}_{qj}^*]$$

with $M$ the mass of the vibrating atom, $a$ the second-quantization destruction operator and $\mathbf{V}$ the polarization vector of the $j$th phonon branch with wavevector $\mathbf{q}$.

This is now substituted into the first equation and the exponentials in the matrix elements are expanded (that is, we assume the atomic vibration amplitudes to be small). Pulling out the term in which one phonon only is involved in the scattering process gives the result

$$\frac{d^2\sigma}{d\Omega dE_1} = \frac{(2\pi)^3}{v}\frac{k_1}{k_0}\sum_{qj}\sum_{\tau}e^{-2w}\frac{\hbar b^2}{2M\omega_{qj}}|\mathbf{k}\cdot\mathbf{V}_{qj}|^2(\bar{n}_{qj} + \tfrac{1}{2} \pm \tfrac{1}{2})$$
$$\times \delta(\mathbf{k} \mp \mathbf{q} + \tau)\delta(\hbar\omega \mp \hbar\omega_{qj})$$

where $e^{-2w}$ is the Debye–Waller factor and $\tau$ a reciprocal lattice vector.

This is the basic equation for the one-phonon scattering process. Most of the important properties are contained in the two $\delta$ functions; the first represents an interference condition for the scattering and the second ensures that energy is conserved. Together they have the effect that the scattered neutron beam in any particular direction just consists of a small number of sharply defined energies. Figure 1 and 2 show typical time-of-flight distributions of scattered neutrons in a given direction from a magnesium single crystal, Fig. 1 being a single well-defined excitation while Fig. 2 shows three partially resolved phonon scattering processes (figures from M.F. Collins, *Proc. Phys. Soc.* 1962). The factors $\mathbf{k}\cdot\mathbf{V}_{qj}$ in the equation can be used as a method of determining the polarization vector of the phonon. For more general lattices with several atoms $k$ per unit cell the scattering equation is the same except that the term

$$e^{-2w}b^2M^{-1}|\mathbf{k}\cdot\mathbf{V}_{qj}|^2$$

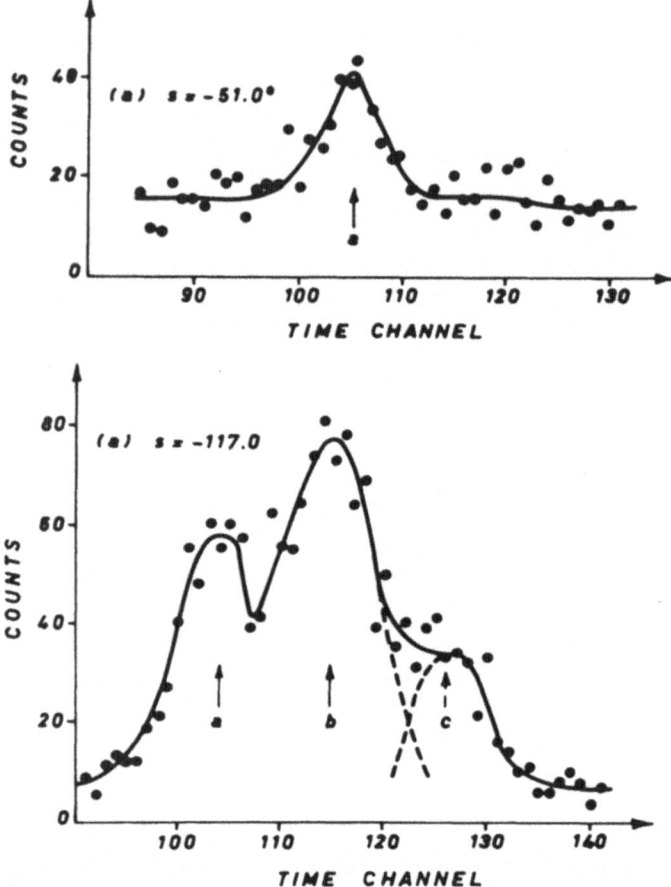

Figs. 1 & 2—Two typical time-of-flight spectra from a magnesium single crystal. Each peak corresponds to a separate phonon mode [M.F. Collins, *Proc. Phys. Soc.* **80**: 302 (1962)].

is replaced by

$$| \sum_k b_k e^{ikR_k} \mathbf{k} \cdot \mathbf{V}_{qjk} M_k^{-1/2} e^{-w_k} |^2$$

This is known as the structure factor and repeats over a larger distance than that of the basic reciprocal lattice. By measuring intensities for more than one $\tau$ value it is possible to solve for the polarization vectors and so positively identify any particular branch $j$. This hasn't been applied really seriously to date since only fairly simple lattices have been studied, and here branch identification is not normally a serious problem.

The scattering by magnons is in many ways rather similar to that by phonons. The matrix elements in the scattering equation are of the type $\langle i \,|\, S^\alpha \,|\, f \rangle$ where $\alpha$ is $x, y$, or $z$. These can be expressed in terms of second quantization creation and destruction operators by use of the Holstein-Primakoff transformation as described in Dr. Elliott's lectures. Evaluation of these matrix elements for a Heisenberg ferromagnet on a Bravais lattice gives the cross section for scattering with the emission or absorption of a single magnon as

$$\frac{d^2\sigma}{d\Omega dE_1} = \left(\frac{\gamma e^2}{mc^2}\right)^2 \frac{k_1}{k_0} \frac{(2\pi)^3}{v} (f(\mathbf{k}))^2 (1 + (\hat{k}_z)^2)\frac{S}{2}$$
$$\sum_{\mathbf{q}\tau} (\bar{n}_\mathbf{q} + \tfrac{1}{2} \pm \tfrac{1}{2}) \delta(\mathbf{k} \mp \mathbf{q} + \tau)\delta(\hbar\omega \mp \hbar\omega_\mathbf{q})$$

where $(\gamma e^2/mc^2)^2$ is a known set of physical constants describing the interaction between an atomic spin and the spin of the neutron; $\hat{k}_z$ is the component of $\mathbf{K}$ along the magnetic $z$-direction and $f(\mathbf{k})$ is the form factor. This is very similar to the phonon formula and so the identical experimental techniques can be used. We can distinguish between phonon and magnon peaks by application of a magnetic field to change the $z$-direction or by heating above the Curie temperature.

For antiferromagnets or complicated ferromagnets a structure factor must be inserted into the cross-section formula analogous to that for phonons. For a simple antiferromagnet this is such that when $\tau$ is an antiferromagnetic reciprocal lattice point the intensity is strong and when it is a nuclear reciprocal lattice point it is weak.

Now what about when the excitations have finite lifetimes? This effectively broadens the energy for a given $\mathbf{q}$ and the $\delta$-functions get smeared out. The width of the neutron line gives a measure of the widths of the phonon or magnon line. The actual cross-section formula for phonons has been worked out by a number of authors assuming small cubic and quartic terms in the Hamiltonian and these show how the lines are broadened and shifted. Unfortunately the experimental techniques are not yet sufficiently advanced for anyone to have had a detailed try at working back from the line widths to deduce the various individual terms in the cubic and quartic equivalent of force constants.

Now, having discussed the basic cross sections, I would like to go on to say a few words about the scope and limitations of the neutron diffraction technique since this is not always appreciated by nonspecialists in the field. Neutrons provide the only experimental method of observing spin waves at general wavevectors. For phonons the only competitor is diffuse X-ray scattering and up to the present this has usually proved to be a more indirect and less satisfactory technique. Thus there are two large fields which have been more or less solely accessible by neutron spectrometry and it is clear that the method is very important.

There are, however, some limitations to the use of neutron spectroscopy. These are:

1. Energies are only obtainable to accuracies of around 1 % with resolution functions in $q$ of around 10 %. These figures are rather low compared to those from optical and infrared data so that those techniques are better in the restricted number of case where they can be used (usually modes at $q = 0$). More powerful neutron sources will of course improve these accuracies but since the fundamental accuracy is defined by the volume in reciprocal space over which phonons are observed, the accuracy only increases by something around the one-fifth power of the source flux at best; large improvements cannot be expected unless some quite-radically new type of source comes along. At the moment the resolution obtained in energy is around 10 % so that lines cannot be resolved unless they are this far apart and, more important, line widths must be this great to be observable. This has the effect that lifetimes in pure materials can only be observed at relatively high temperatures. There are some weak focusing effects in the instruments which may be exploited.

2. About 20 elements have prohibitively large absorption cross sections. These are

| | | | |
|----|----|----|----|
| Li | In | Er | Ir |
| B  | Sm | Tm | Au |
| Rh | Eu | Lu | Hg |
| Ag | Gd | Hf | Pa |
| Cd | Dy | Re | Pu |

though there is no sharp line between possible and impossible materials and some of these may well become accessible as better neutron sources are developed, though others will always be very difficult, e.g., Gd. In some cases the use of separated isotopes gets around the difficulty.

This is not usually a severe limitation since most of the common elements are available and only in the rare earths have serious difficulties arisen.

3. Energies can only be measured up to about 0.1 eV. This does not usually prove to be a serious limitation since a majority of modes have proved to be of lower energy than this. It has however seriously restricted observation of spin waves in transition metals at large values of $q$.

4. Neutron penetration depths in normal crystals are of the order of a few centimeters so that the neutrons see the bulk properties of the material.

5. Neutrons are not particularly good for studying localized modes. This is basically because it is difficult to pick out these modes from the bulk scattering: the neutrons are not specific.

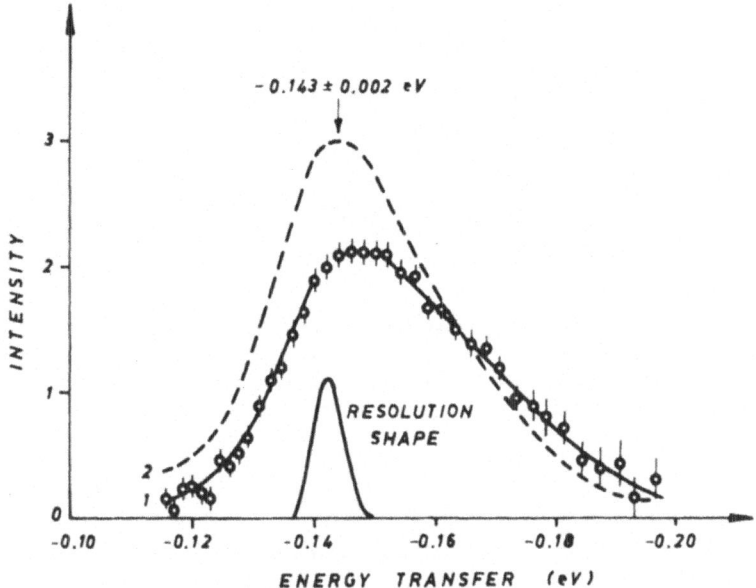

Fig. 3—Titanium hydride. Observed peak of proton vibrating in the tita-nium lattice (by beryllium detector method). Curve (1): observed shape, Curve (2): observed shape multiplied by the factor, exp $\{\hbar\omega/kT\}$.

Perhaps this is an appropriate time to say a few words about the scat-tering of neutrons by localized modes. This scattering obeys energy conser-vation of course, but since $\mathbf{q}$ is no longer a good quantum number, the other $\delta$-function in the cross section is absent. Thus there is a diffuse distribution of scattered neutrons rather than a sharp peak. Only one particular case has proved easily accessible. This is localized vibrations of hydrogen since hydro-gen has an unusually large cross section. Also the modes are often fairly sharp anyway. Figure 3 shows the observed neutron scattering from interstitial hydrogen in titanium (Cocking and Saunderson in *Inelastic Scattering of Neutrons in Solids and Liquids, I.A.E.A.*, Vienna 1963, p 265). The scattering, after correction for a temperature factor, shows a reasonably sharp peak at 0.14 eV. At elevated temperatures the diffusion coefficient of the hydrogen atoms becomes extremely large. This is surprising in view of the high fre-quency of the hydrogen vibrations, which might have been expected to imply that the hydrogen atoms were tightly bound.

To finish this lecture a brief discussion will be given of experimental methods. It was shown earlier that the scattering is defined by two $\delta$-func-tions and that this implies that the energy distribution of the scattered neutron beam should be studied. There are two basic method of measuring or defining neutron energies. These are

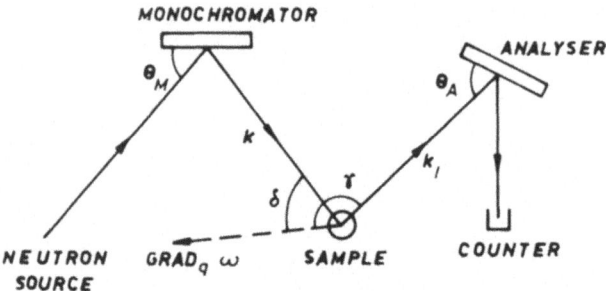

Fig. 4—Schematic diagram of a triple-axis spectrometer. A single-crystal monochromator effects Bragg
scattering of an almost monoenergetic beam on the
sample and a single-crystal analyzer selects those neutrons scattered by the sample with a predetermined
wavelength and direction of motion.

1. Laue scattering from a single crystal.
2. Mechanical pulsing of the beam and subsequent use of time-of-flight
   techniques.

These two methods are embodied in the instruments used for measuring
neutron spectra:

1. *The Triple Axis Spectrometer.* This instrument employs the Laue
scattering technique and has been the principal apparatus used in the study
of elementary excitations.   There are over a dozen in commission about the
world which follow the schematic plan shown in Fig. 4. A special feature
of the instrument is that instead of just analyzing the energy distribution of
the scattered beam, it can be programed to vary other parameters simultaneously (the incident wavelength, the angle of scatter, or the orientation
of the scatterer) such that an energy distribution is obtained at constant
wavevector $\mathbf{q}$. This is especially important in the study of phonons where
the dispersion relations are mainly of use in symmetry directions so that the
dynamical matrix factorizes.

2. *Choppers.* These employ method 2 above. A typical instrument is
shown in Fig. 5 (Dyer and Low in *Inelastic Scattering of Neutrons in Solids
and Liquids*, I.A.E.A., Vienna 1961). A neutron beam passes through two
phased rotating discs which effectively provide a gate at two different points
in space appropriately phased to allow only neutrons of the desired velocity
to pass. The emergent beam is scattered by a specimen and a time-of flight
analysis is performed on the scattered neutron beam. This instrument has
the advantage over the triple-axis spectrometer that it can observe neutrons
scattered in many different directions simultaneously, but the disadvantage
that it cannot be efficiently programed to observe energy distributions at
a given $\mathbf{q}$-vector.

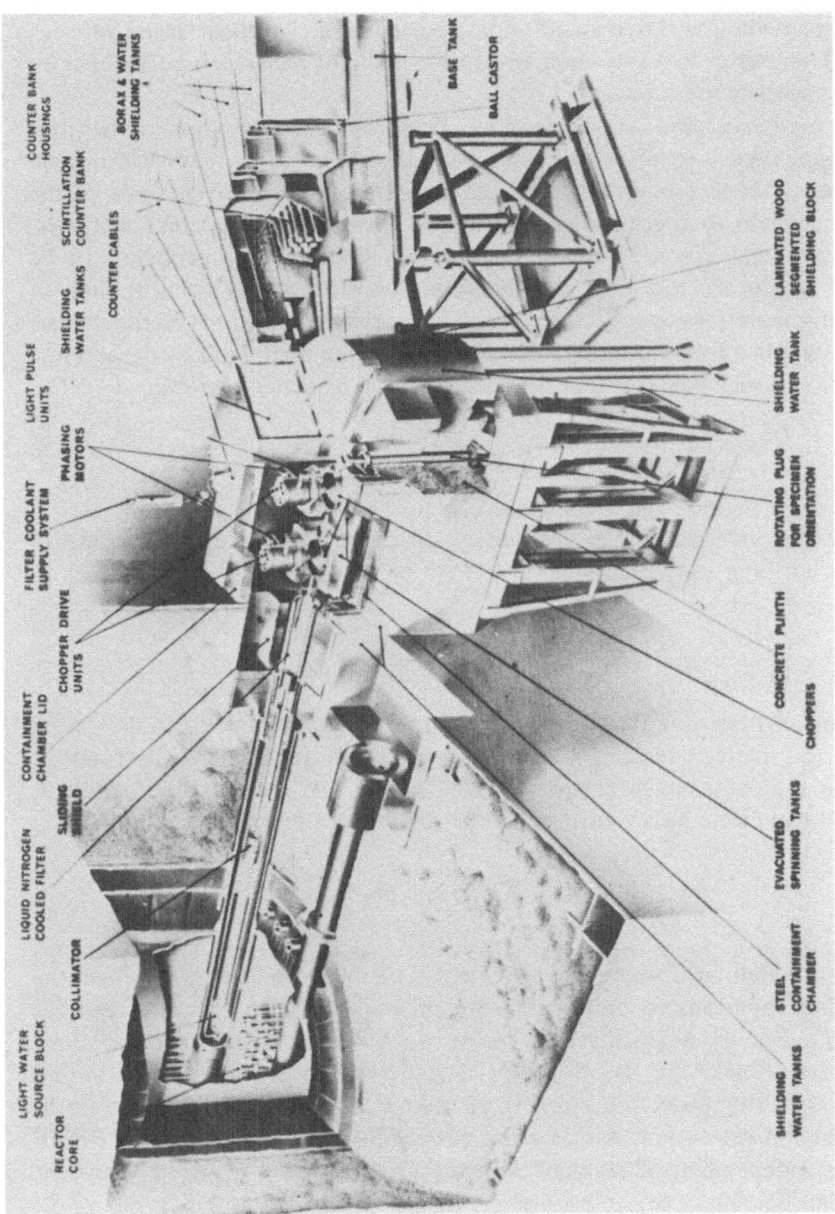

Fig. 5—A neutron time-of-flight spectrometer (Dyer and Low, Inelastic Scattering, of Neutrons in Liquids and Solids, I.A.E.A. Vienna, 1963 p. 179).

3. *The Rotating Crystal Spectrometer.* This hybrid instrument is more akin to the chopper than to the triple-axis spectrometer. A single rotating crystal provides, by Laue scattering, a pulsed monoenergetic beam of neutrons. This passes to a specimen and a time-of-flight analysis is performed on the scattered neutron beam.

4. *Scattering Surface Techniques.* There is a further class of neutron techniques used to study elementary excitations known as *scattering surface techniques.* These are specialized methods which involve no analysis of the scattered beam in energy. They are not so powerful as the other methods, but give higher counting rates, so that they enable some problems to be tackled which are too difficult otherwise. They have only been applied to scattering from ferromagnons to date. There are two variations on the theme:

1. Small angle scattering. This technique, devised by Lowde at Harwell, makes use of the fact that the neutron has a dispersion relation

$$E = \frac{\hbar^2}{2m} k^2$$

while a ferromagnet in the absence of anisotropy fields has the dispersion relation at long wavelengths of

$$E = Dk^2 = \frac{\hbar^2}{2m^z} k^2$$

For typical transition metals $m^z$ is found to be around 20 electron masses or $1\%$ of the neutron mass. The scattering at small angles is of the classical type for that of a heavy particle by a light one, and the maximum angle $\theta_c$ through which the heavy particle can be scattered is given by

$$\theta_c = \frac{m^z}{m} \sim 0.5 \text{ deg}$$

The actual angular distribution of scattered neutrons can be shown from the scattering equations to be independent of angle for $|\theta| < \theta_c$ and zero for $|\theta| > \theta_c$. Figure 6 [Hatherly, Hirakawa, Lowde, Mallett, Stringfellow, and Torrie, *Proc. Phys. Soc.* **84**: 55 (1964)] shows that the observed scattering is indeed of this form, the blurring of the cut off angle being due solely to experimental resolution. A nice point about this technique is that the cut off angle is independent of incident neutron energy so that a "white" incident beam can be employed.

2. Scattering near $\mathbf{k} = \tau$. This technique, mainly pioneered by Riste, is similar to 1 above, except that the small-angle scattering is observed near a Bragg peak by missetting the scattering angle and observing the intensity distribution of scattered neutrons as the crystal is rotated. It has the advan-

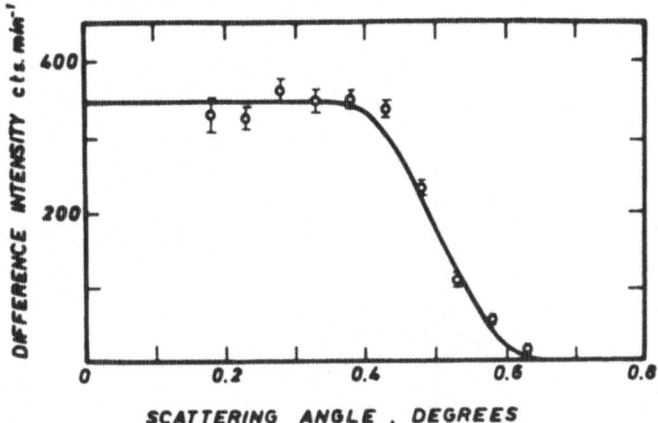

Fig. 6—The small-angle scattering of neutrons by spin waves in iron [Hatherly *et al.*, *Proc. Phys. Soc.* **84**: 55 (1964)].

tage of being able to go to quite large **q** vectors (up to 20% of that at the zone boundary), but the disadvantage of requiring a monoenergetic incident neutron beam. Usually polarized neutrons are used to isolate the spin-wave part of the scattering.

## II. PHONONS

This section will consist of a review of what has been learned in the field of phonons by the use of neutron spectrometry. Complete symmetry-direction dispersion relations have now been measured for around 21 elements and 18 compounds. Clearly, it is not possible to discuss all these measurements in a single talk so I will just describe a selection of these which either served historically to set the pattern for a particular type of material or which have some special feature that I find interesting. It is notable that a large proportion of this experimental work has come from the group at Chalk River mainly associated with the names of Brockhouse, Cowley, Dolling, and Woods.

It is convenient to divide the field up into different types of chemical bond and to start with.

### Ionic Materials

The simplest model here is that of a set of point charges at each of the atomic sites interacting via Coulomb forces together with nearest-neighbour short-range forces. This model (described in more detail in Prof. Balkan-ski's lectures) is known to give about the right binding energy for the lattice. The lattice dynamics so expected was worked out by Kellerman in 1940.

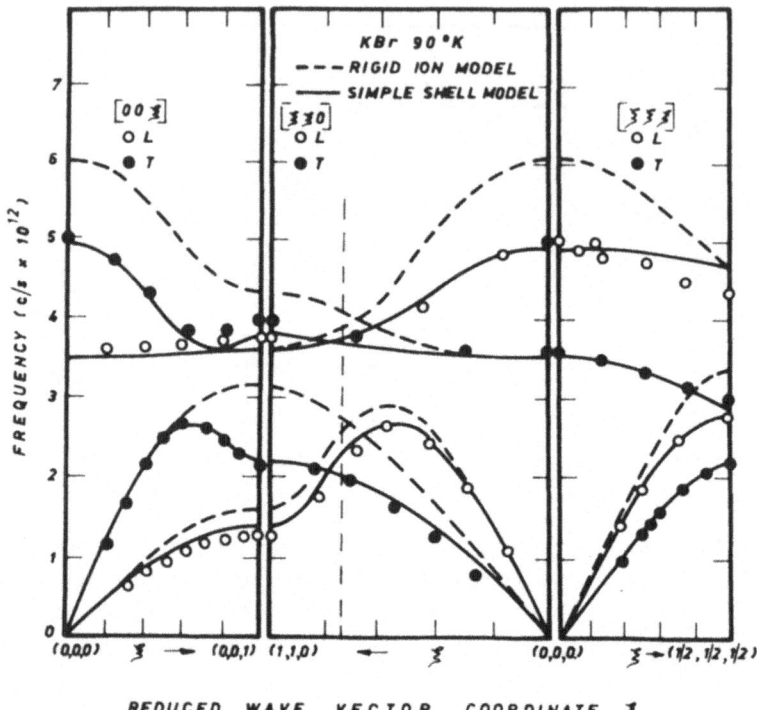

Fig. 7—The measured dispersion curves of potassium bromide (Woods *et al.*, 1963) showing the comparison of the data with the rigid ion and shell models.

The first neutron experiments on NaI and later those on KBr showed that the model wasn't really good enough. Figure 7 shows the data for KBr (Woods, Brockhouse, Cowley, and Cochran 1963), and it is apparent that the simple model, labeled the rigid-ion model in the figure, is very unsatisfactory. However the large splitting between the *TO* and *LO* modes at small *q* does indicate that long-range forces of this general type must be present in the lattice (*cf.* Prof. Balkanski's lectures).

At about the same time as these experimental results were being produced a more satisfactory theory of lattice dynamics of ionic compounds was being developed by many workers, notably Cochran, Cowley, Hardy, Szigetti, Dick, and Overhauser and Mashkevits and Tolpygo. The key idea of all this work is that the ion cannot be regarded as a point charge but that it is deformable. Thus as the ions move in a lattice vibration they deform in some complicated way and this deformation gives rise to long-range electrostatic forces.

These forces are best described by expanding the charge distribution into monopoles, dipoles, and quadrupoles etc. All theories to date have been based on an assumption of just monopole and dipole forces. Probably the simplest is the shell model which pictures an ion as consisting of a positively-charged inner core and an outer (rigid) shell of electrons. Motions of the shell relative to the ion core then give rise to long-range dipole forces. Each is treated as a separate component in working out the lattice dynamics and force constants between these components are assumed to be short-range in nature. Evaluation of the lattice dynamics on these lines gives a much better fit to the experimental data with relatively few disposable parameters as can be seen for KBr in Figure 7. The model used assumed that only the large anion was polarizable.

There are some small remaining discrepancies between theory and experiment. It is currently a matter of controversy whether these should be taken into account by extending the shell model to include more short-range interactions and polarizable cations or by developing a more general theory which takes quadrupole interactions into account. Some of the remaining discrepancies occur for modes where quadrupole effects would be expected to be important.

## Covalent Bonds

Covalent bonds are one case where the simple force-constant approach with nearest-neighbor interactions might be expected to apply so it was a

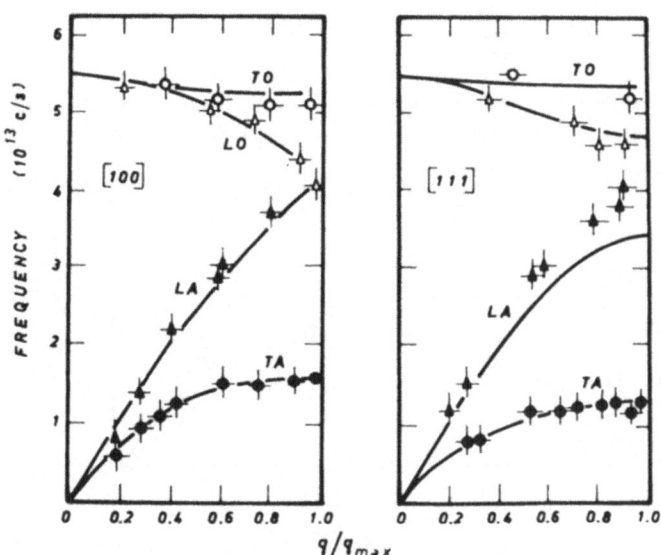

Fig. 8—A comparison of the calculated dispersion curves for the shell model (Cochran 1959) with the observations of Brockhouse and Iyengar on germanium. (a) wavevector along [100]; (b) wavevector along [111].

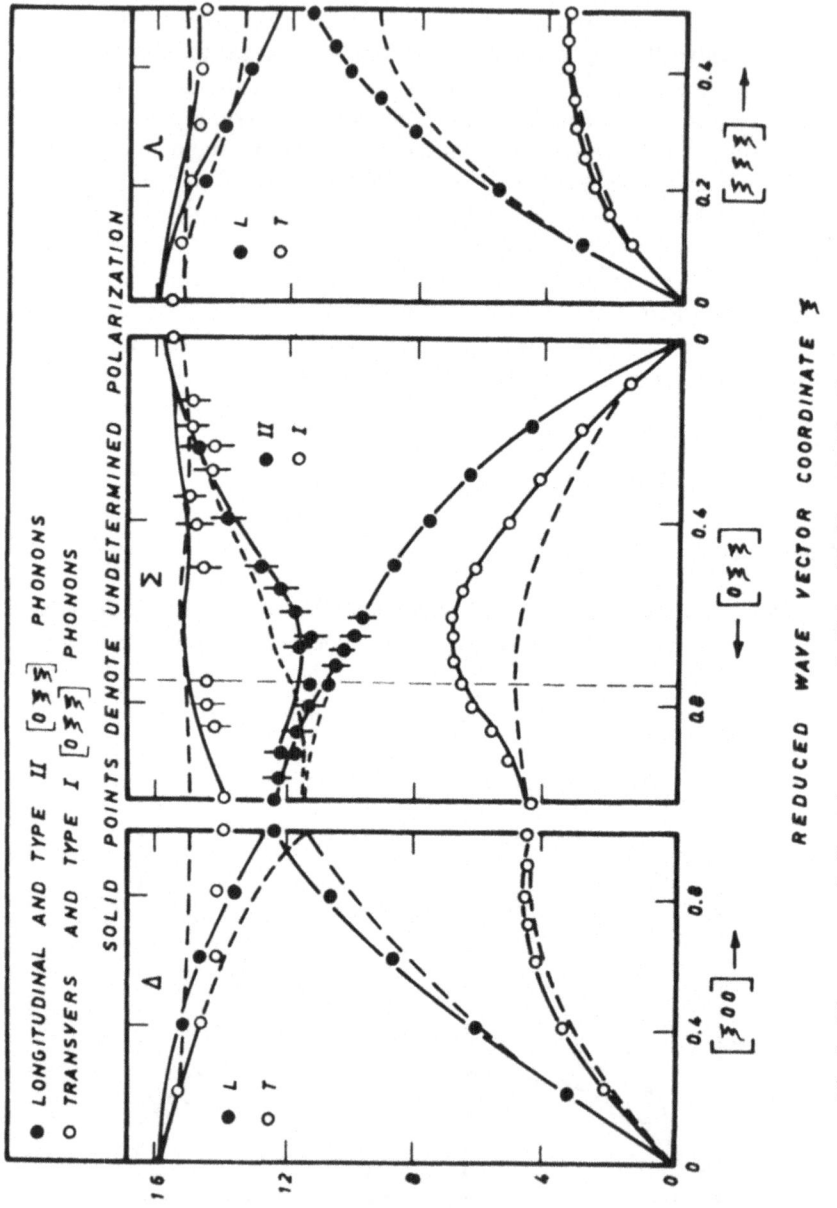

Fig. 9—The measured dispersion curves of silicon (Dolling, 1963). The curves represent various attempts to fit the data on the basis of the shell model.

considerable surprise when the first accurate phonon determination for Germanium by Brockhouse and Iyengar (1958) showed this model to be no good. It was about 20–30% out in predicting frequencies and Hermann, in particular, showed that interactions to at least fifth neighbors were needed to give a fit. Since this involved 15 disposable parameters and there were only 19 pieces of data it was clear that the model was unsatisfactory.

Cochran in 1959 proposed that in these materials as well as in the ionic compounds the ions are not rigid but that dipoles are produced by the vibrations. Describing these dipoles by a shell model as for ionic materials he immediately fitted the experimental data quite reasonably as shown in Fig. 8. Only a simple version of the shell model was used. However, to get really good fits for silicon and germanium quite complicated versions of the shell model must be used. Figure 9 shows a fit for silicon (Dolling 1963). As with the ionic materials it is currently an open question whether such models reflect the real physical situation better than the introduction of quadrupole forces. Figure 10 shows the dispersion law for diamond (Warren, Wenzel, and Yarnell 1965) through here so far as I am aware no detailed fitting to the Shell Model has been performed.

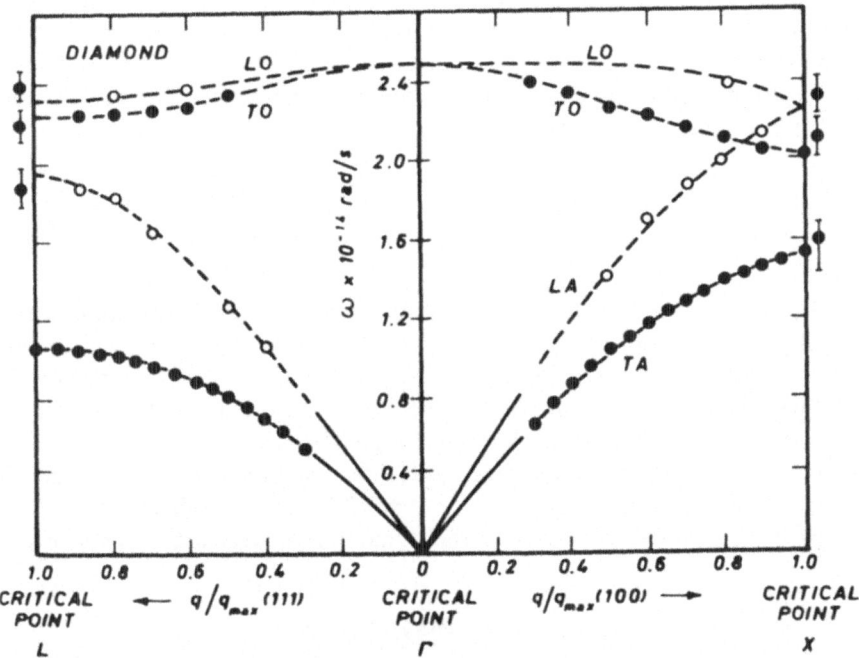

Fig. 10—The measured dispersion curves of diamond (Warren *et al.* 1965).

## Screening of Electric Fields in Ionic Crystals

Lead telluride is a semiconductor in which the existence of ionic charges (monopoles) is confirmed by the splitting of the $LO$ and $TO$ modes at small $\mathbf{q}$. Cowley and Dolling (1965) have observed that in heavily doped $n$-type specimens the conduction electrons are able to screen out the long-range electric fields for optic modes with small $\mathbf{q}$ vector. The small number of conduction electrons might in fact be expected to completely screen out these fields for $q \ll K_F$ so that the $LO$ and $TO$ modes become degenerate while for $q \gg K_F$ the oscillations in space are too rapid to be screened and the $LO$ and $TO$ modes would be the same as in an undoped specimen. The neutron data show a dip in the $LO$ more frequency for $\mathbf{q}$ less than about $2K_F$ though the fact that this wavevector is only of the same order as the instrumental resolution function precluded the production of an accurate shape for the dip.

## Simple Metals

Here it was always clear that the localized force constant model would be inadequate because the forces are coming mainly via itinerant outer electrons. A good theory of lattice vibrations in metals was first produced by Toya in 1958. This theory has since been further developed by Ziman, Sham, and Cochran, and along rather different lines by Harrison. The theory considers three types of interaction

1. Nearest neighbour repulsion.
2. Coulomb forces for positive charges placed in a uniform electron gas.
3. Screening effects of the electrons so as to reduce the Coulomb forces in 2.

The first two effects are fairly straightforward to deal with and all the difficulty lies in the third. Perhaps the simplest method of dealing with this is that of pseudopotentials (Sham, L. J. and Ziman, J., *Solid State Physics* **15** (1963)), though there will not be time to discuss this here. The model assumes that the core and inner electron charge $Z(\mathbf{r})$ is screened out by an electron charge $Z_e(\mathbf{r})$ which has the more widespread spatial extent. For a spherical Fermi surface $Z_e(\mathbf{r})$ behaves like $r^{-3} \cos 2K_F r$ at large distances.

From the model, elements of the dynamical matrix are derived of the form

$$\Phi(\mathbf{k}) = \frac{4\pi^2 e^2 Z(\mathbf{k})}{k^2}\left[1 - \frac{1}{\epsilon(\mathbf{k})}\right]$$

for the screening term where $Z(\mathbf{k})$ is the spatial Fourier transform of $Z(\mathbf{r})$

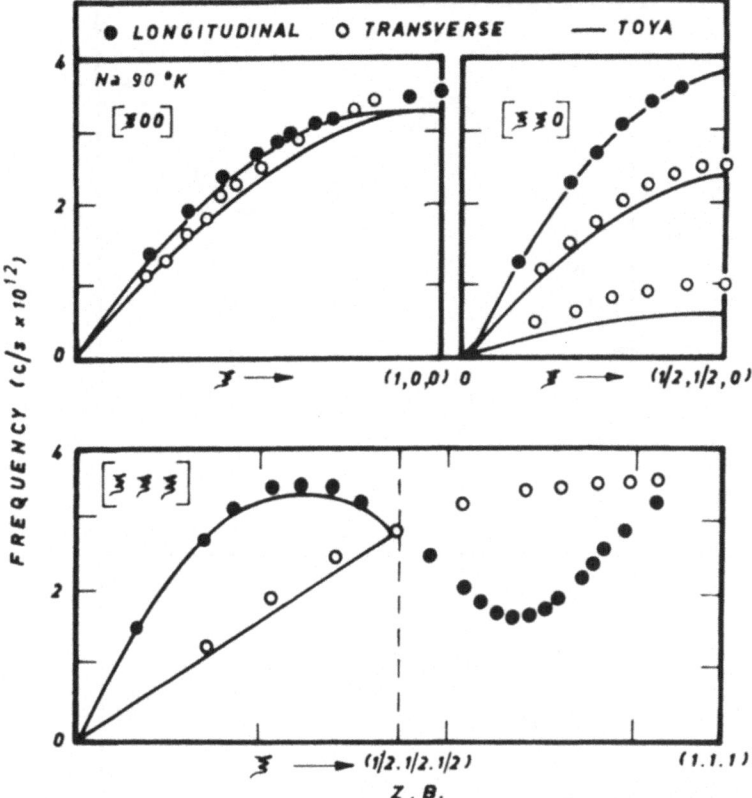

Fig. 11—The dispersion curves for sodium (Woods *et al.* 1962) compared with the theoretical calculations made by Toya. The large filled circles indicate superimposed $L$ and $T$ modes. At $(\frac{1}{2}, \frac{1}{2}, \frac{1}{2})$ and $(1, 0, 0)$ these modes are degenerate by symmetry.

and $\epsilon(\mathbf{k})$ is the dielectric constant. With a spherical Fermi surface $\epsilon(\mathbf{k})$ has a weak logarithmic infinity of slope at $k = 2K_F$. This gives rise to a similar infinity in the slope of the phonon dispersion relation known as a Kohn anomaly which will be discussed more later. Calculations by Toya (1958) with no disposable parameters except the elastic constants gave quite a good fit to the experimental results for sodium. (Woods *et al.* 1962) as shown in Fig. 11. It is not ideal, particularly for the TA mode, but, seeing that the predictions were made before the experiment, it is quite good. This theoretical work seems to be definitely along the right lines but even with the more recent improvements to the theory further work is still necessary to completely fit the data. On going to more complex metals than the alkali metals matters become even more formidable and no satisfactory models to fit the experimental data are available.

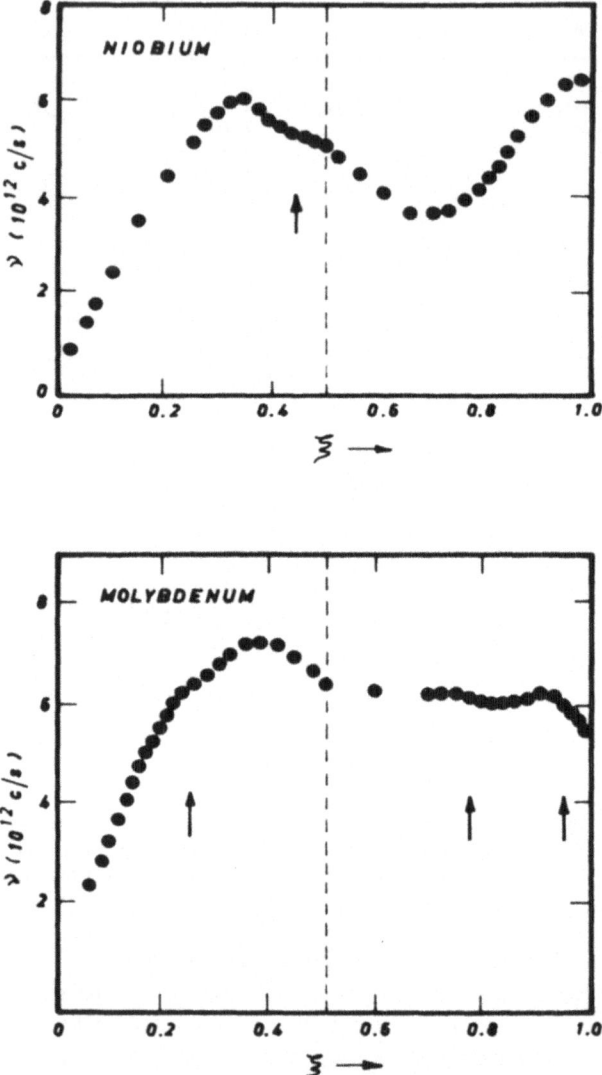

Fig. 12—The upper part shows an expanded plot of the
$[\zeta, \zeta, \zeta] L$ branch in lead at 100°K, showing the pronounced
anomaly (believed to be caused by the Kohn anomaly) near
$\zeta = 0.75$, for a series of measurements under different
experimental conditions. The lower part shows the relation
of observed anomalies to the Fermi surface of lead. The
arrow marked F corresponds to the anomaly shown in the
upper part, that marked G to an obseved anomaly on the $L$
branch. (Brockhouse *et al.* 1962).

## Kohn Anomalies

In the above section it became apparent that there should be Kohn anomalies in metallic phonon dispersion relations when $\mathbf{q} = 2K_F$. In fact more generally such anomalies occur connecting any two parallel bits of

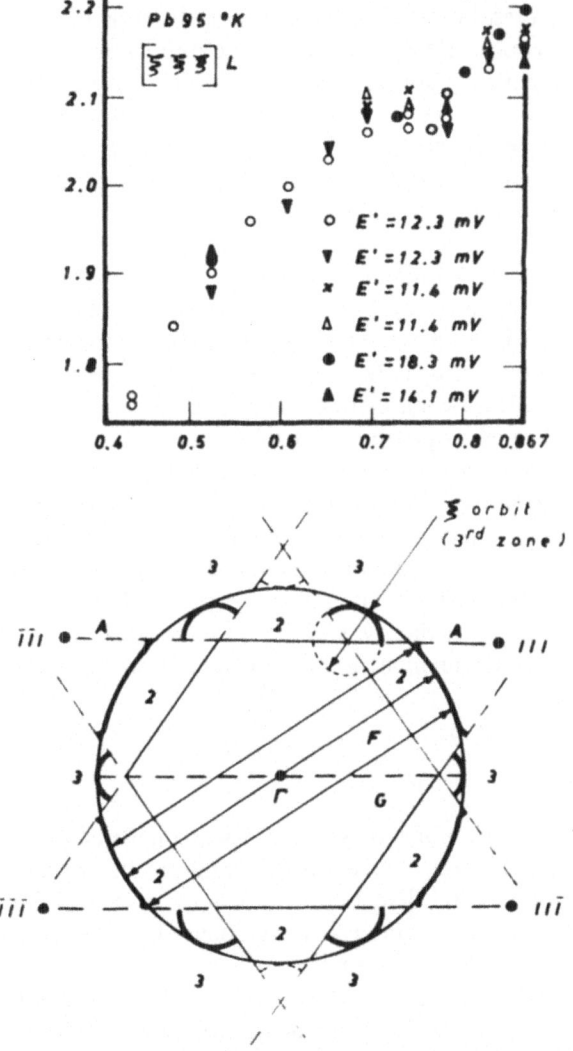

Fig. 13—Comparision of the longitudinal phonon branch in niobium and molybdenum at 296 °K (Woods 1965). The arrows indicate positions of suspected Kohn anomalies.

Fermi surface. Thus the positions of these anomalies can be used to map out Fermi surfaces. Unfortunately the difficulty is that they are rather weak and are only observed in favorable cases. For instance, no anomalies have been seen in the neutron scattering data for sodium.

A case where Kohn anomalies have been seen is lead, shown in Fig. 12 (Brockhouse *et al.* 1961). This seems to be a rather favorable situation and the observed anomalies can be explained in terms of the expected Fermi surface. Incidently, these anomalies have also been seen in the X-ray diffuse scattering and it is possible that this might be quite a good way of observing them, though very little work has been done on the subject. For determining Fermi surfaces this method has the advantages of not requiring the specimen to be at low temperatures or to be particularly pure. It has the disadvantages of not being especially accurate (1 %) and, most serious, it is not easy to assign anomalies unambiguously to specific features of the Fermi surface.

The anomalies are normally under 5 % in energy and their magnitude is not predictable at the moment, though it is generally the case that they show up in materials with large electron–phonon interactions as reflected in the high superconducting critical temperatures or high electrical resistance at high temperatures. It is probable that this is one field where increased experimental accuracies might open up a lot of things that cannot be investigated at the moment. An interesting investigation is that by Woods and co-workers (1965) on Nb–Mo alloys. The dispersion relations for the pure elements is shown in Fig. 13. For a couple of alloys they saw Kohn anomalies at intermediate $q$ values to those in the pure metals showing how a reasonably well-defined Fermi surface is persisting in the alloy phase. Supporters of rigid band approximations in alloy theory will be very gratified by this, though of course with Nb and Mo being neighbors in the periodic system this is rather a favorable case.

### Transition Metals

The above discussion of Nb–Mo alloys leads naturally into a discussion of transition metals. Here there is no fundamental theory available though quite a lot of experimental dispersion curves are on record (Cr, Fe, Ni, Nb, Mo, W, Ta). This has forced interest to be focused on special features of the dispersion law such as Kohn anomalies.

Here two other interesting features will be briefly mentioned: (1) Sharp at Harwell showed that the phonon dispersion law in Nb did not change, certainly by more than a percent or two, on going through the superconducting transition. Velocity of sound mesurements have indicated effects of order 1 part in $10^4$ at long wavelengths and it seems that there is also no large effect at short wavelengths. (2) Moller and Mackintosh in Denmark have measured the phonon dispersion law in chromium. The interesting point was that the dispersion laws were the same above and below $T_N$ and near $T_N$. This is perhaps a little surprising since on going through $T_N$, 0.4 $\mu_\beta$ of moment disappears. This corresponds to quite a sizable shift in the Fermi surface and one might have expected this to alter the screening parameters. More work may be needed on this topic.

### Anharmonic Effects

There are only a small number of experimental results in this field. This is because the linewidths have to be of order 10% to be observable as discussed in Section I. Thus in no pure materials at low temperatures can linewidths be seen. In ionic materials work has been concentrated on the *LO* modes which are found to be especially broad even at quite moderate temperatures. Figure 14 shows these linewidths for KBr (Woods, Brockhouse, Cowley, and Cochran 1963). In the paper these authors show that the large linewidths can be at least qualitatively explained on simple theory. For

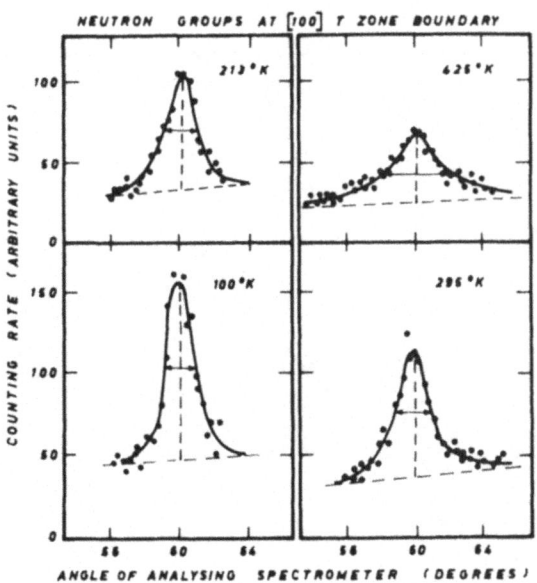

Fig. 14—Neutron groups scattered from potassium bromide at various temperatures (Woods *et al.* 1963). The dashed lines under the peaks represent the background which will preserve the temperature corrected integrated intensities.

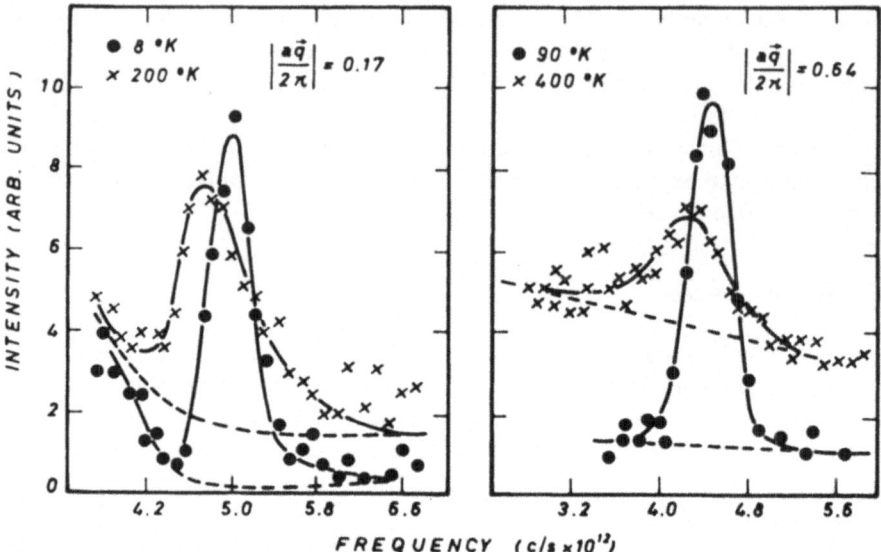

Fig. 15—Neutron groups at four temperatures taken at the point (1, 1, 0) in reciprocal space for lead (Brockhouse *et al.* 1962). The backgrounds have been drawn in to make the areas equal.

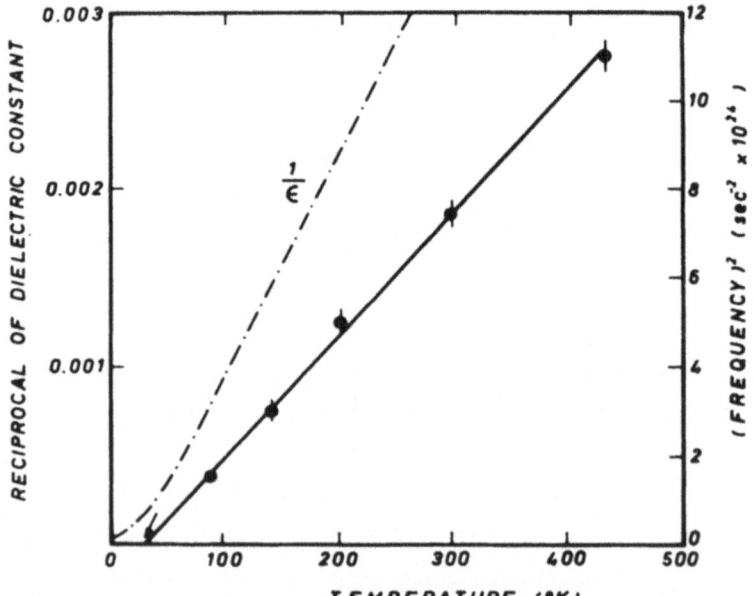

Fig. 16—The temperature dependence of the frequency of a TO mode at $q = 0$ compared with the temperature dependence of the reciprocal of the dielectric constant ($\epsilon$) in $SrTiO_3$ (Cowley 1963).

instance the anharmonic effects can be calculated explicitly for the Coulomb potential.

Data for lead are shown in Fig. 16 (Brockhouse *et al.* 1962). At elevated temperatures the linewidths become very broad and the lifetime is only of order one period of oscillation. A similar situation has been found for aluminium and it appears that near the melting point the simple phonon picture is not really appropriate. However, it is known that phonon frequency spectra are well behaved right up the melting point (though critical points may be smeared out) from neutron scattering and from Debye–Waller factor measurements by, for example, the Mössbauer effect.

A case where really dramatic anharmonic effects have been seen is $SrTiO_3$. This is a ferroelectric material and as the transition is approached one of the optic modes becomes unstable. This mode corresponds to the vibrations of the titanium atom which moves off its cubic site in the ferroelectric state. The temperature dependence of the frequency of this mode is shown in Fig. 16 (Cowley 1963).

## III. MAGNONS

Magnons have been much less investigated than phonons by neutron spectroscopy, though for no clear reason. Only five complete dispersion curves are available, three from Harwell and one each from Chalk River and Riso. This scant amount of data means that the basic outlines have not been investigated experimentally in some areas of the subject. Neutrons are of course the only method currently available for getting complete magnon dispersion curves.

### Antiferromagnets

I am going to start by describing some simple antiferromagnets. At first sight it may seem illogical to start with antiferromagnets rather than ferromagnets since the latter are fundamentally the simpler materials. However, the point is that a number of simple Heisenberg antiferromagnets exist and can be readily experimented on, while simple Heisenberg ferromagnets are rather rare. EuO is probably the prime example of a simple Heisenberg ferromagnet. Unfortunately, europium is one of the elements that has too large a neutron absorption cross section for there to be sufficient scattered neutrons left to do inelastic scattering experiments. The reason for this preponderance of antiferromagnets is that the common mechanism for the exchange interaction in insulators is superexchange, and this interaction is normally antiferromagnetic. Magnetic $d$ electrons at a magnetic atom bond

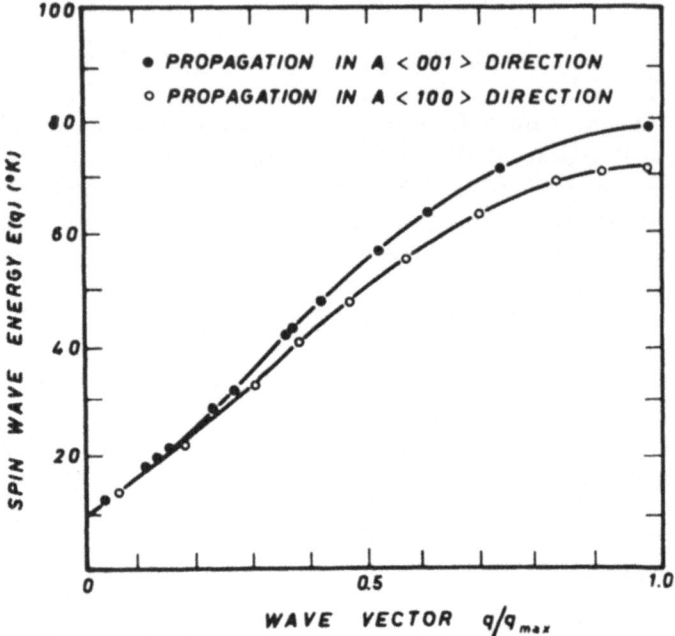

Fig. 17—Spin-wave dispersion relations in $MnF_2$ at 4°K (Tuberfield *et al.* 1964).

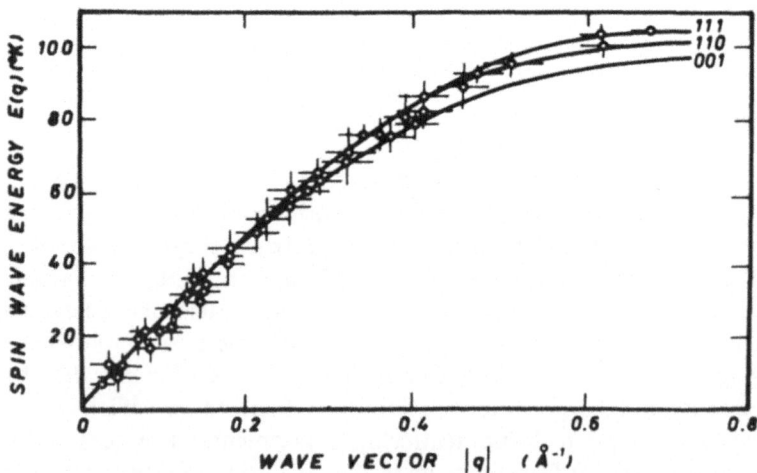

Fig. 18—Spin-wave dispersion relations in $RbMnF_3$ at 4°K (Windsor and Stevenson 1966).

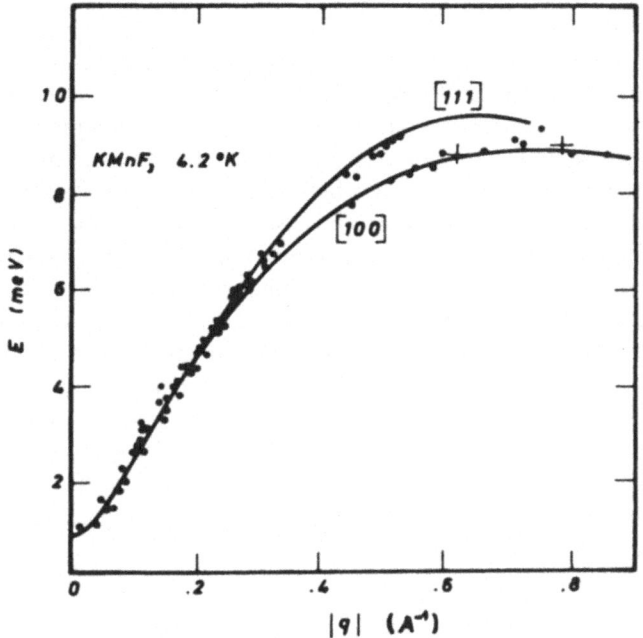

Fig. 19—Spin-wave dispersion relations in KMnF₃ at 4°K (Pickart, Collins, and Windsor 1966).

covalently with a nonmagnetic ligand and some of the moment appears on this ligand. This moment on the ligand couples in with other moments on the ligand from other neighboring magnetic atoms to give the superexchange interaction. Three antiferromagnets of this type have been investigated experimentally. Historically the first was manganese fluoride, $MnF_2$ (Turberfield et al. 1964). The dispersion relations along the $x$ and $z$ axes are as shown in Fig. 17. This gives an excellent fit to a model with exchange to just two sets of magnetic atoms as shown by the solid lines in the figure. The second set of atoms has an exchange energy of around 10% of that to the first set and no further exchange couplings need be postulated.

The other two measurements on antiferromagnets are for $RbMnF_3$ (Windsor and Stevenson 1966) and $KMnF_3$ (Pickart, Collins, and Windsor 1966). The dispersion curves are as shown in Figs. 18 and 19 with the fitted lines corresponding to simple models. For $RbMnF_3$ only nearest-neighbor interactions are needed, in $KMnF_3$ 3% of second-neighbors interactions are needed. The structure in these two cases is such that second neighbors need superexchange through two nonmagnetic atoms and this would be expected to be relatively small on the basis of normal models of superexchange.

Exchange energies have been deduced for all these three materials by a wide variety of techniques and comparisons show good consistency throughout. Exchange energies derived from neutron spectroscopy are found to be of about the same accuracy as are obtained by other techniques. Where the neutron data are essential for an understanding of the magnetic properties is that they show how many exchange interactions are present while other techniques usually give only one combination of exchange constants.

One notable general feature of all these measurements is how good the superexchange model proves to be in predicting the range of the exchange constants. This is in sharp contrast to the phonon situation where in Section II it was shown that no simple model suffices for any type of bond so far investigated. The actual calculation of exchange constants from first principles has so far proved to be too difficult. The magnon energy at $q = 0$ is related to the anisotropy energy so that this anisotropy energy is also measured in a neutron experiment. However, a more accurate way to measure such energies is by antiferromagnetic resonance techniques as will be discussed by Dr. Sievers.

### Terbium

Terbium is the only ferromagnet for which full dispersion curves are available (Moller and Houmann 1966). The magnetic moments here reside on the $4f$ electrons. These are well localized at the terbium atoms and exchange arises indirectly via the conduction electrons. These electrons have a high mobility so that there is no *a priori* reason why the exchange interaction should be of short range. The measured dispersion relations for three symmetry directions of the hexagonal close-packed lattice are shown in Fig. 20. The top part of the figure shows the raw data and the structure in this has the effect that long-range forces must be invoked to describe the data. Dispersion curves in a ferromagnet may be Fourier analyzed to give interplaner force contants in an analogous may to the standard method for phonons first described by Foreman and Lomer. For terbium there are some complications due to the presence of anisotropy energy and to the fact that there are two atoms per unit cell; the bottom half of Fig. 20 shows the processed dispersion curves which it is appropriate to Fourier-analyze. At least four sets of planes are needed to fit the data so it is clear that long-range forces are present.

At higher temperatures (218–230°K) this structure becomes a spiral along $c$. For such a spiral to be stable it is necessary for $J(\mathbf{q}) < J(0)$ where

$$J(q) = \sum_{\mathbf{r}} e^{i\mathbf{q}\cdot\mathbf{r}} J(\mathbf{r})$$

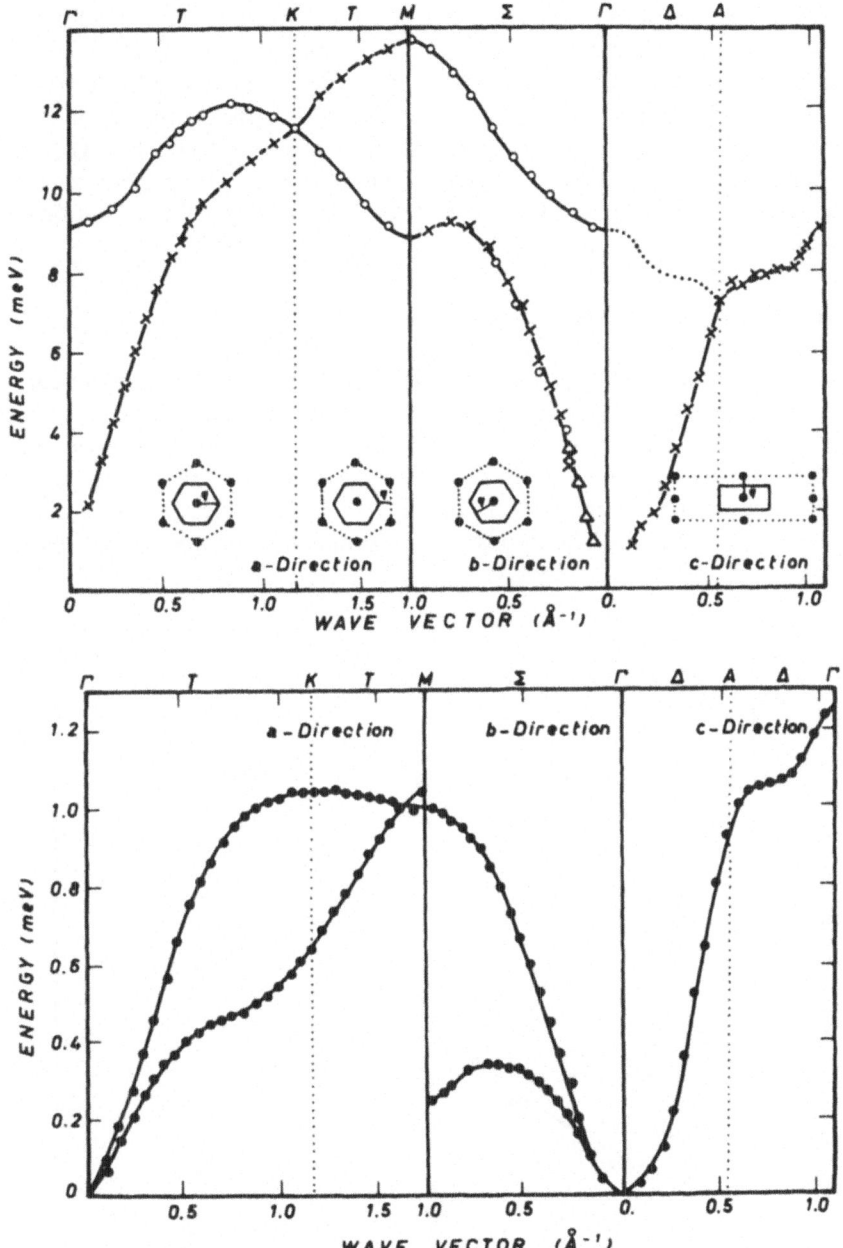

Fig. 20—The top curves show the spin-wave relations in terbium at 90°K, and at the bottom are the derived values of $J(\mathbf{q})$ (Moller and Houmann 1966).

This is not observed to be the case; the spiral wavevector is at a small **q** value in the *c* direction. In fact the set of exchange interactions giving low-temperature dispersion does not support the spiral structure so that *J* itself must be temperature-dependent. At higher temperatures it appears that Moller and Houmann do in fact observe dispersion relations for which $J(\mathbf{q})$ passes through a minimum value (private communication). It is clear that the magnetic properties of terbium are not yet fully understood and that further work is needed.

### Transition Metals and Their Alloys

Owing to the high spin-wave energies involved no full dispersion curves are available for transition metals and the best that has been done is to reach around 20% of the way to the zone boundary. Let us begin with the work on the elements by the Brookhaven group using the scattering surface technique described at the end of Section I. This gives the spin-wave dispersion relation at small **q**. Theoretically for a simple ferromagnet in the absence of anisotropy this is given by

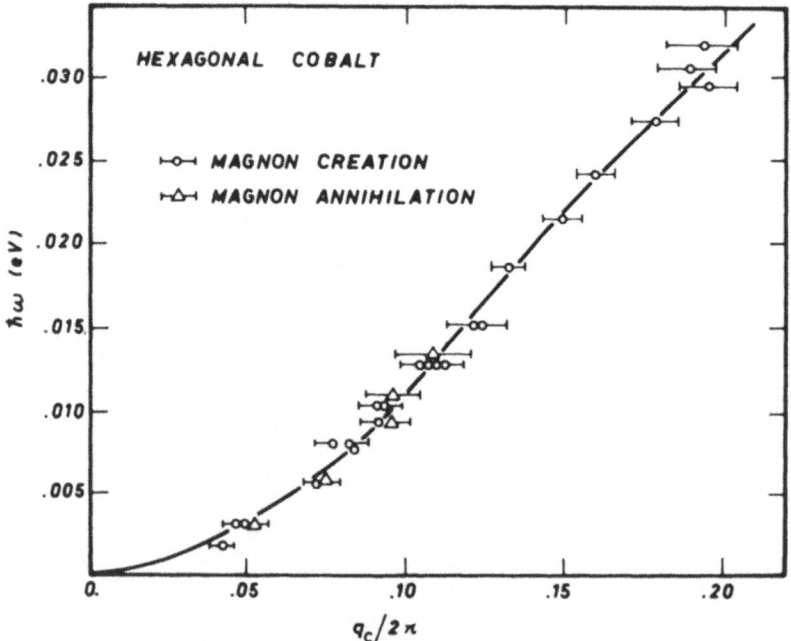

Fig. 21—The acoustic branch of the spin-wave spectrum for hexagonal cobalt (Alperin, H.A. *et al.*, 1966).

$$E = 2S \sum_{\mathbf{r}} J(\mathbf{r})(1 - \cos \mathbf{q} \cdot \mathbf{r})$$

as discussed in Dr. Elliott's lectures. $J(\mathbf{r})$ is the Heisenberg exchange energy between an atom at the origin and one at $\mathbf{r}$. For small $q$ this may be expanded to give

$$E = Dq^2 + Eq^4 + O(q^6)$$
$$= S \sum J(\mathbf{r}) \left[ (\mathbf{q} \cdot \mathbf{r})^2 - \frac{(\mathbf{q} \cdot \mathbf{r})^4}{12} \right] + O(q^6)$$

Thus the term in $q^2$ is related to the second spatial moment of $J(r)$ and that in $q^4$ to the fourth moment. The ratio of the $q^4$ to the $q^2$ term gives the square of a range parameter for the exchange interaction. For both iron [Shirane et al. (1965)] and cobalt this ratio is found to be about ten times larger than would be predicted on the basis of nearest-neighbor interactions only, so that the range parameter is of the order of three times the nearest-neighbor distance. The data for cobalt are shown in Fig. 21; it plainly does not follow just the parabolic form. This is a clear demonstration that long-range interactions are important in the $3d$ ferromagnetic metals.

Another fruitful technique for examining these metals and their alloys is that of the small-angle scattering of neutrons developed by Lowde and described briefly in Section I. This gives the spin-wave energy at rather long wavelengths and measures just the term $Dq^2$ in the dispersion relation quoted above. The technique has been applied to quite a number of alloy systems and some of the results for body-centered alloys are shown in Fig. 22. (Lowde, Shimizu, Stringfellow, and Torrie 1965). These results can be explained only by means of a theory which takes into account the metallic nature of the alloys. For instance, when vanadium with little or no moment [Collins and Low, *Proc. Phys. Soc.* **86**: 535 (1965)] is added to iron the value of $D$ increases, while a simple localized moment picture would predict just the opposite result.

On the right of Fig. 22 is plotted a parameter related to the density of electron states at the Fermi surface for the alloys concerned. The resemblance to the $D$-value is remarkable and fits in well with the theory of Shimizu and Katsuki (*Proceeding International Conference on Magnetism*, Nottingham, 1964) which relates the two directly.

Another point worth making here is that the data imply the existence of reasonably sharp spin-wave modes in these alloys at long wavelengths. That this might be expected theoretically on a localized model has been shown by Murray [*Proc. Phys. Soc.* **89**: 87 (1966)]. She showed that though the spin-wave energy goes as $Dq^2$ at small $q$ the linewidth goes as $q^4$ and so becomes narrow even for a random alloy.

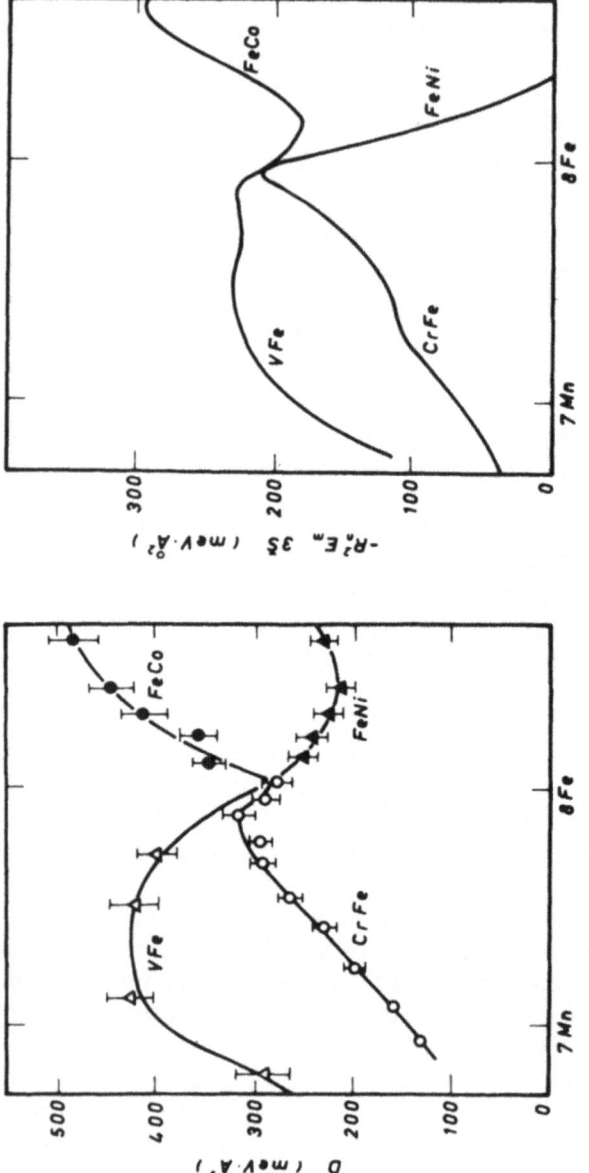

Fig. 22—Left: The observed spin-wave stiffness constant $D$ for some $3d$ alloys plotted against mean electron concentration. Right: The predicted stiffness from the density of states (Lowde *et al.* 1965).

### Ferrimagnets

The only case where any data are available from neutron spectroscopy is $Fe_3O_4$ (Brockhouse and Watanabe 1962). In this inverted spinel structure there are two types of site, the $A$ sites containing $Fe^{3+}$ ions with tetrahedral symmetry and the $B$ sites containing equal numbers of $Fe^{3+}$ and $Fe^{2+}$ ions with octahedral symmetry. The material is ferrimagnetic because the predominant exchange interaction is between the A and B sites so that one $Fe^{3+}$ ion and one $Fe^{2+}$ ion are aligned in one direction on the $B$ sites while one $Fe^{3+}$ ion is aligned in the other direction on the A site. The experimental data are shown in Fig. 23 and parts of two optical branches are apparent as well as the acoustic branch. The spin-wave theory for this lattice has been worked out by Kaplan and the solid line in the figure is fitted from his calculations. The results give a satisfactory fit to the exchange model described above.

### Magnon–Magnon Interactions

The magnon population is too small at low temperatures for this interaction to be observed in pure materials with neutron spectroscopy. However,

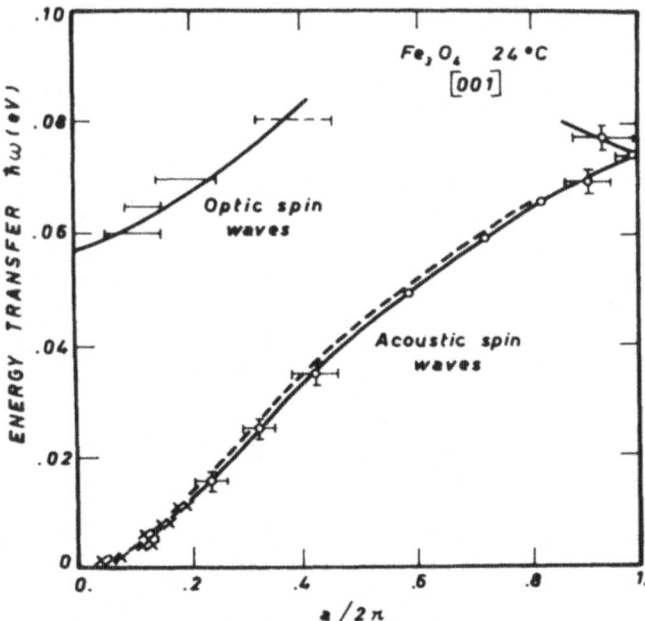

Fig. 23—Experimental curve for spin waves in $Fe_3O_4$ at room temperature. The dashed curve gives the results before correction for resolution (Brockhouse and Watanabe, 1963).

by heating the specimen the population can be made sufficiently large for linewidths to be observable. The point of observability seems to come at somewhere of the order of 0.7 $T_c$. This is such a high temperature that it is not certain that spin-wave theory is applicable; however, it is interesting to investigate what happens and, in fact, spin-wave theory seems to hold up suprisingly well.

The only material for which any extensive work has been done is $MnF_2$ by Turberfield and co-workers (1964). It is convenient to split the problem into two parts; the energy shift due to magnon–magnon interactions and the linewidths. Figure 24 shows measured peak energies of the lines at various temperatures. $T_N$ is 67°K so the data go quite close to $T_N$. The observed energies are compared to energies calculated by Low, who renormalized the spin waves according to the method of Oguchi. The fit is quite good; even amazingly good.

Figures 25 and 27 show the full widths at half height of the spin-wave modes, at 49.5 and 62°K, respectively. The increasing broadening and asymmetry of the spin-wave peaks is quite apparent and at the higher temperature the mode is only propagating for a wavelength or two. Figure 27 shows the change of spin-wave profile as the temperature is raised starting from a sharp $\delta$-function and broadening until in the paramagnetic state it is like a Gaussian centered on an energy of zero. These data were for a mode at

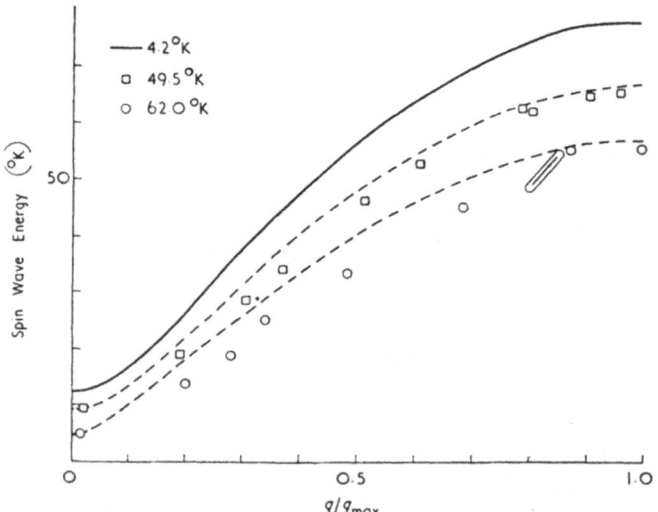

Fig. 24—Spin-wave dispersion curves for propagation in the $\langle 001 \rangle$ direction observed at 4.2, 49.5, and 62°K. The dotted curves are calculated (Low 1965) using a theory of spin-wave interractions (Oguchi 1960).

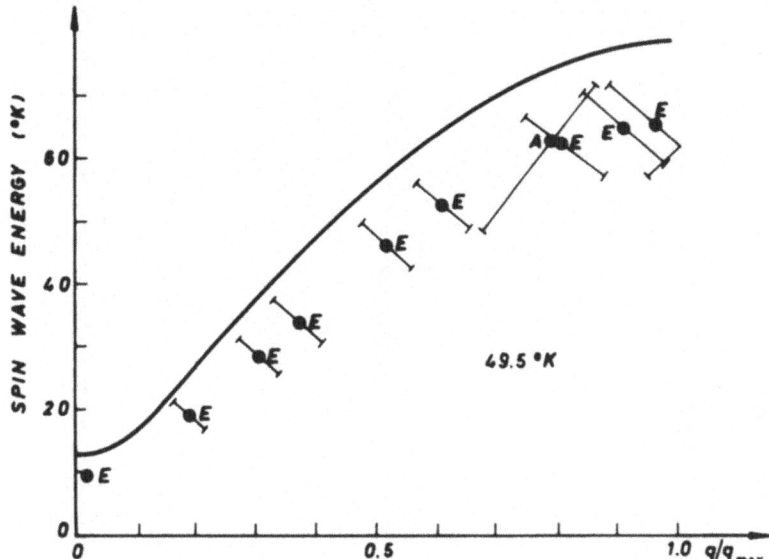

Fig. 25—Line broadening data for spin waves propagating in the $\langle 001 \rangle$ direction in $MnF_2$. The bars indicate the width at half height of the peaks after correction for instrumental resolution. The smooth curve shows the spin-wave dispersion at 4°K (Turberfield *et al.* 1965).

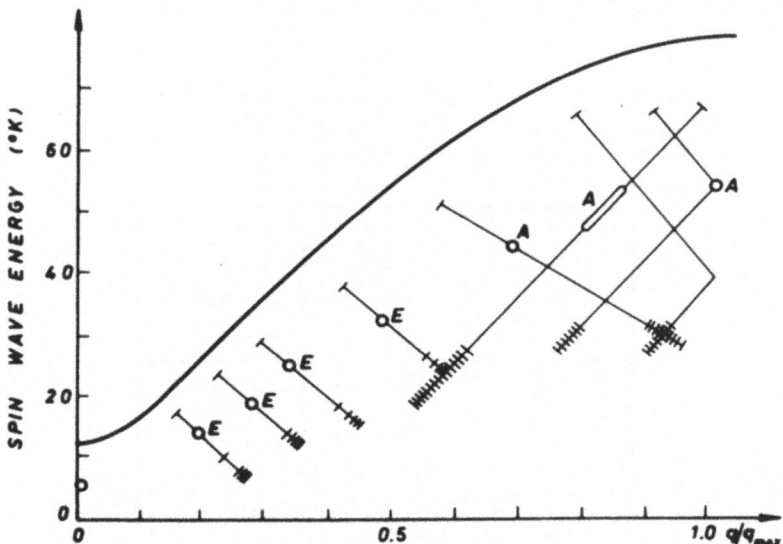

Fig. 26—Line broadening in $MnF_2$ at 62°K. The spin-waves are only propagating for a few wavelengths (Turberfield *et al.* 1965).

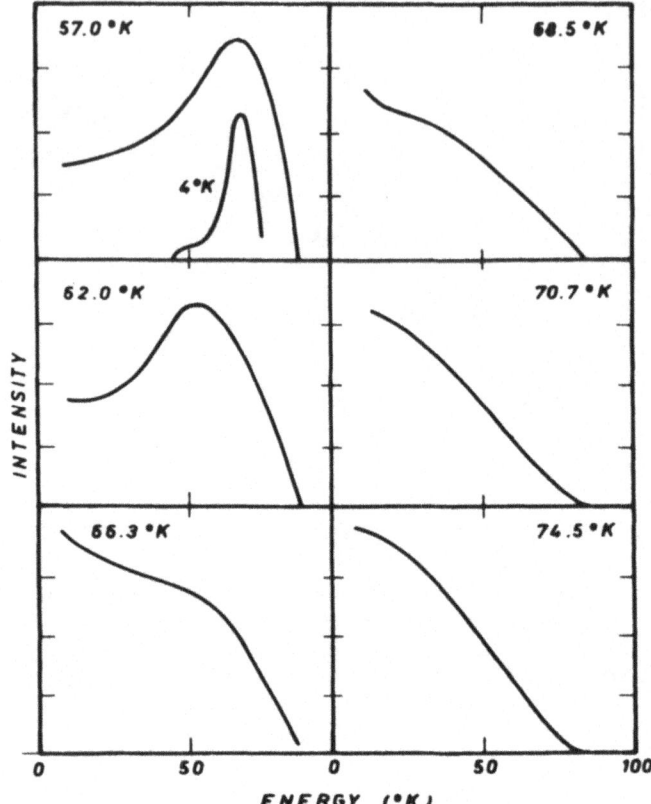

Fig. 27—Spectrum of the magnetic excitations in manganese fluoride at a zone boundary for temperatures below, near to, and above the Neél temperature (67°K) (Turberfield *et al.* 1965).

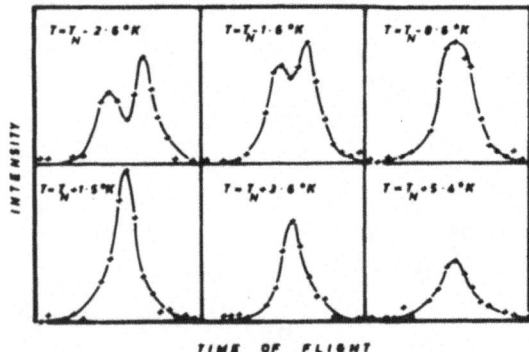

Fig. 28—Temperature dependence of long-wavelength excitations in manganese fluoride near the Neél temperature (Torrie 1966).

the zone boundary; analogous data for a mode near $q = 0$ are shown in Fig. 28 (Torrie, *Proc. Phys. Soc.*, 1966) At $T_N - 1.6°$K, well-defined energy-gain and energy-loss modes are present though these have completely disappeared by $T_N + 1.5°$K.

The narrative has descended to a rather-wooly descriptive level. This is due to the lack of theoretical framework in which to describe the transition from an ordered magnetic state to a paramagnetic state. The experimental data provide a qualitative picture of what is happening and further progress would seem to await the development of theoretical techniques to handle the problem.

## Magnon–Phonon Interactions

This has recently been seen experimentally by Dolling and Cowley (1966) in $UO_2$. The phenomenon has been investigated at very long wavelengths by resonance techniques, but, as in so many other cases, only neutrons can be used at present for general wavelengths. $UO_2$ is an antiferromagnet at low temperatures. The magnon–phonon interaction should be fairly large because the uranium ion is in a state S-1, L-5, J-4 and the large L-value greatly enhances the coupling. There are in fact some complications in describing the magnetic system fully, probably due to crystal field splittings [Blume,

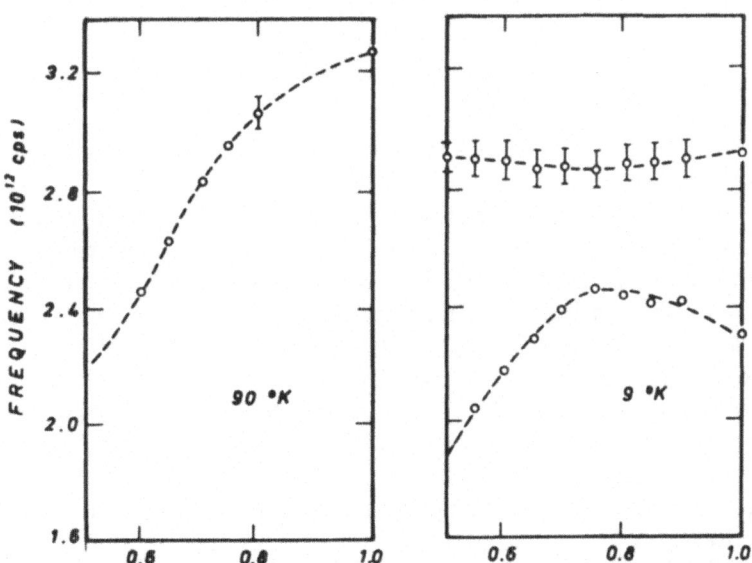

Fig. 29—The dispersion curve for excitations of uranium dioxide at 9°K and 90°K. At 90°K the curve corresponds to the TA phonon branch which is modified at 9°K by magnon–phonon interactions (Dolling and Crowley 1966).

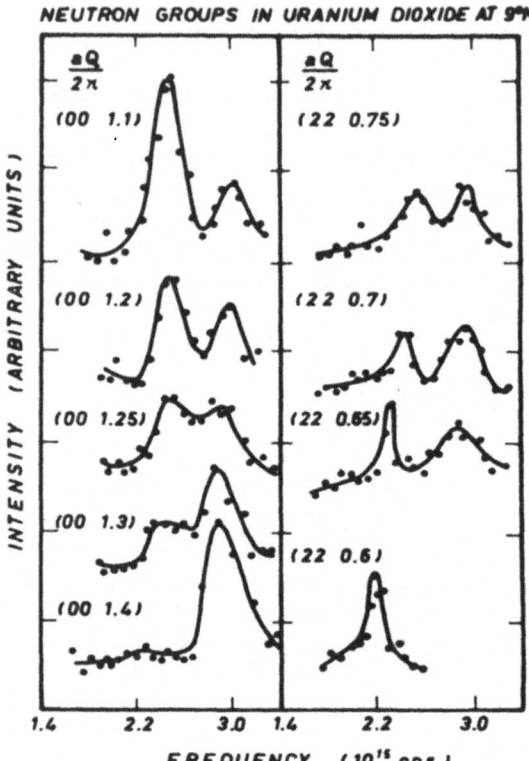

Fig. 30—Neutron groups in UO₂ at 9°K (Dolling
and Cowley 1966).

*Phys, Rev.* **141**: 517 (1966)], but these need not be entered into for the purposes of this discussion.

A section of the dispersion law is shown in Fig. 29 at temperatures well above and well below the Neél temperature of 30°K. It is apparent that there is a spin-wave mode at 9°K which is interacting strongly with the TA phonon mode (seen unperturbed at 90°K). The raw data for these two interacting modes are shown in Fig. 30. On the left the experimental settings are such that the phonon mode is slightly more intense than the magnon mode and on the right vice-versa. It is quite clear that the modes do not cross and that they are each changing in character as they pass through their nearest distance of approach.

If, as Dolling and Cowley postulate, the interaction Hamiltonian is of the quadratic form

$$H = \hbar\omega_q^m a_q^+ a_q + \hbar\omega_q^p b_q^+ b_q + c(a_q b_q^+ + a_q^+ b_q)$$

where $a_q$ and $b_q$ are magnon and phonon destruction operators, then the experimental data give $c/k = 9.6 \pm 1.6°K$ with $k$ as Boltzmann's constant.

## GENERAL REFERENCES

### Section I

*Scattering Cross Sections.*
  Lomer, W.M., and Low, G.G., *Thermal Neutron Scattering* (P.A. Egelstaff—editor) p. 1 (1965).
  Marshall, W., Harvard lecture notes—unpublished (1958).
  Van Hove, L., *Phys. Rev.* **95**: 249 (1954); *ibid* **95**: 1374 (1954).
  Maradudin, A.A. and Fein, A.E., *Phys. Rev.* **128**: 2589 (1962).
*Localized Modes*
  Elliott, R.J., BNL 940 (C-45), 23 (1966)
*Triple-Axis Spectrometry*
  Iyengar, P.K., *Thermal Neutron Scattering* (P.A. Egelstaff—editor) p. 97 (1965).
*Neutron Choppers*
  Brugger, R.M., *Thermal Neutron Scattering* (P.A. Egelstaff—editor) p. 53 (1965).

### Section II

*General*
  Cochran, W., *Repts. Prog. Phys.* **26**: 1 (1963).
  Dolling, G., and Woods, A.D.B., in *Thermal Neutron Scattering* (P.A. Egelstaff— editor) p. 193 (1965).
*NaI*
  Woods, A.D.B., Cochran, W., and Brockhouse, B.N., *Phys. Rev.* **119**: 980 (1960).
*KBr*
  Woods, A.D.B., Brockhouse, B.N., Cowley, R.A., and Cochran, W., *Phys. Rev.* **131**: 1025 (1963); *ibid* **131**: 1030 (1963).
*Ge*
  Brockhouse, B.N., and Iyengar, P.K., *Phys. Rev.* **111**: 747 (1958).
  Cochran, W., *Proc. Roy. Soc.* **A253**: 260 (1959).
*Diamond.*
  Warren, J.L., Wenzel, R.G. and Yarnell, J.L., Inelastic Scattering of Neutrons in Solids and Liquids, I.A.E.A., Vienna, p 361 (1965)
*Na*
  Toya, T., *J. Research Inst. Catalysis, Sapporo* **6**: 161 (1958).
  Woods, A.D.B., Brockhouse, B.N., March, R.H., Stewart, A.T. and Bowers, R., *Phys. Rev.* **128**: 112 (1962).
  Cochran, W., *Proc. Roy Soc.* **A276**: 308 (1963).
*Pb*
  Brockhouse, B.N., Arase, T., Caglioti, G., Rao, K.R. and Woods, A.D.B., *Phys. Rev.* **128**: 1099 (1962).
  Brockhouse, B.N., Rao, K.R. and Woods, A.D.B., *Phys. Rev. Letters* **7**: 93 (1961).
*Nb–Mo*
  Woods, A.D.B., *BNL* 940 (C-45), p 8 (1965)
*Cr*
  Moller, H.B., and Mackintosh, A.R., Inelastic Scattering of Neutrons in Solids and Liquids, *I.A.E.A.*, Vienna (1965)
*SrTiO₃*
  Cowley, R.A., *Phys. Rev. Letters* **9**: 159 (1963).

## Section III

*General.*
Lowde, R.D., *J. Appl. Phys.* **36**: 884 (1965).
*$MnF_2$ and Magnon–Magnon Interactions*
Turberfield, K.C., Okazaki, A., and Stevenson, R.W.H., *Proc. Phys. Soc.* **85**: 743 (1965); Physics Letters, **8**: 9 (1964).
Torrie, B.H., *Proc. Phys. Soc.* **89**: 77 (1966).
*$KMnF_3$*
Pickart, S.J., Collins, M.F. and Windsor, C.G., *J. Appl. Phys.*, **37**: 1054 (1966).
*$RbMnF_3$*
Windsor, C.G. and Stevenson, R.W.H., *Proc. Phys. Soc.* **87**: 501 (1966).
*MnO*
Collins, M.F., *Proc. Int. Conf. Magnetism*, Nottingham (1964).
*b.c.c. alloys*
Lowde, R.D., Shimizu, M., Stringfellow, M.W., and Torrie, B.H., *Phys. Rev. Letters* **14**: 698 (1965).
*Fe*
Shirane, G., Nathans, R., Steinsvoll, O., Alperin, H. A., and Pickart, S.J., *Phys. Rev. Letters* **15**: 146 (1965).
*Co*
Alperin, H.A., Steinsvoll, O., Shirane, G. and Nathans, R., *J. Appl. Phys.* **37**: 1052 (1966).
*$Fe_3O_4$*
Brockhouse, B.N. and Watanabe, H., *Physics Letters*, **1**: 189 (1962).
*Tb*
Moller, H.B., and Houmann, J.C.G., *Phys. Rev. Letters* **16**: 737 (1966).
*$UO_2$*
Dolling, G., and Cowley, R.A., *Phys. Rev. Letters* **16**: 683 (1966).

# Exploring the Excitation Spectra of Crystals Using Far Infrared Radiation*

A. J. Sievers

*Laboratory of Atomic and Solid State Physics*
*Cornell University*
*Ithaca, New York*

## I. FAR INFRARED TECHNIQUES

### a. Introduction

In 1897 far infrared radiation was first isolated by Rubens and Nichols ([1]). They used the selective reflection of alkali halide crystals to separate out the long wavelengths. Later Rubens and Wood ([2]) used focal isolation and a two quartz plate interferometer as a far infrared frequency analyzer and finally Rubens and Von Baeyer ([2]) found that a high pressure mercury arc was a good source in this frequency region. No revolutionary broad band sources have been discovered in the ensuing fifty-five years and today mercury arcs are still used in the frequency region which extends from 100 cm$^{-1}$ to 10 cm$^{-1}$ ($3 \times 10^{12}$ cps to $3 \times 10^{11}$ cps). The development of far infrared interferometers ([3,4]) and liquid helium cooled radiation detectors ([5]), about 1956–1957, opened the way for far infrared solid state spectroscopy. It is easy to demonstrate why these investigations of solids has not taken place earlier. Consider the power available per unit bandwidth for an infrared source used at 1000 cm$^{-1}$ such as a Nernst glower (a rod composed of zirconium and yttrium oxides which is heated to about 2000°K) versus the power available for a mercury arc used in the far infrared at 10 cm$^{-1}$. The mercury arc has an effective noise temperature of about 10,000°K; however, the fused quartz envelope surrounding the arc is relatively opaque to infrared. Only for frequencies below about 100 cm$^{-1}$ does the envelope become transparent

*Work supported in part by the Atomic Energy Commission through Contract AT (30–1)–2391, Technical Report #NYO–2391–95. Additional support has been received from the Advanced Research Project Agency through the use of central facilities and space provided by the Materials Science Center at Cornell, MSC Report #562. The survey of literature pertaining to this review was completed in July 1966.

and the arc itself become an effective source. From the blackbody law one finds that for equal bandwidths the mercury arc emits about 1000 times less power at $10 \text{ cm}^{-1}$ than the relatively cold Nernst glower emits at $1000 \text{ cm}^{-1}$. Keeping this basic energy limitation in mind we now consider some of the present (1966) techniques which are used to sort and detect far infrared radiation.

## b. Grating Monochromator

A typical far infrared monochromator which is used to investigate solid state phenomena is schematically shown in Fig. 1. The instrument is evacuated to eliminate the troublesome atmospheric water vapor vibration rotation absorptions. The optical layout is displayed in a horizontal cross section. The source, labeled S in Fig. 1, is a General Electric UA-2 mercury arc. The radiation is then chopped at approximately 33 cps by a slotted disk mounted directly to a synchronous motor which is powered by a variable frequency supply. The adjustable chopping frequency is important since the laboratory "noise" varies from one run to the next. The radiation is reflected from the off-axis ellipsoid labeled $M_1$. The first filtering of the 200 W of source radiation occurs at $F_1$ which is a reflection grating. The grating grooves are designed so that long wavelengths of interest are specularly reflected from this grating while the short wavelengths are thrown into various grating orders back toward the source. The radiation then focuses on a carbon black impregnated polyethylene window in the center baffle. The beam then

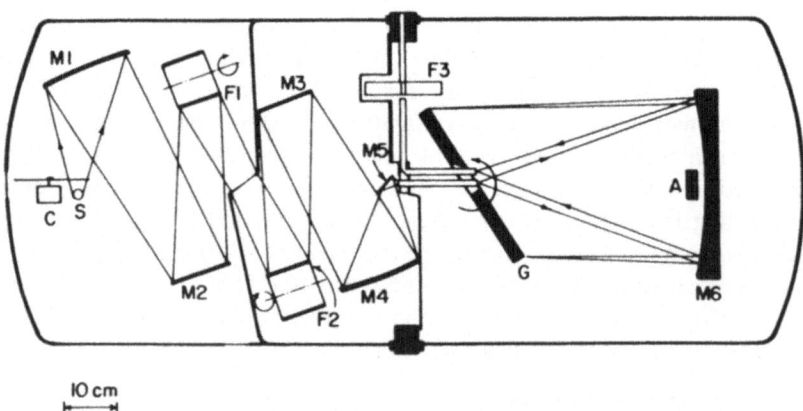

10 cm

Fig. 1—Top view of a far infrared grating monochromator. The instrument is an $f/1.5$ Ebert-type on axis design enclosed in a vacuum tank. The optical path is indicated by the arrows. The two zero-order reflection filters are labeled F1 and F2. The transmission filter wheel is labeled F3.

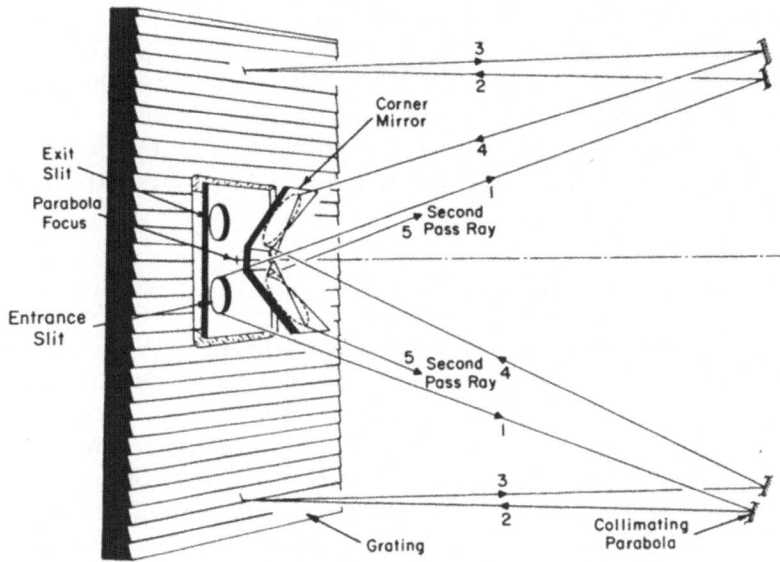

Fig. 2—Top view of the double pass optics. Double passing is achieved by a Walsh-type reimaging corner mirror placed in the focal plane symmetric with the entrance and exit slits.

diverges and the optical path is traced in reverse order first reflecting from filter $F_2$ etc., and finally reflecting from a mirror $M_5$ and focusing on a polished brass tube which transfers the 1.27 cm image to the focal plane of the collimating 36 cm-diameter parabolic mirror $M_6$. The light pipe enables a large radiation cone of 36° to be carried through the grating center with a relatively small loss of grating area. The light pipe plus surrounding baffle also effectively stop most scattered light from the source from reaching $M_6$. In Fig. 1 the channeled radiation diverges from the entrance slit, is collimated by $M_6$, and diffracted by the 30-cm square blazed grating. The first-order radiation (plus higher orders which are not yet completely stopped by filtering) returns to the collimating mirror and is focused on the exit slit. The radiation is then channeled through another brass light pipe, through a 12 position transmission filter wheel ($F_3$) and finally through a carbon black impregnated polyethylene vacuum window to the system to be studied. For a 1% bandwidth the power is now roughly $10^{-10}$ W at about 50 cm$^{-1}$.

The inadequacy of the far infrared source dictates that the resolving power of a monochromator will be limited by just how small the slits can be adjusted and still have adequate power on the detector. Because the power on the detector varies as the square of the slit width, doubling the resolution by halving the slit width reduces the signal by a factor of 4. An alternate means of increasing the resolution is to use a double pass system, which is shown in Fig. 2. Here the resolution is increased by a factor of two while the signal is only halved. Figure 2 shows the double pass optics looking down on the grating. The on-axis design of this instrument enables the double pass feature to be easily incorporated by using a right angle corner mirror similar to that proposed by Walsh for a prism spectrometer ([6]). With 1.2-cm slits (the light pipe diameter) the resolution $v/\Delta v$ of the instrument varies between 72 and 110.

### c. Grating Interferometer

In the spectral range between $100 \ cm^{-1}$ to $40 \ cm^{-1}$ the transmission spectrum of solids usually can be measured with a 1 % bandwidth in a straightforward manner; however, for frequencies below $40 \ cm^{-1}$ only relatively low-resolution studies are possible with a grating monochromator. Following Strong ([4,7]) and Richards ([8]) we have constructed a Strong-type lamellar grating interferometer which can be placed in the optical system as shown in Fig. 3. The lamellar grating, labeled I, consists of two sets of interleaving plates with 50 % of the radiation from the collimating mirror reflecting from the stationary set and 50 % reflecting from a movable set which is mounted on a slide and is advanced in a stepwise fashion by a micrometer screw. This assembly introduces a variable optical path difference between the interfering beams equal to twice the groove depth. The maximum optical path difference determines the maximum resolution and in Fig. 3 is $0.1 \ cm^{-1}$. The intensity on the detector is measured as a function of the path difference between the two sets of plates. The resultant intensity pattern for the central

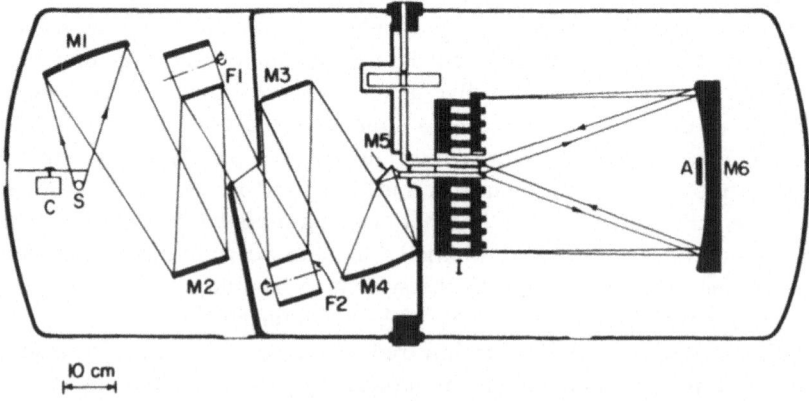

10 cm

Fig. 3—Top view of a far infrared Strong-type lamellar grating interferometer. The two sets of interleaving interferometer plates labeled I replace the grating assembly of Fig. 1.

diffraction order versus distance is called an interferogram. As a function of the interferometer path difference $\Delta$, the intensity can be expressed as

$$E(\Delta) = \int_0^\infty S(v)\, dv + \int_0^\infty S(v) \cos 2\pi v \Delta\, dv \qquad (1)$$

or

$$E(\Delta) - \frac{1}{2} E(0) = \int_0^\infty S(v) \cos 2\pi v \Delta\, dv \qquad (2)$$

where $S(v)$ is the spectral distribution of the radiation and $v$ is the frequency in wavenumbers. We see that $S(v)$ is the cosine Fourier transform of the interferogram function $E(\Delta) - \frac{1}{2}E(0)$. To obtain the transmission spectrum of a solid requires a digital computer to handle the Fourier transform. There are two advantages which offset this disadvantage of having a computer calculation between the experiment and the results.

The first advantage pertains to resolving power. While the maximum resolving power for the monochromator shown in Fig. 1 is determined by the spectral slit widths, the resolution of the interferometer is independent of the slit widths and depends only on the maximum path difference between the two beams. The second advantage (usually known as the Fellgett advantage[8]) concerns signal to noise. The detector at a monochromator output measures only one frequency bandwidth per unit time and the time to cover a given spectral region is the frequency interval divided by the bandwidth per unit time. The detector at an interferometer output measures all components of the spectral region with varying strengths for the total time of the run. Comparing the two methods the larger the spectral region or total time compared to unit time, the larger the Fellgett advantage. The gain in signal to noise with the interferometer is $\frac{1}{2}(\text{TOTAL TIME/UNIT TIME})^{1/2}$.

It is also possible to use an interferometer in a perodic mode, that is, to modulate the optical path difference periodically in time. The computation of the spectrum can then be carried out with narrow band amplifiers rather than a computer. This technique has been used successfully by Genzel and Weber.[9] It does not have the Fellgett advantage unless as many channels are used as there are resolution widths in the spectral range.

The interferometric measurements have their own peculiarities. For instance, a slow drift in detector sensitivity influences only the low-frequency spectral response when measured with an interferometer, whereas a similar effect during a monochromator run would alter the response from a large number of frequencies. On the other hand, sharp noise spikes during a monochromator run only influence a very limited number of frequencies whereas the same noise on an interferogram influences the spectral response at a large number of frequencies.

## d. Detectors

The low-temperature bolometers([10-12]) developed since 1956 can be used most effectively for the far infrared transmission measurements of solids by placing the sample and detector in the same helium cryostat. This feature eliminates transmitting the radiation through a number of cold windows, through the sample, and then through another set of cold windows (with the inherent reflection losses) back to a room temperature detector such as a Golay cell.([13])

The broad-band low-temperature detectors operate on the principle of a rapid variation of resistivity with temperature change. The temperature change is produced by the impinging far infrared radiation. The two types of, bolometers which we have used are the Boyle–Rogers-type carbon bolometer and the Low-type gallium-doped germanium bolometer. A carbon bolometer is made from a 56-$\Omega$ Allen Bradley resistor with final dimensions about 0.5 × 0.5 × 0.05 cm. Leads are attached to evaporated indium strips on the carbon with conducting epoxy. The detectors are then glued with GE 7031 varnish to a 1-milmylar electrical insulator which in turn is glued to a copper block immersed in liquid helium. These detectors are very sensitive the first time they are cooled to helium temperature (comparable to the Low detector) but age rapidly with room temperature cycling. Gallium-doped germanium has proved to be more reliable than the carbon bolometers. Copper leads can be soldered directly to the germanium slab with indium solder and one of these leads also serves as a thermal contact to the liquid helium bath as well as an electrical ground.

## e. Cryostats

These bolometers then are placed at the bottom of an evacuated light pipe assembly which is immersed in liquid helium. A cross section of such a detector can is shown in Fig. 4. The radiation from the grating instrument passes through a polyethylene vacuum window and down the cold light pipe. The radiation passes through a composite alkali halide filter constructed of thin plates of crystal quartz, NaF, NaCl, KCl, and KBr which greatly reduces the intensity of the dc room temperature radiation striking the sample and detector. The radiation next passes through the sample which is attached to a rotatable copper wheel containing five sample positions. This wheel also contains a resistance heater and one junction of a gold–cobalt thermocouple. The other junction of the thermocouple is in contact with the 1.2°K helium bath. The electrical leads are brought up to a room temperature vacuum seal inside the hollow stainless steel axle to which the wheel is screwed. The radiation then passes through another quartz filter and the light pipe image is

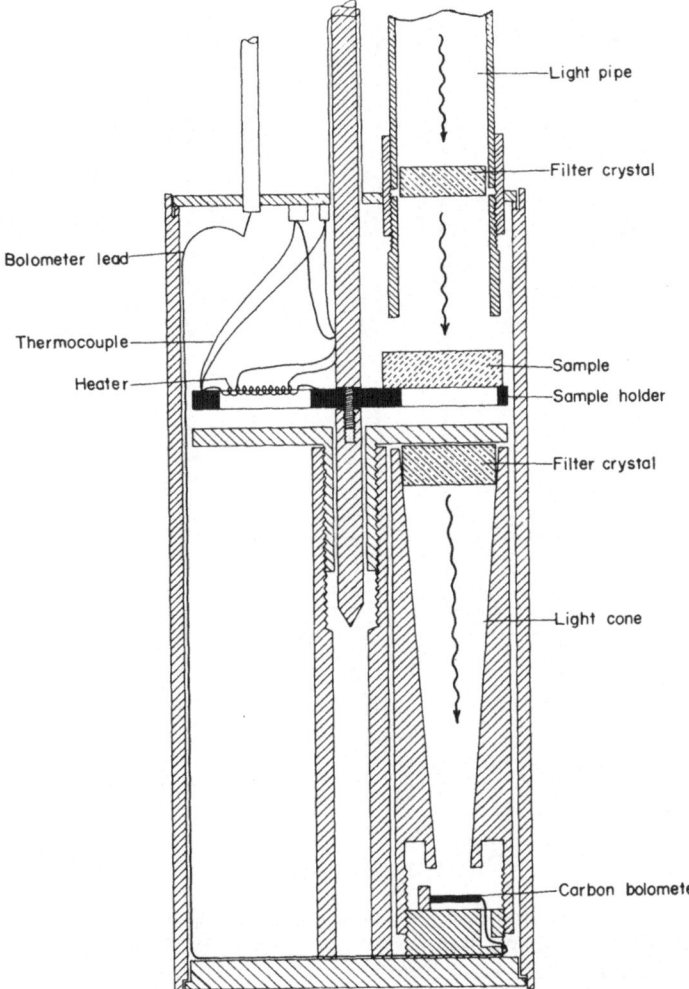

Fig. 4—A cross section of the sample-detector can in a liquid helium cryostat. The transmission of different samples can be compared by rotating the sample holder. The light cone matches the light pipe diameter to the detector diameter.

matched to the detector by a cone which converts the beam from a 1.26 cm diameter and cone angle of $\frac{2}{5}\pi$ to a 4mm diameter and cone angle of $2\pi$. The bolometer responds to the ac heating from the chopped far infrared and the ac voltage is amplified and phase-sensitively detected.

The details of the detector design naturally depend on the experiment to

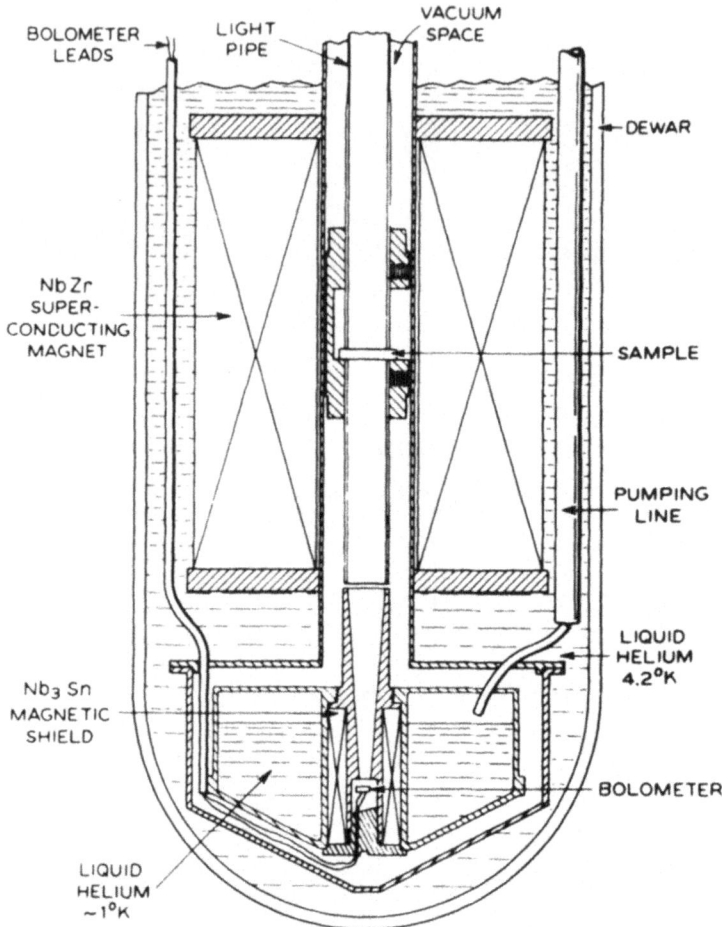

Fig. 5—Sample holder and detector mount for measuring the magnetic field dependence of sample transmission up to 50 KOe. [From P. L. Richards, *Phys. Rev.* **138**: A1769 (1965)].

some extent. A more complex sample and detector assembly used by Richards ([14]) for high-field measurements is shown in Fig. 5. The radiation passes down the light pipe through the sample which is in the center of a 50 kG superconducting magnet and to the bolometer detector. The magnet is in a 4.2° helium bath and the light cone and detector are cooled by a separate pumped liquid helium bath. This feature not only saves helium but the detector can be cooled to a slightly lower temperature, about 1°K.

Improvements in detection techniques are continually being found. For instance we have recently tested a He³ cooled (0.4°K) germanium bolometer and found it to be about three times as sensitive as our $He^4$ cooled (1.2°K) germanium bolometers([15]).

## II. ABSORPTION BY ANTIFERROMAGNETIC CRYSTALS

### a. Introduction

The magnetic ordering temperature $T_N$ for most paramagnetic solids occurs in the temperature range of a few degrees to hundreds of degrees Kelvin. Setting $h\omega = kT_N$ the equivalent frequencies are a few $cm^{-1}$ to hundreds of $cm^{-1}$. We see that the far infrared spectral range is a natural frequency region to measure the interaction energy between spins in magnetically ordered systems.

In the Heisenberg–Dirac–Van Vleck model of ferromagnetism ([16]), one assumes that the quantum mechanical exchange effects couple together the spins of different atoms by a potential of the form

$$V_{ij} = 2J_{ij}\mathbf{S}_i \cdot \mathbf{S}_j \qquad (3)$$

where $J_{ij}$ is the exchange coupling constant and the $\mathbf{S}_i$ and $\mathbf{S}_j$ are the spin operators for the respective ions. This interaction term also appears to describe the coupling in many antiferromagnetic and ferrimagnetic insulators. The interpretation of $V_{ij}$ is very different, however, and was first discussed by Kramers ([17]) and developed by Anderson ([18]) and Van Vleck ([19]). For some insulators the interaction arises from a superexchange mechanism in which the magnetic ions are coupled through electron transfer with intermediate ions. The form of this coupling is the same as that of the Heisenberg direct exchange interaction but the sign of $J$ is usually positive so that the lowest magnetic state corresponds to antiparallel alignment.

A spectroscopic investigation of the exchange coupling might be expected to provide much more detailed information than specific heat or magnetic susceptibility studies because these macroscopic properties represent an average over a large number of spins. Let us consider the possibility of measuring the exchange interaction by inducing an electromagnetic transition in which the relative direction of two neighboring spins is altered with a frequency proportional to $J$. We take two spins, 1 and 2, whch are exchange-coupled and let a uniform external magnetic field oscillating in the $x$-direction act on them. The perturbation can be written as

$$\mathscr{H}' = (g_{1x}\beta S_{1x} + g_{2x}\beta S_{2x})H_x \qquad (4)$$

where $g_{1x}$ and $g_{2x}$ are the spectroscopic splitting factors of the 1 and 2 spin in the $x$-direction. The commutator of the potential and this perturbation is

$$\frac{d}{dt}(\mathbf{S}_1 \cdot \mathbf{S}_2) = -\frac{1}{i\hbar}[\mathcal{H}', \mathbf{S}_1 \cdot \mathbf{S}_2] \tag{5}$$

Performing the commutation, we have

$$\frac{d}{dt}(\mathbf{S}_1 \cdot \mathbf{S}_2) = \frac{\beta H_x}{\hbar}(g_{2x} - g_{1x})(\mathbf{S}_2 \times \mathbf{S}_1)_x \tag{6}$$

If the spins and spectroscopic splitting factors are equivalent, no transitions can be induced. (The classical content of this result is that a uniform oscillating magnetic field acts equivalently on ions with the same gyromagnetic properties; thus it cannot alter the internal coupling between spins.) On the other hand, if the angular momenta are not equivalent, then transitions are possible. This selection rule implies that transitions measuring the coupling constants will not be readily observed for simple isotropic ferromagnets and antiferromagnets. Instead one must turn to the investiagation of anisotropic antiferromagnetic and more complex ferrimagnetic compounds.

## b. One-Magnon Absorption

The uniform precessional modes for a uniaxial two-sublattice antiferromagnet were first described by Kittel[20] and also Nagamiya[21] in 1951. A more complete description of the precessional properties was given by Keffer and Kittel[22]. The experimental testing of these and other theories[23] began around 1956. The largest class of magnetic insulators which have been investigated in the far infrared are the two sublattice antiferromagnetic crystals. Within this class the fluorides of the $3d$-transition group have been studied most completely[24-26].

Many of the general properties of spin wave modes in a two-sublattice array can be demonstrated with the simple example of an antiferromagnetic linear chain with nearest neighbor exchange interactions and magnetic anisotropy energy of orthorhombic symmetry. This is the interaction energy which can arise from long-range magnetic dipole interactions or from the interaction of individual magnetic ions with the local crystalline electric field. Let us consider this model.

The Hamiltonian for the $2n$th and $(2n + 1)$st spins of the linear chain can be written as

$$\mathcal{H}_{2n} = 2J\mathbf{S}_{2n} \cdot (\mathbf{S}_{2n+1} + \mathbf{S}_{2n-1}) + K_x(S_{2n}^x)^2 + K_y(S_{2n}^y)^2 \tag{7}$$

$$\mathcal{H}_{2n+1} = 2J\mathbf{S}_{2n+1} \cdot (\mathbf{S}_{2n+2} + \mathbf{S}_{2n}) + K_x(S_{2n+1}^x)^2 + K_y(S_{2n+1}^y)^2 \tag{8}$$

where $J$ is positive for antiferromagnetic coupling, $K_x$ and $K_y$ describe the appropriate anisotropy energy terms and the even and odd subscripts refer to up and down spins, respectively. Note that for $K_x$, $K_y > 0$ the spins point in the $\pm z$ direction. The equations of motion can be found quantum mechanically by using

$$\frac{d\mathbf{S}_i}{dt} = -\frac{1}{i\hbar}[\mathcal{H}_i, \mathbf{S}_i] \tag{9}$$

However, with our simple model the effective field notation is more instructive. The effective field at each spin is given by

$$g\beta\mathbf{H}_{2n} = -\nabla_{s_{2n}}\mathcal{H}_{2n}$$
$$g\beta\mathbf{H}_{2n+1} = -\nabla_{s_{2n}}\mathcal{H}_{2n+1} \tag{10}$$

where $g$ is the spectroscopic splitting factor, $\beta$ the Bohr magneton, and $\nabla_{s_i}$ is the gradient with respect to the $i$th spin. The effective fields are

$$\mathbf{H}_{2n} = -\frac{\omega_e}{2\gamma}(\mathbf{s}_{2n+1} + \mathbf{s}_{2n-1}) - \frac{\omega_{ax}}{\gamma}(\mathbf{s}_{2n})^x - \frac{\omega_{ay}}{\gamma}(\mathbf{s}_{2n})^y$$

$$\mathbf{H}_{2n+1} = -\frac{\omega_e}{2\gamma}(\mathbf{s}_{2n+2} + \mathbf{s}_{2n}) - \frac{\omega_{ax}}{\gamma}(\mathbf{s}_{2n+1})^x - \frac{\omega_{ay}}{\gamma}(\mathbf{s}_{2n+1})^y \tag{11}$$

where we have set $\hbar\omega_e = 4J\bar{S}$, $\hbar\omega_{ax} = 2K_x\bar{S}$, $\hbar\omega_{ay} = 2K_y S$, and $\mathbf{S}_i = \mathbf{s}_i\bar{S}$, with $\bar{S}$ defined to make $\mathbf{s}_i$, a unity vector of spin; $\gamma = g\beta/\hbar$ is the gyromagnetic ratio. The coupled Bloch equations are

$$\frac{d\mathbf{s}_{2n}}{dt} = \gamma\mathbf{s}_{2n} \times \mathbf{H}_{2n}$$

$$\frac{d\mathbf{s}_{2n+1}}{dt} = \gamma\mathbf{s}_{2n+1} \times \mathbf{H}_{2n+1} \tag{12}$$

For small oscillations about the $z$-axis there are $2N$ linear equations. Because of the translational symmetry under displacement by $2a$, we look for traveling-wave solutions of the form

$$(\mathbf{s}_{2n})^{x,y} = (\mathbf{s}_A)^{x,y}e^{i[\omega t - k 2na]}$$
$$(\mathbf{s}_{2n+1})^{x,y} = (\mathbf{s}_B)^{x,y}e^{i[\omega t - k(2n+1)a]} \tag{13}$$

Two of the four resulting equations are

$$i\omega s_A^x = -\omega_e s_A^y s_B^z + \omega_e s_A^z s_B^y \cos ka + \omega_{ay}s_A^z s_A^y \tag{14}$$

$$i\omega s_A^y = \omega_e s_A^x s_B^z - \omega_e s_A^z s_B^x \cos ka - \omega_{ax}s_A^z s_A^x \tag{15}$$

and the third and fourth equations can be obtained by replacing $A$ by $B$ and $B$ by $A$ in equations (14) and (15). With $s_A^z = -s_B^z = 1$ the eigenfrequencies are

$$\omega_1(ka) = \pm\left\{\left(2\omega_e \cos^2\frac{ka}{2} + \omega_{ax}\right)\left(2\omega_e \sin^2\frac{ka}{2} + \omega_{ay}\right)\right\}^{1/2} \tag{16}$$

$$\omega_2(ka) = \pm\left\{\left(2\omega_e \sin^2\frac{ka}{2} + \omega_{ax}\right)\left(2\omega_e \cos^2\frac{ka}{2} + \omega_{ay}\right)\right\}^{1/2} \tag{17}$$

Because the negative frequencies describe the same modes precessing in the opposite sense we lose no generality by considering only positive frequencies. Setting $ka = 0$, we find the uniform precessional modes which can interact with an rf field:

$$\omega_1(0) = [(2\omega_e + \omega_{ax})\omega_{ay}]^{1/2} \tag{18}$$

$$\omega_2(0) = [(2\omega_e + \omega_{ay})\omega_{ax}]^{1/2} \tag{19}$$

At the magnetic Brillouin zone boundary $ka = \pi/2$ (the Brillouin zone boundary of the chemical cell is $ka = \pi$), the two frequencies are degenerate with

$$\omega_1\left(\frac{\pi}{2}\right) = \omega_2\left(\frac{\pi}{2}\right) = [(\omega_e + \omega_{ax})(\omega_e + \omega_{ay})]^{1/2} \tag{20}$$

For the case where $\omega_e \gg \omega_{ax} \gg \omega_{ay}$ we sketch the dispersion characteristics of the two modes in Fig. 6. By substituting the resonant frequencies into equations (14), etc., the motion of the spins can be determined. For the $k = 0$ mode for $\omega_2(0)$, all spins are in phase, and the two sublattices precess in opposite directions as shown in Fig. 7. In (a) of Fig. 7 we see that the constant-length sublattice vectors precess in opposite directions and trace out elliptical paths on the surface of a sphere. Looking along the $z$-axis in (b) of Fig. 7, the motion of the sulattices produces an oscillating linear magnetic moment in the $y$-direction. The anisotropy of space in the $x, y$ plane causes the two modes to be nondegenerate at $ka = 0$. The application of an external field along the $z$-axis produces a second-order Zeeman shift for each mode. This can readily be understood by considering the oscillating linear magnetic moment associated with the high-frequency mode in Fig. 7. It is the coupling of this moment to the oscillating component of the infrared radiation that produces the resonant absorption. Because there is no time average moment in this mode, the applied field must first create a moment before it can shift the frequency; hence a quadratic shift with magnetic field results. (If $\omega_{ax} = \omega_{ay}$, then circularly polarized modes are produced which are degenerate in frequency.) The application of an external magnetic field along the $z$-

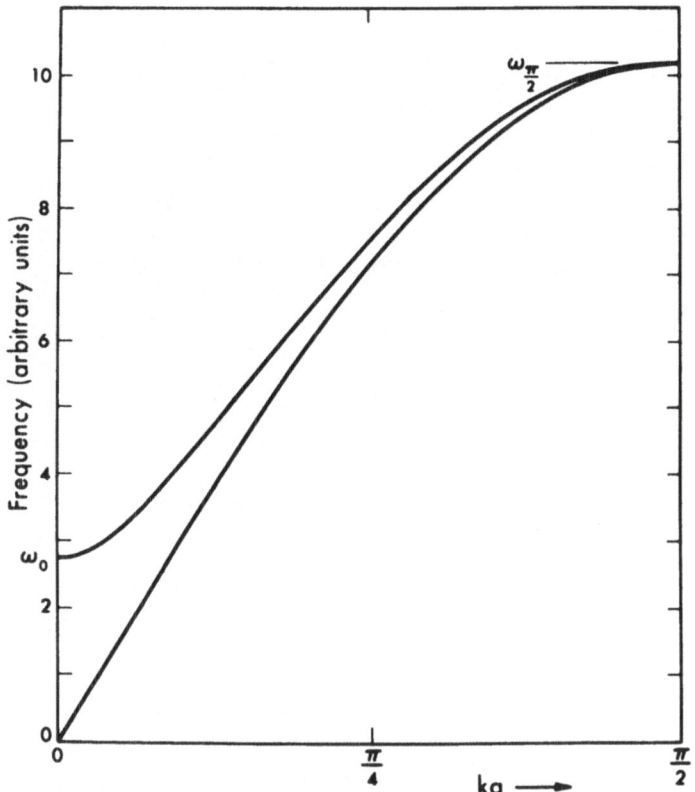

Fig. 6—Dispersion characteristics for a two-sublattice antiferromagnet with orthorhombic anisotropy interaction and $\omega_e \gg \omega_{ax} \gg \omega_{ay}$

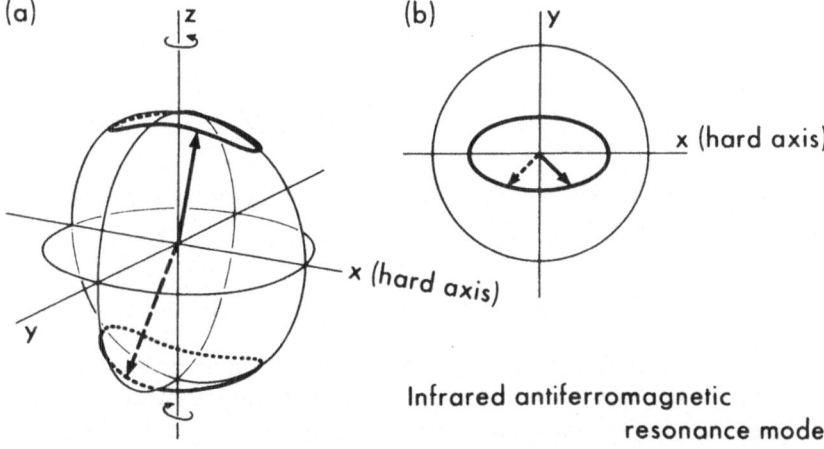

Infrared antiferromagnetic resonance mode

Fig. 7—The $\omega_2(0)$ normal mode. The up and down sublattices precess in opposite directions giving rise to an oscillating $y$ component. The low-frequency mode $\omega_1(0)$ (not shown) has the same form but gives rise to an oscillating $x$ component of magnetization.

axis produces a first-order Zeeman splitting of the two degenerate modes. Both sublattices in each mode give rise to an oscillating magnetic $z$-component; however, at $ka = 0$ there is no net oscillating $z$-component to interact with a $z$-directed rf field.

We can solve for the intensity of the resonance modes $\omega_1(0)$ and $\omega_2(0)$ by introducing an oscillating rf field into the equations of motion and find the rf susceptibility. We obtain for the susceptibility per unit cell

$$\chi_x'(\omega) = \frac{2\gamma^2\hbar\bar{S}\omega_{ay}}{[\omega_1(0)]^2 - \omega^2} = \frac{[\omega_1(0)]^2\chi_x'(0)}{[\omega_1(0)]^2 - \omega^2} \tag{21}$$

$$\chi_y'(\omega) = \frac{2\gamma^2\hbar\bar{S}\omega_{ax}}{[\omega_2(0)]^2 - \omega^2} = \frac{[\omega_2(0)]^2\chi_y'(0)}{[\omega_2(0)]^2 - \omega^2} \tag{22}$$

If we assume the resonance are sharp lines then the integrated absorption can be readily computed by using one of the Kramers–Kronig relations: [27]

$$\chi_x'(\omega) = \frac{2}{\pi}\int_0^\infty \frac{u\chi_x''(u)du}{u^2 - \omega^2} \tag{23}$$

For $\Delta\omega \ll (\omega - \omega_0) \ll \omega_0$ this relation simplifies to

$$\chi_x'(\omega) = \frac{\chi_x''(\omega_0)\Delta\omega}{\pi(\omega_0 - \omega)} \tag{24}$$

The integrated absorption then is

$$\chi_x''[\omega_1(0)]\Delta\omega = \frac{\pi}{2}\omega_1(0)\chi_x'(0) \tag{25}$$

and

$$\chi_y''[\omega_2(0)]\Delta\omega = \frac{\pi}{2}\omega_2(0)\chi_y'(0) \tag{26}$$

The strength of the absorption goes to zero as the anisotropy term in the resonant frequency goes to zero.

### c. Transmission of Magnetic Insulators

In the far infrared region of the spectrum a number of usually valid approximations permit the magnetic susceptibility information to be directly extracted from transmission measurements. Consider a plane wave with wavevector $k = 2\pi\nu(\mu\epsilon)^{1/2}$ passing through a material which is described by

$$\mu = \mu' - i\mu'' = 1 + 4\pi(\chi' - i\chi'') \tag{27}$$

and

$$\epsilon = \epsilon' - i\epsilon'' \tag{28}$$

where $v$ is the frequency in $cm^{-1}$. If multiple reflections are neglected, the transmission through a plate of thickness $d$ for normal incidence is

$$T = \frac{4n}{(n+1)^2 + \kappa^2} e^{-\alpha d} \tag{29}$$

where $\alpha = 4\pi v\kappa$,

$$\mu'\epsilon' - \mu''\epsilon'' = n^2 - \kappa^2 \tag{30}$$

and

$$\mu'\epsilon'' + \mu''\epsilon' = 2n\kappa \tag{31}$$

To develop the approximate transmission expression describing a magnetic insulator, first consider the case where $|\mu| = 1$. Equations (30) and (31) reduce to

$$\epsilon' = n_1^2 - \kappa_1^2 \tag{32}$$

$$\epsilon'' = 2n_1\kappa_1 \tag{33}$$

The lattice vibration electric dipole transitions typically range from 2000 to 100 $cm^{-1}$. The low-frequency wings of these strong lines extend throughout the far infrared region. For pure solids at moderately low temperatures a good approximation is

$$\frac{\kappa_1}{n_1} \ll 1$$

The reflectivity from one surface of dielectric sample then is

$$R = \left(\frac{n-1}{n+1}\right)^2 \tag{34}$$

We now turn on the magnetic interaction so that $|\mu| \neq 1$. Because the intensity of magnetic dipole transitions is typically $10^{-4}$ of the electric dipole intensity, we again consider the case where $\kappa/n, \ll 1$. From equations (27), (30), and (31) the change in index of refraction for $|\mu| \neq 1$ is negligibly small because

$$\frac{\Delta n(v)}{n_1} \cong 2\pi\chi'(v)$$

and $2\pi\chi'_{max}(v)$ for antiferromagnets is typically less than $10^{-2}$. (For a narrow line which has a dissipative component $\chi''$ of the Lorentz form, $\chi'_{max} \approx \frac{1}{2}\chi''_{max}$. The experimental values of $\chi''_{max}$ are typically less than $10^{-3}$ emu/cm³ so that the corresponding $2\pi\chi'_{max}$ are less than $10^{-2}$.) Therefore we take $n = n_1$. From equations (30), (31), (32), and (33) the absorption coefficient can be directly related to $\chi''$. We find that

$$\mu''(v) = 4\pi\chi''(v) = \frac{2}{n_1}[\kappa(v) - \kappa_1(v)] \tag{35}$$

neglecting cubic and higher order terms of $\kappa/n$. Because the maximum value of $\kappa(v)/n$ is usually less than 0.1, expression (35) is fairly accurate. The absorption coefficient [equation (29)] for the magnetic process is $\alpha_m = 8\pi^2 n_1 v\chi''$ (cm⁻¹). A transmission measurement of the sample above and below the ordering temperature is necessary to estimate the lattice contribution to equation (35). Another experimental method is to compare the transmission of the magnetic sample with the transmission of a nonmagnetic material with similar dielectric properties.

### d. Some Experimental Results

A number of antiferromagnetic compounds have been studied spectroscopically. We shall briefly consider the magnetic dynamics observed for some of these compounds.

Ferrous fluoride ($T_N = 78.4°K$) was the first antiferromagnet investigated by far infrared techniques [25]. The $Fe^{2+}$ ions are in a body-centered tetragonal structure. The symmetry of the crystalline electric field upon the $Fe^{2+}$ spins through the spin orbit interaction produces a $c$-directed uniaxial anisotropy which is an order of magnitude larger than the dipolar anisotropy contribution. A doubly degenerate antiferromagnetic resonance mode has been observed at 52.7cm⁻¹ at $T = 2°K$ which splits into two circularly polarized modes with applied magnetic field. As the sample temperature is raised the resonance frequency decreases and the linewidth increases.

The magnetic diple—dipole interaction in the NaCl-structure antiferromagnets, MnO and NiO, produces an unusual uniaxial anisotropy energy in which the uniaxis is the hard axis [28]. Thus the two opposing sublattice moments are directed in an easy (111) plane. The two normal modes for small oscillations of the sublattice magnetizations about a preferred direction within this plane are both nondegenerate and linearly polarized. The high-frequency

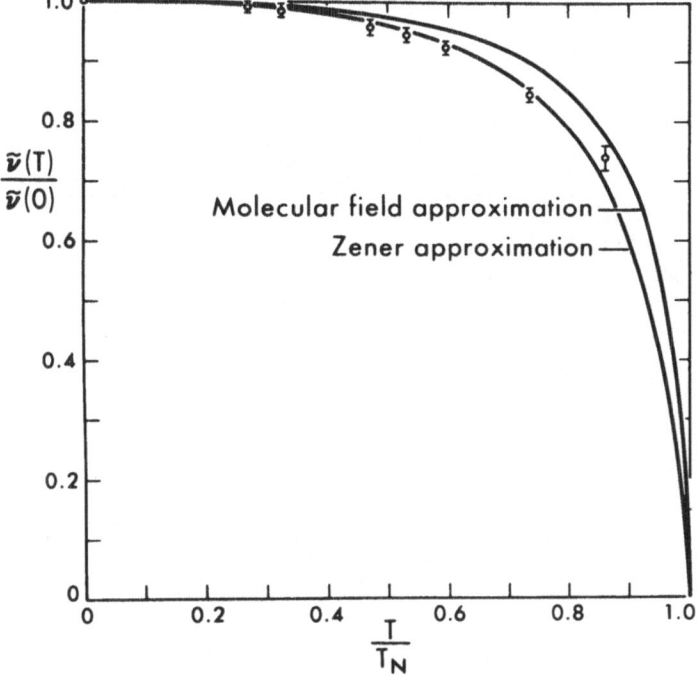

Fig. 8—Temperature dependence of antiferromagnetic resonance in MnO compared with the dependence predicted by the molecular field approximation and the Zener approximation.

mode has been observed at 2°K at 27.5 and 36.6 cm$^{-1}$ in MnO and NiO, respectively. A second-order Zeeman shift is expected for these nondegenerate modes. The temperature dependence of the nondegenerate high-frequency mode has been measured up to 0.87 $T/T_N$. In Fig. 8 the experimental frequencies are compared with the predictions of the Zener theory which considers the temperature dependence starting with correlated neighboring spins and the molecular-field approximation which represents the no-correlation extreme. It is interesting to note that the Zener approximation describes the correct temperature dependence over most of the ordered range. The square root of the neutron diffraction intensity has been used for the temperature dependence of the sublattic magnetization.

Nickel fluoride ($T_N = 73.2°K$) has the same structure as ferrous fluoride but is a weak ferromagnet below the ordering temperature ([14]). The corner and body-center spins are not oppositely directed but canted slightly due to the crystalline field anisotropy. Two nondegenerate spinwave branches are expected for this arrangement. The two modes for $k = 0$ are shown in Fig. 9. The low-frequency resonance of the net moment (mode 1) has been

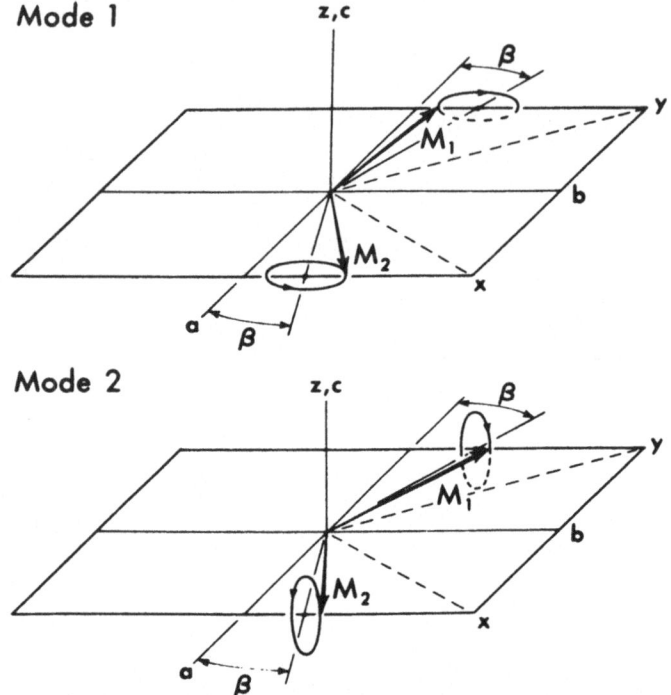

Fig. 9—Normal modes for the canted antiferromagnet $NiF_2$. Mode 1 is the low-frequency ferromagnetic resonance and Mode 2 the high-frequency antiferromagnetic resonance. [From P. L. Richards, *Phys. Rev.* **138**: A1769 (1965)].

observed at 3.33 cm$^{-1}$ and the high-frequency antiferromangetic resonance (mode 2) has been measured at 31.14 cm$^{-1}$.

A more complex canted antiferromagnet is $\alpha$-$CoSO_4$ ($T_N = 12°K$). The large canting angle of 25° is believed to arise from anisotropic superexchange and large anisotropy of the $g$ value for the ground state of the $Co^{2+}$ ion ([29]). No net magnetic moment occurs for this four sublattice antiferromagnet and only three of the four possible $k = 0$ spinwave modes are expected to be infrared active. Three absorptions have been measured at 20.6, 25.4, and 35.8 cm$^{-1}$.

### e. Current Studies

Far infrared absorption associated with two-magnon processes recently has been observed for $FeF_2$ ([30]) and $MnF_2$ ([31]). The absorption coefficient in the frequency region near 100 cm$^{-1}$ is shown in Fig. 10 for $MnF_2$. The observed experimental results are consistent with a process in which the electric vector of the absorbed photon creates a pair of oppositely polarized magnons of wavevector $\mathbf{k}$ and $-\mathbf{k}$. The electric dipole moments associated with two-magnon processes are thought to arise from the combined effects of atomic electric dipole moments and off-diagonal exchange interactions ([32]). Essentially no magnetic field dependence of the absorption frequency is to be expected for this transition because one magnon is emitted from each

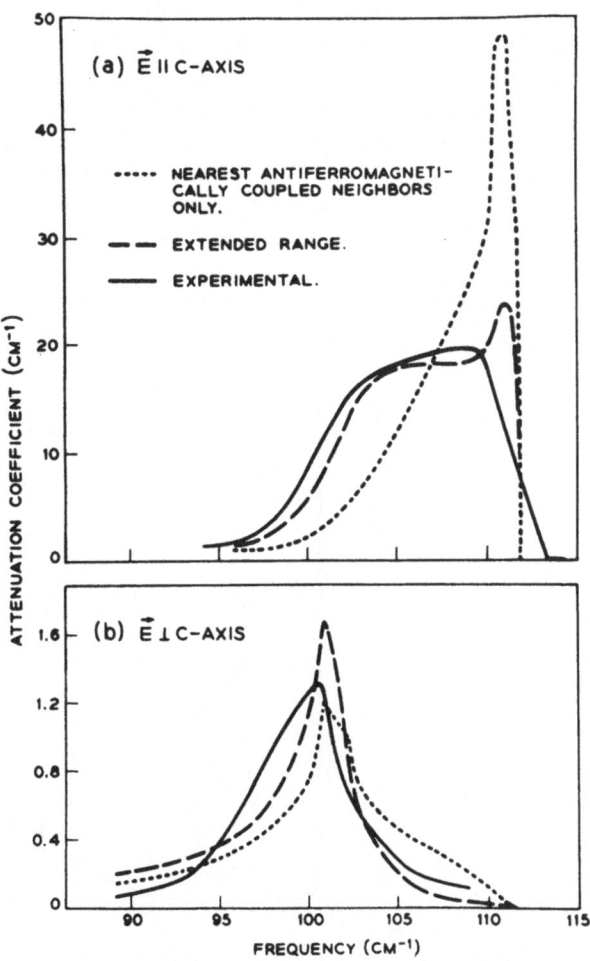

Fig. 10—Theoretical and experimental absorption coefficients for the two-magnon absorption in $MnF_2$ at 4.2°K. (a) electric vector parallel to the C-axis; (b) electric vector perpendicular to the $C$-axis. [From S. J. Allen, R. Loudon, and P.L. Richards, *Phys. Rev. Letters* **16**: 463 (1966)].

of the two normally degenerate spinwave branches at the zone edge. Because a linear splitting of these branches occurs upon the application of an external magnetic field, the frequency of one magnon is decreased and the frequency of the second is increased by the same interval.

One feature which the antiferromagnets from the $3d$-transition group have in common is that their magnetic properties can be understood without considering any magnon—phonon interactions. Recently, an antiferromagnet from the actinide series has been found which does not belong to this class. The magnon and phonon dispersion curves of $UO_2$ have been measured by the inelastic scattering of slow neutrons [33]. The magnon—phonon coupling constant has been found to be about one-third of the ordering temperature indicating that the magnon—phonon interactions are the same order of magnitude as the magnetic interactions. Also a $k = 0$ mode with magnetic character has been located at 18 cm$^{-1}$. Preliminary measurements of the far infrared transmission of $UO_2$ at 2°K have been made at both zero field and also at a field of 56KG. The zero-field spectrum is shown in Fig. 11. The

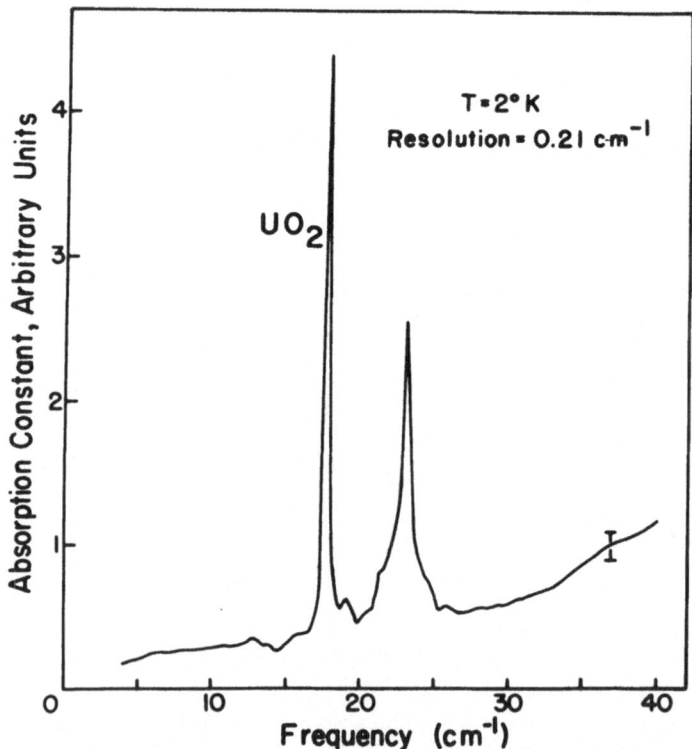

Fig. 11—Far infrared transmission spectrum of antiferromagnetic $UO_2$.

absorption line centered at 17.8 cm$^{-1}$ splits in a magnetic field while the higher-frequency absorption centered at 26 cm$^{-1}$ is field-independent. The strengths of both lines decrease as the temperature is increased. The low-frequency absorption corresponds very closely with the neutron results and is probably an antiferromagnetic resonance absorption. The line is surprisingly narrow. The full width at one-half maximum absorption is 0.28 cm$^{-1}$ slightly larger than the resolution of the interferometer of 0.21 cm$^{-1}$ with which these data were obtained. Because the absorption at 26 cm$^{-1}$ has not been observed in a second crystal obtained from the same supplier three years later, it has been identified tentatively with an electric dipole active lattice impurity mode. Additional absorptions observed at higher frequencies have not been studied in detail but $UO_2$ appears to be an interesting magnetic system. In conclusion the far infrared technique is a useful tool for the investigation of intrinsic magnon processes, and we now turn to the study of lattice impurity modes.

## IIIA. IMPURITY INDUCED ABSORPTION IN ALKALI HALIDE CRYSTALS: GAP MODES

### a. Introduction

Although the ultraviolet absorption spectrum of alkali halide crystals containing substitutional hydride ion impurities ([34]) (U-centers) has been known since 1936, the infrared absorption spectrum associated with the vibrational motion of this impurity was not observed untill 1958. At this time Schäfer ([35]) measured the infrared transmission of a number of alkali halides containing U-centers and found that the resonant frequency occurred far above the maximum frequency of the host lattice and the position of the absorption line depended on the lattice spacing of the host crystal. In particular, it was observed that the Mollwo–Ivey relation, namely,

$$v_0 a^n = \text{constant} \tag{36}$$

which relates (for $n = 1.84$) the absorption frequency $v_0$ of an F-center to the lattice constant a could also apply to the infrared U-center with $n = 2$. Some side band structure on the main line was also found.

Price and Wilkinson then measured a large number of U-center host lattice combinations ([37]) and observed that the exponent $n$ in equation (36) changed for different alkali halide series. In particular, $n = 2.00$ for sodium halides, $n = 2.25$ for potassium halides and $n = 2.50$ for rubidium halides. The deuteride–hydride isotope shift was also measured for a number of host lattices and the frequency ratio was not constant. This ratio was found to be 1.387 for NaCl, 1.398 for KCl, 1.394 for KBr, and 1.378 for KI compared to 1.414 for a simple Einstein oscillator model. This variation in frequency demonstrates that the nearset neighbors participate in the local mode motion by different amounts for the various host lattices. It was also clearly demonstrated that not only did the center frequency of the band shift and the width change with temperature but also that the integrated area under the main absorption line decreased with increasing temperature. Finally Mitsuishi and Yoshinaga ([38]) were able to destroy the infrared U-center absorption by irradiating a crystal at room temperature with ultraviolet light and Fritz ([39]) observed an impurity induced absorption from interstitial hydride ions which was produced by irradiating a U-center crystal with ultraviolet light at low temperatures. These initial experiments demonstrated that an important lattice probe was at hand. For in contrast with the one-phonon absorption spectrum of a pure alkali halide crystal which occurs only for the transverse optic branch at $k = 0$, a number of one-phonon effects are possible in an impure crystal because the impurity destroys the translational symmetry and the lattice wave vector is no longer a good quantum number.

A theory which describes the infrared lattice absorption for crystals with mass defects ([40]) was first given by Wallis and Maradudin ([41]) and also Takeno et al. ([42]). Later a more complete description of the optical properties associated with the mass defect problem for a homonuclear crystal was given by Dawber and Elliott ([43]). Their calculations showed that the induced absorption coefficient associated with a mass defect can be written as

$$\alpha_1(v) = \frac{\pi D}{3\sqrt{\epsilon}\,c}\left(\frac{\epsilon + 2}{3}\right)^2 [m_L^2(v)\delta(v) + m_B^2(v)g(v)] \tag{37}$$

where $D$ is the number of impurities per unit volume, $\epsilon$ is the frequency-dependent dielectric constant of the host crystal, $c$ the velocity of light, $[(\epsilon + 2)/3]^2$ is the local field correction factor to account for the presence of the medium, $m_L(v)$ is the dipole moment associated with the local mode vibration, $\delta(v)$ is the shape function to account for the anharmonicity or randomness in the lattice, $m_B(v)$ is the dipole moment associated with the band modes and $g(v)$ is the density of modes per unit frequency range per unit volume. They also obtained a sum rule which states that

$$\int_0^\infty \alpha_1(v)dv = \frac{\pi e^2}{\sqrt{\epsilon}\,c^2 M'}\left(\frac{\epsilon + 2}{3}\right)^2 Df\,[\text{cm}^{-2}] \tag{38}$$

where $\epsilon$ is the dielectric constant at the resonant frequency, $M'$ is the effective impurity mass, and $f$ is the oscillator strength. (For an allowed electric dipole transition $f = 1$.)

A number of peaks can occur in $\alpha_I(v)$. From the first term in equation (37) the absorption coefficient has a maximum at the local mode frequency above the band of vibrations associated with the perfect lattice. From the second term in equation (37) resonant modes can occur in the band mode region associated with peaking of the functions $m_B(v)$ and $g(v)$ at frequencies where local modes would like to occur but are lifetime-broadened by the finite density of modes. Also, additional structure can appear in the absorption coefficient where maxima occur in the density of normal lattice modes. The experimental observation of localized impurity modes by Schäffer has encouraged experimentalists to search for the corresponding localized gap modes and resonant modes.

To date only the hydride ion has been found to produce a localized lattice mode above the phonon spectrum. However, a number of gap modes have been observed and we intend to discuss the experimental investigations

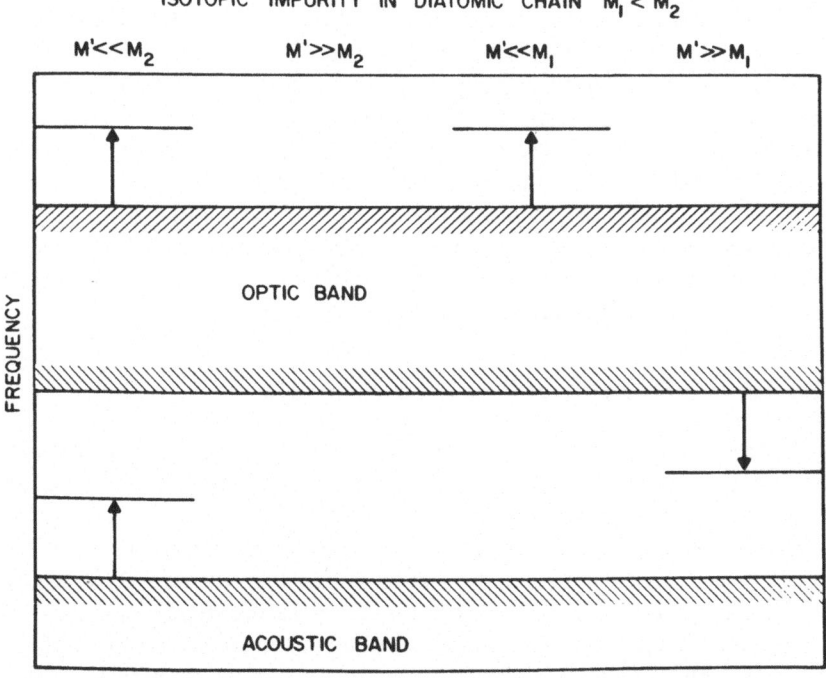

Fig. 12—Isotopic impurity in a diatomic linear chain. The light mass is labeled $M_1$ and the heavy mass $M_2$. Four cases are shown: (1) a light impurity mass replaces the heavy mass of the host lattice ($M' \ll M_2$), (2) a heavy impurity mass replaces the heavy mass of the host ($M' \gg M_2$), (3) a light impurity mass replaces the light host atom ($M' \ll M_1$) and (4) a heavy impurity mass replaces the light mass of the host ($M' \gg M_1$).

of these impurity modes. To introduce these localized modes, we would like to consider a problem solved by Mazur *et al.* ([44]): A simple diatomic linear chain with nearest neighbor force constants and an impurity with mass $M'$ which replaces either the light mass $M_1$ or the heavy mass $M_2$ of the diatomic chain. The nearest neighbor force constants are not altered. The solutions for the impurity problem are shown pictorially in Fig. 12. For $M'$ replacing $M_2$, where $M' \ll M_2$, two impurity modes are generated, a local mode above the top of the optic spectrum and a gap mode between the acoustic and optic branches. Both modes have the same symmetry. In the second case $M' \gg M_2$ and neither local modes or gap modes are predicted. The different behavior for these two cases can readily be understood. At the zone edge of the acoustic branch the heavy atoms $M_2$ vibrate $\pi$ out of phase with each other while the light ions $M_1$ are at rest. The converse is true at the zone edge of the optic branch, the light atoms $M_1$ vibrate $\pi$ out of phase and the heavy atoms $M_2$ are at rest.

Let us consider the first case where $M' \ll M_2$. We start by letting $M' = M_2$ and then increase the mass perturbation by reducing $M'$. Initially the impurity is vibrating out of phase with the other nearest $M_2$ atoms. As $M'$ is decreased the number of atoms participating in this mode decreases and the mode moves up into the gap. A mode moves out of the optic band because of the mutual repulsion between levels of the same symmetry. For the second case $M' \gg M_2$ and the mode at the zone edge acoustic band is depressed. Because the mode is moving away from the optic band no local mode from the optic band is expected. Two more cases are possible if the impurity $M'$ replaces the lighter atom $M_1$ in the diatomic chain and these are also shown in Fig. 12. These modes can be interpreted in a similar manner as above.

### b. Far Infrared Properties of Pure Potassium Iodide

At frequencies below the infrared active transverse optic mode, alkali halide crystals at room temperature are essentially opaque to far infrared radiation. However, as Rubens and Hertz ([45]) first showed in 1912, the crystals become more transparent as the temperature is decreased and the absorption coefficient in this frequency region varies linearly with temperature. The physical process which describes this temperature dependence has been identified ([45,47]) with the anharmonic coupling of the transverse optic phonon $v_T$, $k = 0$, with two other lattice phonons $v_1$, $\mathbf{k}_1$, and $v_2$, $\mathbf{k}_2$ with $\mathbf{k}_1 = \pm \mathbf{k}_2$. The coupling originates from the third-order and higher-order terms which describe the lattice potential energy. The absorption of a photon $v$ can either create two phonons (a summation band with $v = v_1 + v_2$) or create one phonon and destory a lower-energy phonon (a difference band with $v = v_1 - v_2$). In the far infrared frequency region the difference band is important and the absorption coefficient varies as the difference in the population of the two phonon modes, i.e., $n_2 - n_1$.

For the temperature dependence of the absorption coefficient one obtains

$$D(T) = n_2 - n_1 = \frac{1}{e^{(\hbar\nu_2/kT)} - 1} - \frac{1}{e^{(\hbar\nu_1/kT)} - 1} \tag{39}$$

which at high temperature reduces to $D(T) \propto T$. At very low temperatures the difference band is quenched. The frequency dependence of the absorption is expected to contain structure because the absorption band will mainly involve phonons from regions of $k$-space with a high density of phonons.

Of the several alkali halide crystals whose frequency spectra have been calculated ([48]) or measured by inelastic neutron scattering techniques ([49,50]) only five, LiCl, NaBr, NaI, KBr, and KI, have a gap in the spectrum between the optic and acoustic branches. Both NaI and NaBr are hygroscopic and difficult to purify.

Most experimental investigations of gap modes have been made in KI because it has the largest gap of the three remaining crystals. Recently the normal-mode spectrum of KI has been obtained by Dolling *et al.* by inelastic neutron scattering ([50]). Their measured dispersion curves indicate that the separation of the branches at the same zone boundary are

|                      | [100]                  | [111]                  |
|----------------------|------------------------|------------------------|
| LA $\longrightarrow$ LO | 60.7 cm$^{-1}$      | 62.2 cm$^{-1}$      |
| LA $\longrightarrow$ TO | 56.2 cm$^{-1}$      | 25.7 cm$^{-1}$      |
| TA $\longrightarrow$ LO | 81.3 cm$^{-1}$      | 76.2 cm$^{-1}$      |
| TA $\longrightarrow$ TO | 77.0 cm$^{-1}$      | 39.3 cm$^{-1}$      |

With a model which takes into account the polarizability of both ions, axially symmetric short-range forces between first and second nearest neighbors, and a variable ionic charge, they have fit the dispersion curve in the principal directions and calculated the frequency distribution function $g(\nu)$ for KI. Their results are shown in Fig. 13. A well-defined energy gap extends from 69.7 to 95.6 cm$^{-1}$. The two peaks in the density of acoustic modes occur at 62.4 and 53.7 cm$^{-1}$, respectively.

The temperature dependence of the absorption coefficient in pure KI in the frequency region below the fundamental absorption is shown in Fig. 14. A broad maximum in the absorption coefficient at 76 cm$^{-1}$ is clearly visible at 44°K and decreases as the temperature is decreased. With centimeter-thick crystals cooled to 2°K, transmission measurements can be readily obtained up to 85 % of the transverse optic mode frequency. The temperature dependence of the absorption at 76 cm$^{-1}$ can be readily fit with

equation (39). The experimental temperature dependence can be obtained
either by considering the TA $\rightarrow$ LO transition for the [111] direction or the
TA $\rightarrow$ TO transition in the [100] direction. The latter fit is shown in Fig.
14.

In a recent experimental paper Stolen and Dransfeld ([51]) have studied
the far infrared lattice absorption (at lower frequencies than shown in Fig.
14) as a function of temperature for a number of alkali halide crystals. Comparing their experimental results with phonon dispersion curves they found
that two-phonon processes involving one longitudinal and one transverse
branch did not appear. Their results indicate that the KI two-phonon transition shown in Fig. 14 should be identified with the TA $\rightarrow$ TO transition
in the [100] direction.

On the other hand, group theoretical selection rules have been calculated
by Burstein et al.([52]) for the two-phonon infrared absorption in the rocksalt
structure. They find that no infrared-active two-phonon combinations are
allowed at the [100] boundary whereas transition at the [111] boundary are

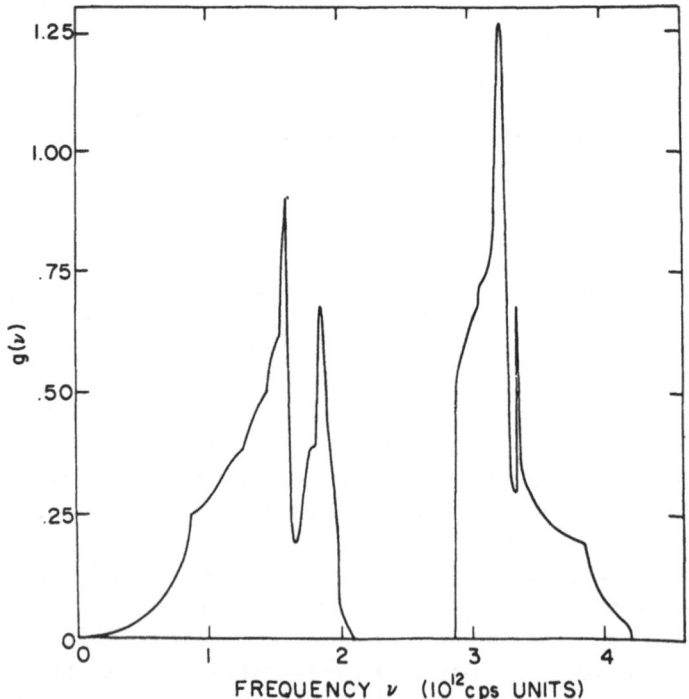

Fig. 13—The frequency distribution for the density of modes in KI. A
well-defined energy gap extends from 69.7 to 95.6 cm⁻¹. [From G.
Dolling et al. Phys. Rev. (1966)].

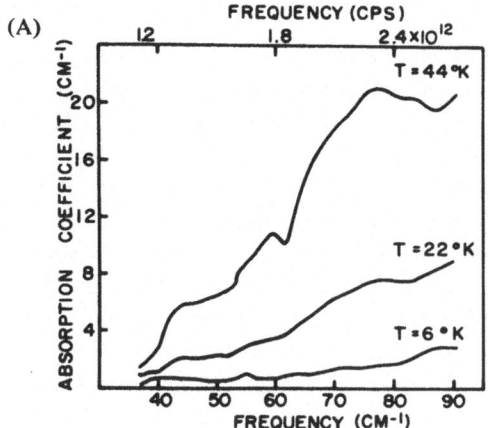

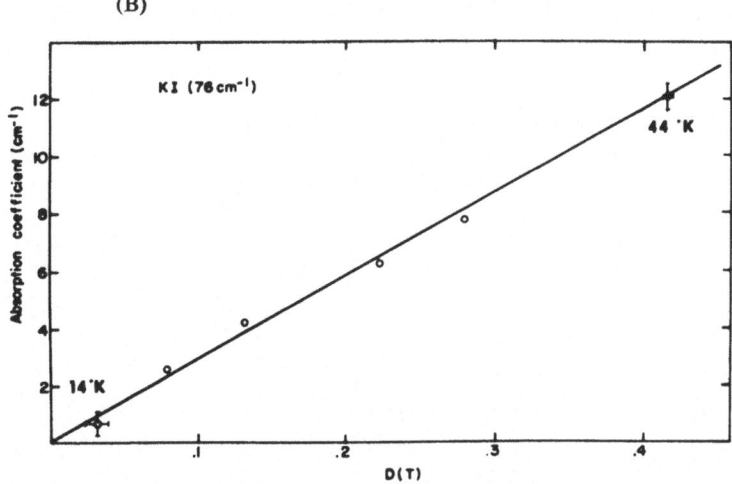

Fig. 14—(A) The temperature dependence of the absorption coefficient in pure KI. The broad maximum in the absorption coefficient occurs at 76 cm⁻¹. (B) The temperature dependence of the absorption coefficient at 76 cm⁻¹ *vs.* $D(T)$. The absorption coefficient varies as the difference in the population of two phonon modes $n_2 - n_1 = D(T)$.

not so restricted. From this point of view one should identify the absorption with the TA $\longrightarrow$ LO transition in the [111] direction for KI. Unfortunately, our experimental results satisfy both proposals equally well. Although a definite low-temperature window does occur in the far infrared region for alkali halide crystals, the details of the intrinsic processes are not completely resolved. Let us now consider spectra which are activated by impurities.

### c. Experimental Studies of Gap Modes from Monatomic Impurities

In this section we shall describe a variety of transmission studies on KI doped with substitutional halogen and metal ion impurities. The experimental results, which are summarized in Table I, will then be used to test a specific lattice model.

Generally the doped crystals are grown in an argon atmosphere from the melt by the Kyropoulos pulling technique. The hydride ion impurity is an exception and we shall consider the experiments with this impurity first. The $U$-center was formed by first additively coloring a KI single crystal by the Van Doorn technique ([53]) and then heating the crystal in a hydrogen atmosphere. The final crystals were colorless. Measurement of the ultraviolet $U$-band in thin platelets gave some indication of the $U$-center concentration (about $10^{18}$ per cm³). The far infrared absorption spectrum measured with a grating monochromator on centimeter-thick crystals cooled to 2°K is shown in Fig. 15A. No sharp absorption lines were observed in the KI gap region and the results are summarized in Table I. In order to identify the band cen-

## TABLE I

### Impurity-Induced Absorption in Potassium Iodide: Gap Modes

| Impurity in KI | Gap mode frequency, cm$^{-1}$ | Band mode frequency, cm$^{-1}$ | Oscillator strength, $f^a$ | Mass ratio $M'/M_{\pm}$ | Force constant $K'/K$ | Ionic radius $r'/r$ |
|---|---|---|---|---|---|---|
| H$^-$ | — | 62.0 | — | 0.078 | — | 0.953 |
| D$^-$ | — | 59.4 | — | 0.156 | — | 0.953 |
| Cl$^-$ | 77.0 | 61 | ln.0.017 tot.0.031 | 0.280 | 0.38 | 0.838 |
| Br$^-$ | 73.8 | 56 60 67 | ln.0.004 tot.0.017 | 0.630 | 0.81 | 0.903 |
| Na$^+$ | — | 53 63.5 | tot.0.015 | 0.588 | — | 0.715 |
| Cs$^+$ | 83.5 | 60 | tot.0.4 | 3.40 | 1.5 | 1.27 |
| Tl$^+$ | — | 55 64.5 | tot.1.0 | 5.20 | — | 1.08 |
| CN$^-$ | 81.2 | | tot.0.070 | 0.204 | 0.312 | 0.547 0.884 |
| NO$_2^-$ | 71.2 79.5 | 63 55 | ln.(0.015) ln.(0.026) tot.0.11 | 0.362 | 0.40 0.56 | 0.585 0.682 |

$^a$ The oscillator strength $f$ is given by Equation (38): ln is the strength of the line absorption. tot. is the strength of the line plus band absorption.

tered at 62 cm⁻¹ with the *U*-center, photochemical conversion was used. By ultraviolet irradiation in the optical *U*-band at room temperature the *U*-centers are converted to *F*-centers plus interstitial hydrogen atoms. The far infrared absorption at 62 cm⁻¹ was decreased by a factor of two. Transmission measurements carried out after the specimen had been heated to 200°C demonstrated that the absorption again increased and the process had been reversed. The absorption coefficient for different stages of the sequence is shown in Fig. 15B. Unfortunately, both *U*-centers and OH⁻-centers behave in this characteristic manner $(^{53,54})$. But additional measurements with a KI: KOH crystal did not show a distinct absorption maximum at 62 cm⁻¹; hence, the impurity induced band is associated with the *U*-center. Another

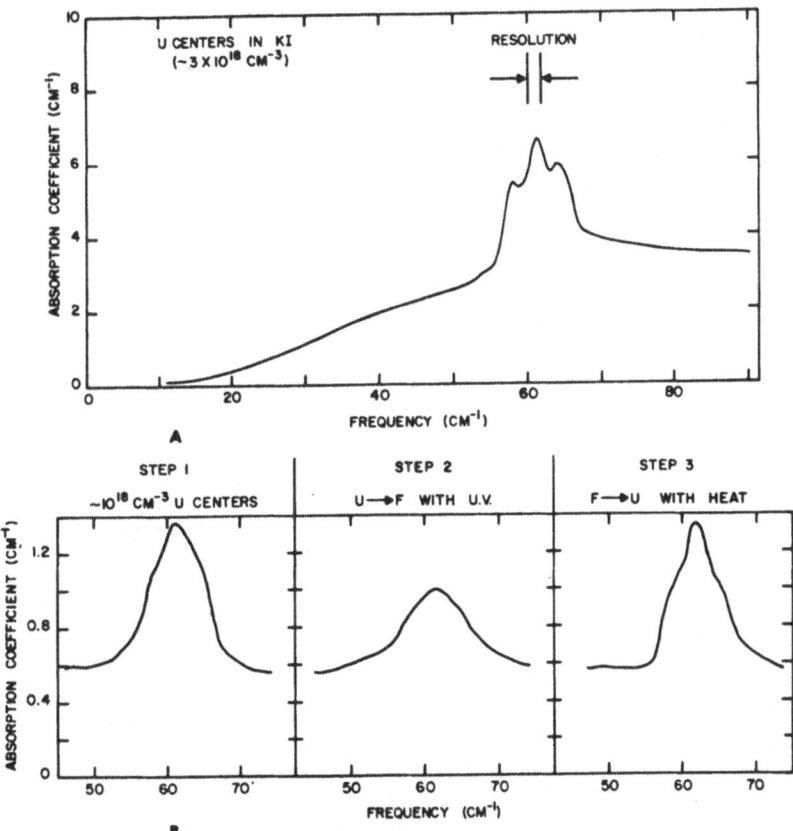

Fig. 15—(A) Impurity-induced absorption from U centers in KI. By comparing the doped crystal with the pure crystal, the intrinsic lattice absorption has been divided out. At 2°K the intrinsic absorption from the infrared active transverse branch becomes appreciable at about 90 cm⁻¹. (B) Photochemical conversion of the U center band.

KI crystal was doped with deuterium ions by the technique described above and a broad peak was found at 59.4 cm⁻¹. No structure was observed for this weak band.

Both sodium ions and thallium ions produce sharp and broad bands in the acoustic phonon spectrum but no sharp lines in the gap region. The impurity induced absorption spectrum for sodium centers is shown in Fig. 16 and the important frequencies are given in Table I. Four concentrations of thallium ions have been studied. The same spectrum was observed for all

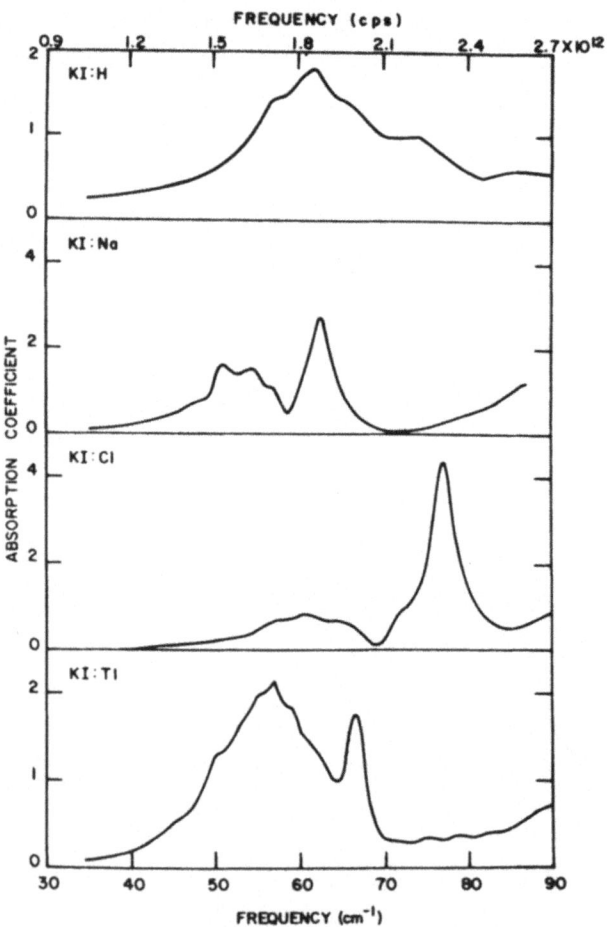

Fig. 16—Impurity-induced absorption for some mona-tomic impurities in KI at 2°K. The impurities are H⁻, Na⁺, Cl⁻, and Ti⁺. Only the Cl⁻ impurity produces a gap mode at 77 cm⁻¹. The absorption frequencies are given in Table I.

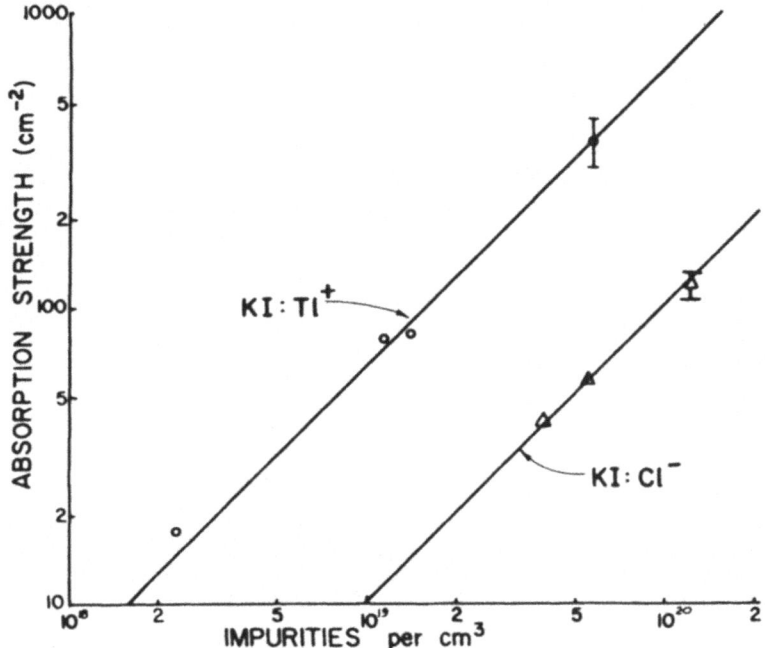

Fig. 17—Impurity-induced absorption strength in KI at 2°K *vs.* impurity concentration for two impurities in KI. A linear dependence is observed for both thallium and chlorine impurities.

crystals and is shown in Fig. 16. The center frequencies for the sharp line and broad band are given in Table I. We have determined the band strength by measuring the area under the absorption curves. For at least an order of magnitude in concentration, the total strength varies linearly with impurity concentration, as shown in Fig. 17.

The only metal ion impurity to date which induces an infrared-active gap mode in KI is $Cs^+$. A sharp absorption line is observed at 83.5 cm$^{-1}$ with an additional broad absorption band occurring near the top of the acoustic spectrum. The spectrum is not shown but the frequencies are recorded in Table I.

The most complete experimental investigation of gap modes has been on the system KI: Cl$^-$([55]). The observed spectrum shown in Fig. 16 consists of a narrow line at 77 cm$^{-1}$ and a band composed of at least three broad lines. Below 30 cm$^{-1}$, the impurity-induced absorption is very small. The total integrated absorption strength has been measured for a number of concentrations and varies linearly with the chemically determined impurity concentration as shown in Fig. 17. The absorption spectra are shown in Fig. 18 for the different impurity concentrations. Higher-resolution studies $v/\Delta v \sim 80$

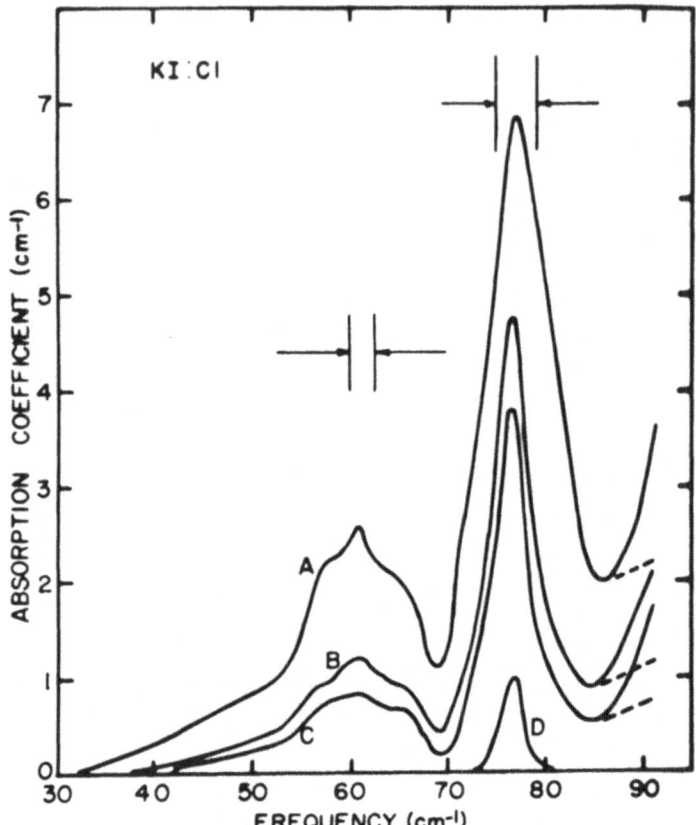

Fig. 18—Impurity-induced absorption strength in KI:KCl at 2°K. The four chlorine concentrations are (A) $1.2 \times 10^{20}$, (B) $5.5 \times 10^{19}$, (C) $3.9 \times 10^{19}$, and (D) $2.8 \times 10^{18}$ Cl ions per cm³. The resolution of the grating monochromator is given by the frequency intervals between the arrows.

with samples C and D in Fig. 18 indicate that another weak absorption occurs at about 73 cm⁻¹. (As we shall see shortly this is probably from an unwanted bromide impurity.) No sharp lines have been found within the broad-band absorption.

The double-pass monochromator previously described has been used to measure the impurity-induced absorption of Br⁻ ions. The absorption coefficient as a function of frequency is given in Fig. 19. The resolution at 74 cm⁻¹ is about $\nu/\Delta\nu \sim 100$. By varying the bromide concentration, the strong absorption at 73.8 cm⁻¹ has been identified with the Br⁻ center as have the three prominent peaks in the acoustic spectrum. The absorption centered at 77 cm$^{-1}$ arises from an appreciable chlorine concentration both in the doping material (KBr) as well as in the host. From a chemical analysis of the sample measured in Fig. 19, a chlorine concentration of 60 ppm was obtained.

Because gap modes should be associated with each chemical isotope of the impurity, neither the spectrum of the Br$^-$-center or the spectrum of the Cl$^-$-center can be regarded as complete. To investigate the isotope shift we have measured the lowest-chlorine-doped KI sample (D) (Fig. 18) with the lamellar interferometer at a resolution of $\nu/\Delta\nu \sim 380$. At this resolution the Cl$^-$ gap mode no longer appears to be a single line but is resolved into two components centered at 76.79 and 77.10 cm$^{-1}$, respectively. The spectrum for the doped crystal D and two pure crystals is shown in Fig. 20. The chlorine impurity is present in all three samples. The lowest absorption curve corresponds to an impurity concentration of about $5 \times 10^{16}$ Cl$^-$ ions/cm$^3$. An impurity-impurity interaction does not appear to be important in this concentration range because the center frequency of the lines does not shift

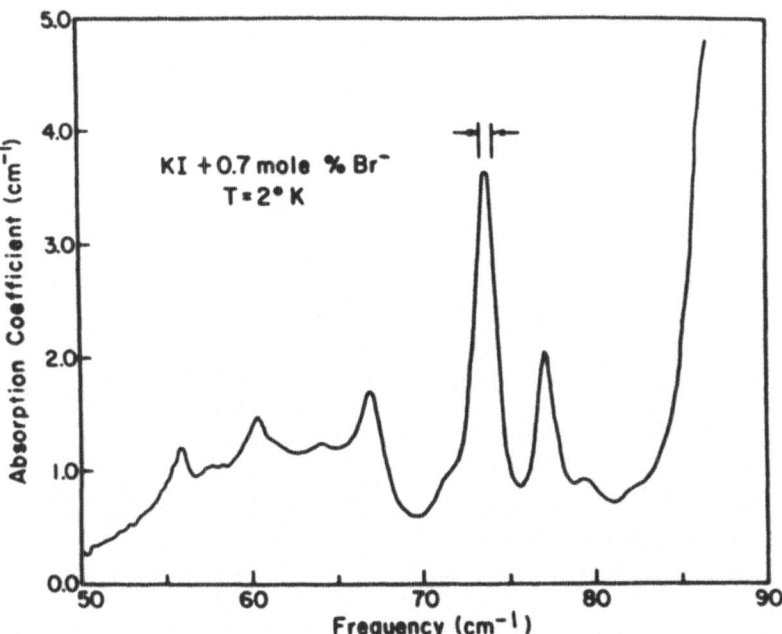

Fig. 19—Impurity-induced absorption coefficient for KI:KBr at 2°K. The strong absorption at 73.8 cm$^{-1}$ has been identified with the bromine impurity and the sharp line at 77 cm$^{-1}$ is from unwanted chlorine impurities in both the dopant and host starting material. The grating monochromator was used in double pass. The resolution is given by the frequency interval between the arrows.

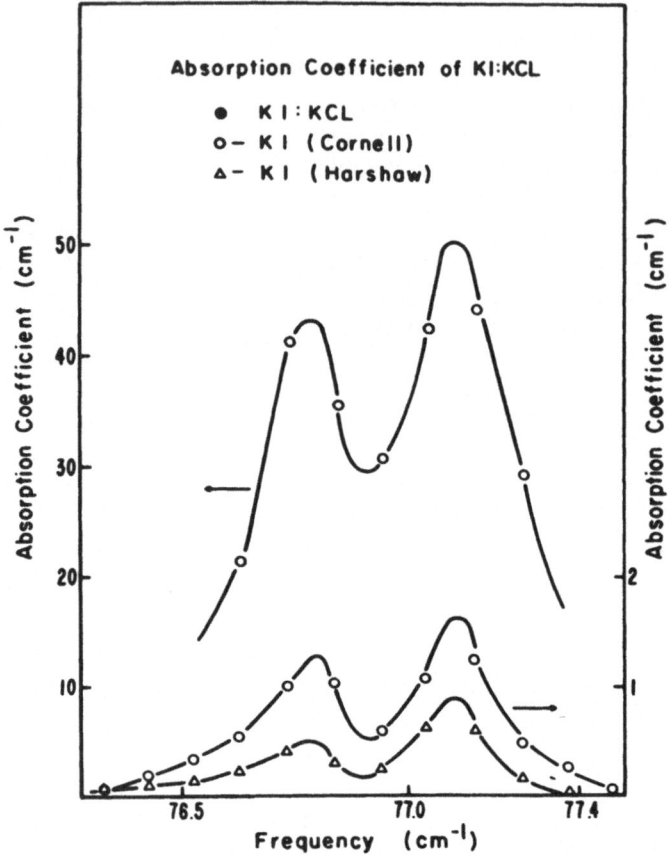

Fig. 20—Isotope shift of Cl⁻(35) and Cl⁻(37) impurities in KI. The impurity-induced absorption coefficient for a Cl⁻ concentration of $2.8 \times 10^{18}$ impurities/cm³ is displayed on the left. The absorption coefficient for two undoped KI crystals is shown on the right.

although the concentration changes by a factor of 50. The width of each lines is still dominated by the instrumental resolution over the entire concentration range and the true width of the gap mode must be less than 0.2 cm⁻¹.

### d. Discussion of Experimental Results

The experimentally determined gap mode frequencies have been summarized in Table I. In addition, the table includes the oscillator strength $f$ as calculated from equation (38), the ionic diameter ratio (impurity ion diameter/host ion diameter) as determined by Pauling[56], the fractional mass change, and the fractional force constant change for a particular model, which will be considered shortly.

From Table I we see that band modes are observed for all impurities whereas gap modes appear for one half of the crystal—dopant combinations. At least some of the broad absorption band in the acoustic spectrum must occur because of the large density of states in this frequency region. Additional absorptions could arise from resonant modes in this frequency region as has been discussed by Maradudin et al [55]. A comparison of the gap mode frequency with the fractional mass change in Table I demonstrates that the experimental results cannot be interpreted with a simple mass defect model. It is not yet clear just how detailed a model is required to describe the experimental results.

Recently Mitani and Takeno (MT)[57] have calculated the frequencies of the infrared-active modes for a system consisting of a point defect with mass and nearest neighbor force constants different from those of the host diatomic lattice of the NaCl type. The long-range interaction forces which introduce coupling between the components of the motion as well as polarizability and lattice distortion are not treated. Taking the interactions between nearest neighbors only and setting central and noncentral force constants equal to each other, they find a solution for the infrared-active impurity mode which is particularly easy to handle. With this model, a one-parameter fit to the experimentally determined frequencies gives the fractional change in coupling constant due to the impurity replacing the host ion. It is the testing of this model which we now consider.

From the paper of Mitani and Takeno ([57]), the eigenfrequency of the $S$-like $A_{2u}$ mode is determined by the solution of the following expression:

$$D(\omega^2) = 1 + \mu + \mu(1 + \lambda)\left(\frac{\omega}{\omega_-}\right)^2$$
$$+ \left[\lambda(1 + \mu) - \mu(1 + \lambda)\left(\frac{\omega}{\omega_-}\right)^2\right]\left(\frac{\omega}{\omega_-}\right)^2 g_\pm(\omega) = 0 \qquad (40)$$

where use has been made of some identity relations satisfied by the Green's functions for this lattice ([58]). The notation in equation (40) is as follows:

$$\mu = \frac{K' - K}{K}, \qquad \lambda = \frac{M' - M}{M}$$

with $K$ and $K'$, respectively, the unperturbed and perturbed nearest neighbor force constants, and $M$ and $M'$ the host ion mass an substitutional impurity mass; $\omega_-$ is the frequency at the top of the acoustic band and $\omega_+$ is the frequency at the bottom of the optic band with $g_\pm(\omega)$ defined below. The $\pm$ sign signifies that the impurity ion replaces either the light (+) ion or the heavy (−) ion.

$$g_+(\omega) = \frac{\omega_+^2}{N} \sum_{k_1 k_2 k_3} \left\{ \omega^2 - \frac{\omega_+^2}{3}[3 - (\cos k_1 + \cos k_2 + \cos k_3)\epsilon] \right\}^{-1} \tag{41}$$

and

$$g_-(\omega) = \frac{\omega_-^2}{N} \sum_{k_1 k_2 k_2} \left\{ \omega^2 - \frac{\omega_-^2}{3}[3 - (\cos k_1 + \cos k_2 + \cos k_3)\epsilon^{-1}] \right\}^{-1} \tag{42}$$

with

$$\epsilon = \left[ \frac{(\omega/\omega_+)^2 - 1}{(\omega/\omega_-)^2 - 1} \right]^{1/2} \tag{43}$$

where $N$ is the number of atoms in the lattice, $k = (k_1 k_2 k_3)$ is the wavevector (the lattice constant is taken to be unity), and the sum extends over the first Brillouin zone. For gap modes the function $g_\pm$ has been tabulated in (MT). The model can be readily forced to fit the experimentally determined KI gap by setting $\omega_- = 69.6$ and $\omega_+ = 95.6$ cm$^{-1}$. The fractional change in the nearest neighbor coupling constant as calculated from equation (40) for the gap mode of Cs$^+$, Cl$^-$, and Br$^-$ is presented in the last column of Table I. A comparison of the ionic diameter ratio (column 7 Table 1) with the force constant change gives consistent results in that a large mismatch in diameters corresponds to a large change in coupling constant. Also if the impurity is larger than the host, the coupling constant is larger than the host value while if the impurity diameter is less than the host diameter the coupling constant is less than the host value.

Another test of this theoretical model can be obtained from the isotope shift data for the Cl$^-$ center. With a value of $K'/K = 0.38$ calculated from the gap mode at 77cm$^{-1}$, the shift in frequency associated with the mass change is calculated to be 1.74 cm$^{-1}$ compared to the measured value of 0.31 cm$^{-1}$. To correctly predict the experimental isotope shift, $K'/K$ must be changed to $K'/K = 0.5$ which then shifts the theoretical gap mode position from 77 to 81 cm$^{-1}$. It appears that the Mitani—Takeno model fails to quantitatively satisfy the isotope shift but it does qualitatively describe many of the experimental results on gap modes.

### e. Absorption Associated with Molecular Impurities

Because of the cubic symmetry ($O_h$) at the monatomic impurity site, infrared-active gap modes are threefold degenerate. A molecular impurity, on the other hand, usually is described by a lower symmetry point group and reduces the local symmetry of the lattice. This more complex defect must be represented by not only a mass change but also an anisotropic coupling to the surrounding lattice. The reduced symmetry of the local lattice then influences the degree of degeneracy of the infrared-active gap mode.

Two molecular impurities have been studied in KI. They are KI: KCN ([59]) and KI: KNO$_2$ ([59-61]). In Fig. 21 low-resolution impurity-induced spectra for the CN$^-(C_{\infty v})$ center and the NO$_2^-(C_{2v})$ center are compared with the Cl$^-$ spectrum. Absorption lines are found in the gap region and broad bands are located at the edge of the acoustic spectrum. For the CN$^-$ center only one infrared-active mode has been observed in the gap although two are to be expected from the impurity symmetry. The center frequency is given in Table I. For the NO$_2^-$-center two modes are found in the gap region and two broad bands in the acoustic spectrum as shown in Fig. 21. The center frequencies are given in Table I. The NO$_2^-$-center has been studied more exten-

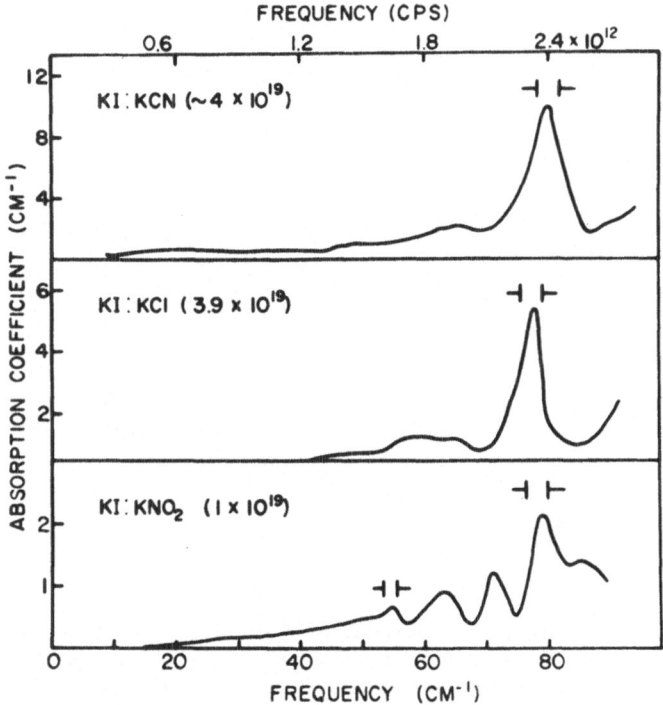

Fig. 21—Impurity-induced absorption coefficient for two molecular impurities in KI at 2°K. The absorption spectra for KI: KCN (4 × 10$^{19}$ impurities/cm$^3$) and for KI: KNO$_2$(1 × 10$^{19}$ impurities per cm$^3$) are compared with the absorption spectrum for a monatomic impurity KI: KCl (4 × 10$^{19}$ impurities/cm$^3$). The resolution for the single pass grating monochromator is shown.

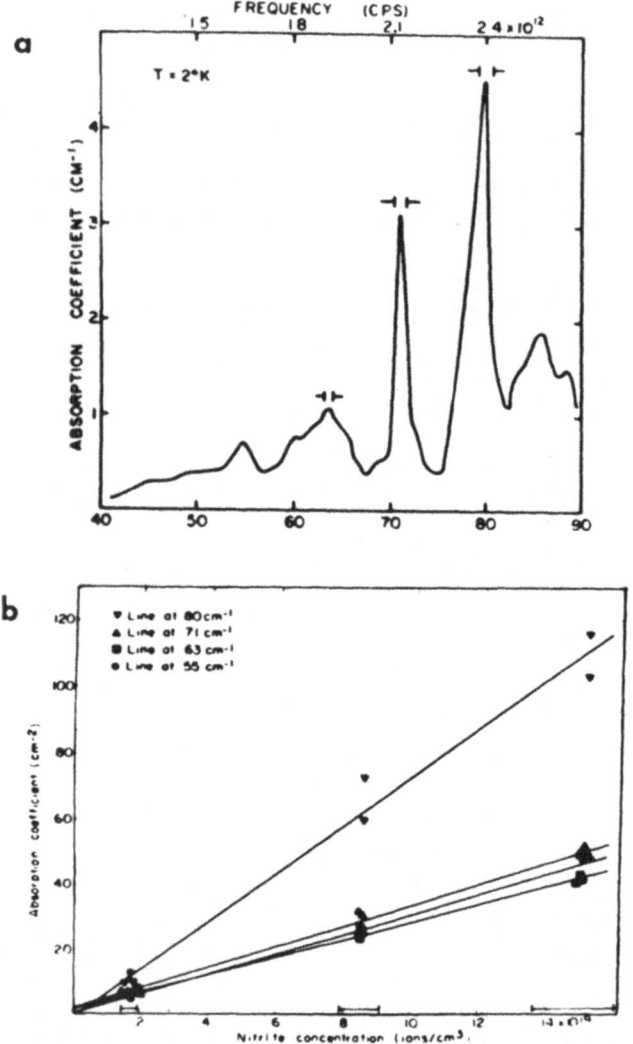

Fig. 22—The far infrared absorption spectrum of KI: KNO₂ (A) A medium resolution spectrum of KI: $KNO_2(1.5 \times 10^{19}$ $NO_2^-$ ions/cm³, $1 \times 10^{18}$ $NO_3^-$ ions/cm³). (B) The linear dependence of the absorption strengths for the absorption peaks at 80, 71, 63, and 55 cm⁻¹ with $NO_2^-$ concentration. (C) Top trace: high-resolution spectrum displaying the two nitrite gap modes at 79.5 nad 71.2 cm⁻¹. Bottom trace: high-resolution spectrum of an air-grown KI: KNO₃ crystal which contains $5 \times 10^{17}$ $NO_3^-$ ions and $7 \times 10^{17}$ $NO_2^-$ ions/cm³. None of the lines can be unambiguously identified with the $NO_3^-$ impurity. The $NO_2^-$ lines can be readily seen. (D) Temperature dependence of the impurity-induced nitrite absorption. The temperature dependent host lattice absorption has been divided out.

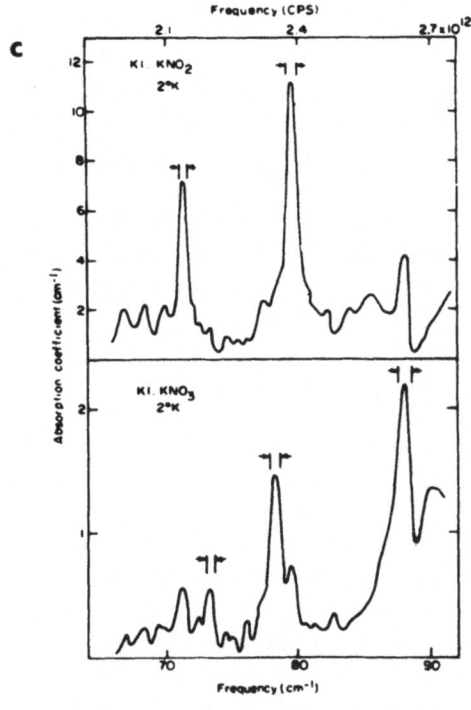

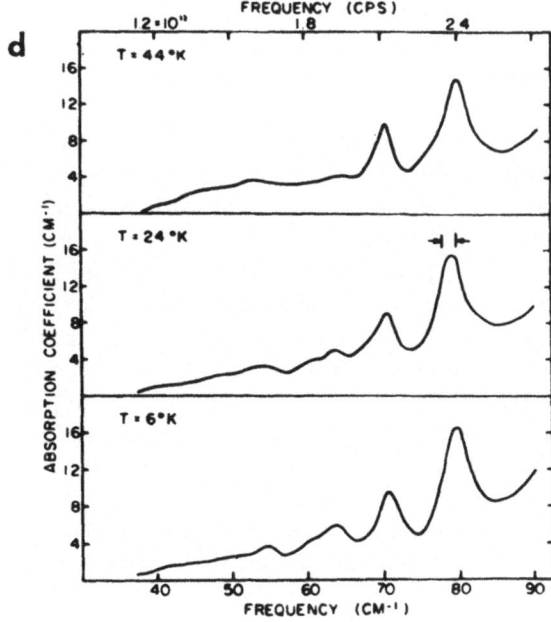

sively and we shall describe a number of experiments, the results of which are summarized in Fig. 22A, B, C, and D.

In Fig. 22A a higher-resolution spectrum of KI: $KNO_2$ crystals at 2°K is presented. The nitrite concentration is 1.5 $\times$ $10^{19}$ $NO_2^-$ ions $cm^3$ and the nitrate concentration is 1 $\times$ $10^{18}$ $NO_3^-$ ions/$cm^3$. To determine whether the four measured absorptions are due to the nitrite impurity and not nitrate, which is also present in all the crystals, the integrated absorption coefficient for each line has been plotted versus the concentration for the three KI: $KNO_2$ crystals studied. In Fig. 22B the absorption strength versus nitrite concentration is shown to vary linearly. In the upper half of Fig. 22C a higher-resolution measurement of Fig. 22A in the gap mode region is displayed. In the bottom of C the spectrum of an air-grown KI:$KNO_3$ crystal which contained about 5 $\times$ $10^{17}$ $NO_3^-$ ions and about 7 $\times$ $10^{17}$ $NO_2^-$ ions/$cm^3$ is shown. The frequencies of the three prominent lines are given in Table I. The $NO_2^-$ gap modes are also visible. An unambigous identification of the three lines with the $NO_3^-$-center is not possible because the absorptions are weak. For example, the two higher-frequency lines correspond very closely to the frequencies of the $OH^-$ ion absorption lines measured by Renk ([62]). Also chlorine impurities are probably present and this would introduce an uncertainty about any line found near 77 $cm^{-1}$. Finally, in Fig. 22D the temperature dependence of the impurity-induced nitrite absorption is shown at low resolution. The broad bands wash out as the temperature is increased while the gap modes are relatively temperature-independent.

A number of attempts have been made to explain the dynamics of the nitrite ion in KI. Since all present models are only partially successful only a qualitative description which focuses mainly on the far infrared results is presented here.

Timusk and Staude ([63]) first reported phonon structure in the $NO_2^-$ electronic absorption at 400 m$\mu$ in alkali halide crystals at 4.2°K. For the $No_2^-$-center, sharp structure was obsered in absorption on the high-frequency side of the electronic (or electronic plus molecular vibration) transition. It was suggested that the lines which appeared at 63 and 70 $cm^{-1}$ from the no-phonon line might arise from band absorption and gap mode absorption. The low symmetry of the $NO_2^-$ ion ensures that *only* modes which transform according to the $A_1$ representation (both infrared-active and Raman-active) couple with the nondegenerate electronic transitions. A possible interpretation is to identify the two far infrared absorptions at 71.2 and 63 $cm^{-1}$ with the infrared-active modes transforming according to the $A_1$ representation. To do this we must assume the $C_{2v}$ symmetry of the impurity center lifts the threefold degeneracy of the infrared-active $T_{1u}$ mode and produces three infrared-active nondegenerate modes transforming as the $A_1$, $B_1$, and

$B_2$ representations of the $C_{2v}$ point group. From the dielectric constant measurements of Sack and Moriarty ([64]) and the near infrared work of Narayanamurti et al. ([65]) the $NO_2^-$ ion is definitely displaced from the normal ion equilibrium position in KI and also the $C_2$ axis is probably oriented along the equivalent $\langle 110 \rangle$ directions. The near infrared measurements dictate that the molecule is almost freely rotating about this twofold axis, but not about either of the other two principal axes of the molecule. The effective coupling of the impurity to the lattice will be very different in the plane perpendicular to the $C$-axis as compared to along the $C$-axis. To explain the far infrared results we try the following model: If, because of the free rotation, an average potential is used for vibrations within the plane, the $B_1$- and $B_2$-type modes would appear degenerate. In this approximation the $B_1 + B_2$ absorption would be twice as strong as the $A_1$ absorption. A comparison of the experimental oscillator strengths for the two nitrite gap modes in Table I gives a ratio of 1.7. Also, the total oscillator strength of these two lines is the same order of magnitude as found for the $Cl^-$ gap mode. In addition, using the (MT) model, the effective perturbed force constants can be estimated for each frequency and compared with the ratio of the ionic diameters ([66]). A geometric mean of the diameters along the two principal axes perpendicular to the twofold axis is used for one effective diameter in Table I. The mismatch variation is $50\%$ and a force constant variation of $30\%$ is required to fit both gap mode frequencies. The results are consistent with the $Cl^-$ results in Table I, although the details of the molecular motion of the $NO_2^-$-center in KI are no doubt complex.


## IIIB. IMPURITY-INDUCED ABSORPTION IN ALKALI HALIDE CRYSTALS: LOW-LYING RESONANT MODES

### a. Introduction

Impurity-induced one-phonon absorption which is associated with large peaks in the phonon density of modes has been observed by Angress et al. ([67]) and also by Balkanski and Nusimovici ([68]). At this time we shall not deal with this band-mode absorption but focus our attention on some additional sharp absorption lines which have been observed in a number of doped alkali halide crystals at still lower frequencies by using far infrared optical techniques. For these crystals substitutional defects can give rise to discrete absorptions which are electric dipole in nature. The observed center frequencies occur in a phonon frequency region where the host density of lattice modes is expected to increase monotonically with increasing frequency.

The absorption process has been identified with the excitation of a quasi-localized mode or resonant lattice mode. Brout and Visscher ([72]) first calculated the frequency and lifetime of such a mode for the mass defect. Later Visscher ([73]) considered the possibility of lattice resonant modes arising from both mass and coupling constant changes. With the aid of a computer he calculated some relevant properties in a simple cubic lattice with impurity mass and nearest neighbor force constants different from those of the host lattice. The calculations indicated that either a large impurity mass or a weak coupling constant could give rise to a lattice resonant mode. The optical properties of resonant modes associated with a mass defect perturbation for a homonuclear crystal was developed later by Dawber and Elliott ([43]).

The first observations of resonant modes were made by Pohl ([74]) for molecular defects and Walker and Pohl ([75]) for monatomic defects in alkali halide crystals. By means of low-temperature thermal conductivity measurements, they measured the phonon scattering associated with resonant modes (see R. O. Pohl, this summer school). This work plus the optical absorption measurements on $U$-centers ([35]) was the stimulus for the far infrared investigations. The description of these latter studies can best be divided into four sections: In the first two sections some optical measurements are described, line widths are determined, and the results are compared with the (MT) theory described earlier. The comparison suggests that for many dopant —lattice combinations the impurity is only weakly coupled to the lattice. In the third section the temperature-dependent properties of the optical absorption lines are described and a phenomenological description is presented. Finally in the fourth section strain measurements are described and interpreted.

## b. Silver-Activated Resonant Modes

The first optically active resonant modes were identified with the substitutional silver ion impurity in alkali halide crystals ([69,70]). The far infrared absorption spectra found for silver-activated potassium chloride and bromide were quite similar. The spectra are characterized by an absorption line superimposed on the low-frequency wing of a broad absorption band. For KCl: AgCl the low-frequency absorption line has been observed at 38.6 cm$^{-1}$ and is shown in Fig. 23. The frequency corresponds to about one quarter of the Debye temperature. Only one silver doping has been studied; however, a number of silver dopings have been measured in a KBr host crystal. In this case a strong low-frequency absorption is observed at 33.5 cm$^{-1}$. The integrated absorption strength varies linearly with silver concentrations. In addition to this line, broad absorption bands are observed at higher frequen-

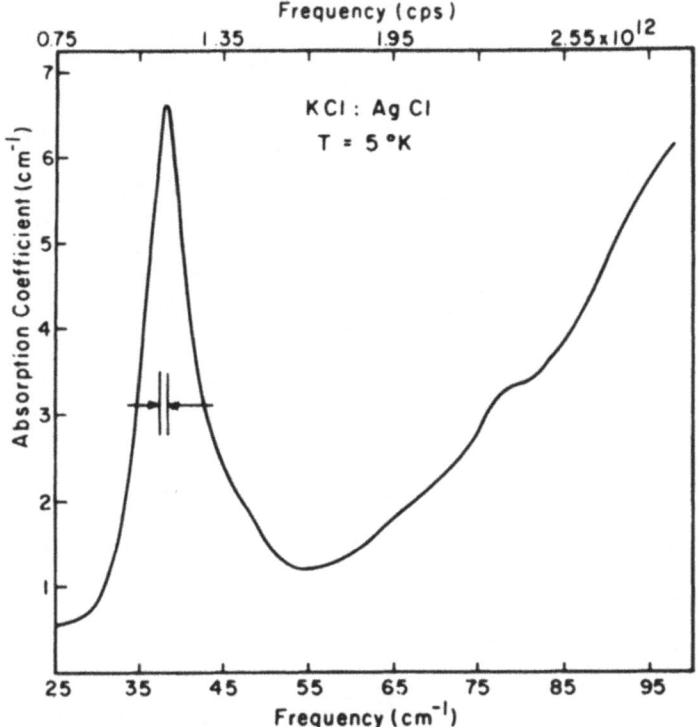

Fig. 23—Absorption coefficient for the silver-activated resonant mode
in KCl:AgCl. The instrumental resolution is indicated.

cies. The absorptions probably arise from the peaked density of states
although no direct one-to-one correlations with the dispersion curves, found
by Woods *et al.* ([76]), have been obtained.

Silver concentrations of $10^{18}$ ions/cm³ in potassium iodide yield a single
sharp absorption line at 17.4 cm⁻¹. The full width at one-half maximum
absorption is 1.3 cm⁻¹. Correcting for the effect of the monochromator
(single pass) slit width of 0.7 cm⁻¹, we estimate the true width to be approximately 0.6 cm⁻¹. In Fig. 24 the resonant mode in KI : AgI is shown for
a temperature of 6°K. This particular dopant—lattice combination has the
curious property of not being stable at room temperature. The strength of
the infrared absorption line slowly decreases and disappears in about a three-
month period. The conversion process can be increased by heating the crystal
to 140°C. The results from such a temperature cycling are also shown in
Fig. 24. Although the kinetics have not been studied in detail, one possible
interpretation of these results is that AgI is being formed in the crystal.

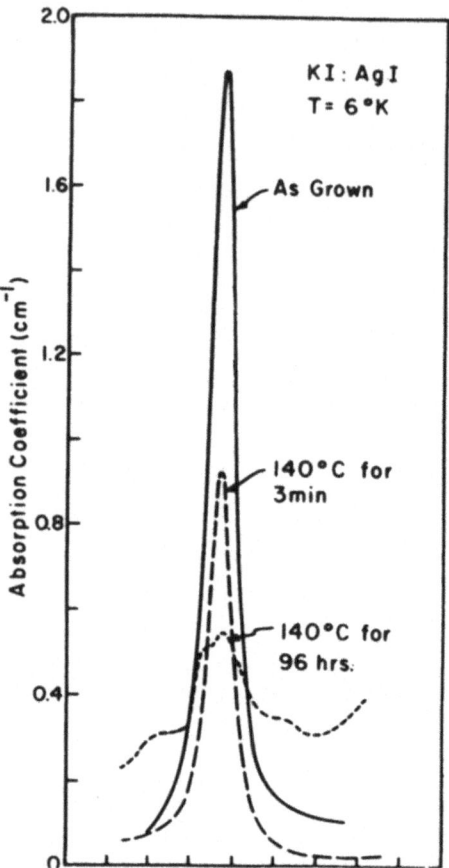

Fig. 24—Instability of the silver-activated resonant mode in KI:AgI. The transmission was measured with the crystal at 6°K between each cycle to 413°K.

The other alkali halide crystals doped with silver which have been measured appear to be stable and a summary of the resonant mode frequency for a number of crystals is shown in Fig. 25. The center frequencies are displayed on a log-frequency—log-lattice constant plot and are also given in Table II. Silver impurities in five different host lattices produce absorptions whose center frequencies can be fit by the following Ivey-type relation

$$va^3 = \text{constant} \tag{44}$$

The exponent $n = 3$ is not the same as has previously been observed for either the $F$-center or the infrared $U$-center. No physical significance has been found for this expression, and the good fit for most of the salts could either be accidental or imply that the impurity—lattice coupling is different for RbCl and KI. For the present, the fit to equation (44) is assumed to be accidental. A simple mass defect perturbation cannot account for the decrease of the resonant mode frequency with increasing lattice constant and a more complex model must be considered.

To determine the significance of these experimental results we again turn to the model of Mitani and Takeno (MT) ([57])—the specific model against which some of the properties of the gap modes were tested. With the appropriate approximations their formalism can be used to obtain resonant mode frequencies in terms of the coupling constant and mass perturbation parameters. To solve equation. (40), the secular equation which determines the vibration frequencies of the $A_{2_\mu}$ mode, the function $g_+(\omega^2, 000)$ must be evaluated. For host crystals which have a small gap between the optic and acoustic branches, that is $[(\omega_+ - \omega_-)/\omega_-] \ll |$ so that $\epsilon \approx 1$, $g_\pm(\omega^2,000)$ can be evaluated by taking the limit $N \to \infty$, replacing $\omega^2$ by $\omega^2 + i\delta$ where $\delta$ is posi-

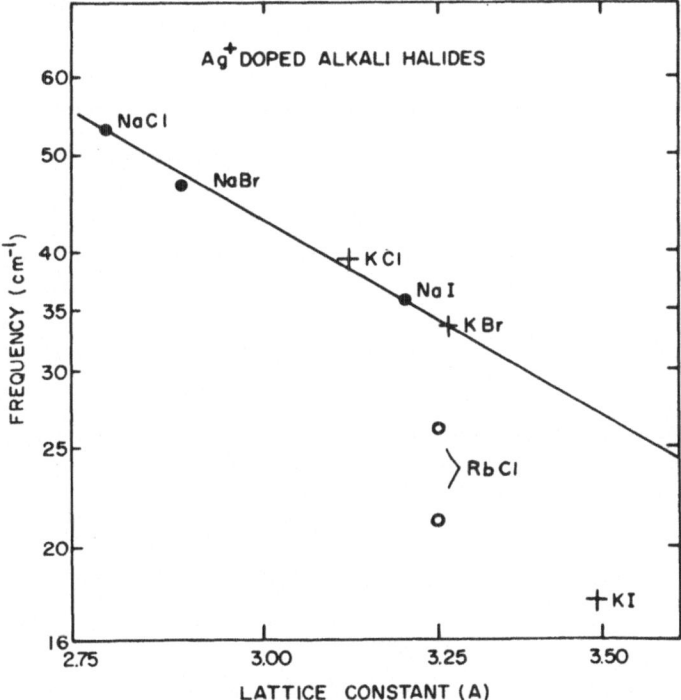

Fig. 25—The resonant mode frequency *vs.* the host lattice constant for an array of silver-doped alkali halide crystals. The solid line has a slope-3 on the log–log plot.

## TABLE II

### Impurity-Induced Absorption in Alkali Halide Crystals: Resonant Modes

| Lattice: impurity combination | Resonant mode frequency $\nu_0$, cm$^{-1}$ | Experiment, frequency line width $\omega_0/\Gamma$ | Theory, frequency line width $\omega_r/\Gamma_r$ | Mass ratio $M'/M_+$ | Force constant ratio $K'/K$ | Ionic radius ratio $r'/r$ |
|---|---|---|---|---|---|---|
| NaCl:Li$^+$ | 44.0 | 60 | 30 | 0.304 | — | 0.632 |
| NaCl:F$^-$ | 59.0 | | | | | |
| NaCl:Cu$^+$ | 23.7 | 39 | 100 | 2.76 | 0.04 | 1.0 |
| NaCl:Ag$^+$ | 52.5 | 5 | 4 | 4.7 | 0.6 | 1.33 |
| NaBr:Ag$^+$ | 48.0 | — | 2 | 4.7 | 1.1 | 1.33 |
| NaI:Ag$^+$ | 36.7 | — | 2 | 4.7 | 1.1 | 1.33 |
| KCl:Li$^+$ | 42 | — | 16 | 0.179 | — | 0.450 |
| KCl:Ag$^+$ | 38.8 | 7 | 9 | 2.77 | 0.30 | 0.947 |
| KBr:Li$^+$(6) | 17.9 | 18 | 100 | 0.154 | 0.0052 | 0.450 |
| KBr:Li$^+$(7) | 16.3 | 20 | 140 | 0.179 | 0.0051 | 0.450 |
| KBr:Ag$^+$ | 33.5 | 7.5 | 7 | 2.77 | 0.36 | 0.947 |
| KI:Ag$^+$ | 17.4 | 19 | 22 | 2.77 | 0.172 | 0.947 |
| RbCl:Ag$^+$ | 21.4 | — | 15 | 1.26 | — | 0.852 |
| | 26.4 | | | | | |
| | 36.1 | | | | | |

tive and infinitesimal, and using the Debye approximation with a frequency cutoff at $\omega_m^2 = 2\omega_\pm^2$. To the lowest order in $\omega/\omega_\pm$, Re $g_\pm(\omega^2 + i\delta,000)$ $= -\frac{3}{2}$ and Im $g_\pm(\omega^2 + i\epsilon,000) = -i(3\pi/4\sqrt{2})(\omega/\omega_\pm)$. Equation Re $D(\omega^2 + i\delta) = 0$ determines the resonant-mode frequency of the $A_{2\mu}$ mode (the resonant mode solution is designed by $\omega_0$) with a width given by

$$\frac{\Gamma}{2} = \left| \frac{\text{Im } D(\omega_0^2 + i\delta)}{[(d/d\omega) \text{ Re } D(\omega^2 + i\delta]_{\omega=\omega_r}} \right| \tag{45}$$

These facts are used with equation (40) to determine the resonant mode frequency $\omega_r$ of the $S$-like mode and its width $\Gamma_r$.

The final expressions are

$$\left(\frac{\omega_r}{\omega_-}\right)^2 = \frac{2(1 + \mu)}{3\lambda + \lambda\mu - 2\mu} \tag{46}$$

and

$$\frac{\Gamma}{\omega_r} = \frac{3\pi}{4\sqrt{2}} |\lambda| \left(\frac{\omega_r}{\omega_\pm}\right)\left(\frac{\omega_r}{\omega_-}\right)^2 \tag{47}$$

where $\lambda = M'/M_\pm - 1$, $\mu = k'/k - 1$, and $M_+$ and $M_-$ are the light and heavy host masses as previously defined. In the isotope approximation where $M' \gg M_\pm$ and $K' \sim K$ these relations reduce to

$$\omega_r^2 \approx 4\left(\frac{M_\pm}{M_-}\right)\left(\frac{K}{M'}\right) \tag{48}$$

and

$$\frac{\Gamma}{\omega_r} \approx \frac{\pi}{2\sqrt{2}}\left(\frac{\omega_r}{\omega_\pm}\right) \tag{49}$$

for the weak coupling constant approximation $K' \ll K$ and $M' \sim M_\pm$ and the two expressions are

$$\omega_r^2 \approx 6\left(\frac{M_\pm}{M_-}\right)\left(\frac{K'}{M'}\right) \tag{50}$$

and

$$\frac{\Gamma}{\omega_r} = \frac{3\pi}{4\sqrt{2}}|\lambda|\left(\frac{\omega_r}{\omega_\pm}\right)\left(\frac{\omega_r}{\omega_-}\right)^2$$

The resonant frequency has the simple Einstein oscillator form with an effective mass equal to $M'(M_-/M_\pm)$. A comparison of these two different limits indicates that weak coupling constants are more effective than heavy masses in decreasing the linewidth of the resonant mode. This conclusion is physically plausible. The frequency dependence of the linewidth in the isotope approximation merely reflects the density of phonon modes into which the resonant mode decays. The lower the resonant mode the lower the density of phonon modes into which the resonant mode decays. In the weak-coupling case the density of phonon modes again comes into play but the resonant-mode linewidth is also decreased owing to the additional uncoupling of the resonant mode from the phonon modes by small force constant $K'$. At low frequencies the two limiting cases have different properties. The heavy mass approximation gives rise to a resonant mode with an amplitude which is peaked around the impurity but at large distances looks more like a phonon, while the weak coupling approximation produces a resonant mode which has a peaked amplitude at the impurity site but at large distances looks more like a local mode.

A strigent test for the model is contained in equation (47). With an experimentally determined frequency for the resonant mode $\nu_0$ and also for the top of the acoustic spectrum $\nu_-$, the ratio of the line width to resonant-mode frequency $\Gamma_r/\omega_r$ can be calculated and compared with the experimentally measured ratio $\Gamma/\omega_0$. The comparison is given in columns four and three of Table II. Considering the simplicity of the model, the agreement between theory and experiment is surprisingly good for the silver ion impurity. The good agreement encourages one to go a step further and determine the coupling constant ratio $K'/K$, from equation (46), which fits the experimental data. The values obtained for this parameter are displayed in column 6 of Table II. For the sodium halide salts the force constants are approximately the same as for the host crystal while for the potassium salts the coupling between the silver ion and the host is definitely weaker. The ratio of the impurity ionic radius to the host ionic radius is given in column 7 of Table II. The ratio is seen to vary in a consistent manner with the coupling constant parameter.

Because both the optical experiments and the model demonstrate that the impurity mass plays only a secondary role in producing lattice resonant modes, far infrared measurements have been carried out for a number of impurity ions with small ionic radius and small mass. The most fruitful of these studies has been on lithium-doped alkali halide crystals.

### c. Lithium-Activated Resonant Modes

Far infrared absorption lines have been observed to occur far down in the acoustic continuum for the lithium impurity in NaCl, KCl, and KBr host lattices ([71]). The measured frequencies are given in Table II. The induced absorption spectrum for KCl: Li$^+$ is shown in Fig. 26. Although a distinct absorption band does occur, the line width is much larger than can be obtained with the (MT) model.

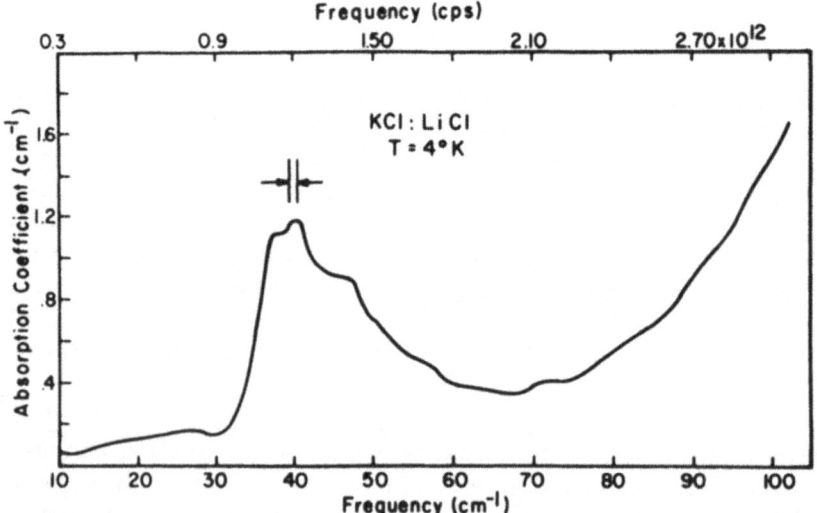

Fig. 26—Induced absorption spectrum for KCl:Li+. Only this one broad band has been observed in the spectral region from 3 to 100 cm$^{-1}$.

Recent studies have shown that low-lying tunneling states are associated with a translational instability of the lithium impurity in the KCl lattice ([77]). Some observed properties of this defect system are well described by a model in which the lithium ion moves in a multiwell potential formed by the neighboring ions with a repulsive barrier at the normal lattice site ([78]). However, the far-infrared transmission of KCl:LiCl single crystals does not permit an identification of the impurity-induced energy-level scheme since no sharp absorption lines and only one broad band at 40 cm$^{-1}$ have been observed in the spectral region from 3 to 100 cm$^{-1}$.

Similar measurements have been carried out for KBr:Li$^+$. In this case a sharp low-frequency lattice mode is found with additional structure at higher frequencies. Again the measured width for the resonant mode absorption, although small, is still much larger than given by the (MT) model (see Table II). The sharp absorption line does allow the lithium 6,7 isotope shift to be measured. The absorption coefficients found for both the Li$^6$ isotope and Li$^7$ isotope are shown in Fig. 27. For Li$^6$ ions the absorption is observed at $17.9 \pm 0.2$ cm$^{-1}$ with a full width at half-maximum absorption of approximately 1.4 cm$^{-1}$. In this frequency region, the bandwidth from our monochromator is approximately 0.7 cm$^{-1}$. The true width is estimated to be between 0.8 and 1.0 cm$^{-1}$. Two concentrations have been measured: for curve A there are $1 \times 10^{19}$ Li$^6$ ions/cm$^3$ and for curve B there are $1.5 \times 10^{18}$ Li$^6$ ions/cm$^3$. With the higher concentration, A, a prominent band is observed at 45.5 and at 83 cm$^{-1}$. For Li$^7$ a strong absorption is located at $16.3 \pm 0.2$ cm$^{-1}$ with a full width at half-maximum absorption of 1.2 cm$^{-1}$. We estimate the true width here to be between 0.6 and 0.8 cm$^{-1}$. Again, two concentrations have been investigated. For curve C there are $1.2 \times 10^{19}$ ions/cm$^3$ and for curve D there are $1.3 \times 10^{18}$ ions cm$^3$. A prominent band is observed at 43 cm$^{-1}$ and another at 83 cm$^{-1}$ in the more concentrated sample.

The measured isotope shift ($\nu_6/\nu_7 = 1.10$) for the low-frequency absorption is larger than predicted with a simple Einstein oscillator model [$(\frac{7}{6})^{1/2} = 1.08$]. Because the isotope shift will be large when the ion in question has a large amplitude in a normal mode, we picture for the first excited state of the resonant mode a threefold degenerate $T_{1u}$ state in which the lithium ion is undergoing much larger excursions than the nearest neighbor ions in the crystal. Most of the higher frequency absorption in Fig. 27 probably corresponds to the one phonon absorption spectrum from the impurity-activated phonons. The absorption coefficient should vary as a weighted one-phonon density of states; however, sufficient detail is not observed to verify this prediction. One unusual feature is the small but observable isotope shift of the band at 43 cm$^{-1}$ for Li$^7$ to 45.5 cm$^{-1}$ for Li$^6$. Also the band appears to be slightly stronger for Li$^6$ than for Li$^7$. This absorption can either be identified with another slightly infrared-active impurity mode or as the second overtone

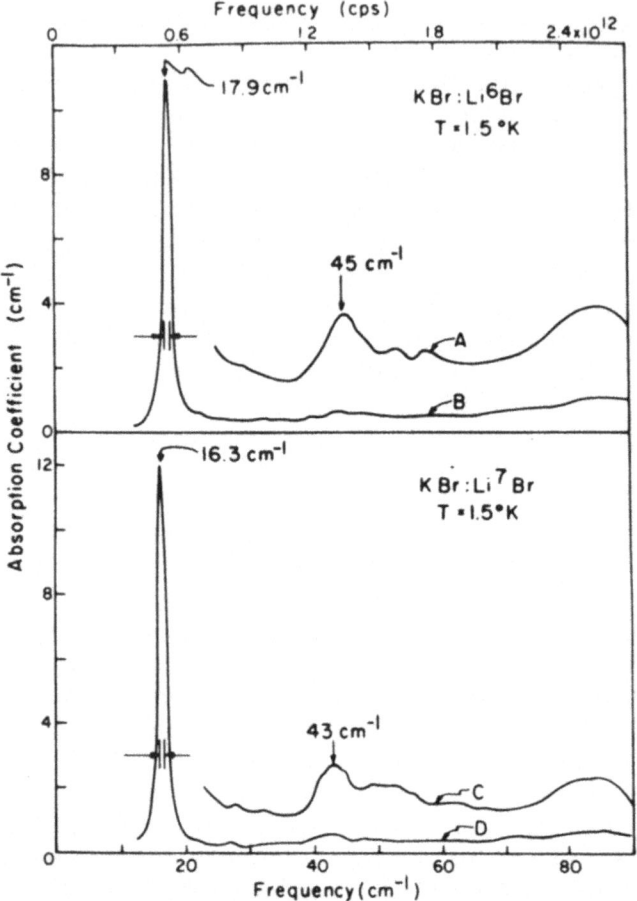

Fig. 27—Impurity induced absorption coefficient *vs.* frequency for the two lithium isotopes in KBr. For curves A and B the concentrations of KBr:Li⁶Br are $1 \times 10^{19}$ and $1.5 \times 10^{18}$Li⁶ ions/cm³, respectively. For curves C and D the concentrations of KBr:Li⁷Br are $1.2 \times 10^{19}$ and $1.3 \times 10^{18}$Li⁷ ions/cm³, respectively.

of the low-frequency absorption. If anharmonic forces are important for this system, the second overtone can become infrared-active in an octahedral environment. If the four experimentally determined frequencies for the resonant mode and the second overtone for the two isotopes are fit to an anharmonic Einstein oscillator of the form

$$\Omega_7(n) = v_e[n + \tfrac{1}{2}] + v_e X_e[(n + \tfrac{1}{2})^2 + \tfrac{1}{4}] \tag{51}$$

$$\Omega_6(n) = \rho v_e[n + \tfrac{1}{2}] + \rho^2 v_e X_e[(n + \tfrac{1}{2})^2 + \tfrac{1}{4}] \tag{52}$$

where $v_e$ is the harmonic oscillator frequency, $X_e$ the anharmonic contribution, and $\rho = (M_6/M_7)^{1/2}$, the square root of the mass ratio. The data can be fit with $v_e = 18.4 \text{ cm}^{-1}$, $X_e = -0.058$ and $\rho = 1.11$. The large linewidths can be reconciled with the small widths predicted by the (MT) model for the harmonic approximation as shown in Table II by introducing anharmonic coupling to band phonons. Assuming the anharmonic contribution dominates the linewidth, the coupling constants for both isotopes are calculated to be $K'/K \approx 0.005$. This appreciable softening of the forces between the Li ion and its surroundings is also consistent with an anharmonic potential model.

In the harmonic approximation the resonant-mode absorption would be temperature-independent; however, the anharmonic coupling is expected to alter this description and we now turn to this problem.

### d. Temperature-Dependent Properties

In contrast with temperature-dependent studies on gap modes which are hindered by the strong temperature dependence of the host-lattice absorp-

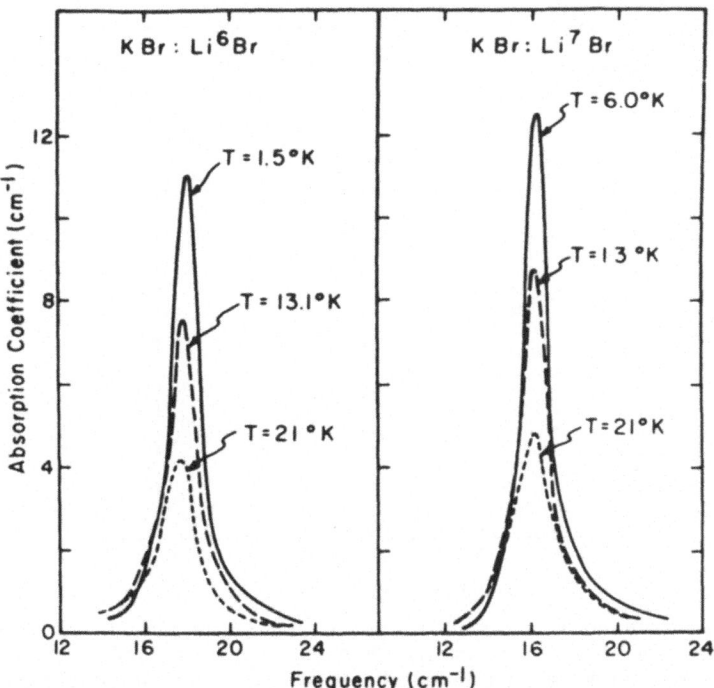

Fig. 28—Temperature dependence of the absorption coefficient in the neighborhood of the resonant mode absorption for Li⁶ and for Li⁷ in KBr.

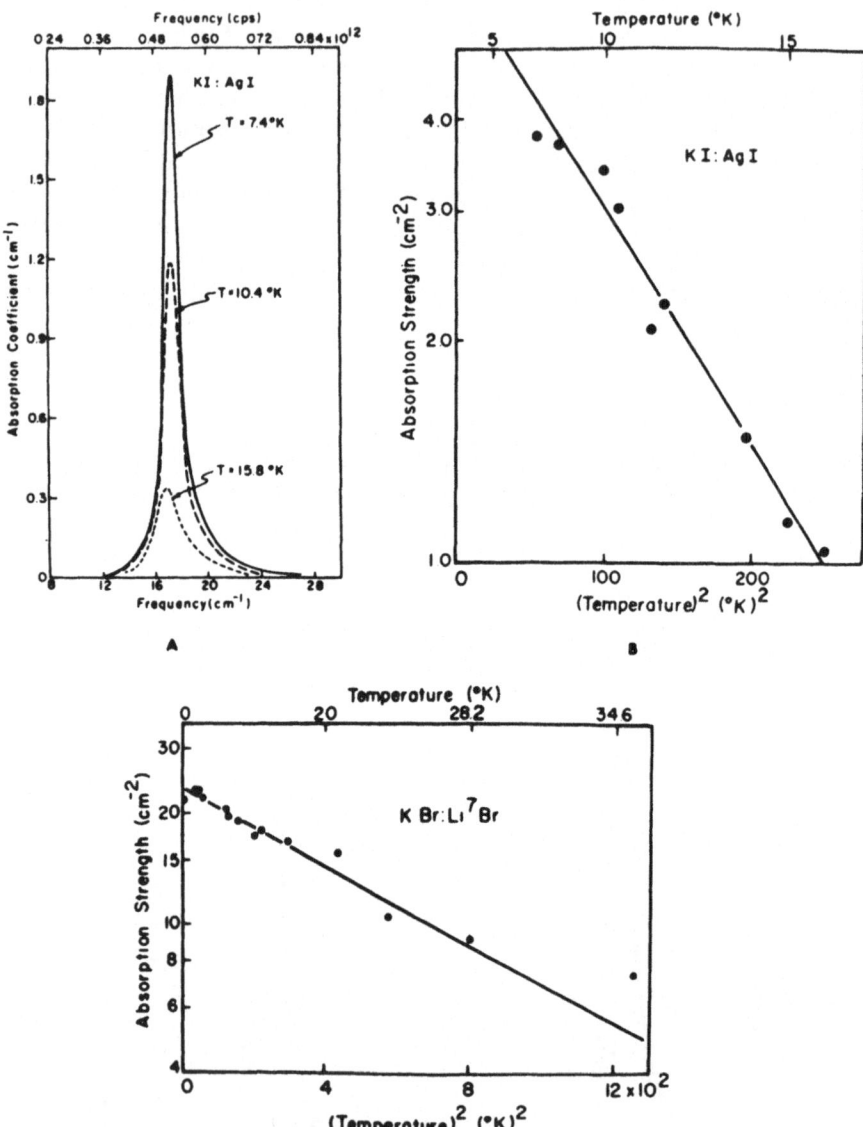

Fig. 29—Temperature dependence of the line strength for resonant modes. (A) Temperature dependence of the absorption coefficient in the neighborhood of the resonant mode absorption for KI:AgI. (B) Temperature dependence of the integrated absorption coefficient for KI: AgI (C) for KBr:Li Br.

tion, temperature-dependent studies on resonant modes at moderately low temperatures are relatively straightforward. One of the most striking features which has been observed with these low-frequency absorptions is the rapid temperature dependence of the line strength ([79]). In Fig. 28 the absorption coefficient in the neighborhood of the resonant mode absorption for $Li^6$ and also for $Li^7$ in KBr are shown for three different temperatures. As the sample temperature is increased, the eigenfrequency decreases slightly and the half width increases. The largest change occurs in ($\int \alpha(\nu)d\nu[cm^{-2}]$) the line strength. The temperature dependence of the line strength has also been measured for $KI: Ag^+$. The exponential form of the temperature dependence of the absorption strength is shown in Fig. 29 where a semilog plot of the strength vs. temperature can be fit to a straight line. Similar straight-line fits have been obtained for both isotopes of Li in KBr. The temperature dependence for $Li^7$ is also shown in Fig. 29.

Recently the strength of the $U$-center mode has also been observed to have this characteristic temperature dependence ([79,80]). These results are shown in Fig. 30. This system is to some extent more straightforward and the description of the temperature dependence follows in direct analogy with the temperature dependence of some zero phonon electronic transitions ([81]). The necessary approximations are the following:

The local mode is assumed to be represented by an Einstein oscillator and the adiabatic approximation is invoked; that is, the local-mode transition is assumed to occur much faster than the other phonon transitions in the crystal. In this manner the coordinates of the local mode and the normal lattice modes are separated. Because of anharmonic coupling the local mode transition is strongly influenced by the position of the neighboring ions and one assumes that the important coupling mechanism is determined by the relative displacement of these neighboring atoms. The modulation of the local-mode levels then is assumed to be a function of the local strain at the defect. For modulation which is a linear function of strain, the probability of finding a local mode transition with no transfer of energy to the lattice phonons can be calculated. From the quadratic coupling terms can be obtained the temperature dependence of the center frequency and also the temperature dependence of the line width of the local mode. For a continuum model, this modulation can be written as

$$\nu(t) = \nu_0 + \alpha S(t) + \beta[S(t)]^2 \tag{53}$$

where $S(t)$ represents the instantaneous lattice strain field of the initial and final oscillator energy levels. The instantaneous frequency varies with time as a result of the time-dependent strain modulation. In a classical approximation the function representing the complex amplitude associated with the frequency $\nu(t)$ can be written as ([81])

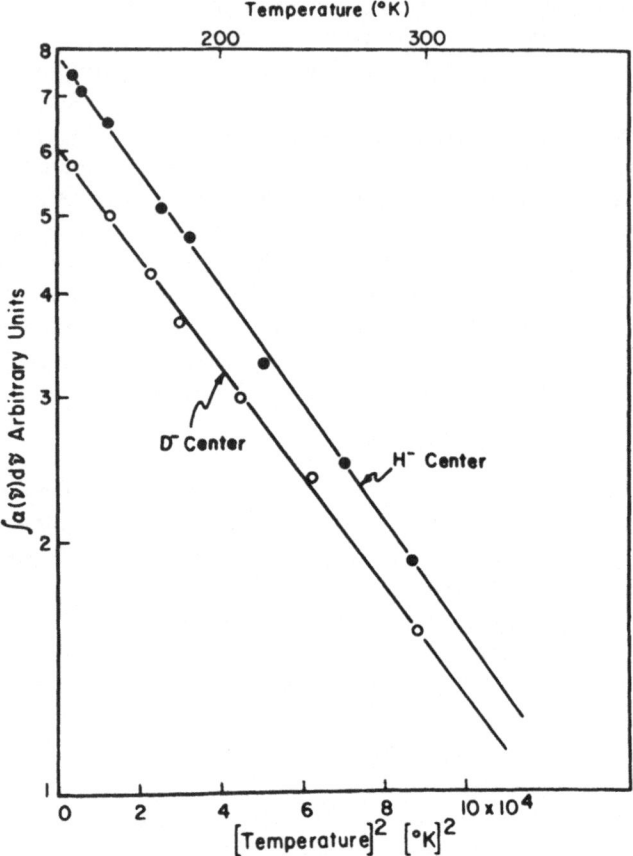

Fig. 30—Temperature dependence of the absorption strength for the hydride center and also the deuteride center in KCl [From W. C. Price and G.R. Wilkinson([37])].

$$A = A_0 \exp[i \int^t v(t)\, dt] \tag{54}$$

In the Fourier analysis of this function two characteristic features appear in the resulting spectrum: A sharp peak at the frequency

$$v = v_0 + \langle \beta[S(t)]^2 \rangle \tag{55}$$

and a continuous spectrum extending over a range of frequencies comparable to those describing the time variations of $S(t)$. The linear term in equation (53) does not contribute to the average value of the frequency because positive and negative excursions are equally likely, but the amplitude of the linear term does determine the strength of the sharp spike.

Let us proceed to the phonon description. For the modulation which is a linear function of strain, the probability of finding a local-mode transition with no transfer of energy to the lattice phonons can be written in the Debye approximation as ([79,80])

$$\frac{I_L}{I_B} = \exp\left[-\alpha'\left(1 - 4\left(\frac{T}{\theta}\right)^2 \int_9^{\theta/T} \frac{x\,dx}{e^x - 1}\right)\right] \tag{56}$$

where $I_L$ represents the strength of the sharp spike and $I_B$, the total strength of the sharp spike plus broad multiphonon structure. This expression states that the probability of a local-mode transition is proportional to the Debye—Waller factor.

The frequency shift with temperature can be written as ([82,83])

$$\Delta v_0(T) = \delta \int_0^\infty dv \rho(v) n(v) \tag{57}$$

and the line width from the Raman scattering of phonons by the impurity can be written as

$$\Gamma(T) = \gamma \int_0^\infty dv [\rho(v)]^2 n(v)[1 + n(v)] \tag{58}$$

Here $n(v) = [\exp(hv/kT) - 1]^{-1}$ and $\rho(v)$ is the effective density of phonons, that is, the density of states weighted by the coupling parameters. Assuming a Debye spectrum of acoustic phonons, these expressions reduce to

$$\Delta v_0(T) = \delta'\left(\frac{T}{\theta}\right)^4 \int_0^{\theta/T} \frac{x^3 dx}{e^x - 1} \tag{59}$$

and

$$\Gamma(T) = \gamma'\left(\frac{T}{\theta}\right)^7 \int_0^{\theta/T} \frac{x^6 e^x dx}{(e^x - 1)^2} \tag{60}$$

Using these two expressions, with a $\theta$ slightly smaller than the Debye $\theta$, good agreement with experiment has been found for U-centers in a number of alkali halide crystals ([80]) but poor agreement with experiment has been noted for the temperature dependence of the line strength for the hydride ion in the alkaline earth fluorides ([83]).

In the case of resonant modes it is not possible to separate the coordinates of the local mode from the normal lattice using the adiababitc approximation. Although this problem has not been solved satisfactorily, clearly the solution must rely on the weak coupling of the impurity ion to the rest of the lattice. Because of the apparent similarities of the temperature-depend-

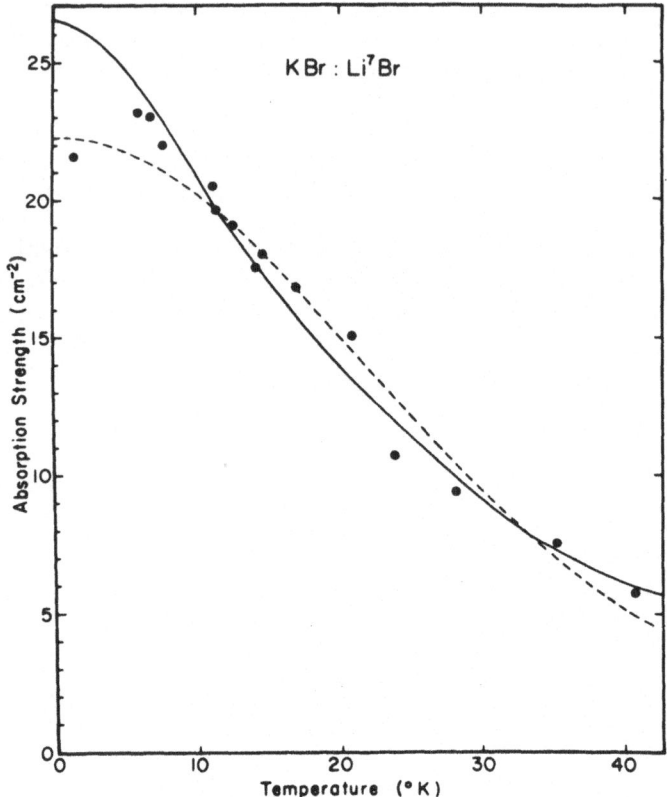

Fig. 31—The temperature dependence of the absorption strength
for KBr:Li⁷Br with a linear temperature scale. The solid and
dashed curves are "best fits" of equation (56), see text.

ent properties of the local mode and the weakly coupled resonant mode,
we study the experimental results for the resonant mode in more detail.

In Fig. 31 the temperature dependence of the intensity of the resonant
mode in KBr: Li⁷Br is considered. A linear temperature scale is used in this
case in order to compare the experimental results with equation (56). The
dashed curve represents the temperature variation of equation (56) with
$\theta = 174°K$. The solid curve indicates the temperature variation of equation
(56) with a phonon cutoff at a much lower frequency equivalent to $\theta = 25°K$.
The expression with the low-frequency cutoff perhaps fits the experimental
results more closely.

The temperature dependence of the linewidth for KI: AgI and KBr:
Li⁷Br and the temperature dependence of the resonance frequency for both
KBr: Li⁶Br and KBr: Li⁷Br and also for KI:AgI are shown in Fig. 32. The

large errors in the experimental points occur mainly because a low-resolu-
tion-grating instrument was used for these measurements.

Although only the $T^7$ Raman relaxation term is necessary to describe
the local mode linewidth results [equation (60)], both a direct one-phonon
process and a Raman two-phonon process are to be expected here for
resonant modes. The one-phonon process is postulated to account for relaxation processes to band phonon modes with energies very close to the resonant
mode energy. Neglecting the higher excited states of the resonant mode the
temperature-dependent linewidth in this case is then

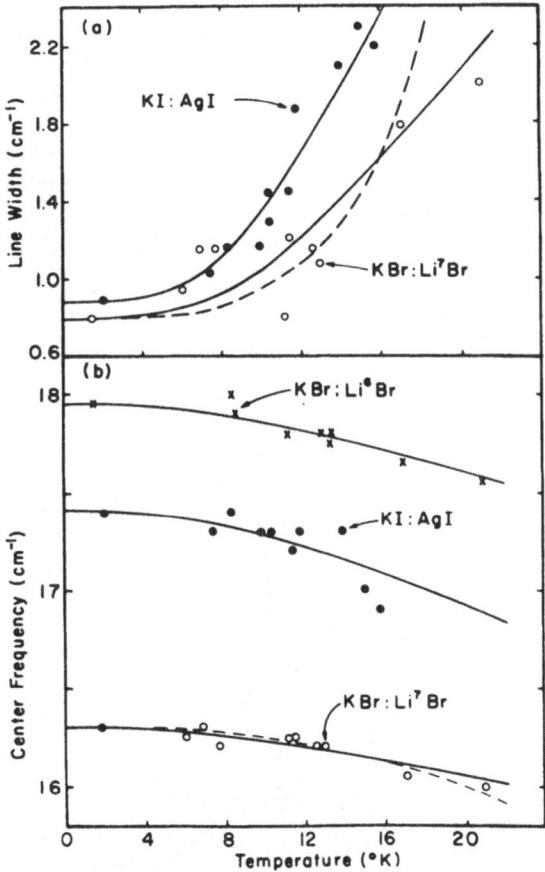

Fig. 32—Temperature-dependent properties of the
linewidth and resonant mode center frequencies:
(A) temperature dependence of linewidth for KI:
AgI and KBr:Li⁷Br; (B) temperature dependence of
the resonant frequency for KBr:Li⁶Br, KBr:Li⁷Br and
KI:AgI. The solid curves are described in the text.

$$\Gamma(T) = \Gamma_D \coth \frac{T_0}{2T} + \Gamma_R \left(\frac{T}{\theta}\right)^7 \int_0^{\theta/T} \frac{x^6 e^x dx}{(e^x - 1)^2} \tag{61}$$

where $\Gamma_D$ is the coefficient of the direct process, $T_0$ the resonant mode frequency in °K, and $\Gamma_R$ the coefficient for the Raman process. The three curves in Fig. 32 are obtained by assigning the width at 0°K to $\Gamma_D$ and then fitting $\Gamma_R$ to the experimental results at 16°K. The dotted curve is for KBr with a $\theta = 174$°K and the two solid curves are obtained from equation (61) with the phonon cutoff $\theta$ reduced to 25°K. The frequency shift of the absorption lines illustrated in the lower half of Fig. 32 can be fitted equally well with a $\theta$ of 25° or 174°K in equation (59).

The low phonon cutoff which fits the experimental results probably indicates that low-frequency phonons couple more strongly to the resonant mode than do high-frequency phonons.

Recently Weber and Nette ([84]) have observed a resonant mode associated with NaCl: Cu$^+$. Their results are given in Fig. 33. As the temperature increases the oscillator strength of the absorption remains almost constant and the halfwidth increases by a factor of 20 on going to 79°K. At low tem-

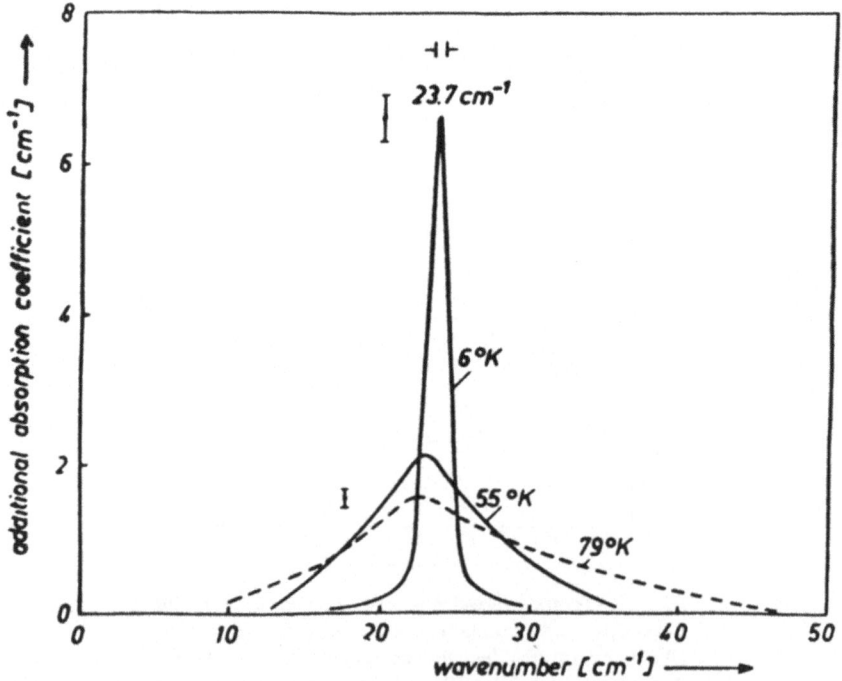

Fig. 33—Temperature dependence of the resonant mode for NaCl:Cu$^+$ [From R. Weber and P. Nette, *Physics Letters* **20**: 493 (1966)].

peratures the theoretical line width agrees fairly well with the experimental value as given in Table II. Our model can explain the temperature-dependent results only if the linear strain coupling is much weaker than the quadratic coupling. Just why the quadratic term should be more important here than for the resonant modes considered earlier is an interesting problem.

For all the resonant modes, the temperature dependence of the absorption coefficient can give rise to a temperature-dependent contribution to the static dielectric constant. From the Kramers—Kroning relations the impurity-induced contribution associated with the resonant-mode absorption at 0°K is

$$\Delta \epsilon_1^i = \frac{2}{\pi} \int_0^{\nu_c > \bar{\nu}} \frac{\epsilon_2^i(\nu')d\nu'}{\nu'} \tag{62}$$

where $\epsilon = \epsilon_1 - i\epsilon_2$ and $\epsilon_2^i$ describes the low-frequency impurity-induced absorption contribution. The integral in equation (62) encompasses both the sharp resonant line at $\nu_0$ and also any low-frequency one-phonon absorption at $\bar{\nu}$ associated with the resonant mode. The intrinsic lattice absorption from the transverse optic branch is far removed from the low-frequency region considered here and the integral is cut off at $\nu_c$ below the TO mode. As the temperature of the system increases, the center of gravity of the total impurity-induced absorption shifts, and equation (62) guarantees that the impurity-induced contribution to the static dielectric constant is also temperature-dependent. This effect is quite modest for KI: Ag$^+$ and KBr: Li$^+$ but should become larger for systems with lower-frequency impurity modes.

### e. Stress-Induced Frequency Shift

With the application of uniaxial stress along different crystallographic directions, the coupling of the resonant mode excited state to lattice distortions of different symmetries can be distinguished. Consider the electric dipole transition between the two vibrational states $\psi$ which transform according to the $A_{1g}$ and $T_{1u}$ irreducible representations of the $O_h$ point group appropriate to the defect site. The local symmetry of the lattice is lowered by the uniaxial elastic deformation and a splitting of the degenerate $T_{1u}$ state can occur. The lattice distortions which will perturb the resonant-mode excited state can be obtained as follows. The matrix elements associated with the stress perturbation $H'$ are for the ground state,

$$\Delta E_g = \langle \psi(A_{1g}) | H' | \psi(A_{1g}) \rangle \tag{63}$$

and for the excited state,

$$\Delta E_e = \langle \psi(T_{1u}) | H' | \psi(T_{1u}) \rangle \tag{64}$$

For these matrix elements to be invariant under all symmetry operations of the octahedral group $O_h$ they must transform as the $A_{1g}$ representation. Thus the stress perturbation must transform according to the representations contained in the direct product of

$$T_{1u} \times T_{1u} = A_{1g} + E_g + T_{2g} \qquad (65)$$

and the only modes which interact with the excited state of the resonant mode are the long-wavelength acoustic distortions of $A_{1g}$ (spherical), $E_g$ (tetragonal and orthorhombic), and $T_{2g}$ (trigonal) symmetry. The anharmonic coupling of these modes to the resonant mode can be estimated from the slopes of the absorption frequency vs. applied stress experiment.

If the direction cosines of the pressure $P$ with respect to the crystal axes are labeled by $\alpha$, $\beta$, and $\gamma$, and the direction cosines of the polarization vector of the radiation relative to the crystal axes by $\alpha'$, $\beta'$, and $\gamma'$ then the slope of the frequency vs. stress curve for a cubic crystal is given by ([85])

$$\frac{dv}{dP} = D(S_{11} + 2S_{12}) + E(S_{11} - S_{12})$$
$$\times [(2\gamma'^2 - \alpha'^2 - \beta'^2)(2\gamma^2 - \alpha^2 - \beta^2) + 3(\alpha'^2 - \beta'^2)(\alpha^2 - \beta^2)]$$
$$+ 2FS_{44}(\alpha'\beta'\alpha\beta + \beta'\gamma'\beta\gamma + \gamma'\alpha'\gamma\alpha)$$

where $D$, $E$, and $F$ are the coupling coefficients associated with strain components which transform as the representations $A_{1g}$, $E_g$, and $T_{2g}$, respectively, and $S_{11}$, $S_{12}$, and $S_{44}$ are the compliances appropriate to the local region around the impurity.

The measurement of the shift of the resonant mode with polarized radiation requires some experimental techniques which have not been considered previously. A schematic diagram of the uniaxial stress cryostat is presented in Fig. 34. The far infrared radiation from the interferometer enters through a light pipe and passes through the cold radiation filter labeled $F$, reflects from a 45° mirror, and then passes through a fine wire grid labeled $P$ which polarizes the radiation. Two grids of orthogonal orientation can be inserted in the radiation path. The radiation passes through the sample labeled $S$ which is placed between two thin pieces of cardboard (to equalize the stress) which in turn are in contact with the stress assembly. The radiation is then reflected from a second right-angle bend and finally reaches the germanium detector labeled D. Compared with simple transmission measurements, the intensity of radiation at the detector in Fig. 34 is reduced by a factor of four (the data gathering time is increased by a factor of 16) because of the two right-angle bends and the wire polarizer. The experiments would be extremely difficult without an interferometer.

## UNIAXIAL STRESS CRYOSTAT

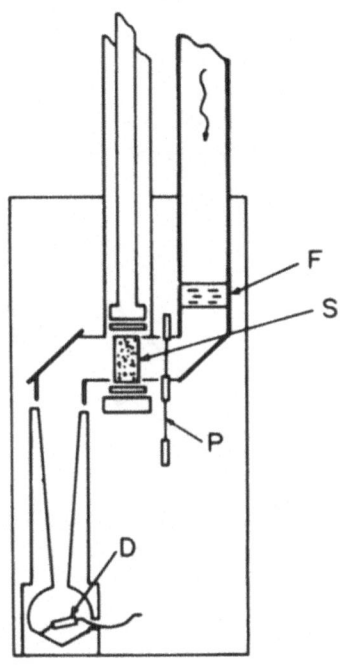

Fig. 34—Sample holder and detector
mount for stress measurements: F is a
low temperature transmission filter, P
the polarizer, S the sample, and D the
detector.

The KBr: Li⁶Br samples used in these measurements were single crystals
approximately 13 × 10 mm in a section normal to the beam and 2.5 mm
thick. A sample temperature of 2°K was maintained during the experiment
by exchange gas coupling to the surrounding pumped helium exchange bath.
Static weights were used to produce the uniaxial compressional stress in the
sample up to about 2 kg/mm². For larger stresses a manually operated
hydraulic pump and piston arrangement was used.

The experimentally determined absorption frequency as a function of
applied stress is shown in Fig. 35 for two crystallographic orientations. For
stress applied along the [100] axis and also along the [111] axis, two absorption lines which shift linearly with applied stress were measured. For stress
along the [110] direction, three absorption lines have been observed. Frequency shifts as large as 2.5 cm⁻¹, equivalent to a relative shift of 14%, have
been observed for the *E*-vector polarized parallel to the stress in the [100]

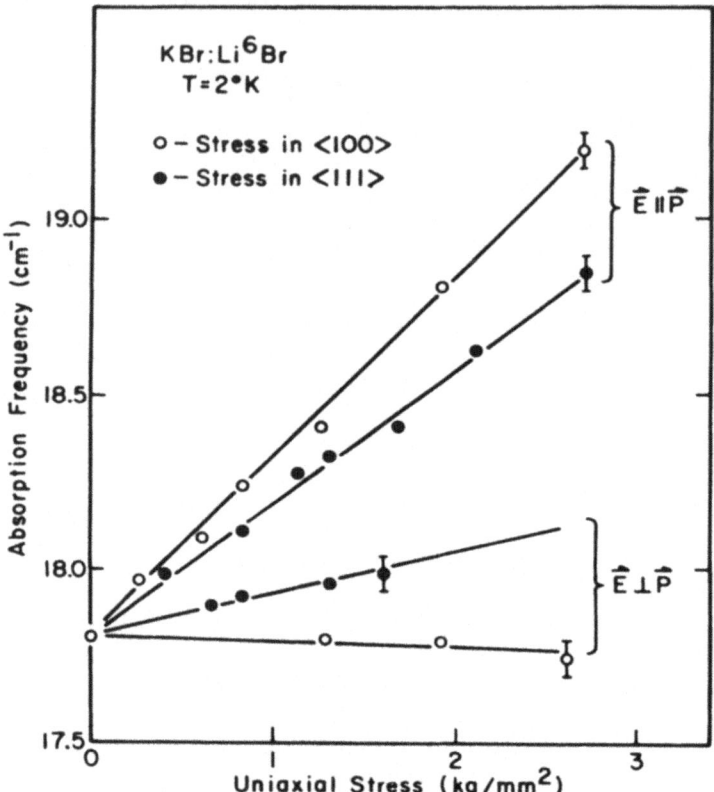

Fig. 35—Absorption-frequency centroid versus applied uniaxial stress: E represents the electric vector of the polarized radiation and P the applied uniaxial stress.

direction. In all cases, the relative aborption strengths for different polarizations remain constant as a function of stress.

From our experiments ([86]) the coefficients of equation (66) are found to be, in units of cm$^{-1}$,

$$D = 880 \pm 160$$
$$E = 360 \pm 70$$
$$F = 140 \pm 45$$

if the compliances are taken to be those appropriate to the unperturbed crystal. On this basis we deduce that the resonant mode in this system is most strongly coupled to long-wavelength, spherically symmetric, acoustic modes of the surrounding lattice.

## IV. SUMMARY

Impurity-induced absorption in crystals in many cases can be studied by far infrared spectroscopy. Whereas only the $H^-$ local mode has been observed above the phonon spectrum of the pure crystal, a large number of gap modes between the optic and acoustic phonon bands have been observed in the far infrared. To quench the two-phonon absorption band of the host crystal the transmission measurements must be carried out on crystals cooled to low temperatures. Only limited temperature-dependent studies are possible. Gap modes are extremely sharp with widths less than $0.2 \text{ cm}^{-1}$. The isotope shift of a gap mode is a sensitive probe of the impurity lattice coupling constants. Some of the experimental observations can qualitatively be explained with a simple impurity—lattice model. With low-symmetry centers such as polyatomic molecules the spectroscopic techniques appropriate to other frequency regions can often be used to investigate these far infrared gap modes.

A number of impurity-induced resonant modes have been observed in alkali halide crystals. Sharp absorption lines have been identified with the weak coupling of the impurity atoms to the host lattices. The large isotope shift for the $Li^6 - Li^7$ absorption lines indicates that the resonant mode can be approximated by an Einstein oscillator. By measuring the temperature dependence and strain dependence of the absorption, the anharmonic coupling of the resonant mode to the phonon modes has been observed. Both type of measurements indicate that a strong coupling exists. Although the characteristic tunneling states associated with off-center impurity ions have not yet been detected in the far infrared, only a few dopant—lattice combinations have been studied.

In conclusion, with far infrared techniques a wealth of detailed information can be obtained about the properties of localized and resonant lattice states in crystals.

At this time I should like to note that a number of my students have contributed to the experimental work which has been reported. The development of the interferometer and monochromator was carried out by I. Nolt, C. D. Lytle, and R. Alexander. Both the stress splitting of the resonant mode and the isotope shift of the $Cl^-$ gap mode were measured by I. Nolt. C. D. Lytle and R. Westwig measured the molecular impurity mode spectrum and the $Br^-$ gap-mode spectrum, respectively. The far infrared properties of $UO_2$ have been studied by K. Aring. I would also like to express my sincere appreciation to Prof. S. Takeno for his close cooperation with some of the theoretical problems. Also discussions with Professors J. A. Krumhansl, R. O. Pohl, A. A. Maradudin, and G. F. Nardelli have been particularly helpful.

Finally I would like to thank Professor Nardelli for the pleasant accommodations provided at this summer school.

This work has been supported by the U. S. Atomic Energy Commission. Additional support has been received from ARPA through the use of central facilities and space provided by the Materials Science Center.

## REFERENCES

1. H. Rubens and E. F. Nichols, *Phys. Rev.* **4**: 314 (1897).
2. H. Rubens and R. W. Wood, *Phil. Mag.* **21**: 249 (1911); H. Rubens and O. Von Baeyer, **21**: 689 (1911).
3. H.A. Gebbie and G.A. Vanasse, *Nature*, **178**: 432 (1956).
4. J. Strong, *J. Opt. Soc. Am.* **47**: 354 (1957).
5. W.S. Boyle and K.F. Rodgers, *J. Opt. Soc. Am.* **49**: 66 (1959).
6. A. Walsh, *J. Opt. Soc. Am.* **42**: 96 (1952).
7. J. Strong and G.A. Vanasse, *J. Opt. Soc. Am.* **49**: 844 (1959).
8. P.L. Richards, *J. Opt. Soc. Am.* **54**: 1474 (1964).
9. L. Genzel and R. Weber, *Z. Angew. Phys.* **10**: 127 (1957); **10**, 195 (1958).
10. F.J. Low, *J. Opt. Soc. Am.* **51**: 1300 (1961).
11. D. Bloor, T.J. Dean, G.O. Jones, D.H. Martin, P.A. Mawer, C.H. Perry, *Proc, Roy. Soc. London* **261**: 10 (1961).
12. M.A. Kinch and B.V. Rollin, *Brit. J. Appl. Phys.* **14**: 672 (1963).
13. M.J.E. Golay, *Rev. Sci. Instr.* **18**: 347 (1947).
14. P.L. Richards, *Phys. Rev.* **138**: A1769 (1965).
15. H.D. Drew and A.J. Sievers, to be published.
16. J.H. Van Vleck, *The Theory of Electric and Magnetic Susceptibilities*, Oxford Univ. Press, 1932.
17. H.A. Kramers, *Physica* **1**: 182 (1934).
18. P.W. Anderson, Phys. Rev. **79**: 350 (1950); *ibid.* **115**, 2 (1959).
19. J.H. Van Vleck, *J. Phys. Radium* **20**: 124 (1959).
20. C. Kittel, *Phys. Rev.* **82**: 565 (1951).
21. T. Nagamiya, *Prog. Theoret. Phys. (Kyoto)* **4**: 342 (1941).
22. F. Keffer and C. Kittel, *Phys. Rev.* **85**: 329 (1952).
23. T. Nagamiya, K. Yosida, and R. Kubo, *Adv. in Phys.* **4**: 14 (1955).
24. F.M. Johnson and A.H. Nethercot. *Phys. Rev.* **114**: 705 (1959).
25. R.C. Ohlmann and M. Tinkham, *Phys. Rev.* **123**: 425 (1961).
26. P.L. Richards, *J. Appl. Phys.* **35**: 850 (1964).
27. M. Tinkham, *Phys. Rev.* **124**: 311 (1961).
28. A.J. Sievers and M. Tinkham, *Phys. Rev.* **129**: 1566 (1963).
29. I.F. Silvera, J.H.M. Thornley, and M. Tinkham, *Phys. Rev.* **136**: A695 (1964).
30. H.W. Halley and I. Silvera, *Phys. Rev. Letters* **15**: 654 (1965).
31. S.J. Allen, R. Loudon, and P. L. Richards, *Phys. Rev. Letters* **16**: 463 (1966).
32. T. Tanabe, T. Moriya and S. Sugano, *Phys. Rev. Letters* **15**: 1023 (1965), also see T. Tanabe, *J. Phys. Soc. Japan* **21**: 926 (1966).
33. G. Dolling and R.A. Cowley, *Phys. Rev. Letters* **16**: 683 (1966).
34. R. Hilsch and R.W. Pohl, *Gött Nachr. Math. Phys. Neue Folge*, **2**: S. 139 (1936). R. Hilsch and R.W. Pohl, *Trans. Faraday Soc.* **34**: 883 (1938).

35. G. Schäfter, *J. Phys. Chem. Solids* **12**: 233 (1960).
36. E. Mollwo, *Z. Physik* **56**: 85 (1933) and H. Ivey, *Phys. Rev.* **72**: 341 (1947).
37. W.C. Price and G.R. Wilkinson, University of London King's College, Molecular and Solid State Spectroscopy Technical Report No. 2 (1960). (unpublished).
38. A. Mitsuishi and H. Yoshinaga, *Prog. Theoret. Physics (Kyoto) Suppl.* **23**: 241 (1962).
39. B. Fritz, *J. Phys. Chem. Solids* **23**: 375 (1962).
40. For a complete list of references on the mass defect problem see: A.A. Maradudin, E.W. Montroll, and G.H. Weiss, *Theory of Lattice Dynamics in the Harmonic Approximation* Solid State Physics Suppl. 3, F. Seitz and D. Turnbull, Editors, Academic Press, 1963.
41. R.E. Wallis and A.A. Maradudin, *Prog. Theor. Phys.* **24**: 525 (1960).
42. S. Takeno. S. Kashiwamura and E. Teramoto, *Prog. Theoret. Phys. (Kyoto) suppl.* **23**: 124 (1962).
43. P.G. Dawber and R.J. Elliott, *Proc. Roy. Soc.* **273**: 222 (1963); *Proc. Phys. Soc.* **81**: 453 (1963).
44. P. Mazur, E.W. Montroll, and R.B. Potts, *J. Wash. Acad. Sci.* **46**: 2 (1956).
45. H. Rubens and G. Hertz, *Berlin, Ber.* **14**: 256 (1912).
46. H. Bilz and L. Genzel, *Z. Physik* **169**: 53 (1962).
47. E. Burstein, *Phonons and Phonon Interactions*, T.A. Bak, ed., W.A. Benjamin Inc., New York, 1964, p. 276.
48. A.M. Karo and J.R. Hardy, *Phys. Rev.* **129**: 2024 (1963).
49. B.N. Brockhouse, *Phonons and Phonon Interactions*, T.A. Bak, ed, W.A. Benjamin Inc., New York, 1964, p. 221.
50. G. Dolling, R.A. Cowley, C. Schittenhelm and I.M. Thorson, *Phys. Rev.* **147**, 577 (1966).
51. R. Stolen and K. Dransfeld, *Phys. Rev.* **139**: A 1295 (1965).
52. E. Burstein, F.A. Johnson, and R. Loudon, *Phys. Rev.* **139**: A1239 (1965).
53. J.H. Schulman and W.D. Compton, *Color Centers in Solids*, MacMillan Comapny, New York, 1962, p. 49.
54. J. Rolfe, *Phys. Rev. Letters* **1**: 56 (1958).
55. A.J. Sievers, A.A. Maradudin, and S.S. Jaswal, *Phys. Rev.* **138**: A272 (1965).
56. L. Pauling, *Nature of the Chemical Bond*, Cornell University Press, Ithaca, 1945.
57. Y. Mitani and S. Takeno, *Prog. Theoret. Phys. (Kyoto)* **33**: 779 (1965).
58. The general expression for the eigenfrequency of the $A_{2u}$ mode is obtained from the (M-T) paper but the simple form of equation (40) has been obtained by S. Takeno and the author.
59. C.D. Lytle, Thesis, Materials Science Center Report #390, Cornell University, Ithaca, (1965).
60. A.J. Sievers and C.D. Lytle, *Physics Letters* **14**: 271 (1965).
61. K.F. Renk, *Physics Letters* **14**: 281 (1965).
62. K.F. Renk, *Physics Letters* **20**: 137 (1966).
63. T. Timusk and W. Staude, *Phys. Rev. Letters* **13**: 373 (1964).
64. H.S. Sack and M.C. Moriarty, *Solid State Comm.* **3**: 93 (1965).
65. V. Narayanamurti, W.D. Seward, and R.O. Pohl, *Phys. Rev.* (in press).
66. W.C. Price, W.F. Sherman, and G.R. Wilkinson, *Spectroch. Acta* **16**: 663 (1960).
67. J.F. Angress, S.D. Smith, K.F. Renk, *Proc. Int. Conf. Lattice Dynamics Copenhagen* 1963, p. 467.
68. M. Balkanski and M. Nusimovici, *Phys. Stat. Sol.* **5**: 635 (1964).
69. A.J. Sievers, *Phys. Rev. Letters*, **13**: 310 (1964).
70. R. Weber, *Phys. Letters* **12**: 311 (1964).

71. A.J. Sievers and S. Takeno, *Phys. Rev.* **140**: 1030 (1965).
72. R. Brout and W.M. Visscher, *Phys. Rev. Letters* **9**: 54 (1962).
73. W.M. Visscher, *Phys. Rev.* **129**: 28 (1963).
74. R.O. Pohl, *Phys. Rev. Letters* **8**: 481 (1962).
75. C.T. Walker and R.O. Pohl, *Phys. Rev.* **131**: 1433 (1963).
76. A.D.B. Woods, B.N. Brockhouse, R.A. Cowley, and W. Cochran, *Phys. Rev.* **131**: 1025 (1963).
77. H.S. Sack and M.C. Moriarty, *Solid State Comm.* **3**: 93 (1965); G. Lombardo and R.O. Pohl, *Phys. Rev. Letters* **15**: 291 (1965); J.A.D. Matthew, *Solid State Comm.* **3**: 365 (1965).
78. S.P. Bowen, M. Gomez, J.A. Krumhansl, and J.A.D. Matthew, *Phys. Rev. Letters* **16**: 1105 (1966).
79. S. Takeno and A.J. Sievers, *Phys. Rev. Letters*. **15**: 1020 (1965).
80. S.S. Mitra and R.S. Singh, *Phys. Rev. Letters* **16**: 694 (1966).
81. R.H. Silsbee and D.B. Fitchen, *Rev. Mod. Phys.* **36**: 432 (1964).
82. G.F. Imbusch, W.M. Yen, A.L. Schawlow, D.E. McCumber and M.D. Sturge, *Phys. Rev.* **133**: A1029 (1964).
83. R.J. Elliott, W. Hayes, G.D. Jones, H.F. Macdonald, and C.T. Sennett, *Proc. Roy. Soc.* **A289**: 1 (1965).
84. R. Weber and P. Nette, *Physics Letters* **20**: 493 (1966); R. Weber (private communication).
85. W. Gebhardt and K. Maier, *Phys. Stat. Sol.* **8**: 303 (1965).
86. I.G. Nolt and A.J. Sievers, *Phys. Rev. Letters* **16**: 1103 (1966).

# Study of Localized Excitations Caused By Point Defects Using Thermal Conductivity*

R. O. Pohl

*Laboratory of Atomic and Solid State Physics*
*Cornell University*
*Ithaca, New York*

## I. DETERMINATION OF PHONON SCATTERING RATES FROM THERMAL CONDUCTIVITY MEASUREMENTS

The thermal motion of atoms in a soild can be described with traveling elastic waves. Their density of states is disturbed by the introduction of lattice defects. Some of the waves will disappear, new ones, more or less localized excitations, or impurity modes, will appear. Some of these impurity modes will be infrared- or photon-active. Others can be observed best through their interaction with the elastic running waves or phonons. Obviously, the best method for the study of these latter modes is ultrasonic techniques. Since, however, the generation and detection of monochromatic elastic waves in the frequency range of interest, typically above $10^{10}$ sec$^{-1}$ is still a major experimental problem ([1]), measurements of the lattice thermal conductivity have been used extensively for the detection of impurity modes in dielectric solids, in which heat is carried exclusively by phonons.

Figure 1 uses a simple analogy to describe this method. Suppose we wish to measure an optical extinction band (extinction: absorption and scattering) as shown at the top of Fig. 1a, without the use of a monochromator. In principle this can be done with a polychromatic light source ("white light") of known spectral distribution $dE/dv$ as shown for three temperatures $T_1$, $T_2$, $T_3$. If we measure the total energy transmitted by the sample as a function

*Work supported in part by the U. S. Atomic Energy Commission through Contract AT(30-1)-2391, Technical Report #NYO-2391-35. Additional support was received through the Advanced Research Projects Agency by the use of the Report Facility of the Materials Science Center at Cornell, MSC Report #530. The survey of literature pertaining to this review was completed in July 1966.

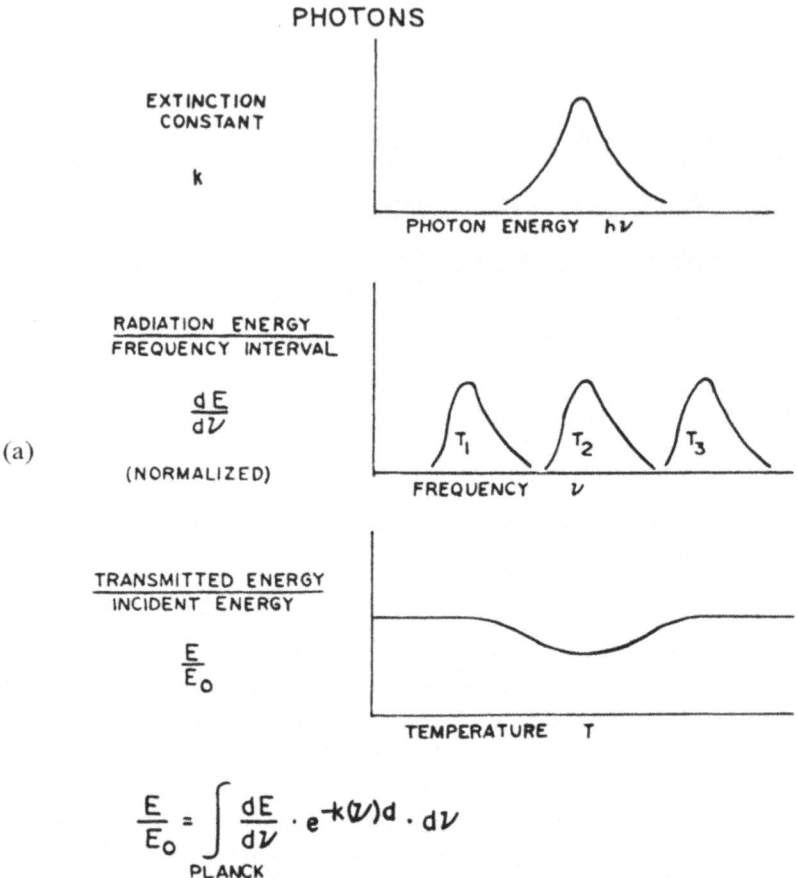

Fig. 1—a, b—Spectroscopy using broad bands of radiation. See text.

of the temperature of the light source, we observe a transmission spectrum as shown in the third line of Fig. 1a: in the temperature range in which most of the photons emitted by the lamp are on "speaking terms" with the extinguishing centers, the transmission (transmitted energy $E$)/(incident energy $E_0$) will have a minimum. The determination of the shape of the extinction band $k(v)$ involves fitting the expression

$$\frac{E}{E_0} = \int_{\text{Planck}} \frac{dE}{dv} \cdot e^{-k(v)} \cdot dv \qquad (1)$$

to the experimental data by choosing the proper $k(v)$. It is precisely this technique that is used in thermal conductivity. Instead of the extinction constant $k$ we want to know the phonon scattering relaxation rate or the inverse relaxation time $\tau^{-1}$, i.e., instead of the distance over which the beam gets attenuated

## PHONONS

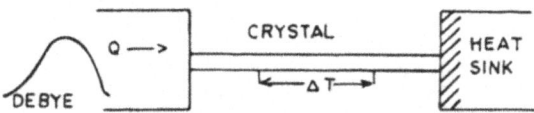

$$k \stackrel{?}{=} (\text{RELAXATION TIME } \tau)^{-1}$$

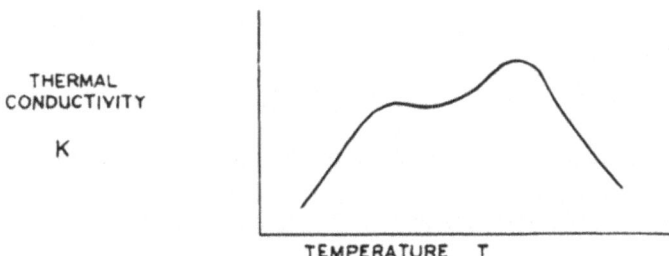

THERMAL CONDUCTIVITY

$$K = 1/3 \, C \cdot v \cdot v \cdot \tau$$
$$C = \text{SPECIFIC HEAT}$$
$$v = \text{SOUND VELOCITY}$$

(b)

THERMAL
CONDUCTIVITY

K

TEMPERATURE  T

$$K = \frac{1}{3} \int_{\text{DEBYE}} \frac{dC}{d\nu} \cdot v^2 \cdot \tau(\nu) \cdot d\nu$$

to 37% of its initial energy we ask for the time (Fig. 1b). A heater, attached to one end of the crystal, emits polychromatic phonons of total power $Q$ into the solid. Some of them get scattered by the defects, some by scattering processes inherent to the pure crystal (this is, of course, identical to light; in Fig. 1a we simply ignored the "fundamental extinction"). The resulting temperature gradient $T$ is measured. The thermal conductivity $K$

$$K = Q(A\nabla T)^{-1} \tag{2}$$

where $A$ is the cross section of the crystal, can be written as

$$K = \frac{1}{3} \cdot C_v \cdot v_{\text{ave}} \cdot l_{\text{ave}} = \frac{1}{3} C_v v_{\text{ave}}^2 \tau_{\text{ave}} \tag{3}$$

where $C_v$ is the specific heat of the crystal at constant volume, and $v$ and $l$ are velocity of sound and phonon mean free path $l = v\tau$, respectively.

The constant factor follows from gas kinetic considerations ([2]). Since we are interested in the frequency (and possibly also the temperature) dependence of $\tau$, we write $K$ in its integral form

$$K = \frac{1}{3} \int_q \frac{dC}{d\omega} v^2 \tau(\omega) d\omega \tag{4}$$

which we call the Debye model of thermal conductivity; $\omega$ is the phonon angular frequency $\omega = 2\pi v$. The integral has to be extended over all normal modes $q$ of the crystal. In order to keep the number of adjustable parameters manageably small, a number of important simplifications must be made, which we list below:

1. Debye density of states (see Fig. 2), including:
2. No distinction of phonon polarization.
3. Constant sound velocity.
4. The combined relaxation time $\tau_{\text{total}}$ is determined from the individual relaxation time $\tau_i$ describing the various scattering processes with

$$\tau_{\text{total}} = \left( \sum_i \tau_i^{-1} \right)^{-1} \tag{5}$$

These simplifications are so profound that only the experiment can tell whether they are justifiable. In the following we shall list four experiments which were used to test this model. A detailed description of one such test is given in Ref. 3.

In all four cases the conductivity of an undoped sample was compared with that of crystals containing known amounts of defects with phonon scattering cross sections preferably known from theory. A Debye integral was then fitted to the undoped sample. In view of the considerable number of more or less adjustable parameters available for the choice of $\tau_{\text{pure}}^{-1}$ (boundary, phonon—phonon, isotopic, and residual defect scattering), good fits to the undoped sample can always be obtained with little effort.

The next step consists of trying to fit the conductivity of the doped sample using the same integral with a relaxation rate $\tau_{\text{total}}^{-1} = \tau_{\text{pure}}^{-1} + \tau_{\text{defect}}^{-1}$, where $\tau_{\text{defect}}^{-1}$ should be known from theory, such that this fit contains no further adjustable parameters. If such a fit is successful in a number of different hosts containing different defects in varying concentrations, then one may conclude that such an analysis can actually be used to find an unknown defect scattering rate $\tau_{\text{defect}}^{-1}$, which is, of course, our goal.

## 1. Ge Isotopes([4])

The "pure" crystal contained enriched $Ge^{74}$ (*ca.* 95%), the "doped" crystal consisted of the normal isotopic mixture. According to theory ([5,6]),

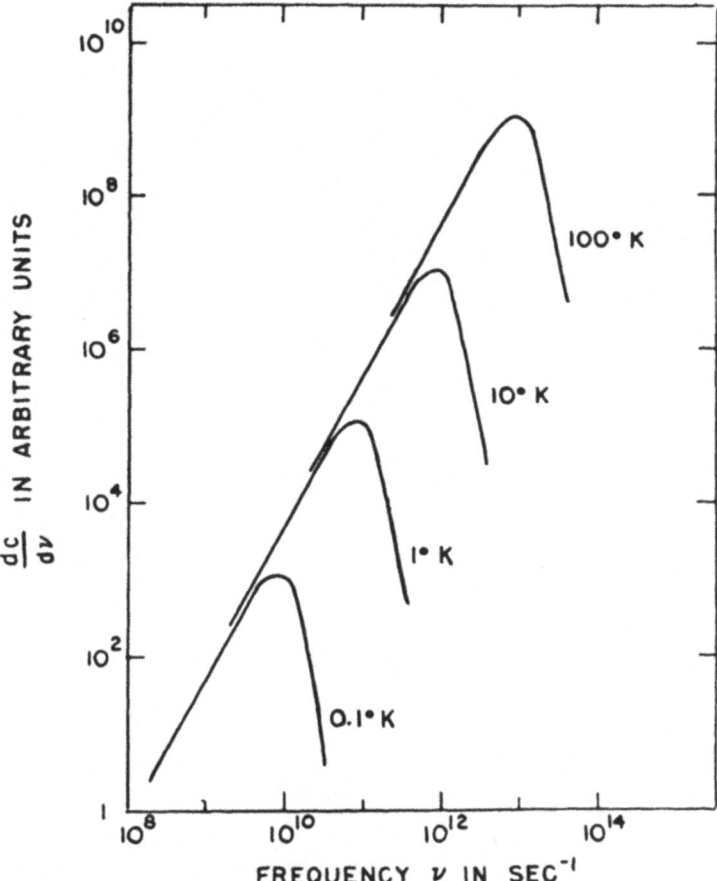

Fig. 2—Spectral distribution of the specific heat $c$ for a Debye density of states with the temperature as parameters. $v = \omega/2\pi$ is the phonon frequency.

the perturbation of the lattice periodicity should result in a phonon scattering rate

$$\tau_{\text{defect}}^{-1} = \frac{\Omega_0 \Gamma \omega^4}{4\pi v^3} \qquad \Gamma = \sum_i f_i \left(1 - \frac{M_i}{\bar{M}}\right)^2 \qquad (6)$$

i.e., a Rayleigh-type scattering law. Here $\Omega_0$ is the volume of the unit cell, $f_i$ the fraction of unit cells having mass $M_i$, $\bar{M}$ the average mass, and $v$ the velocity of sound.

The circles in Fig. 3 represent experimental data, the lines were calculated ([7]) following the description given above. After a fit to the enriched sample had been produced by properly adjusting $\tau_{\text{pure}}^{-1}$, a very good fit to the normal

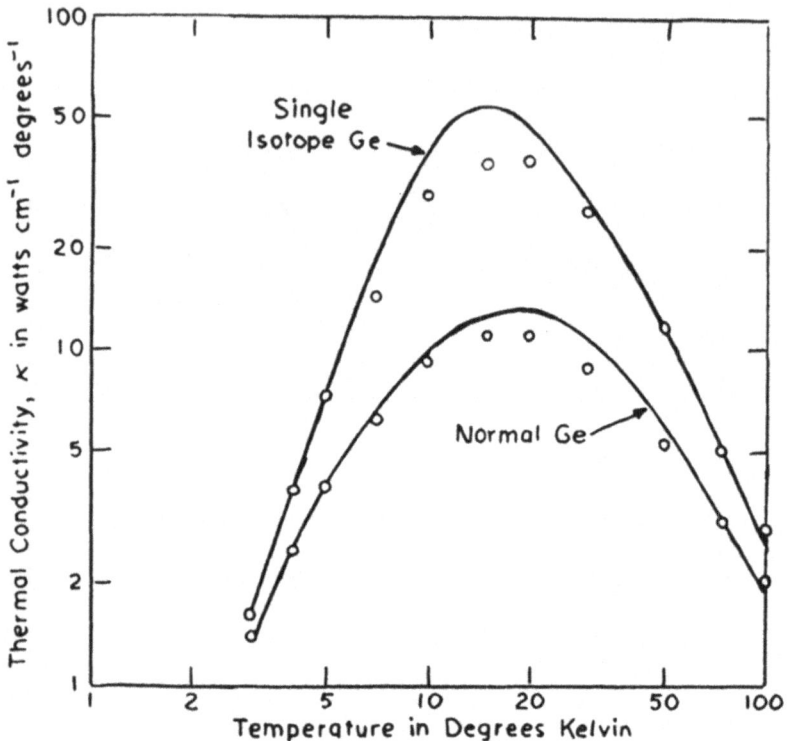

Fig. 3—Isotopic scattering in germanium. Circles represent experimental data, [4] Lines are machine fits using the Debye model [7].

Ge was obtained by including the Rayleigh term from equation (6) corresponding to the increased isotopic concentration.

## 2. Molecular Defects [3]

Here we have no theoretical prediction of the scattering cross section. Therefore the test had to proceed in a slightly different way: Fig. 4 shows the very strong reduction produced by $NO_2^-$ ions substituted for $Br^-$ in KBr. After curve fitting the conductivity of the undoped KBr, a scattering rate had to be determined that resulted in a fit to the lowest $NO_2^-$ doping. This was accomplished using $\tau_{\text{defect}}^{-1} = 8 \times 10^{-23} \text{ cm}^3 \; n_{NO_2}\omega$. It was then shown that a good fit to the more highly doped samples could be obtained by scaling $\tau_{\text{defect}}^{-1}$ according to the nitrite concentration $n_{NO_2}$, (see curves in Fig. 4). From this it was concluded that once a defect relaxation time was determined for one concentration, it was possible to fit the curves regardless of the rela-

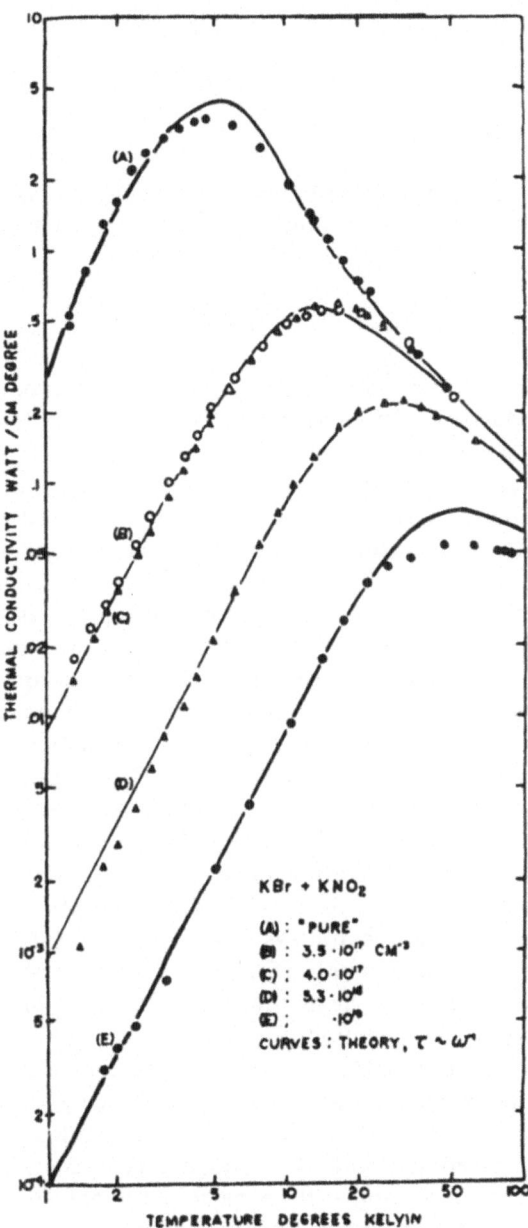

Fig. 4—Thermal conductivity of KBr:NO$_2^-$. Curves are machine fits. Nitrite concentrations in the crystals were: (A): 0; (B): $3.5 \times 10^{17}$cm$^{-3}$; (6): $4 \times 10^{17}$ cm$^{-3}$; (D): $5.3 \times 10^{18}$ cm$^{-3}$; (E): $5.7 \times 10^{19}$ cm.$^{-3}$ $n_{KBr} = 1.4 \times 10^{22}$ cm$^{-3}$.

tive strength of defect *vs.* intrinsic scattering by simply scaling the defect scattering according to the concentration and by using the Debye model.

## 3. Boundary Effect in LiF ([8])

According to theory ([9]) the phonon scattering by crystal boundaries in the regime of ballistic phonon flow is given by the frequency-independent scattering time

$$\tau_b = 1.12 \frac{d}{v} \tag{7}$$

where $d$ is the diameter of the crystal of square cross section. Figure 5 shows the thermal conductivity of four samples of LiF of high isotopic and chemical purity of different diameter. They were all cleaved off the same boule and their surfaces had been roughened by sand blasting to ensure applicability of the theory which requires absence of specular phonon scattering at the surfaces. Since the strength of the boundary scattering can be changed without changing the other physical and chemical properties of the crystal, we can also use the boundary effect to test the Debye model, using $\tau_b^{-1}$ as the theoretically known defect scattering term. This test proceeded in two steps. First the conductivity of the largest crystal was fitted. This was accomplished with an (arbitrarily chosen) term $\tau_u^{-1}$ describing phonon—phonon and residual defect scattering plus the boundary term as given in equation 7. The curve thus calculated is shown as the solid line labeled (A) in Fig. 5 ([10]). Below 6°K ($\theta_{LiF} = 722°K$) a least-squares fit to the data is proportional to $T^n$, where $n$ is equal to 3 as predicted by theory [equations (7) and (3)] to better than 1 %. Therefore in this temperature range $\tau_u^{-1}$ is entirely negligible. These data therefore represent a test of the Debye model in its very simplest form, namely, equation (3), which follows from equation (4) for a frequency-independent $\tau(\omega)$. Similar results are found for smaller crystals, and the quantity needed to fit the data was again found to be equal to the crystal diameter (except for the same 15 % mentioned in Ref. 10).

The second step proceeds in exactly the same way as above under 1 and 2. We increase the boundary scattering rate by reducing the sample size, and ask whether the experimental data can be correctly fitted by combining the previously determined $\tau_u^{-1}$ with the changing $\tau_b^{-1}$ according to equations (5) and (4). This comparison will be meaningful only in the temperature range where both intrinsic and boundary scattering processes scatter about equally strongly, i.e., near the peak of the curves. Again the agreement between the Debye model and experiment is very good. This is of particular interest. Boundary scattering occurs at the surface only, the other scattering processes occur in the bulk. Treating them as equal by adding both scattering

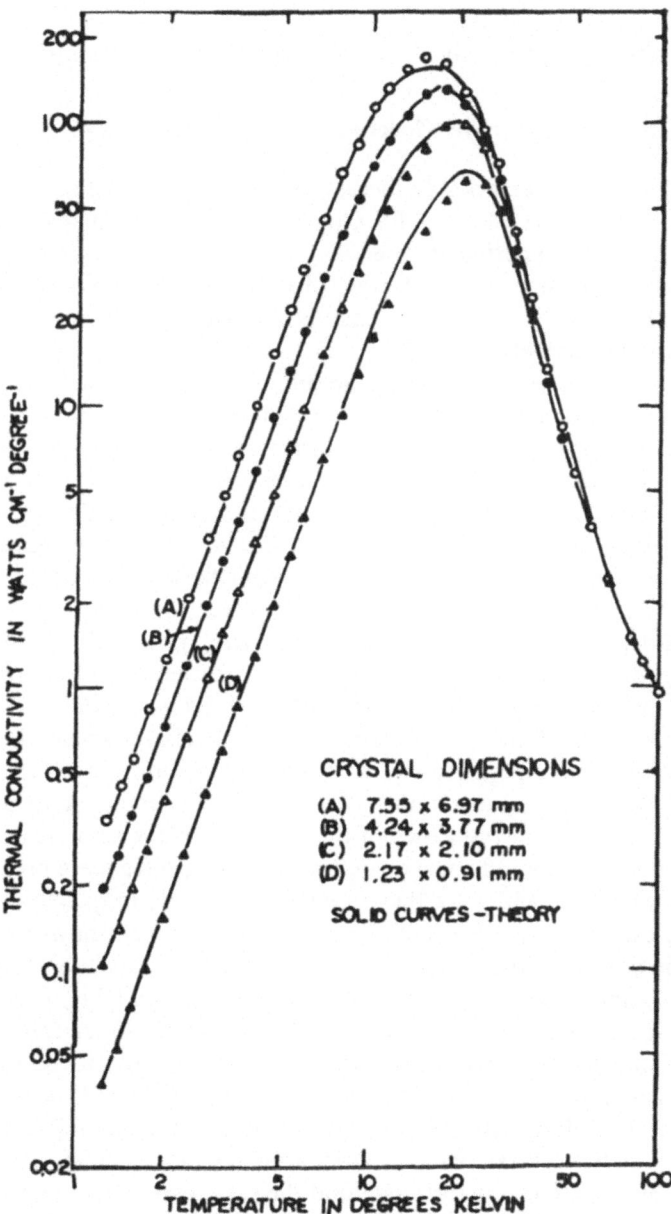

Fig. 5—Thermal conductivity of high-purity LiF crystels (99.99% Li⁷) for different crystal diameters. Curves are machine fits.

rates as done in the Debye model is questionable ([11]), in particular if heat is carried through the crystal in a Poiseuille type flow. We refer to Krumhansl's lectures. Apparently in our crystals these problems do not occur.

## 4. Limitations of the Debye Model of Thermal Conductivity.

In spite of the remarkable success of the Debye model as demonstrated above, we known that the simplifications it contains eventually must lead to disagreement with experiment, or else we must conclude that thermal conductivity is such a coarse tool that none of the fine points so far postulated by theory are noticeable. The next example, however, will indeed point out a limitation of the Debye model.

So far we have been talking about phonon—phonon three-quantum processes without distinguishing between normal, ie., momentum conserving, and Umklapp, i.e., momentum destroying, processes ([12]). The role of U-processes in causing a thermal resistance is fairly easily understood through the Bragg reflection which the resultant phonon undergoes. The N-processes by themselves, however, do not lead to a thermal resistance, since the total crystal momentum is conserved. Nevertheless they are expected to play a very important role in the transfer of heat by replenishing phonon states which had been depleted through momentum destroying processes. Thus, if certain scattering processes were particularly strong for certain phonons, like for instance in the case of U-processes, where strict selection rules must be expected because of the complicated dispersion relations near the edges of the Brillouin zones, N-processes, by funneling phonons from less strongly scattered states into states which scatter strongly, could contribute significantly to the thermal resistance of the crystal. For the same reason N-processes should have an influence on defect scattering, since these processes also destroy momentum and since they usually are strongly frequency-dependent. One such example is the Rayleigh-type scattering mechanism which we know is found in crystals with isotopic mass fluctuations. LiF crystals of high chemical purity and different isotopic composition ($Li^6$ and $Li^7$, is isotopically pure) have recently become available. The influence of isotopic mixtures on the conductivity in these crystals had to be expected to be quite small (the unit cell in equation (6) of the diatomic LiF lattice is formed by the LiF molecule). Therefore extreme care had to be taken to avoid unintentional impurities in the crystals which might simulate the effect we were looking for. For this reason it was fortunate that the same experiment was performed independently in two different laboratories with differently prepared samples ([8,13]). Practically identical results were obtained in both investigations. In Fig. 6 are shown those reported in Ref. 8. The solid surves were computed with the Debye integral, using the relaxation rate $\tau_{pure}^{-1}$ adjusted to fit

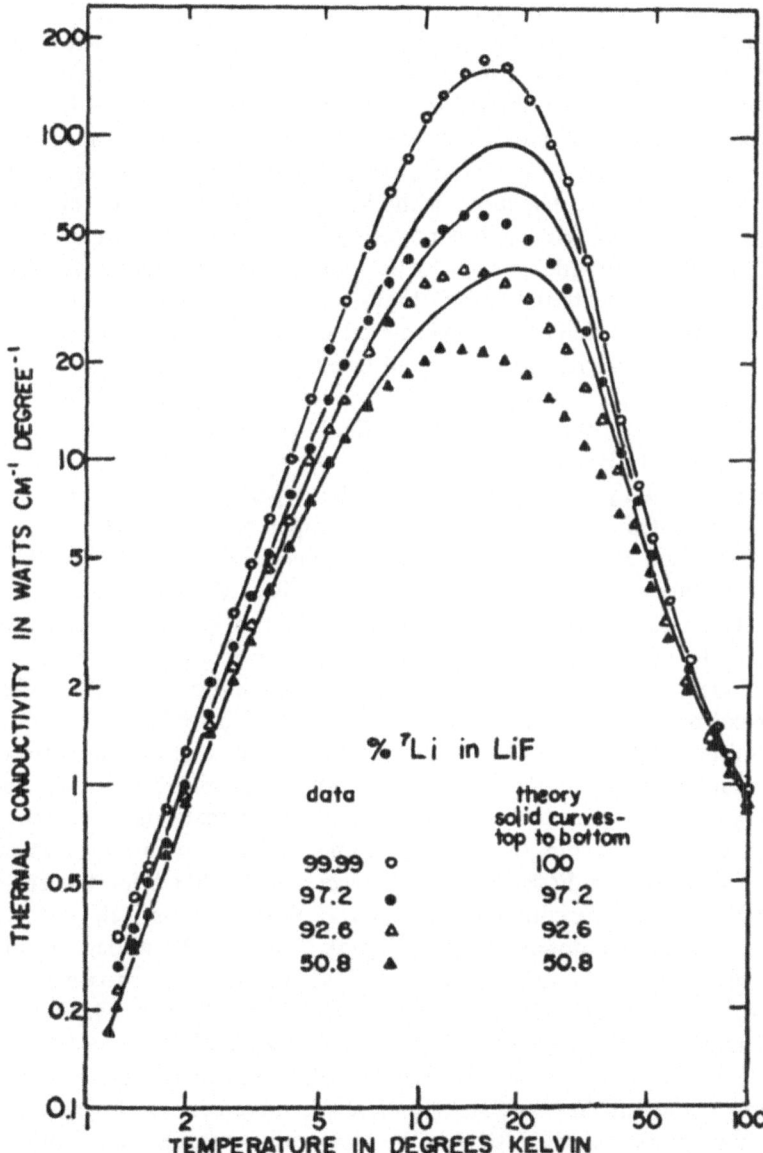

Fig. 6—Isotope effect in LiF. Solid curves computed with the Debye integral and the Rayleigh scattering term from equation (6). The crystals had slightly different diameters.

the pure crystal data plus a Rayleigh term (equation 6) as given by theory for the three isotopic concentrations (for the four crystals slightly different $d$'s had to be used because the samples had different diameters). The measured curves fall way below the computed ones in the temperature range around and above the maximum of the thermal conductivity curves. This is the range in which strong N-processes are most likely to occur: at low temperatures all anharmonic processes should vanish, at high temperatures U-processes should become dominant because most of the heat will then be carried by phonons of large wavevectors. Hence, it is presently believed that the additional scattering is indeed caused by N-processes funneling low- frequency phonons into the strongly scattered ($\propto \omega^4$) high-frequency states. This view is supported by the fact that the Debye model can describe the boundary effect in the same high-purity LiF quite well (see above under 3). Since boundary scattering is frequency-independent, it does not deplete certain phonon states more than others, and hence the N-processes cannot produce an enhancement of the "defect" scattering.

It has actually been possible to fit the experimental data obtained on isotopic mixtures of LiF over the entire temperature range ([13]) by using a theory designed specifically to take proper account of N-processes ([7]), but this procedure required a rather large number of additional adjustable parameters. Consequently, the agreement between calculation and experiment has to be considered with some caution. For our purpose, however, it is sufficient to note that there is indeed a strong indication of the influence of normal processes on the thermal conductivity. This raises two questions: Why didn't the N-processes shown up in case 1 and 2 of our test? In which cases should we expect N-processes to be of importance? From the theory it follows that in order to become important, their scattering strength must be large enough in comparison to the momentum destroying (U- and other) processes. These conditions apparently were not fulfilled for the enriched Ge with a maximum conductivity of ca 35 W/cm-deg as well as for KBr of natural isotopic composition with a peak at *ca*. 3.5 W/cm-deg. It appears furthermore, that if the LiF crystal with 92.6 % Li[7] had been chosen as "pure" and some scattering rate $\tau_{pure}^{-1}$ been adjusted to fit its conductivity (the open triangles in Fig. 6), then the Debye model would have indeed been able to describe the 50.8 % crystal. In other words: Even in LiF the N-processes lose importance in the less pure samples. We therefore conclude that in all the other crystals studied so far the impurity scattering was sufficiently strong to cover the influence of the N-processes. The only other material where influence of N-processes has been observed to-date is solid helium. Even in fairly imperfect crystals of He[3] and He[4] and mixtures thereof, the Debye model has failed in very much the same way as described above for LiF ([14]). Again we refer to Krumhansl's lectures in this volume.

From the discussion presented in this chapter we conclude that in most cases the Debye model gives an adequate description of the experimental results. We have therefore become confident that we can use this model to extract unknown defect scattering cross sections from thermal conductivity data.

## II. PHONON SCATTERING BY LOW-FREQUENCY IMPURITY MODES. MOLECULAR DEFECTS AND THE LI⁺ ION.

Thermal conductivity was among the first experimental tools through which the existence of impurity modes inside the acoustic continuum of solids was demonstrated ([15]). In the course of many attempts to understand these modes in detail it was learned that additional independent experimental techniques had to be employed, because it was found that frequencies of the impurity modes as well as their phonon scattering cross sections were in disagreement with existing theories. Much has since been learned about the photon-active impurity modes through measurements of optical absorption in the far infrared, and we refer to the lectures by Sievers, Balkanski, and Nardelli. There is, however, no good evidence that any one mode couples to both phonons and photons, which means that the two techniques cannot be used to study the same mode. In this dilemma help has come from an unexpected source, namely, from the study of molecular impurities in solid solution.

These molecular impurities have particularly simple phonon scattering cross sections which could be extracted with confidence from experiment using the Debye model, and near-infrared techniques could be used to search for the impurity modes associated with these defects. Again it was found that the modes which couple to the phonons do not show up in optical absorption, but using a simple model it was possible to deduce them from the modes which are visible in the near-infrared spectrum. It is therefore advantageous to begin our studies of localized excitations by studying molecular rather than monatomic defects.

### 1. The CN⁻ Ion in Solid Solution ([16])

The thermal conductivity of potassium halides is very strongly reduced in the presence of small concentrations of CN⁻ (see Fig. 7). Without any analysis of the data it is apparent that the phonons which are the dominant carriers of heat at about 0.6 and 6°K are scattered more strongly than the other phonons, and a resonant scattering mechanism appears likely. A quantitative description of the data was obtained using the Debye model and a defect scattering rate containing two resonance expressions of Lorentzian form:

$$\tau_{\text{defect}}^{-1} = \frac{nA}{\omega_0^2} \cdot \frac{(\omega/\omega_0)^2}{[1 - (\omega/\omega_0)^2]^2} + \frac{nA'}{\omega_0'^2} \cdot \frac{(\omega/\omega_0')^2}{[1 - (\omega/\omega_0')^2]^2} \tag{8}$$

Here $n$ is the spectroscopically determined concentration of $CN^-$ ions, the $A$-s are adjustable parameters containing the strength of the scattering processes, and the $\omega_0$-s are the resonant frequencies.

The solid curves in Fig. 7 are the best fits to the experiment using the scattering strength and resonant frequencies as given in Table I. Within the accuracy of the determination of $n$, $A$ was found to be independent of concentration. Naturally, one will try to associate the scattering with some new energy states connected with the $CN^-$ ion in the halogen vacancy, and one will ask about the nature of these states and also how they couple to the phonons. Progress has been achieved concerning the first part of this question by studying the stretching vibration of the $CN^-$ in the near infrared. It was found that at sufficiently high temperatures the infrared spectrum is very nearly identical to that of a freely rotating $CN^-$ ion, as shown by the characteristic rotation—vibration spectrum known for diatomic molecules in free space. At low temperature ($T < 80°K$ in the potassium salts) certain preferred orientations for the CN dumbbell were found and from stress-induced alignment these directions were determined to be in the $\langle 100 \rangle$ directions of the host lattice. In the liquid helium range the spectrum showed in addition to a very narrow central line two distinct sum and difference satel-

## TABLE I

**Resonant Frequency $\omega_0$ and Scattering Strength $A$ Determined for Several Defects in Alkali Halides, Using the Debye Model and an Inverse Phonon Scattering Relaxation Time of a Lorentzian form [see equation (8) and text].**

| System | $\omega_0$, in wave-Numbers, $cm^{-1}$ | $\omega_0$ (rad/sec) | $A$ (cm/sec)$^3$ | Reference |
|---|---|---|---|---|
| KCl :CN | 1.6 | $3 \times 10^{11}$ | $1.04 \times 10^{13}$ | 16 |
| KBr :CN | 1.6 | $3 \times 10^{11}$ | $1.04 \times 10^{13}$ | 16 |
| KI :CN | 1.6 | $3 \times 10^{11}$ | $2 \times 10^{13}$ | 16 |
| KCl :CN | 18.0 | $3.54 \times 10^{12}$ | $9.38 \times 10^{15}$ | 16 |
| KBr :CN | 18.0 | $3.54 \times 10^{12}$ | $9.38 \times 10^{15}$ | 16 |
| RbCl:CN | .68 | $1.28 \times 10^{11}$ | $4.9 \times 10^{11}$ | 19 |
| KCl :NO$_2$ | 20.0 | $3.8 \times 10^{12}$ | $4.77 \times 10^{15}$ | 18 |
| KCl :Li | 1.2 | $2.26 \times 10^{11}$ | $1.86 \times 10^{12}$ | 21 |
| KBr :Li | 3.2 | $6 \times 10^{11}$ | $1 \times 10^{11}$ | 21 |

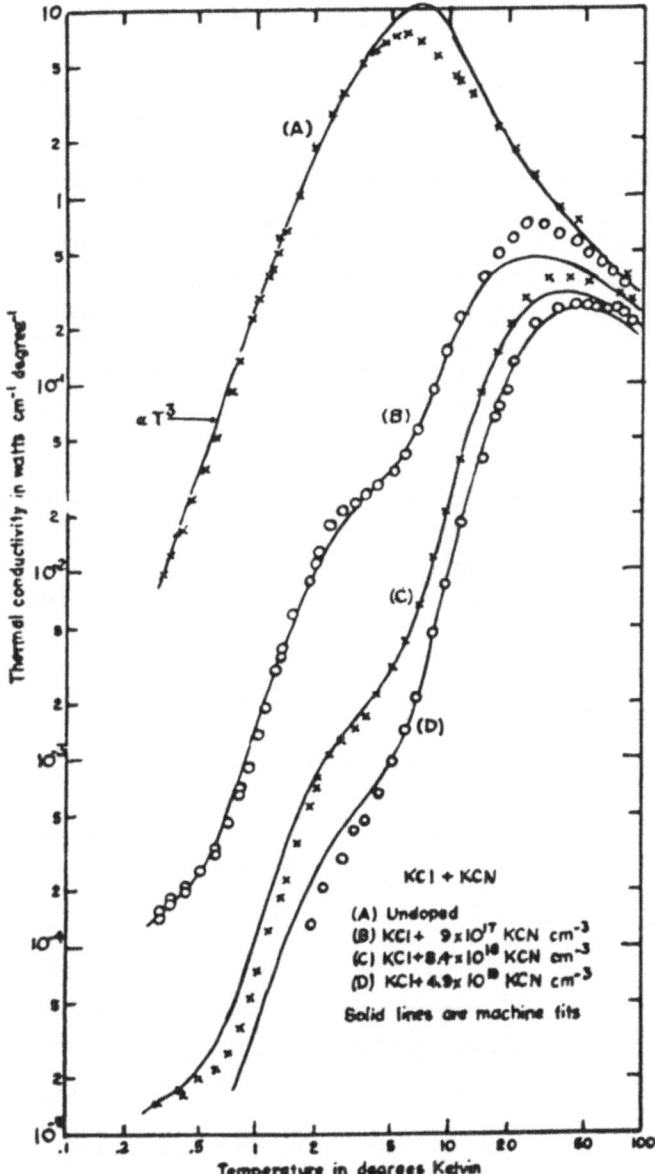

Fig. 7—Thermal conductivity of KCl containing different concentrations of CN$^-$ ions. The concentration of NCO$^-$ ions, inevitable as a result of a contamination of the starting material, varied between 1–3 × 10$^{18}$cm$^{-3}$. Solid lines are machine fits using the Debye model.

lites, indicating that another harmonic oscillator was coupled to the stretching vibration. This oscillation was identified as a libration about the six $\langle 100 \rangle$ equilibrium orientations. All of these results could be described very satisfactorily assuming that the molecule ion was subjected to a cosine potential of octahedral symmetry ([17]). Thus the lowest motional states, those with energies well below the potential barrier could be described as simple harmonic motion of the $CN^-$ ion around the $\langle 100 \rangle$ equilibrium orientations, and the states with energies far above the potential barrier, completely unaffected by this potential, were those of a freely rotating ion.

The barrier parameter $K$ and thus the magnitude of the barrier height can be computed from the observed librational frequency and the rotational inertia of the molecule. Furthermore a quantum mechanical calculation shows that the librational levels are split. The reason for this is that the eigenstates of a particle subjected to $N$ equivalent harmonic potentials are $N$-fold degenerate. If, however, the barrier height between these potentials is finite so that some overlap between wavefunctions in neighboring wells becomes possible, the degeneracy becomes at least partially lifted. This splitting, known as the tunnel splitting, is a measure of the frequency with which the particle will shuttle back and forth between the different wells without being excited to states above the barrier, which are of course shared by all wells.

Figure 8 shows a two-dimensional sketch of the potential and the eigen-

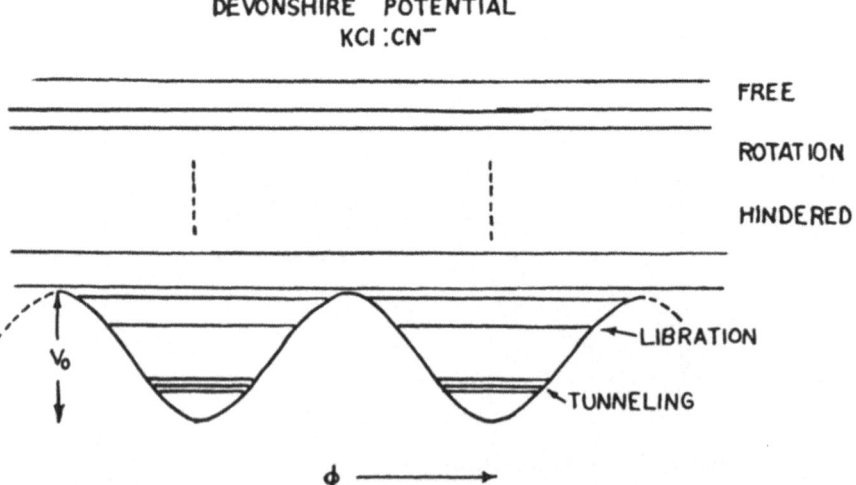

Fig. 8—Energy levels for a diatomic molecule subjected to a three-dimensional cosine potential of octahedral symmetry. Schematic. The curve indicating the potential as function of angle represents a cross section through the potential surface in a (100) plane. The minima occur in $\langle 100 \rangle$ directions, the maxima in the $\langle 110 \rangle$ directions.

values for our system. If we picture the potential as a cross section through a (100) plane, then the potential maxima occur in the $\langle 110 \rangle$ orientations, and $V_0 = 1.25$K.

The same study was also made with RbCl, NaCl, and NaBr as hosts. Figure 9 shows the conductivity of RbCl:CN[18]. The solid curves (below 1°K) were computed using the Debye model and a resonance scattering term of Lorentzian form. The dip at high temperatures could not be fitted with such a resonance expression. The resonance angular frequency $\omega_0$ connected with it was therefore derived in the following way: The temperature $T_0$ at which the dips occur that can be fitted with a Lorentzian scattering cross section can be determined as the temperature at which the two tangents drawn to the conductivity curve below and above the dip intersect with each other. A comparison of this—somewhat arbitrary—temperature $T_0$ with the resonance frequency $\omega_0$ derived from the machine fit resulted for altogether six different resonances in a simple connection:

$$\hbar\omega_0 = 4.25kT_0 \tag{9}$$

to within 10%. Therefore 1°K corresponds to $\cong$ 3 cm$^{-1}$ in the wavenumber

## TABLE II

**Energy States of Rotational Motion and Potential Barriers of the CN$^-$ Ion in some Alkali Halides, as Derived from Near Infrared Spectroscopy and Specific Heat Measurements, Compared with Resonant Frequencies Obtained from Measurements of Low Temperature Thermal Conductivity.** All energies are given in wavenumbers (1 cm$^{-1}$ $\cong$ 1.24 × 10$^{-4}$ eV).

| Host | Libration frequency (cm$^{-1}$) | $V_0$ (cm$^{-1}$) | Four lowest states. Symmetries, degeneracies, and energies above bottom of potential well (cm$^{-1}$) | | | | Resonant frequencies $\omega_0$ from $R(T)$ (cm$^{-1}$) |
|------|------|------|------|------|------|------|------|
| | | | $A_{1g}(1)$ | $T_{1u}(3)$ | $E_g(2)$ | $T_{2g}(3)$ | |
| KCl | 12 | 25 | 6.0 | 7.4 | 8.4 | 19.5 | 1.6 | 18 |
| KBr | 12 | 25 | 6.0 | 7.4 | 8.4 | 19.5 | 1.6 | 18 |
| KI | 12 | 25 | 6.0 | 7.4 | 8.4 | 19.5 | 1.6 | not determined |
| RbCl | 19 | 44 | 9.5 +.625 | 9.5 +.75 | 9.5 | 28.5 | .68 | ~33[a] |
| NaCl | not observed | >100 | tunnel splitting $\ll 10^{-10}$cm$^{-1}$ | | | | not observed $\ll 10^{-2}$cm$^{-1}$ | ~140[a] |
| NaBr | not observed | >100 | not determined | | | | | |

[a]This energy was not determined by curvefitting, but rather using relation (9), see text.

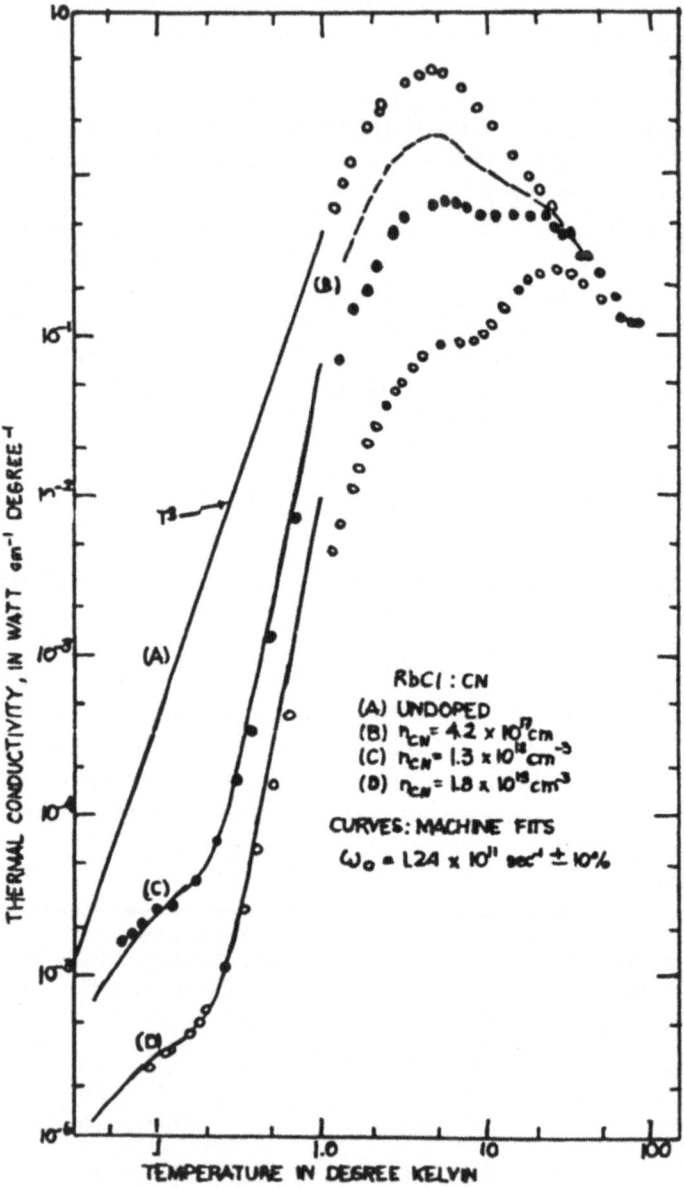

Fig. 9—Thermal conductivity of RbCl:CN. The solid curves (below 1°K) were computed using the Debye model and a resonance scattering term of Lorentzian form.

measure $(1 \text{ cm}^{-1} \triangleq 3 \times 10^{10} \text{ sec}^{-1}$ for frequency $f$; $1 \text{ cm}^{-1} \triangleq 1.42°K$ for $\hbar\omega = kT$, and $1 \text{ cm}^{-1} \triangleq 1.24 \times 10^{-4} \text{ eV}$ in energy).

Relation (9) was then used also for dips occurring at higher temperatures, like for the high-temperature dip in RbCl:CN. In this way $\omega_0 = 33$ cm$^{-1}$ was determined from $T_0 = 11°K$. It must be emphasized that such a determination is questionable, because we do not know the scattering cross sections for these high temperature dips, but we shall see in the following that this choice appears reasonable.

In Table II we summarize the information about the low-lying states as obtained for the systems studied. In addition to these spectroscopic studies, specific heat measurements were performed ([16]). At low temperatures the phonon density of states in the host lattice becomes quite small, and impurity induced states should become noticeable as anomalies in the specific heat. These anomalies were indeed observed (in KCl:CN). They could be explained quantitatively as a set of Schottky anomalies with the energies and degeneracies predicted for the tunnel states from the infrared work. It must be added at this point that the tunnel states and the states of free rotation were not resolved in the infrared spectrum, although their presence was indicated through the width and shape of the spectral lines and bands. We refer to Ref. 16 for these spectra.

Now that it has been established that the CN$^-$ ion is remarkably mobile in some of the alkali halide host lattices we return to the question as to the states which cause the strong phonon scattering observed in thermal conductivity. From a comparison of the last two columns in Table II with the motional energy levels the answer suggests itself quite readily: there is a very close agreement between the energy of the tunnel splitting (columns 4–6) and the energy of the low-temperature resonance in all hosts. In NaCl, where it was concluded that the CN$^-$ is frozen-in and hence that the tunnel splitting is very small, no low-temperature scattering was observed, from which an upper limit for the frequency of the low-temperature resonance could be deduced. The high-temperature resonances agree with the energy difference between the tunneling states (columns 4–6) and the states close and above the potential barrier (column 3), whereas the energy between the tunneling states and the librational state $T_{2g}$ is not observed in thermal conductivity in any of the crystals. Note that by using equation (9) for RbCl and NaCl barrier heights are obtained which are consistent with the ones determined spectroscopically. We take this as evidence that our method of determining frequencies of phonon resonances occurring at high temperatures is correct.

The close agreement between the energies derived from spectroscopy and specific heat on the one side and from thermal conductivity on the other side has been interpreted with the model that the tunneling states and the states of free rotation couple strongly to the phonons, whereas the librational ones do not. This picture has been confirmed by the observation that molecules which are known to be frozen in but do librate (e.g., $NO_3^-$ or $NCO^-$), have no noticeable influence on the thermal conductivity [3,16]. Furthermore it explains why only the librational satellites could be resolved in the IR, but not the satellites belonging to tunneling and rotation: These states are probably lifetime broadened because of their interactions with the phonons.

## 2. The $NO_2^-$ Ion in Solid Solution [19]

Measurements of infrared absorption and thermal conductivity on $NO_2^-$ doped samples have been explained in the same way as the observations on $CN^-$ presented above. Since the $NO_2^-$ is an asymmetric top, the low-lying motional states are considerably more complex and cannot be reproduced here. All the observations, however, can be interpreted with the picture that the rotational and the tunnel states scatter phonons, whereas the librational ones do not.

It may seem that the study of the motional states connected with the rotation of impurity molecule ions has sidetracked us from our real goal, namely, the study of impurity modes. One can, of course, stretch the definition of the word "impurity mode" to encompass these motional states, but we shall see that the reason for our discussing them in this connection lies deeper. This will become clear when we discuss the small $Li^+$ ion.

## 3. The $Li^+$ Ion in Solid Solution

The first observation of the peculiar properties of the small $Li^+$ ion came from thermal conductivity measurements [20] They have since been extended and will be published shortly [21]. In the following we shall summarize the results. Figure 10 shows the experimental data obtained on KCl:Li. The LiCl used as doping was reagent grade material of normal isotopic composition, i.e., 92.6% $Li^7Cl$. The prominent feature is the strong reduction of the thermal conductivity at low temperatures. Neither a rapid quench nor a slow cool after a high-temperature anneal changed the essential features of these curves. It was found that in samples which had been kept at room temperature for about one year the thermal conductivity below $1°K$ had increased in such a way that for these samples the conductivity was exactly like that observed in crystals with smaller $Li^+$ doping. Quenching restored the original low conductivity. The data shown in Fig. 10 therefore were obtained on quenched crystals. The solid curves below $1°K$ were computed, using the Debye model

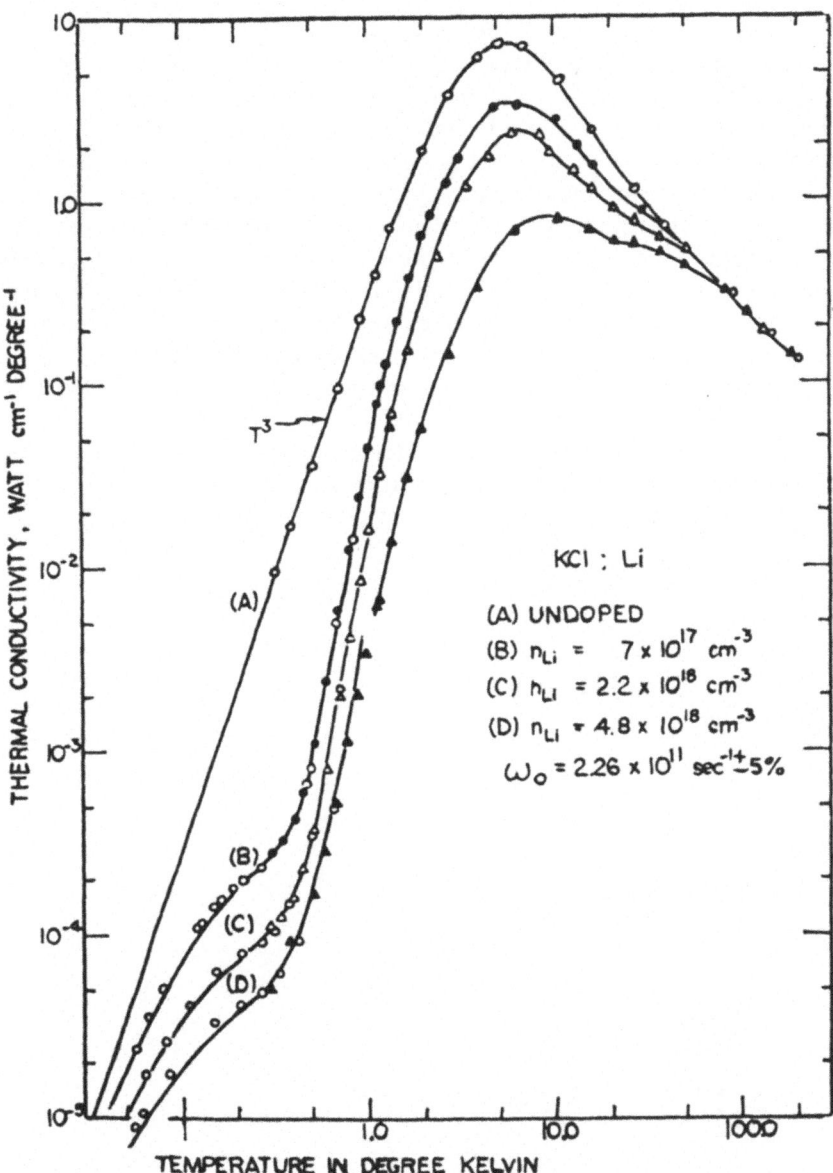

Fig. 10—Thermal conductivity of KCl containing several Li⁺ concentrations as determined from flame photometry. The solid curve through the pure KCl data is a machine fit. In the doped crystal the machine fit (solid curves) extends only up to about 1°K.

and a relaxation rate of Lorentzian form. Resonant angular frequency $\omega_0$ and scattering strength $A$ are listed in Table I. Note how well the effect of the $Li^+$ on the thermal conductivity scales with the $Li^+$ concentration as determined by flame photometry. An interesting comparison of $A$ and $\omega_0$ is shown in Fig. 11, where $A\omega_0^{-2}$ is plotted versus $\omega_0$ for the molecular defects and also for the $Li^+$. It follows that in first approximation $A\omega_0^{-3} = 1.8 \times 10^{-22}$ cm$^3$, independent of the resonant frequency and the scattering center. The striking similarity between the scattering caused by the monatomic and the molecular impurities suggested that the $Li^+$ in KCl was causing an impurity mode similar to that found with the molecular defects. There the low-lying resonance was attributed to a tunneling motion between equivalent equilibrium orientations. Since no rotational inertia is associated with the monatomic $Li^+$ ion, the closest similarity with the molecular case is found if the small $Li^+$ is not stable in the center of the cavity, and the potential to which it is subjected has several equivalent off-center positions. This model was tested with measurements of the low-temperature dielectric constant ([22]) and by the observation of an electrocaloric effect ([23]): Lithium-doped KCl crystals

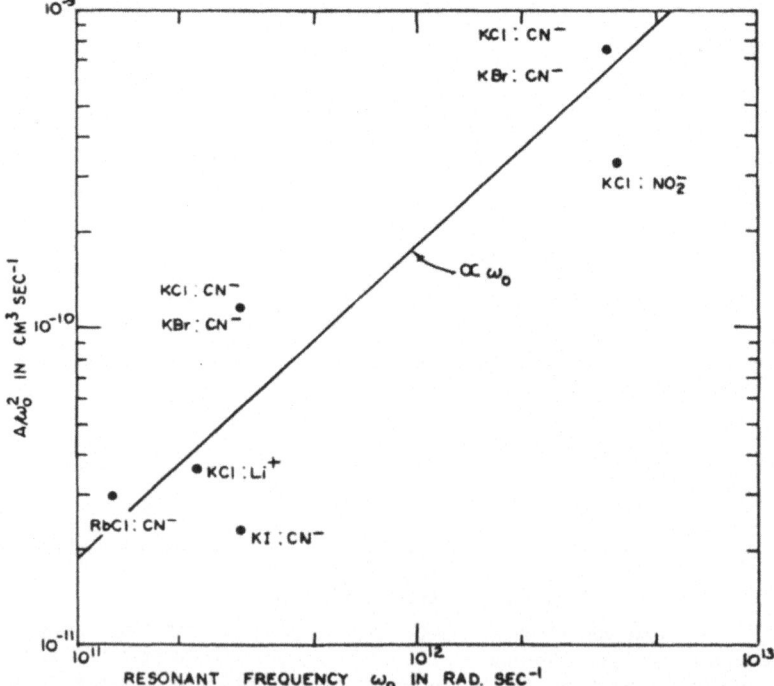

Fig. 11—Dependence of the scattering strength $A$ (equation 8) on the resonance frequency $\omega_0$. In first approximation $A$ appears to be proportional to $\omega_0^3$ for the systems studied. The only exception found so far is for KBr:Li, see text.

showed an increase in dielectric constant which increased with $T^{-1}$ as expected for a Langevin type polarizability. These results were identical to those obtained on molecular defects, which had been explained through an alignment of the permanent electric moment of the molecule ion[22,24]. In KCl:Li such an electric moment could come about through the positive lithium ion not coinciding with the negative effective charge of the cavity, which for reasons of symmetry is expected to be located in the cavity center. In the electrocaloric effect the alignment of the Li$^+$ ion in the presence of an external electric field was used to decrease the entropy of the system. Adiabatic removal of the electric field resulted in a randomization of the ions, thus increasing the entropy of the Li system and decreasing the temperature of the sample. The results obtained here were in qualitative agreement with previous measurements of the electrocaloric effect on KCl containing OH$^-$ ions [25,26], and the electric dipole moment found for the Li$^+$ was 2.5 debyes $= 8.4 \times 10^{-30}$ A-sec-m $= 0.53\ e$Å, assuming a Lorentz local field $E_{loc} = (\epsilon + 2)E_{app}/3$, with $\epsilon = 4.5$.

In the language of our classical model this means that the $x$-coordinates of the equilibrium positions of the Li$^+$ are 0.53 Å away from the cavity center. A schematic two-dimensional model of the potential is shown in Fig. 12. The higher energy levels are those of a three-dimensional oscillator, whereas the lower states, in particular the ground state, are split due to tunneling in much the same way as those of the rotational motion of impurity molecules. In this model phonon scattering would occur by excitation of the Li$^+$ ion between the tunnel states.

This model has since been refined with a quantum mechanical calculation and estimates for the shape and the barrier heights for the various translational motions have been given [27-29]. From inelastic measurements [30] on KCl:Li it has been concluded that the absolute minima of the potential surface lie in the $\langle 111 \rangle$ directions. Recent calculations [31] tend to support these results: It was found that in the KCl lattice in which the nearest neighbor Cl$^-$ ions were allowed to relax around the Li$^+$ the energy of the lattice had a minimum for the Li$^+$ displaced in the $\langle 111 \rangle$ directions, whereas maxima were found for $\langle 110 \rangle$ displacements. The difference in energy was found to be 40 cm$^{-1}$. For a motion from $\langle 111 \rangle$ to $\langle \bar{1}\bar{1}\bar{1} \rangle$ a barrier height of 2000 cm$^{-1}$ was found. In $\langle 100 \rangle$ the potential is assumed to have a saddle-point [32]. According to these calculations (and also the ones discussed in Ref. 28) the ion should avoid the center of the cavity, and its motion should be a quasirotation along the edges a cube about 1 Å wide whose center coincides with the center of the cavity. From these considerations it appears that the striking similarity between the observed phonon scattering by the Li$^+$ and that found for molecular defects is perhaps not too surprising. We conclude that the impurity mode associated with the tunneling of the Li$^+$

MONATOMIC  IMPURITY  WITH
CENTRAL  INSTABILITY

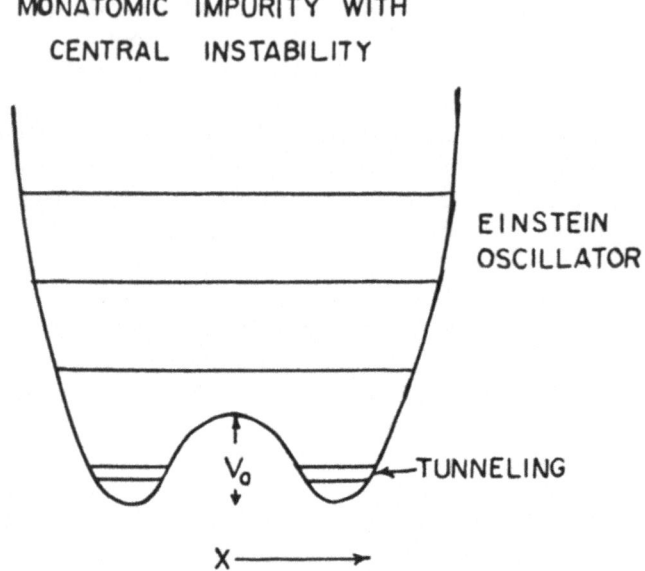

Fig. 12—Schematic one-dimensional potential for an off-center
ion. Note that the x-coordinate describes a translation, not a rotation; as a result the potential barrier rises high near the cavity wall.
The number of tunneling levels will depend on the number of
equivalent wells. See text and footnote 34. There is some evidence
for the existence of the higher states described as Einstein oscillator
levels. In the far infrared a broad absorption centered at about 40
cm$^{-1}$ has been observed (A.J. Sievers, private communication). It
may be caused by transitions between these infrared-active states.

ion, which has been called the "tunneling mode" has a very low energy,
scatters phonons with a scattering cross section of Lorentzian resonance
form, and its quantum mechanical description appears to be similar to that
of the tunneling of a molecular defect.

When an electric field is applied to the crystal the level splitting of the
tunnel states ($\hbar\omega_0$) should increase, and consequently the phonon scattering
should shift to higher frequencies (we consider a one-dimensional case, like
Fig. 12 with the field applied in the x-direction). Consequently, the thermal
conductivity should decrease at temperatures above the dip $T_0$, and increase
below $T_0$. This has indeed been observed ([21]). The results are in qualitative
agreement with one resonance frequency shifting to higher frequency $\omega_0$
+ $\Delta\omega$ in an applied field. More careful studies of this effect presently under-
way at Cornell ([33]) have shown that the picture of a two-level system causing
the phonon scattering is not adequate to describe the observations and that
the realistic three-dimensional potential must be considered. In a potential

of $n$ equivalent wells the lowest vibrational state consists of $n$ states which will be partly degenerate ([34]). Off-hand it is not clear why their phonon scattering is described with one resonance frequency alone, and it is conceivable that the $\omega_0$ found for zero electric field is an average of several resonances spaced perhaps within a factor of two in frequency. An electric field should change the levels, lift degeneracies, and furthermore influence the matrix elements for the coupling to the phonons. At least some of these effects are observed in the studies mentioned ([33]), and it is hoped that they together with specific heat and dielectric measurements will shed more light on the impurity modes of the tunneling type.

In addition to the dip in the thermal conductivity curves observed at low temperatures (Fig. 10) a dip at high temperatures is observed. Although it appears to be rather weak it must be kept in mind that it occurs at a tem-

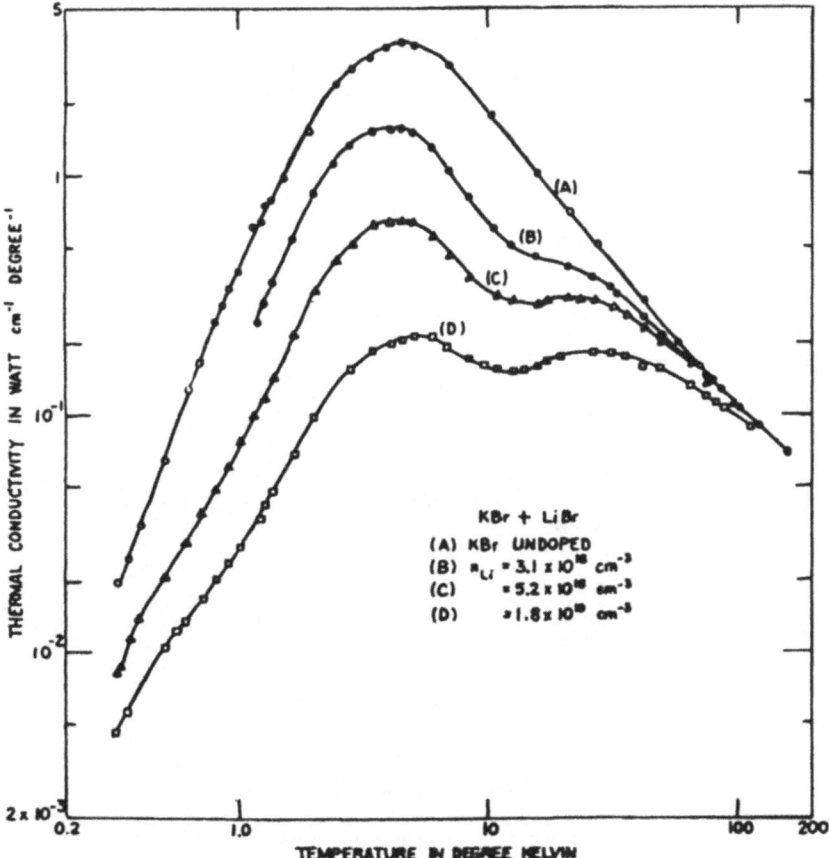

Fig. 13—Thermal conductivity of KBr:Li.

perature where intrinsic processes considerably reduce the phonon mean
free path, whereas the low-temperature dip occurs at a temperature where
the intrinsic mean free path is equal to the sample diameter, *ca.* 0.5 cm.
The shape of the high-temperature dip is similar to that observed for RbCl:CN
at high temperatures. Hence we suspect another resonance scattering process
and we try to analyze it in the same way the RbCl:CN data were analyzed:
From $T_0 = 21°K$, $\omega_0 \triangleq 63 \text{ cm}^{-1}$ is determined. It is tempting to associate
this resonance with an excitation of the Li$^+$ ion across the barrier height
estimated to be $\sim 40 \text{ cm}^{-1}$ ([31]) (see above). We shall see in Section III,
however, that very similar dips in the thermal conductivity of doped alkali
halides are observed in cases where a central instability of the impurity ion
can almost certainly be excluded. We therefore postpone the discussion of
the possible nature of this high-energy resonant mode until later.

Fig. 13 shows the thermal conductivity of KBr:Li for three different Li$^+$
concentrations. Again two dips are observed in the doped samples, but the
scattering at low temperatures is now considerably weaker. The high-tempera-

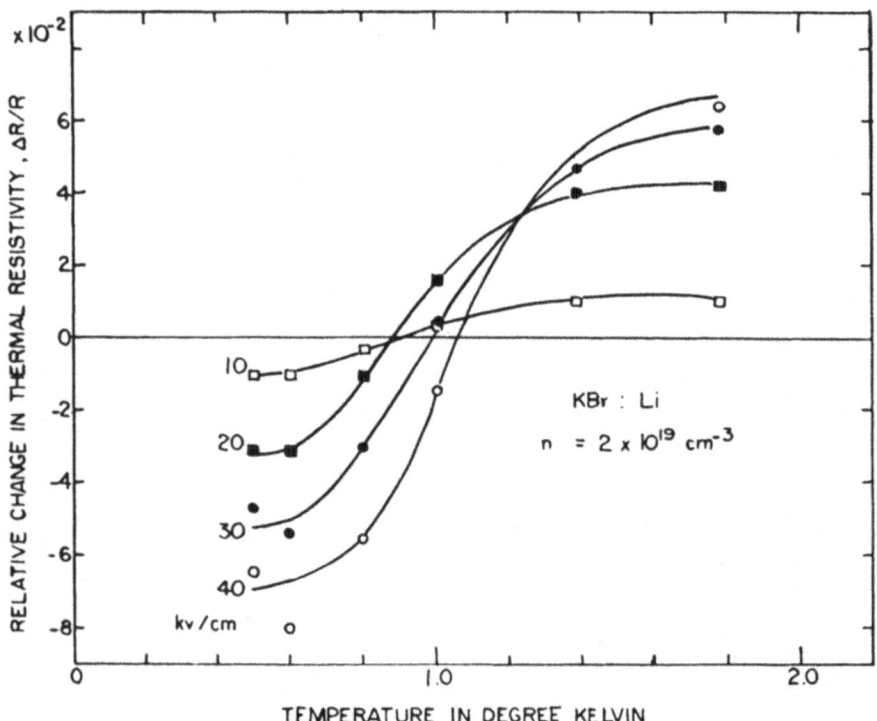

Fig. 14—Influence of an electric field on the thermal conductivity of KBr:Li for four
different fields applied along [100] perpendicular to the heat flow. Experiment. Thermal
resistivity = (Thermal conductivity)$^{-1}$.

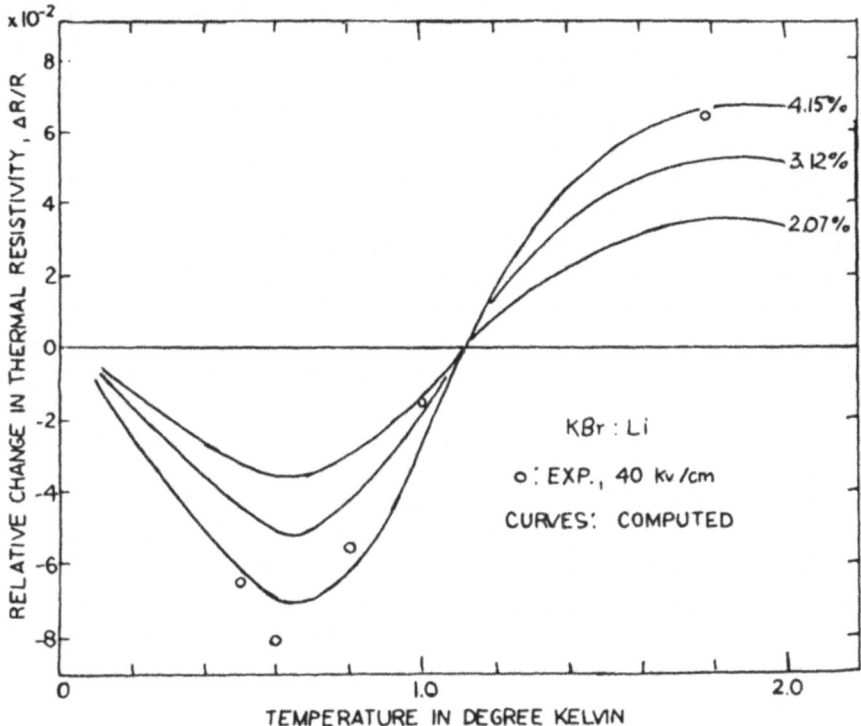

Fig. 15—Comparison of experimental results shown in Fig. 14 with theory. Solid curves calculated for a shift in resonant frequency by amount indicated. Zero field resonant frequency $\omega_0 \triangleq 3.2$ cm$^{-1}$.

ture scattering is about comparable to that found in KCl:Li, considering the higher concentrations dissolved in the KBr host lattice. Due to its weakness the low-temperature resonance appears partly washed out by the high-temperature resonance. It existence, however, could be clearly established by measuring the influence of dc electric fields applied perpendicular to the heat flow on the thermal conductivity ([21]) (see Figs. 14 and 15). From the excellent agreement between experiment and calculation we conclude that the zero-field low-temperature scattering can indeed be ascribed to a resonance scattering process with a cross section of Lorentzian form. Its parameters $A$ and $\omega_0$ are listed in Table I. A comparison with the scattering strengths observed for all the other resonance scattering strengths determined so far shows that $A$ falls about 300 times below the expected value. It therefore appears that the nature of the scattering process in KBr:Li is quite different from the scattering processes discussed so far. This is confirmed by the absence of an electrocaloric effect ([35]) and by far-infrared studies of the infrared-active impurity mode found at 17.8 cm$^{-1}$ in KBr:Li$^6$ ([36]). Stress-induced frequency shifts of this absorption line were found to be in agreement with the assumption that the $Li^+$ ion in KBr sits in a environment of cubic symmetry [37]. In view of the fact that the reduction of the thermal conductivity scales well with the $Li^+$ concentration, it seems unlikely that the scattering be caused by some Li aggregates. Quenching studies have not been performed yet, but the crystals used for the far-infrared work had exactly the same history as the ones used for thermal conductivity. The nature of the impurity mode at $3.2 \ cm^{-1}$ is therefore not understood. No similar modes seem to have been observed yet with molecular or other monatomic impurities. The high-temperature dip in KBr:Li occurs at $13°K$. Using relation (9) $\omega_0 = 39 \ cm^{-1}$ is determined. It does not show an isotope effect; two samples, one doped with LiBr of normal isotopic composition, the other doped with enriched $Li^6$-Br in almost identical concentration were measured between 4 and $100°K$. Within the experimental accuracy no effect of the different isotopic mass was observed. It is believed that an effect of $10\%$ as found in the far-infrared work should definitely have been observable. We postpone the discussion to Section III. For completeness' sake we mention results obtained on the only other lithium-doped alkali halide on which the thermal conductivity has been studied. In NaCl:Li no low-temperature depression is noticed. A high-temperature dip with $T_0 \sim 45°K$ corresponding to $\omega_0 = 135 \ cm^{-1}$ is similar to those to be discussed below.

## III. PHONON SCATTERING BY POINT DEFECTS.

### 1. Theory.

The introduction of point defects into the lattice disturbs its periodicity and thus gives rise to phonon scattering. The simplest such case is the isotope effect, where the perturbation is caused by the difference in mass only. If one disregards any changes in the vibrational spectrum of the lattice and merely asks how a phonon from the state $q$ gets elastically scattered into the state $q'$, one obtains the well-known Rayleigh type scattering law given in equation (6). Its validity has been tested on several materials with different isotopic compositions, as explained in Section I.

If the perturbation is caused by different force constants or a strain field resulting from a mismatch of the atomic radii of host and impurity atoms the scattering cross section also is expected to obey a Rayleigh-type scattering law [6], although there is still some incertainty about the numerical factor and even the frequency dependence, resulting largely from our ignorance of the actual displacement field [38]. Very little is known experimentally about the phonon scattering by such strain fields. One important point, however, seems to have been established: mass *and* spring changes around the imperfection may give compensating contributions to the scattering amplitude. The scattering power of the center may thereby be reduced below the value calculated on the basis of pure isotopic scattering ([39,40]). Quantum mechanically it means that the matrix elements describing the two scattering processes can cancel each other partially and therefore have to be added before they are squared. The situation is similar to impedance matching for sound waves traversing the interface between two media, where the amount of energy reflected can be very small, although the density mismatch is large, because the difference in elastic constants compensates for the different densities. Evidence for this effect has been found in KCl: Ag and KBr: Ag (see Fig. 17).

The introduction of point defects also produces impurity modes in the lattice. They can take up energy from the traveling waves, but they do not contribute to the heat flow. Their influence on the thermal conductivity has been described with two different techniques which shall be sketched in the following. In the presence of the defects a certain part of the phonon spectrum is replaced by impurity modes. We assume that this part is described by one average frequency $\omega_s$ and a certain width $\Delta\omega$. Each impurity mode has a spatial damping factor describing the localization of the mode. Since no traveling waves of frequency $\omega_0$ are left in the crystal only three-quantum processes have to be considered in a manner similar to Umklapp processes. The relaxation rate for this process ([15]) has a resonance form and depends on the phonon frequency as well as the temperature. The weakness of this model lies in the large number of adjustable parameters and also the rather strict requirements about how the impurities affect the vibrational spectrum. The second approach ([41-47]) was aimed at eliminating some of these shortcomings by actually calculating the response of the defect lattice to a periodic driving force. An incoming plane wave will cause a wave to go out from the defect in much the same way as scattered light will go out from a scattering particle (Huyghens principle). The scattered amplitude was calculated in the harmonic approximation using Green's functions techniques. We refer to Nardelli's lectures. The results obtained in all of these investigations are of the following basic form

$$\tau^{-1} = \frac{A\omega^4}{(\omega_0^2 - \omega^2)^2 + \Gamma\omega^6} \tag{10}$$

This (elastic) scattering process has resonant character. The resonant frequency $\omega_0$ characterizes the frequency range in which the response of the lattice will be a maximum, and will be determined by the specific defect system. We shall write down the relaxation rate for the case of isotopic defects using the Debye approximation for the phonon density of states for the umperturbed lattice, using the nomenclature as used in Ref. 46:

$$\tau^{-1} = \frac{N_s \beta^2 Q \omega^3}{N(1 - \beta \omega^2 P)^2 + \beta^2 \omega^4 Q^2} \tag{11}$$

where $N_s$ is the concentration (number density) of defects, $N$ is the concentration of host atoms, $\beta = \Delta m/m$, $m$ is the mass of host atom, $m + \Delta m$ is the mass of the defect, $\omega_D$ is the Debye limiting angular frequency, and

$$Q = \frac{3\pi\omega}{2\omega_D^3}$$

$$P = \frac{3}{2\omega_D^3}\left(2\omega_D + \omega \ln\left|\frac{\omega_D - \omega}{\omega_D + \omega}\right|\right)$$

In this case

$$\omega_0^2 = \frac{1}{\beta P} = \frac{m}{\Delta m} \frac{2\omega_D^3}{3\left(2\omega_D + \ln\left|\frac{\omega_D - \omega}{\omega_D + \omega}\right|\right)}$$

For sufficiently large $\Delta m$ only low frequncies have to be considered, and the logarithmic term turns out to be a small correction. Then

$$\omega_0^2 = \frac{m}{\Delta m} \frac{\omega_D^2}{3} \tag{13}$$

On the other hand, if a heavy atom of mass $m + \Delta m$ is introduced into the lattice, the simplest oscillation, namely, that in which the impurity is assumed to oscillate with the nearest neighbors held fixed (note that we are talking about an isotopic defect, hence no change in force constant and volume), has the angular frequency $\omega_E$

$$\omega_E^2 = \frac{m}{m + \Delta m} \cdot \frac{3\omega_D^2}{5} \tag{14}$$

which is quite similar to the $\omega_0$ determined above. Hence we see that the resonant frequency in this special case at least is comparable to that of the Einstein oscillator caused by the impurity.

In the long-wavelength limit, $\omega \ll \omega_0$, equation (11) transforms into the original expression for the isotopic scattering. For real isotopes $\Delta m \ll m$, and therefore $\omega_0$ should be quite large. It is for this reason that the observations of the isotope effect so far have not revealed any resonant scattering. Finally the phonon scattering by low-lying states caused by the introduction of molecular impurities has been calculated using the same techniques [48]. Consequently a resonant scattering of the form of equation (10) has been found, and in particular in the low-frequency limit a scattering proportional to $\omega^4$.

## 2. Comparison with Experiment

Figure 16, 17, and 18 show examples of the influence of monatomic monovalent impurities on the thermal conductivity in KCl ([40]). Similar results were obtained also on KBr ([3,40]) and NaCl ([50]). In Fig. 19 the influence of monatomic divalent impurities is shown ([51]). Divalent ions tend to coagulate at room temperature. The crystals used in these experiments were therefore quenched from high temperatures. Then every divalent ion can be assumed to be associated with one cation vacancy on a site of a next-nearest neighbor.

In all these systems the thermal conductivity curve shows a pronounced indentation in the temperature range where in the undoped samples phonon–phonon processes dominate. A defect scattering process with a monotonic frequency dependence certainly cannot describe these data. Scattering by resonances appears to be a likely explanation, but here we encounter the difficulty mentioned in the beginning of Section II. In none of these crystals were such modes detected in far infrared absorption except in KCl:Ag([52]). This is perhaps not too surprising because of the different coupling involved for phonon and photon interaction, but it means that our entire knowledge about

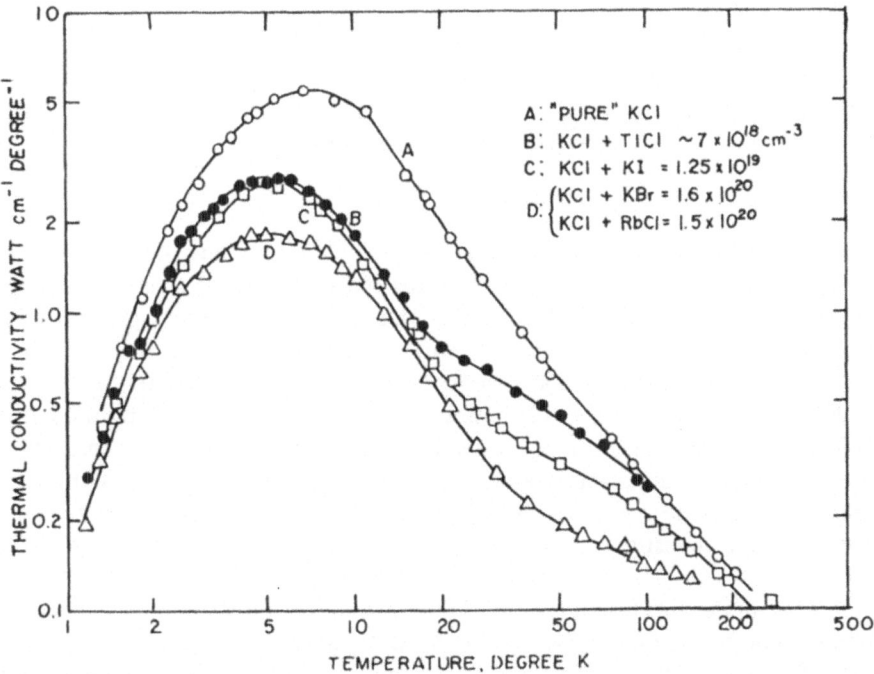

Fig. 16—Thermal Conductivity of KCl containing Tl+, I−, and Br− in the concentrations indicated ([40]).

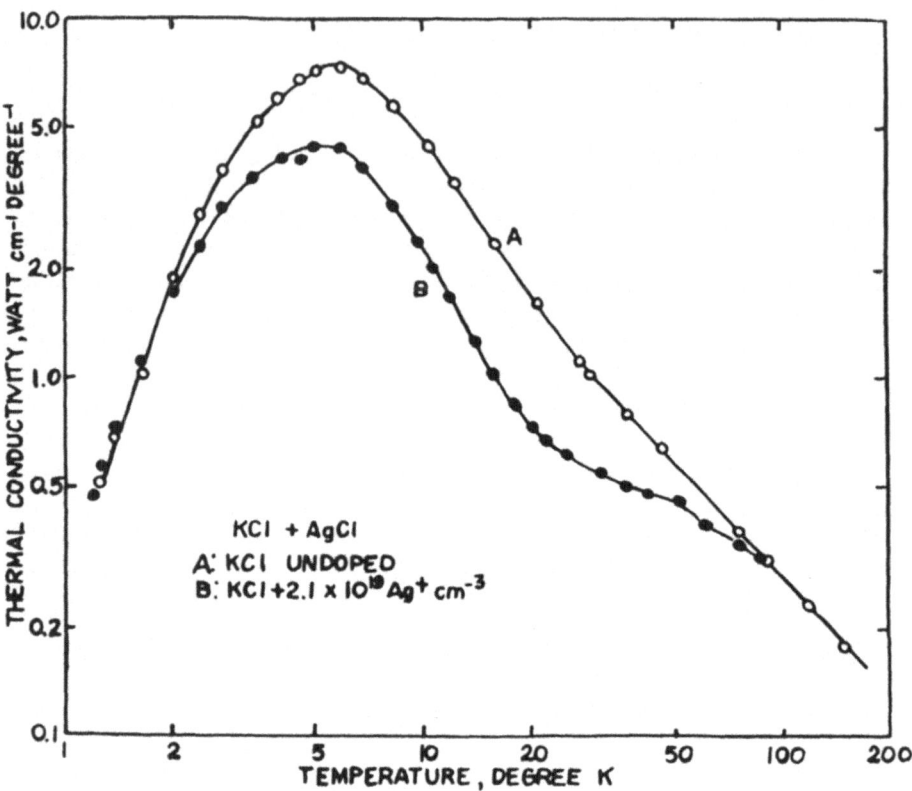

Fig. 17—Thermal conductivity of KCl:Ag+ ([39,40]).

these modes has to be derived from the thermal conductivity data. Attempts in this direction have been made using the inelastic and the elastic scattering mechanism as discussed before. With the inelastic scattering process ([15]), reasonable fits to the experimental curves were obtained for KCl:I⁻, Na⁺, Ca⁺⁺, and the resonant mode frequencies were determined to be 69, 85, and 90 cm⁻¹, respectively. The difficulty with this model, as pointed out before, is the large number of adjustable parameters. Therefore even a good fit to the experimental data cannot be considered to be a proof of the correctness of this model. We shall came back to this scattering process later, but shall first investigate how well the elastic scattering model can explain the data.

This model was tested on KCl:I ([45,46]), and without using any adjustable parameters good qualitative fits to the data were obtained involving only the difference in mass between the Cl⁻ and the I⁻ ions, and the resonance frequency was determined to be 57 cm⁻¹. This was very encouraging because it showed that the elastic scattering model was in principle able to explain the observed temperature dependence of the thermal conductivity. It has since been found, however, that for all other systems studied, like the ones shown in these lectures, the fits turned out to be considerably poorer ([40]). We can only summarize these efforts by stating that it was never possible to produce the sharpness of the dip and very rapid recovery of the thermal conductivity toward higher temperatures by using a resonant-mode scattering cross section. One can perhaps go even a little further: a thermal conductivity curve varying as rapidly with temperature as observed in these experiments can never be described with a scattering cross section depending on $\omega$ alone. Even rapid damping ($\Gamma$) or complete cut-offs of the relaxation time above a certain phonon frequency can never produce such dips, the reason being that the frequency distribution of the phonons involved in the transport of heat at a given temperature varies only slowly with the latter, as shown in Fig. 2. Hence we must conclude that only a strongly temperature-dependent phonon scattering cross section will be capable of properly describing experimental results. A similar difficulty is observed with CN⁻ in RbCl. CN⁻ in KCl shows two distinct resonances, both describable with a Lorentzian scattering rate.

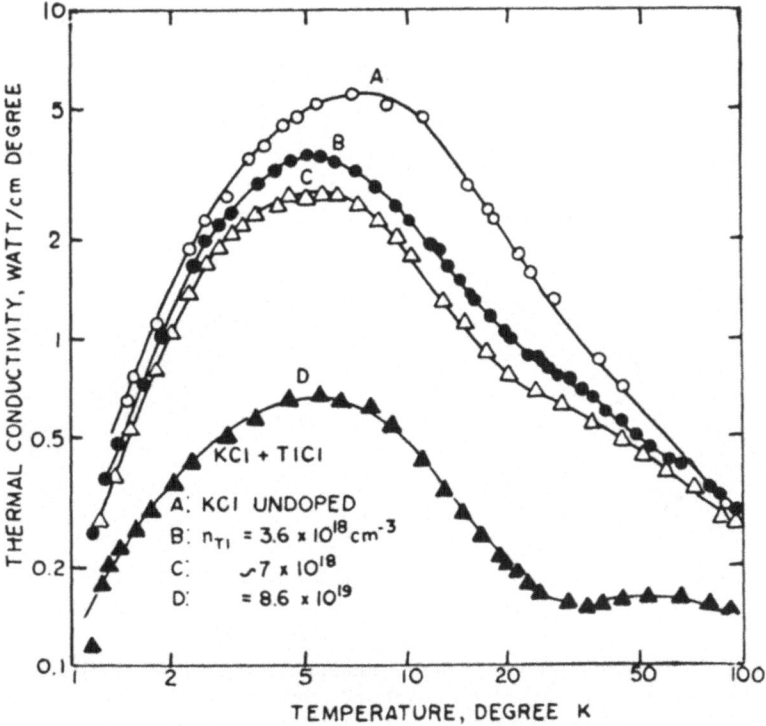

Fig. 18—Thermal conductivity of KCl:Tl⁺ for different concentrations ([40]).

We associate the high-frequency resonance with the onset of free rotation. Similar results are obtained for CN⁻ in RbCl, but here the onset of rotation occurs at a higher energy, as found from infrared spectroscopy. We ascribe the high-temperature dip in the thermal conductivity curve of RbCl:CN to a phonon scattering by these rotational states. The thermal conductivity in this temperature range however cannot be described with a Lorentzian scattering cross section. What is the cause for this difference? In KCl the high resonance occurs in a region where intrinsic processes are unimportant; in RbCl, however, the resonance occurs above the maximum, where intrinsic processes are strongly active. Hence we may conclude that in the presence of intrinsic phonon-phonon scattering an otherwise "simple" scattering process gets greatly modified—or vice versa. The shape of the thermal conductivity curves in crystals containing monatomic impurities is similar to that observed with RbCl:CN. Hence the scattering caused by these defects may by itself be very simple indeed, say, of a resonance form as given in equation

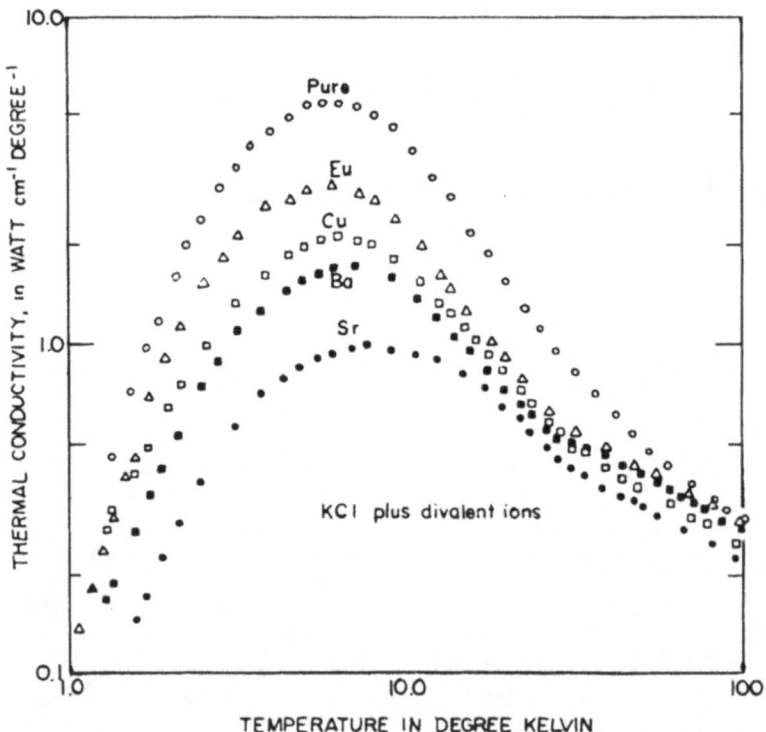

Fig. 19—Thermal conductivity of KCl containing several divalent impurities ([49]). The samples were well quenched and contained therefore vacancies in a concentration equal to the divalent ion concentration, between 200–300 ppm.

(10), but in a temperature region where phonon–phonon collisions are important, the observed scattering is greatly modified. This means a breakdown of the relaxation time approximation as stated in equation (5): the phonon scattering rates are not additive. If this were indeed true, another description of the thermal conductivity in doped crystals would be more appropriate. Let us re-inspect the experimental curves: At low temperature, a rise compatible with a Rayleigh cross section is observed; then, above the maximum, the curves drop again, and the *temperature dependence* in this range is very similar to that observed in the undoped host, (see in particular Fig. 18, also Figs. 10 and 13). Above $T_0$ a rapid recovery of the conductivity toward that of the undoped crystal is observed, and in most crystals the effect of the defects has disappeared above 100°K. We assume that the introduction of the defects causes pseudolocalized modes in the acoustic continuum which strongly couple to the phonons via three-quantum processes. The situation is similar to that of Umklapp scattering, except that instead of a Bragg-reflected wave a pseudolocalized mode gets excited. The temperature dependence of the phonon scattering cross section and hence of the thermal conductivity is therefore similar to that of pure crystals, but the conductivity is considerably reduced. As the temperature rises, the energy of the dominant phonons increases, which reduces the possibility of exciting the pseudolocalized mode via a three-quantum process, and hence the conductivity is now dominated again by the Umklapp processes as in the pure crystal. Before taking refuge to this very drastic model one must certainly investigate all other possible explanations which do not require abandoning the additivity of relaxation rates, like the inelastic three-quantum process ([15]), or the resonance on phonon–phonon collisions proposed by Carruthers ([53]). However, with the experimental information presently available it appears important to also consider the model presented above.

In Section II we associated the high-temperature dip in the conductivity of RbCl:CN with the onset of free rotation, and this had let us to a relation between $T_0$, the temperature of the dip, and the resonance frequency of the scattering process [see equation (9)]. Because of the resemblance between the conductivity observed with monatomic defects to that found with RbCl:CN, we tentatively determine the resonance frequencies associated with monatonic defects by using equation (9). In this way we find the following values (in wavenumbers): KCl + Tl$^+$: 57 cm$^{-1}$; Li$^+$: 63 cm$^{-1}$; Ag$^+$: 66 cm$^{-1}$; I$^-$: 72 cm$^{-1}$; Rb$^+$, Br$^-$: 105 cm$^{-1}$; (for comparison $\omega_D = 163$ cm$^{-1}$). In KBr: Li we obtain $\omega_0 = 39$ cm$^{-1}$.

Next we want to discuss the low-temperature scattering cross sections for molecular impurities. They were found to be of Lorentzian form, increasing proportional to $\omega^2$ below the resonance frequency $\omega_0$ and decreasing proportional to $\omega^{-2}$ above $\omega_0$. Such a relaxation rate is presented in Fig. 20.

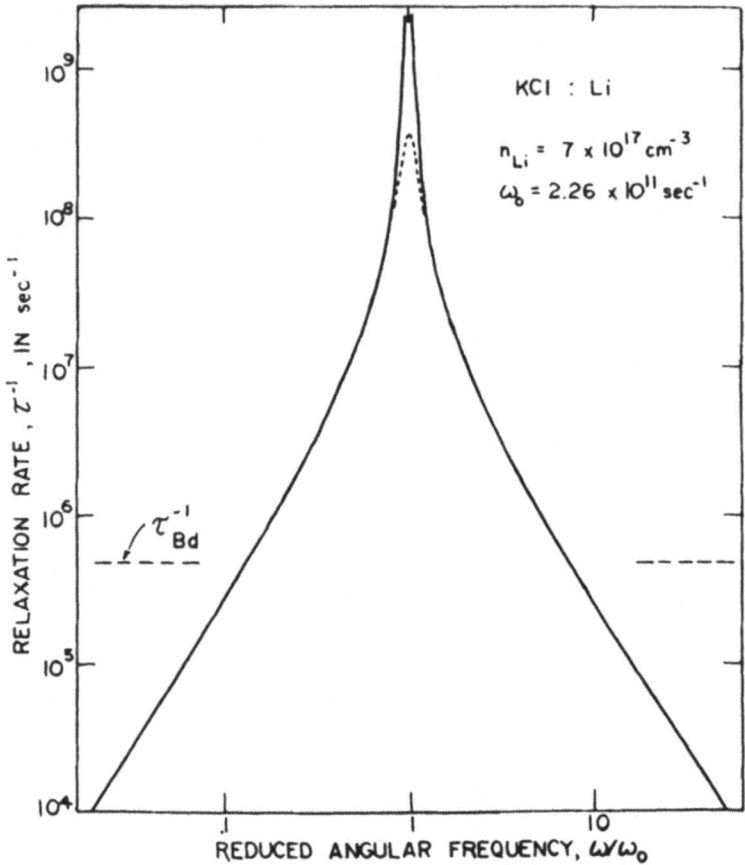

Fig. 20—Resonance relaxation rate of Lorentzian form. The curve shown
was used to fit the thermal conductivity of KCl:Li [curve (B) in Fig. 10].

Sketched into the graph is the effect of a rather large damping, of form $\Gamma\omega^2$
as an additional term in the resonance denominator in equation (8). $\Gamma$ for the
case shown is such that the energy of the classical oscillator should decrease
to 37% in about one period. As can be seen, the effect of this damping
is quite small indeed except over a small frequency interval, which is not
noticeable in thermal conductivity because of the broad phonon spectrum
involved in the heat transport (Fig. 2). The relaxation rate proposed for scattering by molecular defects ([48]) varies proportional to $\omega^4$ below $\omega_0$, goes
through a resonance maximum, and then decreases proportional to $\omega^{-2}$ above
$\omega_0$, depending on the damping. Therefore the relaxation rate is not symmetric
around $\omega_0$. Now it is conceivable that by considering several scattering processes as expected for the scattering by the several tunneling states of the

molecule (or $Li^+$ ion) an equally good fit to the experiment could be obtained. This is presently being investigated ([21]). Such an effort unfortunately is not too promising. The experimental data require a decrease in the relaxation rate proportional to $\omega^2$ above the resonance frequency, and therefore the damping term $\Gamma\omega^6$ is important for such a fit. This increases the number of adjustable parameters. One might hope that $\Gamma$ can be obtained from a best fit to the data. This unfortunately is wrong. An increases in $\Gamma$ will flatten only the resonance peak, which remains unnoticed in the thermal conductivity, and eventually tends to shift the maximum of the relaxation rate to lower frequencies. Hence the use of a relaxation rate of the form given in equation (10) will only increase the number of adjustable parameters and will not provide the information about $\Gamma$ which is needed so badly for an understanding of the interaction between the phonons and the defects. Therefore the reason we prefer to use the Lorentzian term (equation (8) rather than the one predicted by theory to describe the low-temperature resonances—apart from the very good agreement with the experiment—is the fact that it contains a minimum number of adjustable parameters.

## IV. SUMMARY AND OUTLOOK

We have demonstrated the use of thermal conductivity measurements for the study of changes in the vibrational spectrum of dielectric solids containing defects. This method appears to work well for low-lying motional states, as demonstrated for the case of tunneling states of molecular and monatomic defects. These states appear to scatter phonons with a resonance scattering process, and a simple relaxation rate depending only on the phonon frequency has resulted in very satisfactory fits to the experimental data. Because of the broad frequency band of phonons used in thermal conductivity no information can be obtained about the line shape near the center of the resonance and the total area under the curve, and therefore one important quantity, the lifetime of the excited state, or the damping of the harmonic oscillator, cannot be determined from this experiment. From the scattering rate in the tails of the curve one can, however, determine a quantity $A$, in this article called the scattering strength, and this quantity contains the matrix element of the phonon scattering, and lifetime, and possible also the width of the individual lines caused by random internal inhomogeneities in the local crystal field. More work will be needed to separate these quantities, and specific heat and paraelectric resonance appear to be particularly promising for this purpose.

Many monatomic defects in alkali halides produce characteristic indentations on the high-temperature side of the conductivity curve, typically above 10°K. They can be qualitatively described with elastic phonon scattering by pseudolocalized impurity modes. Extensive experimental studies have shown, however, that such a theory cannot describe the shape of the "dips" correctly. It is therefore suspected that inelastic processes play a very important role in these crystals, and that the thermal conductivity is more adequately described with a model which considers a change in three-quantum processes brought about by the quasi-localized modes. Because of the narrow temperature range over which these effects occur it seems logical to assume that one of the many impurity modes caused by the defect accounts for this phenomenon.

The low-temperature scattering by molecules in alkali halide crystals (and also by Li in KCl) tells us that these defects cause resonant states in the lattice. On the other hand, we have seen that the resonant frequencies of these states are identical with the frequency of the motional states of the defects in their vacancies. We can say that when the defect performs a rotational or quasi-rotational motion it periodically kicks the lattice with a frequency equal to that of its rotation, thus causing a large outgoing scattered wave, or a resonance. The fact that the resonance can be associated with a rotational state of the defect which itself is of course highly localized in the lattice leads to a tempting analogy: the high-temperature resonances observed with monatomic defects also have their origin in some simple motion of the defect, which has a rotational character. One motion which would be very similar to that of a rotating molecule is a translational motion of the monatomic defect which takes place in not one, but two or even three perpendicular directions, just like the tunneling motion of the small $Li^+$. If a larger ion performs such a motion, it will probably move the nearest neighbor ions, pushing away the ion to which it comes closest. This motion could be best described as a slushing around of the impurity in a deformable shell formed by its six nearest neighbors and would bear some resemblance to a rotation of an impurity ion. It is reasonable that such a mode should be relatively little changed if the mass of the central atom is changed, as was indeed observed in $KBr:Li^6$ and $KBr:Li^7$. The absence of an effect of the next-nearest neighbors, as observed with the divalent impurities, is also quite likely.

This article attempts to summarize our present understanding of the interaction between phonons and impurity modes. Many questions are still unanswered. We want to mention only a few of those which have not been brought up so far: (a) What is the coupling mechanism? So far we simply assume it is through the stress which changes the cavity in which the defect is located and which causes the defect to rearrange. But what are the details and can one explain the frequency dependence of the relaxation process, the Lorentzian resonance form? (b) Obviously, longitudinal and transverse phonons must get scattered differently. Assume that the transverse phonons get scattered by the defects. Then the low-temperature conductivity should not decrease below the value characteristic for heat transfer by longitudinal phonons alone. This is not observed. Hence some other mechanism must continuously mix longitudinal and transverse phonons. At low temperatures this mechanism can only be boundary scattering. We have evidence that most of the so-called boundary scattering occurs in a thin layer under the surface ([8]) containing a high dislocation density as a result of the surface treatment necessary to produce a Casimir-type boundary effect. Can these defects mix longitudinal and transverse phonons? (c) Finally, in this article we have restricted ourselves to alkali halides. Do other host lattices with less ionic bonding show similar impurity mode scattering?

Little is presently known about these questions, but we are convinced that the studies of thermal conductivity will help to contribute to their answers—and moreover, will bring up new problems which will deepen our understanding of lattice vibrations.

I would like to acknowledge numerous stimulating conversations with my colleagues at Cornell University, in particular, with Dr. J. A. Krumhansl and Dr. A. J. Sievers. The financial support of the U. S. Atomic Energy Commission and the Advanced Research Projects Agency in preparing this manuscript is gratefully acknowledged.

## REFERENCES AND FOOTNOTES

1. We refer to a recent paper by Jlukor and Jacobsen, dealing with the generation and detection of elastic waves with frequency $f = 1.14.10^{11} \mathrm{sec}^{-1}$. In print.
2. P. Debye, *Vortraege ueber die Kinetische Theorie der Materie und der Elektrizitaet*, p. 17–60. Berlin B.G. Teubner, 1914.
3. R.O. Pohl, *Z. Physik* **175**: 358 (1963).
4. T.H. Geballe and G.W. Hull, *Phys. Rev.* **110**: 773 (1958).
5. I. Pomeranchuk, *J. Phys. (U.S.S.R.)* **6**: 237 (1942).
6. P.G. Klemens, *Proc. Phys. Soc. (London)* **A68**: 1113 (1955).
7. J. Callaway, *Phys. Rev.* **113**: 1046 (1959).
8. P.D. Thacher, Ph. D. Thesis, Cornell 1965. To appear in *Phys. Rev.*
9. H.B.G. Casimir, *Physica* **5**: 495 (1938). For a discussion of the numerical factor see also Po. Carruthers, *Rev. Mod. Physics*, **33**: 92 (1961). For crystals of finite length, end corrections have to be applied, see R. Berman, F.E. Simon, and J.M. Ziman, *Proc. Roy. Soc. (London)* **A220**: 171 (1953) and R. Berman, E.L. Foster, and J.M. Ziman, *Proc. Roy. Soc. (London)* **A231**: 130 (1955).
10. This is not quite true: The boundary relaxation time had to be chosen 15% longer than given by equation (7). Incorrect averages over the sound velocities for the various phonon branches in LiF may be the reason, i.e., the Debye velocity of sound may be at fault. We have, however, not been able to remove the discrepancy in spite of numerous attempts to compute "correct" velocity averages (see Ref. 8). We believe therefore that the discrepancy has a deeper reason, connected with the actual phonon

scattering process at the crystal surface, but because of the smallness of the effect we shall ignore it completely in these lectures.

11. P. Carruthers, *Rev. Modern Physics* **33**: 92 (1961) Ch. VII.

12. R.E. Peierls, *Quantum Theory of Solids*, Oxford, Clarendon, Press, 1956).

13. Berman and J.C.F. Brock, *Proc. Roy. Soc. (London)* **A289**: 46 (1965).

14. B. Bertman, H.A. Fairbank. R.A. Guyer, and C.W. White, *Phys. Rev.* **142**: 79 (1966) and R. Berman, C.L. Bounds, and S.J. Rogers, *Proc. Roy. Soc. (London)* **A289**: 66 (1965).

15. C.T. Walker and R.O. Pohl, *Phys. Revue* **131**: 1433 (1963), interpretation by C.M. Wagner, *Phys. Rev.* **131**: 1443 (1963).

16. V. Narayanamurti, *Phys. Rev. Letters* **13**: 693 (1964), and W.D. Seward and V. Narayanamurti, *Phys. Rev.* **148**: 463 (1966).

17. A.F. Devonshire, *Proc. Roy. Soc. (London)* **A153**: 601 (1936). The foundation of this work had been laid by L. Pauling, *Phys. Rev.* **36**: 430 (1930).

18. J.P. Harrison, P.P. Peressini, and R.O. Pohl, to be published.

19. R.O. Pohl, *Phys. Rev. Letters* **8**: 481 (1962), and V. Narayanamurti, W.D. Seward, and R.O. Pohl, *Phys. Rev.* 1966. **148**: 481.

20. F.C. Baumann, *Bull. Am. Phys. Soc.* **9**: 644 (1964).

21. F.C. Baumann, J.R. Harrison, R.O. Pohl, and W.D. Seward, to be published.

22. H.S. Sack and M.C. Moriarty, *Solid State Commun.* **3**: 93 (1965).

23. G. Lombardo and R.O. Pohl, *Phys. Rev. Letters* **15**: 291 (1965).

24. W. Känzig, H.R. Hart, and S. Roberts, *Phys. Rev. Letters* **13**: 543 (1964).

25. U. Kuhn and F. Luty, *Solid State Commun.* **3**: 31 (1965).

26. I. Shepherd and G. Feher, *Phys. Rev. Letters* **15**: 194 (1965).

27. J.A.D. Matthew, *Solid State Commun.* **3**: 365 (1965).

28. S.P. Bowen, M. Gomez, J.A. Krumhansl, and J.A.D. Matthew, *Phys. Rev. Letters* **16**: 1105 (1966).

29. M.E. Baur and W.R. Salzman, to be published.

30. N.E. Byer and H.S. Sack, Cornell University Materials Science Center Report No. 473, 1966 (to be published).

31. R.D. Hatcher and W. Wilson, private communication.

32. Previous calculations (G.J. Dienes, R.D. Hatcher, R. Smoluchowski, and W. Wilson, *Phys. Rev. Letters* **16**: 25 (1966) for the $\langle 100 \rangle$ directions were found to be unreliable, and hence the question whether the potential in $\langle 100 \rangle$ has a saddle point cannot be decided with certainty. R.D. Hatcher, private communication.

33. J.P. Harrison and P.P. Peressini, private communication.

34. For potential minima along the $\langle 111 \rangle$ directions one would expect four levels (see, for instance, Ref. 28). The lowest one, of symmetry $A_{1g}$, singly degenerate; the next one up, $T_{1u}$, triply degenerate, the next one, $T_{2g}$, triply degenerate, and the highest one, $A_{2u}$, singly degenerate. With the potential assumed in Ref. 28, the energy differences between adjacent levels are all equal. This, however, is probably not true. From microwave absorption experiments (A. Lakatos and H.S. Sack *Solid State Commun.* **4**: 315 (1966) a level splitting $\delta = 0.81$ cm$^{-1}$ was deduced, in contrast to the splitting of $1.2 \pm 10\%$ cm$^{-1}$ found in the thermal conductivity.

35. G. Lombardo and R.O. Pohl, *Bull. Am. Phys. Soc.* **11**: 212 (1966). In an earlier study of the dielectric constant of KBr:Li at low temperatures (Ref. 22) an effect of the Li$^+$ ion had been noted. Its magnitude, however, has since been found to be considerably smaller than in KCl:Li. H. Bogardus and H.S. Sack, private communication.

36. A.J. Sievers and S. Takeno, *Phys. Rev.* **140**: A1030 (1965).

37. I.G. Nolt and A.J. Sievers, *Phys. Rev. Letters* **16**: 1103 (1966).

38. Ref. 11, pg. 109 *ff.*
39. J.A. Krumhansl and J.A.D. Matthew, *Phys. Rev.* **140**: A1812 (1965).
40. F.C. Baumann and R.O. Pohl, to be published.
41. S. Takeno, *Progr. Theoret. Phys. (Kyoto)* **29**: 191 (1963); **30**: 144 (1963).
42. J. Callaway, *Nuovo Cimento* **29**: 883 (1963); *J. Math. Phys.* **5**: 783 (1964).
43. J.A. Krumhansl, *Proceedings of the 1963 International Conference on Lattice Dynamics*, Copenhagen. Pergamon Press 1964, p. 523.
44. M.V. Klein, *Phys. Rev.* **131**: 1500 (1963); **141**: 716 (1966).
45. R.J. Elliott and D.W. Taylor, *Proc. Phys. Soc. (London)* **83**: 189 (1964).
46. C.W. McCombie and J. Slater, *Proc. Phys. Soc. (London)* **84**: 499 (1964).
47. K. Thoma and W. Ludwig, *Phys. Status Solidi* **8**: 487 (1965).
48. Max Wagner, *Phys. Rev.* **131**: 2520 (1963); **133**: A750 (1964).
49. Such an expression is obtained quite generally if a wave hits an obstacle capable of resonating. As an example we cite the scattering of sound waves in water containing gas bubbles (P.M. Morse and H. Feshbach, *Methods of Theoretical Physics*, McGraw-Hill, New York, 1953, Ch.11. 3). The scattering cross section has exactly the form of equation (10) except for the damping term whose frequency dependence depends on details of the liquid and the enclosed gas.
50. R. Caldwell and M.V. Klein, private communication.
51. J.W. Schwartz and C.T. Walker, *Phys. Rev. Letters* **16**: 97 (1966).
52. A.J. Sievers, *Phys. Rev. Letters* **13**: 310 (1964).
53. P. Carruthers, *Phys. Rev.* **125**: 123 (1962); *Phys. Rev.* **126**: 1448 (1962).

# Elementary Excitations
# and
# Their Observations

## R. J. Elliott

*Department of Physics*
*University of Oxford*
*Oxford, England*

## I. INTRODUCTION

The most important similarities among elementary excitations arise from the symmetry of the crystal. Because of the translational symmetry every excitation must vary across the crystal from unit cell to unit cell only by a phase factor $e^{i\mathbf{k}\cdot\mathbf{z}}$. For a crystal of $N$ unit cells there are $N$ possible values of $k$ uniformly spread over the first Brillouin zone. In group theoretical language each $k$ corresponds to a different representation of the translation group.

In addition, there are extra symmetry elements in a space group which reflect the symmetry of the unit cell. These control the form of the excitation inside each unit cell so that the general form is

$$u_k(z)e^{i\mathbf{k}\cdot\mathbf{z}} \tag{1}$$

or

$$\sum_{R_n} \phi(z - R_n)e^{i k \cdot \mathbf{R}_n} \tag{2}$$

In the presence of impurities all the translational symmetry is lost and $\mathbf{k}$ is no longer an adequate description. Only the small group of point symmetry elements is left, and the reduced symmetry makes the calculation of properties much more difficult.

## II. TYPES OF EXCITATION

### a. Phonons

Phonons are primarily the excitations associated with the heavy nuclear motion. However, there is always a complementary electronic motion which affects the energy $\epsilon(k)$ of the excitation as well as the coupling to an external stimulus through the form of the $\phi(z - R_n)$. The usual approximation for discussing the electronic motion is that of the shell model ([1]).

### b. Electronic Modes.

If an adiabatic approximation is assumed, electronic modes may usually be considered with the nuclei fixed. The detailed description differs with the basic approximation which provides a satisfactory description of the electronic ground state.

*1. Metals.* Here the ground state may be described by the Fermi distribution of electrons in single-particle Bloch states which themselves have wavefunctions in the form of (1). The excitations in the system correspond to the promotion of an electron across the Fermi surface leaving behind a hole. These two particle excitations do not have a unique $\epsilon(\mathbf{k})$ curve but cover a continuous energy range for each $k$. For a simple case these are shown in Fig. 1. The interelectronic interactions give rise, in addition, to collective modes with unique $\epsilon(\mathbf{k})$ excitation curves. In metals these are electron density fluctuations [plasmons ([2])] and magnons ([3]) in those materials showing magnetic order.

*2. Semiconductors.* Again the ground state is adequately described in terms of single particle bands. But now, because of the energy gap, the

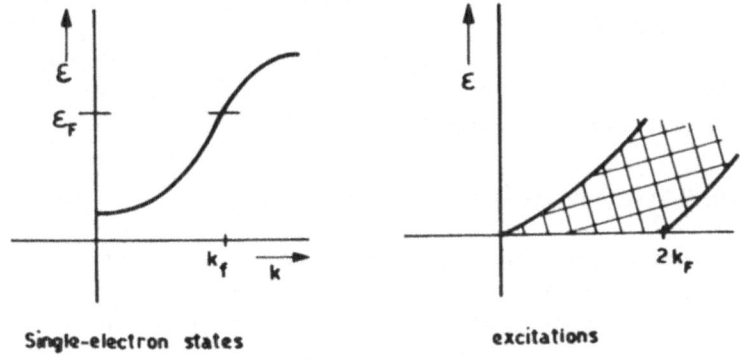

Single-electron states                                    excitations

Figure 1

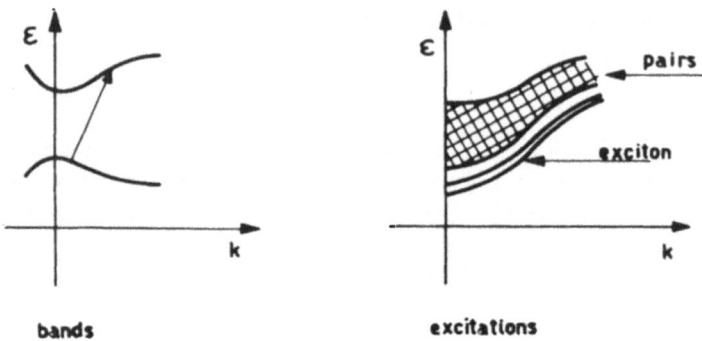

Figure 2.

continuum of excitations does not stretch down to zero (see Fig. 2). The residual Coulomb interaction may now bind the pairs into excitons which have a separate $\epsilon(\mathbf{k})$ curve for each relative motion ([4]).

## c. Insulators

If the electron correlation energy is high so that it takes a lot of energy to transfer electrons between atoms creating ionized states, the best simple description of a system is in terms of atomic wavefunctions. Then the excited states are approximately those with an excited atom, where the wavefunction may be denoted by $\psi_i$ when the excitation is centered at $\mathbf{R}_i$. By symmetry there will be a band of such states with

$$\psi(k) = \sum_i \psi_i e^{i\mathbf{k}\cdot\mathbf{R}_i}$$

$\epsilon(\mathbf{k})$ will tend to be flat in these cases since if electron transfer is easy the band picture of (b) would be more appropriate. This picture works well for paramagnetic salts when each crystal field splitting gives a separate exciton, and molecular crystal like anthracene. Rare-gas solids and alkali halides may be described in this way—the first exciton arising from the atomic transition $p^6 \longrightarrow p^5 s$—but are really intermediate between (b) and (c)([5]).

## d. Spin Waves (Magnons)

In paramagnetic ions the ground atomic state is normally degenerate. Exchange interaction between the ions may cause a magnetically ordered ground state. Excitations between those states which were previously degenerate now have bands $\epsilon(\mathbf{k})$ of excitations which spread over a range of energy comparable with the exchange. If it is a simple spin degeneracy these excitations correspond to spin deviations; in the presence of orbital momentum they may be more complex. The exchange interaction will also split the excited crystal field levels causing a modification of those excitons.

## III. TYPES OF EXPERIMENTAL INFORMATION

It is clear from the above that most types of elementary excitations have unique $\epsilon(\mathbf{k})$ curves, although there may sometimes be continua. At low $T$, so that the system is more or less in its ground state, few excitations will be thermally excited. Excitation processes through external stimulation can normally be expanded in the number of quanta produced. Because of the form of equations (1) and (2) the response to a stimulus of wavevector $\mathbf{k}$, frequency $\omega$ will be

$$\chi(\mathbf{k}, \omega) = A(\mathbf{k}, \omega)\delta[\omega \pm \omega(\mathbf{k})] \tag{3}$$

which will allow a direct determination of $\epsilon(\mathbf{k}) = \hbar\omega(\mathbf{k})$. The function $A$ will reflect the detailed form of $u_k(z)$ and the strength of the interaction. The point symmetry may cause additional selection rules and occasionally make $A = 0$

Two quantum processes will require

$$\mathbf{k} = \mathbf{k}_1 + \mathbf{k}_2$$
$$\omega = \omega(\mathbf{k}_1) + \omega(\mathbf{k}_2) \tag{4}$$

which gives a continuum response over a finite range rather than the $\delta$-function of (3). It is therefre more difficult to unravel the experimental information to obtain details of the elementary excitations, although in some cases it can be a very useful method ([6]).

The type of experimental information obtained will depend on the $\omega(\mathbf{k})$ relationship of the exciting quanta.

### a. Neutrons

Neutrons are particularly useful having energies of a few tens of meV for $\mathbf{k}$'s of typical Brillouin zone size. At present an upper limitation is about 70 meV but hopefully this will be extended. A scattering experiment $\mathbf{q} \rightarrow \mathbf{q}'$ has conservation rules

$$\mathbf{q} - \mathbf{q}' = \mathbf{k}$$
$$\frac{\hbar}{2m_0}(\mathbf{q}^2 - \mathbf{q}'^2) = \pm\epsilon(\mathbf{k}) \tag{5}$$

and gives a direct measurement of $\epsilon(\mathbf{k})$. They interact direclty with nuclei and are ideal for the study of phonons. They also interact with electrons through a magnetic interaction and may be used to study spin waves and crystal field splittings at low energies.

## b. Light

Light is particularly conveient in that it is readily usable over a wide energy range. Over most of the energy region of interest, however, (say up to 10 eV.), the $\mathbf{k}$ values are very small on the Brillouin zone scale, and may often be considered zero. The direct absorption and scattering (Raman) experiments therefore only observe the $\mathbf{k} \sim 0$ modes, in single-quantum excitation. It interacts through its electric field with the charged ions to excite phonons, and with the electron polarizability to excite excitons. This interaction may be so strong as to excite polaritons ([7]), mixed excitations. It also interacts more weakly through its magnetic field with spin waves and paramagnetic excitons.

A wealth of two quantum effects have also been studied. Two-phonon absorption and Raman scattering has been extensively studied ([9]). Recently two-mangnon effects of similar form have been observed ([10]), but phonon—magnon pairs have yet to be found. Exciton—phonon interactions tend to be large in ionic crystals and experiments indicate that many phonons may accompany an excition transition. They are correspondingly difficult to interpret. With weaker coupling in a semiconductor, exciton–phonon pairs are more clearly seen. Recently exciton–magnon pairs have been observed ([11]).

## c. Other types of Excitations

Other types of excitations tend only to be used for special situations. High-energy electrons may be used to study plasmons. X-ray scattering can give some information about phonons. Mechanically created accoustic waves will interact with low-energy spin waves.

## REFERENCES

1. See the article by Ludwig this volume p. 93.
2. See the article by Resibois this volume p. 340.
3. See the article by Elliott this volume p. 306.
4. R.S. Knox "Excitons," Suppl. 5 "*Solid State Physics*" (Academic Press) R. J. Elliott in *Excitons and Polarons*, (Oliver and Boyd, 1963)
5. J.C. Phillips and F. Hermanson, *Phys. Rev.* (to appear)

6. See the article by Balkanski this volume p. 113.
7. See the article by Collins this volume p. 156.
8. See the article by Hopfield this volume p. 413.
9. See the article by Burstein this volume p. 367.
10. Allen, Loudon, and Richards, *Phys. Rev. Letters* **16**: 463 (1966).
    Woods-Halley and Silvera, *Phys. Rev. Letters* **15**: 654 (1965).
    Tanabe, Moriya and Sugarno, *Phys. Rev. Letters* **15**: 1023 (1965).
11. Schawlow et al., *Phys. Rev. Letters* **15**: 656 (1965).

# Spin Waves

## R. J. Elliott

*Department of Physics*
*University of Oxford*
*Oxford, England*

## I. INTRODUCTION

The elementary excitations of an ordered magnetic system are called spin waves or magnons. We shall begin with a pedagogical account of the simplest situation and indicate how the theory may be refined and extended to more complicated situations. The magnetic ordering arises from the exchange interaction between electrons. In those cases which can be treated on an atomic model and where the ground state has only spin degeneracy this has the Heisenberg form. Between atoms on site $i$ and $j$ it takes the form

$$-2J_{ij}\mathbf{S}_i \cdot \mathbf{S}_j \tag{1}$$

If orbital electron motion is important it takes a much more complicated form ([1]) involving tensor operators of angular momentum $Y_l^m(\mathbf{S})$. In addition, single-ion anisotropy in the form

$$\sum_l^m A_l^m Y_l^m(\mathbf{S}_i) \tag{2}$$

is normally present, together with dipole–dipole interactions, and applied magnetic fields. These effects will only be discussed qualitatively. For a complete treatement see the extensive reviews of Walker ([2]) and Keffer. ([3])

## II. BLOCH THEORY

For a Hamiltonian consisting entirely of terms like (1) and positive $J_{ij}$, the ground state is one with each spin aligned

$$S_i^z = S \qquad S_t = \sum_i S_i^z = NS \tag{3}$$

The ground state energy is

$$E_0 = -\sum_{ij} J_{ij} S^2 \tag{4}$$

In fact, the ground state is degenerate. Since

$$[\mathbf{S}_i \cdot \mathbf{S}_j, \mathbf{S}_i + \mathbf{S}_j] = 0 \tag{5}$$

$$[\mathcal{H}, \mathbf{S}_i] = 0 \tag{6}$$

and the ground state is one of $|\mathbf{S}_t| = NS$

The states of $S_t^z = NS - 1$ may be treated exactly. For if we call $\psi_i$ the state with the deviation of the $i$th spin ($S_i^z = S - 1$), a correct state may be expected by symmetry to be

$$\psi_{\mathbf{k}} = N^{-1/2} \sum_i e^{i\mathbf{k} \cdot \mathbf{R}_i} \psi_i \tag{7}$$

This is confirmed by considering $\mathcal{H}\psi_i$. If equation (1) is rewriten as

$$S_i^z S_j^z + \tfrac{1}{2}(S_i^+ S_j^- + S_i^- S_j^+) \tag{8}$$

the first operator is diagonal on the $\psi_i$ while the second transfers a deviation from $i \rightarrow j$. Using

$$S^- |S_z\rangle = |M\rangle = [(S + M)(S - M + 1)]^{1/2} |M - 1\rangle \tag{9}$$

we find

$$\mathcal{H}\psi_i = E_0 + 2S \sum_j J_{ij}\psi_i - 2S \sum_j J_{ij}\psi_j \tag{10}$$

i.e.

$$(\mathcal{H} - E_0)\psi_{\mathbf{k}} = 2S \sum_j J_{ij}(1 - e^{i\mathbf{k} \cdot (\mathbf{R}_i - \mathbf{R}_j)})\psi_{\mathbf{k}} = \epsilon(k)\psi_{\mathbf{k}} \tag{11}$$

The expression on the right is the excitation energy $\epsilon(\mathbf{k})$. It is seen that $\epsilon(\mathbf{k}) = 0$ for $\mathbf{k} = 0$, i.e., the completely symmetric excitation to $S_t = NS$, $S_t^z = NS - 1$. At small $\mathbf{k}$, $\epsilon$ is quadratic

$$\epsilon(\mathbf{k}) = Dk^2 \tag{12}$$

where in a cubic crystal

$$D = \tfrac{2}{3}S \sum_j J_{ij}(\mathbf{R}_i - \mathbf{R}_j)^2 \tag{13}$$

While these expressions are exact for single spin deviations the method leads to complications for multiple reversal. For the energy is modified if

two spins are reversed inside the range of $J_{ij}$ using the first term (8). Two reversals on the same atom are different because of (9), and the total number of
reversals on any atom is restricted to $2S + 1$. These two effects may be regarded as giving rise to interactions between the spin waves—called dynamical
and kinematical ([4]). We may reasonably expect them to be small if only a
small number of excitations are present. Thus to first approximation, the spin
waves can be taken as noninteracting. There is no restriction on the number
in any $\mathbf{k}$ value so they are bosons in this approximation. At low $T$ the average
number of excitations

$$NS - \langle S_i^z \rangle = \int_{\text{B.Z}} \frac{V}{(2\pi)^3} \frac{d\mathbf{k}}{(e^{\epsilon(\mathbf{k})/kT} - 1)} \propto T^{3/2} \tag{14}$$

The average energy

$$\langle E \rangle = \int_{\text{B.Z}} \frac{V}{(2\pi)^3} \frac{\epsilon(\mathbf{k})d\mathbf{k}}{(e^{\epsilon(\mathbf{k})/kT} - 1)} \propto T^{5/2} \tag{15}$$

## III. EQUATION OF MOTION METHOD

The basic property of spin waves in any system is the $\epsilon(\mathbf{k})$, which can
be measured by neutron scattering etc. From it detailed information about
tha range of $J_{ij}$ and the form of more complex interactions can be determined. It is therefore convenient to find a more general method to calculate
$\epsilon(\mathbf{k})$. A clue to the construction of such a method can be obtained by noting
that the excitations of §I are obtained by the operation

$$\psi_k = N^{-1/2} \sum_i e^{i\mathbf{k} \cdot \mathbf{R}_i} S_i^- \bar{\psi}_0 \tag{16}$$

and involve the components of $\mathbf{S}$ perpendicular to the ordered moment. In
the classical limit they in fact consists of a circular precession of the perpendicular components with a phase variation $e^{i\mathbf{k} \cdot \mathbf{R}}$ across the crystal. The
excitation energies could therefore be obtained from the equation of notion
of these perpendicular components.
In the system of §I using the commutators

$$[S_i^z; S_j^{\pm}] = \pm S_i^{\pm} \delta_{ij}$$
$$[S_i^+, S_j^-] = 2S_i^z \delta_{ij} \tag{17}$$

in the equation

$$-i\hbar \frac{\partial S_i^-}{\partial t} = [\mathcal{H}, S_i^-] \tag{18}$$

gives

$$= -2 \sum_j J_{ij}(S_i^z S_j^- - S_j^z S_i^-) \tag{19}$$

Near $T = 0$, $S_i^z = S$ and the equations may be linearized Performing the Fourier transform to

$$S^-(\mathbf{k}) = N^{-1/2} \sum_i e^{i\mathbf{k} \cdot \mathbf{R}_i} S_i^- \tag{20}$$

gives the same characteristic energy as (11)

This method may be generalized to any form of Hamiltonian and any spin order (which can be determined by elastic neutron scattering). The equations may be linearized by replacing the ordered component of each spin (which may not necessarily be parallel) by its static value, and neglecting terms which as higher than first order in the perpendicular components.

As another simple example we consider a simple antiferromagnet where some of the $J_{ij}$ in the Hamiltonian of type (1) are negative. If we further assume an ordering of the $\uparrow \downarrow \uparrow \downarrow$ form

$$S_i^z = Se^{i\mathbf{w} \cdot \mathbf{R}_i} \tag{21}$$

we can proceed in the same way. This ground state is not an eigenstate of $S_t$ as required by (6), but we shall show latter that it is nearly so and (21) is a good enough approximation to determine $\epsilon(\mathbf{k})$. On taking the Fourier transform of (19) using (21) we find, as may be expected from the reduction of the Brillouin zone associated with the spin ordering, that $S^-(\mathbf{k})$ is mixed to $S^-(\mathbf{k} + \mathbf{w})$. In fact

$$\hbar\omega S^-(\mathbf{k}) = 2S[J(\mathbf{w}) - J(\mathbf{k} + \mathbf{w})]S^-(\mathbf{k} + \mathbf{w})$$
$$\hbar\omega S^-(\mathbf{k} + \mathbf{w}) = 2S[J(\mathbf{w}) - J(\mathbf{k})]S^-(\mathbf{k}) \tag{22}$$

so that

$$(\hbar\omega)^2 = 4S^2[J(\mathbf{w}) - J(\mathbf{k} + \mathbf{w})][J(\mathbf{w}) - J(\mathbf{k})] \tag{23}$$

and $\hbar\omega$ is proportional to $k$ at low $k$. The basic excitations are mixtures with different amplitudes on the sites of different $S_i^z$.

In more complicated crystals the form of $\epsilon(\mathbf{k})$ will be more complex. If there are $n$ magnetic ions in a magnetic unit cell there will be $n$ branches to the spectrum. The form of this spectrum may be discussed in terms of the magnetic space group ([5]). However, if the Hamiltonian has only Heisenberg interactions and point anisotropy there is more symmetry than this and the additional degeneracies have been discussed in terms of a "spin space group"([6]). Figure (1) shows a complicated example, yttrium iron garnet a ferrimagnet with space group $O_h^{10}$ and 20 magnetic ions per cell. Some extra degeneracies arise because of time reversal symmetry. For example the

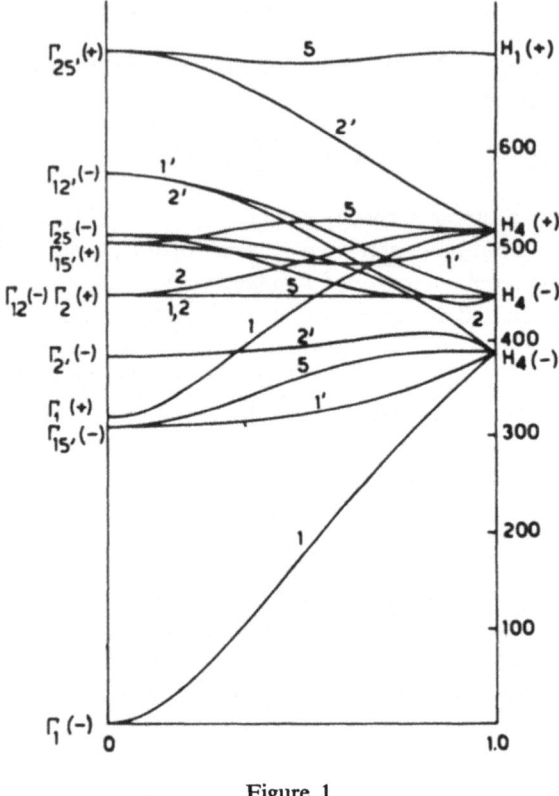

Figure 1

simple antiferromagnet discussed above has a doubly degenerate spectrum over the whole zone which is only resolved by application of a magnetic field.

## IV. IMPROVEMENTS TO SPIN-WAVE THEORY

### a. Low $T$ Expansion

We have shown how to obtain the spin-wave spectrum under the approximation of saturation of each ordered spin. This is correct at $T = 0°$K in a ferromagnet but does not even describe the proper ground state in other cases. It becomes progressively worse as $T$ increases. Early attempts to improve this used the thermal average $\langle S_i^z \rangle$ in linearizing (19). This corresponds to a simple decoupling procedure in the Green's function method [7]. The Green's function $\langle\langle S_i^-(t); S_i^+(0) \rangle\rangle$ has an equation of motion like (18) and

(19) with an additional source term $2\pi\hbar\delta(t)[S_i^-, S_j^+]$ on the right. However, the simple decoupling gives spin-wave energies renormalized by a correction proportional to $T^{3/2}$ at low $T$ from (14). This is incorrect; Dyson ([4]) showed the correction to be like $T^{5/2}$ and this is confirmed by better decoupling schemes. The physical origin of the difference is that the important quantity is not the angle a spin makes with a fixed magnetization axis, but rather the relative angle between spins. This latter goes like the energy (15) in a simple nearest-neighbor coupling model.

A more direct way of obtaining the low-$T$ result uses the creation and destruction operators of Holstein and Primakoff ([8]). If we transform the operator $S^-$ given by (9) into a creator of spin deviations by defining $S - M = n$ we find

$$S^-|n\rangle = \sqrt{(2S - n)(n + 1)}\,|n + 1\rangle \tag{24}$$

It can then be related to a creation operator $a^+$ by

$$S^- = a^+(2S - a^+a)^{1/2} \tag{25}$$

When $n$ is small the correction from the number operator $a^+a$ may be neglected in (25) and

$$S^- = \sqrt{2S}a^+ \qquad S^+ = \sqrt{2S}a \qquad S^z = S - a^+a \tag{26}$$

It may be noted that $a$, $a^+$ operate in an infinite space where $S^-$ operates in one of only $2S + 1$ dimensions. This is related to the kinematical interaction mentioned above and turns out to lead to negligible corrections. Working to second order in the $a$'s gives a Hamiltonian

$$\mathscr{H} = -2\sum_{ij} J_{ij}(S^2 - Sa_i^+a_i - Sa_j^+a_j + Sa_i^+a_j + Sa_j^+a_i) \tag{27}$$

which may be diagonalized by a Fourier transformation

$$a_k = N^{-1/2}\sum_i e^{i\mathbf{k}\cdot\mathbf{R}_i}a_i \tag{28}$$

to

$$\mathscr{H} = \sum_k \epsilon(\mathbf{k})a_k^+a_k \tag{29}$$

where $\epsilon(\mathbf{k})$ has the form (11.) Also to this order the Bose commutation rules hold

$$[a_k, a_k^+] = 1 \tag{30}$$

Going to fourth order in the $a$'s in (25) and (26) we get some extra terms in

$\mathscr{H}$ representing spin-wave interactions. Those which give a contribution to first order in perturbation theory have the form

$$\sum_{k,q} C(\mathbf{k}, \mathbf{q}) a_k^+ a_k a_q^+ a_q \tag{31}$$

and to that order give a corrected spin-wave energy

$$\epsilon'(\mathbf{k}) = \epsilon(\mathbf{k}) + \sum_q C(\mathbf{k}, \mathbf{q}) n_q \tag{32}$$

where $n_q$ is the boson excitation number.

Since at low $k$   $\epsilon = Dk^2$, $C$ must by symmetry have the form $dk^2q^2$. Thus

$$\epsilon'(\mathbf{k}) = k^2 (D + \sum_q dq^2 n_q) \tag{33}$$

and at low $T$ the correction is of order $T^{5/2}$. The magnitude of the correction depends on the detailed from of $J_{ij}$ ([9]).

## b. Antiferromagnetic Ground State

The problem of the antiferromagnetic ground state can also be treated by a similar method ([10]). Operators like (26) apply to the up spin sites. For down spin sites the appropriate deviation operators are

$$S_j^+ = \sqrt{2S}\, a_j^+ \qquad S_j^- = \sqrt{2S}\, a_j \qquad S_j^z = -S + a_j^+ a_j \tag{34}$$

For coupling between spins on the same sublattice the Heisenberg Hamiltonian has the form (27); for spins on different sublattices the terms quadratic in $a$ are

$$-2SJ_{ij}(a_i a_j + a_i^+ a_j^+ + a_i^+ a_i + a_j^+ a_j)$$

On Fourier transforming using (28) the form of $\mathscr{H}$ is more complicated; the terms which survive because of translationsl symmetry are

$$\mathscr{H} = \tfrac{1}{2} \sum_k A_k(a_k^+ a_k + a_{-k}^+ a_{-k}) + \tfrac{1}{2} \sum_k B_k(a_k a_{-k} + a_k^+ a_{-k}^+) \tag{35}$$

which may be diagonalized by a transformation

$$C_k = u_k a_k + v_k a_{-k}^+ \tag{36}$$

to give

$$\mathscr{H} = \sum_k (A_k^2 - B_k^2)^{1/2}(C_k^+ C_k + C_k C_k^+) - A_k \tag{37}$$

Unlike (29) the second term in (37) gives a zero point motion and shows that we have assumed an incorrect ground state. For the simple model of §III

$$A_k = 2S[J(w) - \tfrac{1}{2}J(q) - \tfrac{1}{2}J(w + q)]$$
$$B_k = S[J(w + q) - J(q)] \tag{38}$$

so that the excitation energy is the same as (23).

The deviation from the fully aligned state

$$\Delta = S - \langle S_i^z \rangle = \langle a_i^+ a_i \rangle = \frac{1}{N} \sum_k u_k^2 \langle C_k^+ C_k \rangle + v_k^2 \langle C_k C_k^+ \rangle \tag{39}$$

$$= \frac{1}{N} \sum_k \frac{A_k}{\epsilon_k} \left( n_k + \frac{1}{2} \right) - \frac{1}{2} \tag{40}$$

This turns out to be small for simple nearest-neighbor coupling and gives, for example, in a bcc lattice at $T = 0$

$$\frac{\Delta}{S} = \frac{0.059}{S} \tag{41}$$

Experimentally ([11]) in Mn $F_2$ the result is much smaller still; nuclear resonance gives $0.004 \pm 0.003$, while the best theoretical calculation is $0.018$. Presumably anisotropy and other forces help to stabilize the ordered ground state.

## V. SPIN WAVES IN METALS ([13])

The methods so far described have applied specifically to insulators. However, spin waves are observed in magnetic metals and may be accounted for theoretically if some correlation is included in the band theory. The simple molecular field theory of Stoner and Slater does not include such correlation effects. In terms of operators

$$\mathscr{H} = \sum_{k\sigma} (E(k) + \tfrac{1}{2}VM\sigma)b_{k\sigma}^+ b_{k\sigma} \tag{42}$$

where the $b$ are Fermion operators and $\sigma$ has the value $\pm 1$ for the two spin orientations. $E(k)$ is the usual band energy and $VM$ the molecular field term proportional to the difference in population

$$M = \sum_k (n_{k+} - n_{k-}) \tag{43}$$

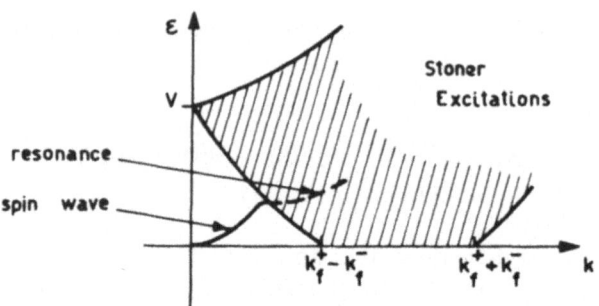

An ordinary electron excitation without spin change is created by $b^+_{k+q\sigma}b_{k\sigma}$ and a typical spectrum is shown in Fig. 1 of the previous paper (See p. 301). Spin deviation excitations are given by

$$S^-_k(\mathbf{q}) = b^+_{k+q-}b_{k+} \tag{44}$$

with energy $E(\mathbf{k} + \mathbf{q}) - E(\mathbf{k}) + VM = \hbar\omega_k(\mathbf{q})$.

The spectrum of these so-called Stoner excitations is shown in Fig. 2 The interaction energy between electrons may be written

$$\frac{1}{2}\sum_{\mathbf{k}_1,\mathbf{k}_2\mathbf{q}} Vb^+_{\mathbf{k}_1+\mathbf{q},\sigma_1}b^+_{\mathbf{k}_2-\mathbf{q},\sigma_2}b_{\mathbf{k}_2,\sigma_2}b_{\mathbf{k}_1,\sigma_1} \tag{45}$$

where the matrix element $V$ has been taken as a constant although it should have some dependence on $\mathbf{k}_1$, $\mathbf{k}_2$, and $\mathbf{q}$. Putting $\mathbf{q} = 0$ and using first-order perturbation theory will give the molecular field correction term in (42). A better result may be obtained from the equation of motion for $S^-_k(\mathbf{q})$ [13].

After simplification by replacing $b^+_{k\sigma}b_{k\sigma}$ by its average $n_{k\sigma}$ in the quartic terms one finds

$$-i\hbar\frac{\partial}{\partial t}S^-_k(\mathbf{q}) = \hbar\omega_k(\mathbf{q})S^-_k(\mathbf{q}) + \frac{1}{2}V(n_{k+} - n_{k+q-})\sum_{k'}S^-_{k'}(\mathbf{q})$$

There is a collective mode solution of this equation in the form

$$\sum A_k(\mathbf{q}, \omega)S^-_k(\mathbf{q})$$

The eigenvalues are given after some manipulation by the equation

$$1 + \frac{V}{N}\sum_k \frac{n_{k+} - n_{k+q-}}{\hbar[\omega - \omega_k(\mathbf{q})]} = 0$$

It is found that $\omega \propto q^2$ at small $q$. As the collective modes approach the Stoner excitations there is an anomaly and a resonance passes into the band which broadens rapidly. So far these effects have not been detected experimentally. These spin waves may be regarded as a form of exciton somewhat similar to those in semiconductors. The electron–hole pairs are not confined to a single atom but have larger extent in the crystal.

## REFERENCES

1.  P.R. Levy, *Phys. Rev.* **135**: A155 (1964).
2.  L.R. Walker, *Magnetism Vol. I*, Ed. Rado and Suhl, (1965).
3.  F. Keffer, *Handbuch der Physik* (to appear).
4.  F.J. Dyson, *Phys. Rev.* **102**: 1230 (1956).
5.  J. Dimmock and R.G. Wheeler, *Phys. Rev.* **127**: 391 (1962).
6.  W. Brinkman and R.J. Elliott, *J. App. Phys.* **37**: 1475 (1966); *Proc. Roy. Soc.* **294**: 343 (1966).
7.  C.W. Haas and H.S. Jarett, *Phys. Rev.* **135**: A 1809 (1964).
8.  T. Holstein and H. Primakoff, *Phys. Rev.* **58**: 1098 (1940).
9.  W. Marshall, *Proc. Low Temp. Conf. London* (1962).
10. R. Kubo, *Phys. Rev.* **87**: 568 (1952).
11. E.D. Jones and K.B. Jefferts, *Phys. Rev.* **135**: A 1277 (1964).
12. H.L. Davies, *Phys. Rev.* **120**: 789 (1960).
13. F. Englert and M. Antonoff, *Physica* **30**: 429 (1964).
14. T. Izuyama, *Prog. Theor. Phys.* **23**: 969 (1960).
15. T. Izuyama, R. Kubo and D. Kim, *Prog. Theor. Phys.* **28**: 1025 (1963).

# Theory of Scattering in Solids and Localized Spin Waves

J. Callaway

*Department of Physics*
*Louisiana State University*
*Baton Rouge, Louisiana*

In these lectures, we will discuss the properties of localized spin waves in solids using the techniques of solid state scattering theory. These techniques will be developed as required.

It has been known for many years that if an impurity atom, for instance, one of different mass, is present in a vibrating lattice a localized vibrational mode may appear under suitable circumstances. Similarly, localized electronic states associated with imperfections are observed. It would therefore be expected that in a magnetic ordered system imperfections could disturb the spin wave excitation spectrum in such a way as to produce a localized spin wave. We will see how this comes about mathematically; however, at the present time, not much is known about these objects from experiment. Our considerations will be based on the Heisenberg Hamiltonian for a ferromagnet. We will consider a hypothetical system in which only nearest neighbor interactions are important. We will write this Hamiltonian as

$$H = -\sum_{m,\Delta} J(\mathbf{R}_m, \mathbf{R}_m + \Delta)S(\mathbf{R}_m)S(\mathbf{R}_m + \Delta) \tag{1}$$

We will not include either external fields nor dipole–dipole interactions. In writing $H$, we use the following notation: $\mathbf{R}_m$ is a lattice vector. At each site $\mathbf{R}_m$ there is an atom with a spin operator $S(\mathbf{R}_m)$. The vector $\Delta$ connects an atom with one of its nearest neighbors. The spins on neighboring atoms are coupled by an exchange integral $J(\mathbf{R}_m, \mathbf{R}_m + \Delta)$ The system will be said to contain a magnetic defect if at some site $\mathbf{R}_0$ (which we choose as the origin) there is an atom whose spin $S'$ is coupled to its neighbors by an exchange integral $J'$, and either or both these quantities differ from those referring to the rest of the crystal. The exchange for the rest of the crystal will be $J$, the magnitude of the spin $S$. Thus, $S^2(\mathbf{R}_m)$ is diagonal with elements $S(S + 1)$

if $\mathbf{R}_m \neq \mathbf{R}_0$, and with elements $S'(S'+1)$ if $\mathbf{R}_m = \mathbf{R}_0$. We will further suppose that $J$ is positive so that the perfect crystal is ferromagnetic.

We next introduce the usual raising and lowering operators $S_\pm$ through the definition

$$S_\pm(\mathbf{R}_m) = S_x(\mathbf{R}_m) \pm iS_y(\mathbf{R}_m) \tag{2}$$

These operators obey the usual commutation relations

$$[S_+(\mathbf{R}_m), S_-(\mathbf{R}_n)] = 2\delta_{mn}S_z(\mathbf{R}_m), \tag{3a}$$

$$[S_z(\mathbf{R}_m), S_-(\mathbf{R}_n)] = -\delta_{mn}S_-(\mathbf{R}_m), \tag{3b}$$

$$[S_z(\mathbf{R}_m), S_+(\mathbf{R}_n)] = \delta_{mn}S_+(\mathbf{R}_m) \tag{3c}$$

These operators are introduced in the Hamiltonian which then becomes

$$H = -\sum_{m,\Delta} J(\mathbf{R}_m, \mathbf{R}_m + \Delta)[S_z(\mathbf{R}_m)S_z(\mathbf{R}_m + \Delta) + S_+(\mathbf{R}_m)S_-(\mathbf{R}_m + \Delta)] \tag{4}$$

The ground state of a ferromagnet is described by the ket $|0\rangle$ which characterizes the condition of complete spin alignment. Thus, in the ferromagnetic state, no spin can be raised:

$$S_+(\mathbf{R}_m)|0\rangle = |0\rangle \tag{5}$$

for all $\mathbf{R}_m$.

Our interest will focus on the states in which there is a single spin deviation. These give rise to the spin waves. An orthonormal basis for the one-spin deviation states is furnished by the vectors we will denote by $|m\rangle$, corresponding to the presence of a single spin deviation at $\mathbf{R}_m$:

$$|m\rangle = [2S(\mathbf{R}_m)]^{-1/2}S_-(\mathbf{R}_m)|0\rangle \tag{6}$$

Using the commutation relations for the spin operators, we verify easily that

$$\langle m|n\rangle = [4S(\mathbf{R}_m)S(\mathbf{R}_n)]^{-1/2}\langle 0|S_+(\mathbf{R}_m)S_-(\mathbf{R}_n)|0\rangle = \delta_{mn} \tag{7}$$

An arbitrary state $|\Psi\rangle$ containing a single spin deviation may be expanded in the kets $|m\rangle$:

$$|\Psi\rangle = \sum_m \phi(\mathbf{R}_m)|m\rangle \tag{8}$$

Evidently $\phi(\mathbf{R}_m)$ is the projection of $|\Psi\rangle$ on $|m\rangle$,

$$\phi(\mathbf{R}_m) = \langle m|\Psi\rangle \tag{9}$$

and we can interpret $|\phi(\mathbf{R}_m)|^2$ as the probability that there is a spin deviation on site $m$. (An interesting calculation concerning the spread of an excitation giving an initially localized spin disturbance—an initial state $|m\rangle$—is given by Huber ([5]). Thus $\phi(\mathbf{R}_m)$ may be interpreted as the "wave function" for a spin wave.

The function $\phi$ obeys an effective Schrödinger equation. We want the state vector $|\Psi\rangle$ to be an eigenstate of the Hamiltonian (4).

$$H|\Psi\rangle = E|\Psi\rangle$$

We insert equation (8) and form the scalar product with the bra $\langle n|$. With the use of equation (9) this leads to

$$\sum_m \langle n|H|m\rangle \phi(\mathbf{R}_m) = E\phi(\mathbf{R}_n) \tag{10}$$

It is a straightforward matter to evaluate the matrix elements of the Hamiltonian on this basis with the aid of the commutation relations for the spin operators. We have

$$
\begin{aligned}
\langle n|H|m\rangle &= -[4S(\mathbf{R}_n)S(\mathbf{R}_m)]^{-1/2}\sum_{j\Delta}\langle 0|\,S_+(\mathbf{R}_n)[S_z(\mathbf{R}_j)S_z(\mathbf{R}_j+\Delta)\\
&\quad + S_+(\mathbf{R}_j)S_-(\mathbf{R}_j+\Delta)]S_-(\mathbf{R}_m)|0\rangle J(\mathbf{R}_j,\mathbf{R}_j+\Delta)\\
&= -[4S(\mathbf{R}_n)S(\mathbf{R}_m)]^{-1/2}\{\sum_{j\Delta}J(\mathbf{R}_j,\mathbf{R}_j+\Delta)\langle 0|[-\delta_{jn}S_+(\mathbf{R}_n)S_z(\mathbf{R}_j+\Delta)\\
&\quad - \delta_{m,j+\Delta}S_z(\mathbf{R}_j)S_+(\mathbf{R}_n) + 2S_+(\mathbf{R}_j)\delta_{j+\Delta,n}S_z(\mathbf{R}_n)]S_-(\mathbf{R}_m)|0\rangle\\
&\quad + \langle 0|\,HS_+(\mathbf{R}_n)S_-(\mathbf{R}_m)|0\rangle\}\\
&= E_0\delta_{mn} - [4S(\mathbf{R}_m)S(\mathbf{R}_n)]^{-1/2}\sum_{j\Delta}J(\mathbf{R}_j,\mathbf{R}_j+\Delta)\{-\delta_{jn}S(\mathbf{R}_j+\Delta)\cdot\\
&\quad \cdot\langle 0|\,S_+(\mathbf{R}_n)S_-(\mathbf{R}_m)|0\rangle - \delta_{n,j+\Delta}S(\mathbf{R}_j)\langle 0|\,S_+(\mathbf{R}_n)S_-(\mathbf{R}_m)|0\rangle\\
&\quad + \delta_{j,n}\delta_{m,j+\Delta}\langle 0|\,S_+(\mathbf{R}_n)S_-(\mathbf{R}_m)|0\rangle\\
&\quad + \delta_{m,j}\delta_{n,j+\Delta}\langle 0|\,S_+(\mathbf{R}_j)S_-(\mathbf{R}_m)|0\rangle\\
&\quad + 2\delta_{n,j+\Delta}S(\mathbf{R}_n)\langle 0|\,S_+(\mathbf{R}_j)S_-(\mathbf{R}_m)|0\rangle\\
&\quad - 2\delta_{mn}\delta_{n,j+\Delta}\langle 0|\,S_+(\mathbf{R}_j)S_-(\mathbf{R}_m)|0\rangle\}\\
&= E_0\delta_{mn} - \sum_{j,\Delta}J(\mathbf{R}_j,\mathbf{R}_j+\Delta)\{-S(\mathbf{R}_j+\Delta)\delta_{jn}\delta_{mn} - S(\mathbf{R}_j)\delta_{mn}\delta_{n,j+\Delta}\\
&\quad + [4S(\mathbf{R}_m)S(\mathbf{R}_n)]^{-1/2}2\delta_{n,j+\Delta}\delta_{jm}S(\mathbf{R}_n)S(\mathbf{R}_m)\\
&= E_0\delta_{mn} + 2\delta_{mn}\sum_\Delta J(\mathbf{R}_n,\mathbf{R}_n+\Delta)S(\mathbf{R}_n+\Delta)\\
&\quad - 2\sum_\Delta J(\mathbf{R}_m,\mathbf{R}_m+\Delta)\delta_{n,m+\Delta}[S(\mathbf{R}_m)S(\mathbf{R}_m+\Delta)]^{1/2} \tag{11}
\end{aligned}
$$

With the use of equation (11) our Schrödinger equation becomes

$$
\begin{aligned}
2\sum_\Delta J(\mathbf{R}_n,\mathbf{R}_n+\Delta)[S(\mathbf{R}_n+\Delta)\phi(\mathbf{R}_n) - \{S(\mathbf{R}_n)S(\mathbf{R}_n+\Delta)\}^{1/2}\phi(\mathbf{R}_n+\Delta)]\\
= (E - E_0)\phi(\mathbf{R}_n) \tag{12}
\end{aligned}
$$

Here $E_0$ is the energy of the ferromagnetic state $|0\rangle$.

We can separate the matrix elements of the Hamiltonian into a part $H_0$ corresponding to the perfect crystal and a part (which we will denote as $V$ and refer to as the scattering potential) due to the imperfection. To determine $H_0$, we merely set all the spins in equation (11) equal and do the same for the exchange integrals; then

$$\langle m\,|\,H_0\,|\,n\rangle = (E_0 + 2JSZ)\delta_{nm} - 2JS\delta_{m-n,\,\Delta} \tag{13}$$

where $z$ is the-number of nearest neighbors (coordination number). The final $\delta$ in (13) indicates that a contribution to $H_0$ is obtained whenever $\mathbf{R}_m - \mathbf{R}_n$ is equal to *any* vector connecting an atom to its nearest neighbors. To determine $\langle n\,|\,V\,|\,m\rangle$ we subtract equation (13) from equation (11), remembering that only a single defect is present and it is at the origin.

$$\langle n\,|\,V\,|\,m\rangle = \delta_{nm}[2SZ(J' - J)\delta_{n,\,0} + 2(J'S' - JS)\delta_{n,\,\Delta}$$
$$- 2[J'(SS')^{1/2} - JS]\delta_{m-n,\,\Delta}(\delta_{m,\,0} + \delta_{n,\,0}) \tag{14}$$

Let us solve the Schrodinger equation in our formalism for the perfect crystal. This is, from equation (12),

$$2JS[Z\phi^{(0)}(\mathbf{R}_n) - \sum_{\Delta} \phi^{(0)}(\mathbf{R}_n + \mathbf{\Delta})] = (E - E_0)\phi^{(0)}(\mathbf{R}_n) \tag{15}$$

Let us try as a solution

$$\phi^{(0)}(\mathbf{R}_n) = \lambda e^{i\mathbf{R}\cdot\mathbf{R}_n} \tag{16}$$

where $\lambda$ is a normalization constant to be determined later. When (16) is substituted into (15) we find that it is indeed a solution, provided that

$$E - E_0 = -2JS[Z - \sum_{\Delta} e^{i k \cdot \Delta}] \tag{17}$$

This is the standard result for the spin wave energy. The normalization constant $\lambda$ can be chosen so that the $\phi$ from an orthonormal set. Note that there is a different $\phi$ for each $\mathbf{k}$, so we should really write $\phi^{(0)} = \phi^{(0)}(\mathbf{k}, \mathbf{R}_m)$.

We will require that

$$\sum_n \phi^{(0)}(\mathbf{k}', \mathbf{R}_n)^*\phi^{(0)}(\mathbf{k}, \mathbf{R}_n) = \delta(\mathbf{k} - \mathbf{k}') \tag{18}$$

We have, however, the identity ([3])

$$\sum_n \exp[i(\mathbf{k} - \mathbf{k}')\cdot\mathbf{R}_n] = \frac{(2\pi)^3}{\Omega} \sum_l \delta(\mathbf{k} - \mathbf{k}' - \mathbf{K}_l) \tag{19}$$

where $\mathbf{K}_l$ is a reciprocal lattice vector and $\Omega$ is the volume of the unit cell.

However, if we restrict $\mathbf{k}$ and $\mathbf{k}'$ to lie inside the first Brillouin zone, the summation over reciprocal lattice vectors disappears, and we obtain equation (18), provided that

$$\lambda = \frac{\Omega^{1/2}}{(2\pi)^{3/2}} \tag{19a}$$

With this choice for the normalization, we also have that

$$\int \phi(\mathbf{k}, \mathbf{R}_n)^* \phi(\mathbf{k}, \mathbf{R}_m) d^3k = \delta_{mn} \tag{20}$$

where the integral includes the first Brillouin zone.

Let us denote the eigenstates of $H_0$, which are the unperturbed spin wave states, by $|\mathbf{k}\rangle$. From equation (8) we see that

$$|\mathbf{k}\rangle = \frac{\Omega^{1/2}}{(2\pi)^{3/2}} \sum_n e^{i\mathbf{k}\cdot\mathbf{R}_n} |n\rangle \tag{21a}$$

$$|m\rangle = \int \phi^{(0)}(\mathbf{k}, \mathbf{R}_m)^* |\mathbf{k}\rangle d^3k = \frac{\Omega^{1/2}}{(2\pi)^{3/2}} \int e^{-i\mathbf{k}\cdot\mathbf{R}_m} |\mathbf{k}\rangle d^3k \tag{21b}$$

The spin wave states are also orthonormal. With the use of equation (19), we find that

$$\langle \mathbf{k} | \mathbf{k}' \rangle = \delta(\mathbf{k} - \mathbf{k}') \tag{22}$$

We thus have two sets of orthonormal basis functions for the perfect crystal: (1) The states $|n\rangle$ corresponding to a localized spin deviation, and (2) the spin wave states $|\mathbf{k}\rangle$.

We will now turn to a discussion of the scattering problem, and to localized spin waves. The scattering of spin waves by an imperfection can be completely described by the $t$-matrix, which we require on the basis of spin wave states: that is, we want to determine the elements $\langle \mathbf{k} | t | \mathbf{k}' \rangle$. If we know $t$, we can compute the transition probability per unit time; hence the scattering cross section can be obtained. The poles of $t$ determine the localized spin wave states.

Our discussion will be quite general at this point. It applies to any crystal in which we have some system of excitations: electrons, phonons, magnons, governed by a Hamiltonian $H_0$ and a potential $V$ representing an imperfection. We will assume that there is some orthonormal basis in which $V$ is a finite matrix or can reasonably be approximated by a finite matrix; otherwise, there are no essential approximations. However, since our major concern here is with spin waves in simple ferromagnets, we will make an "unessential" approximation which is valid for the spin wave system described that only one band of energy levesl is to be considered. For such a general system as we have described there are two sets of states for the perfect crystal; wavelike states, which we will denote by $|\mathbf{k}\rangle$ and localized states $|n\rangle$, and these are connected by equation (21). For electrons in a simple band for instance the $|\mathbf{k}\rangle$ (or more exactly the representation coefficients $\langle \mathbf{r}|\mathbf{k}\rangle$) are Bloch functions $\psi(\mathbf{k},\mathbf{r})$; and the localized states $|n\rangle$ (or $\langle \mathbf{r}|n\rangle$) are Wannier functions $a(\mathbf{r} - \mathbf{R}_n)$. To describe a scattering process, we require the matrix element of the scattering potential between a free particle, i.e., perfect crystal state and a scattering state, say $|\mathbf{k}\rangle_+$, which is a solution of the Lippmann–Schwinger equation

$$|\mathbf{k}\rangle_+ = |\mathbf{k}\rangle + \frac{1}{E^+ - H_0}V|\mathbf{k}\rangle_+ \tag{23}$$

The superscript $+$ on the $E$ indicates that the energy has an infinitesimal positive imaginary part. The transition probability per unit time between two perfect crystal states, $k \longrightarrow q$ is proportional to

$$|\langle \mathbf{q}|V|\mathbf{k}\rangle_+|^2 \tag{24}$$

However, we can write

$$\langle \mathbf{q}|V|\mathbf{k}\rangle_+ = \langle \mathbf{q}|t|\mathbf{k}\rangle \tag{25}$$

where on the right-hand side we have two perfect crystal states and the $t$-operator or $t$-matrix which satisfies the "integral" equation

$$t = V + V\frac{1}{E^+ - H_0}t \tag{26}$$

We will now consider the determination of the $t$-matrix element $\langle \mathbf{q}|t|\mathbf{k}\rangle$. The essential observation is that equation (26) is much easier to solve on the basis of the localized states $|n\rangle$ than on the states $|\mathbf{k}\rangle$, for in the former case, our assumption that $V$ is a finite matrix means that (26) can be solved by ordinary matrix methods. Thus we have

$$\langle \mathbf{q}|t|\mathbf{k}\rangle = \sum_{m,n} \langle \mathbf{q}|n\rangle\langle n|t|m\rangle\langle m|\mathbf{k}\rangle$$

$$= \sum_{m,n} \phi^{(0)}(\mathbf{q}, \mathbf{R}_n)^*\langle n|t|m\rangle\phi^{(0)}(\mathbf{k}, \mathbf{R}_m)$$

$$= \frac{\Omega}{(8\pi)^3} \sum_{n,m} e^{i(\mathbf{k}\cdot\mathbf{R}_m - \mathbf{q}\cdot\mathbf{R}_n)}\langle n|t|m\rangle \tag{27}$$

Let us consider equation (26) on the localized basis

$$\langle n|t|m\rangle = \langle n|V|m\rangle + \sum_{j,l} \langle n|V|j\rangle\Big\langle j\Big|\frac{1}{E^+ - H_0}\Big|l\Big\rangle\langle l|t|m\rangle \qquad (28)$$

We will refer to the matrix elements of the operator $(E^+ - H_0)^{-1}$ as Green's functions. We can write

$$\mathcal{G}(\mathbf{R}_j - \mathbf{R}_l) \equiv \Big\langle j\Big|\frac{1}{E^+ - H_0}\Big|l\Big\rangle$$

$$= \frac{\Omega}{(2\pi)^3}\iint d^3k\,d^3q\,e^{-i\mathbf{q}\cdot\mathbf{R}_j + i\mathbf{k}\cdot\mathbf{R}_l}\Big\langle \mathbf{q}\Big|\frac{1}{E^+ - H_0}\Big|\mathbf{k}\Big\rangle$$

$$= \frac{\Omega}{(2\pi)^3}\iint e^{-i\mathbf{q}\cdot\mathbf{R}_j + i\mathbf{k}\cdot\mathbf{R}_l}\frac{\delta(\mathbf{q} - \mathbf{k})}{E^+ - E(\mathbf{k})}d^3k\,d^3q$$

$$= \frac{\Omega}{(2\pi)^3}\int \frac{e^{-i\mathbf{k}\cdot(\mathbf{R}_j - \mathbf{R}_l)}}{E^+ - E(\mathbf{k})}d^3k \qquad (29)$$

Thus the Green's functions can be expressed as integrals over the energy spectrum. For a general $E(\mathbf{k})$, the evaluation can be quite difficult. The asymptotic form for large $|\mathbf{R}_j - \mathbf{R}_l|$ can be determined by the method of stationary phase (see Callaway [2]). In this way it can be shown that if the energy $E$ is within the continuum, $\mathcal{G}(\mathbf{R})$ is proportional to $\exp(i\mathbf{P}\cdot\mathbf{R})/R$ (we won't discuss the determination of $\mathbf{P}$ here: see the reference cited); but if $E$ lies outside the continuum, then $\mathcal{G}(\mathbf{R})$ falls off exponentially with $R$. In addition certain tricks are available for cubic lattices with the simple energy spectrum given by equation (17). We may use some of these later. At this point, however, let us look briefly at $\mathcal{G}(0)$. Using the definition of the density of states (which we denote as $G(E)$), we can write

$$\mathcal{G}(0) = \frac{\Omega}{(2\pi)^3}\int d^3k\,\frac{1}{E^+ - E(k)}$$

$$= \int \frac{G(E')}{E^+ - E'}dE'$$

$$= P\int \frac{G(E')}{E - E'} - i\pi G(E) \qquad (30)$$

To obtain equation (30), we have used the identity

$$\lim_{\epsilon\to 0}\frac{1}{x + i\epsilon} = P\Big(\frac{1}{x}\Big) - i\pi\delta(x)$$

and the symbol $P$ indicates that the integral is a Cauchy principal value. Let us now return to equation (26), with the understanding that the quantities in it are expressed as matrices on the localized basis. Afterward, we will consider the transformation of equation (27). Let us also denote the operator $(E^+ - H_0)^{-1}$ by $\mathscr{G}$. Then we see that (26) possesses the formal solution

$$t = V \frac{1}{1 - \mathscr{G}V} \tag{31}$$

However, equation (31) is not merely formal but is a practical and useful expression. Its utility rests upon the possibility of evaluating the inverse matrix in (31). To see how this may be done, it is convenient to consider the matrices $\mathscr{G}$ and $V$ in a block form. If there are $N$ cells in the crystal, and the scattering potential $V$ is nonzero in $n$ of these cells, we find it convenient to write the matrices $V$ and $\mathscr{G}$ in the form shown below with dimensions indicated:

$$V = \begin{array}{c} \\ n \\ N-n \end{array} \overset{\begin{array}{cc} n & N-n \end{array}}{\left( \begin{array}{c|c} V & 0 \\ \hline 0 & 0 \end{array} \right)}$$

$$\mathscr{G} = \begin{array}{c} \\ n \\ N-n \end{array} \overset{\begin{array}{cc} n & N-n \end{array}}{\left( \begin{array}{c|c} g_{aa} & g_{ab} \\ \hline g_{ba} & g_{bb} \end{array} \right)}$$

Then the matrix $1 - \mathscr{G}V$ becomes

$$1 - \mathscr{G}V = \left( \begin{array}{c|c} 1 - g_{aa}V & 0 \\ \hline - g_{ba}V & I \end{array} \right) \tag{32}$$

in which the lower right-hand box contains an $(N - n) \times (N - n)$ dimensional unit matrix. To invert $1 - \mathscr{G}V$, it will be seen that if we construct the inverse of the left-hand $n \times n$ matrix $1 - g_{aa}V$ the inverse of the full matrix can be found immediately. Let us denote this inverse as $P/D$

$$(1 - g_{aa}V)^{-1} = \frac{P_{aa}}{D} \tag{33a}$$

where $D$ is the determinant of the $n \times n$ matrix $1 - g_{aa}V$

$$D = \det(1 - g_{aa}V) \tag{33b}$$

and $P$ is the adjugate matrix. (It is convenient for future use to single out the determinant $D$). Then

$$[1 - \mathscr{G}V]^{-1} = \begin{pmatrix} \dfrac{P_{aa}}{D} & \vdots & 0 \\ \hdashline \dfrac{g_{ba}VP_{aa}}{D} & \vdots & I \end{pmatrix}$$

To construct $t$, we multiply on the left by $V$. Then we get

$$t = \begin{pmatrix} \dfrac{V_{aa}P_{aa}}{D} & \vdots & 0 \\ \hdashline 0 & \vdots & 0 \end{pmatrix} \tag{34a}$$

The nonzero portion of $t$ has the same range as the potential. Our result can be expressed as follows: Suppose

$$\langle n | V | m \rangle = \begin{cases} 0 \text{ if } R_m > R_{\max} \\ \quad\quad \text{or} \\ 0 \text{ if } R_n > R_{\max} \end{cases}$$

(definition of finite range). Then

$$\langle n | t | m \rangle = D^{-1} \sum_j \langle n | V | j \rangle \langle j | P | m \rangle \text{ if } R_n, R_m > R_{\max}$$
$$= 0 \text{ if either } R_m \text{ or } R_n \text{ are greater than } R_{\max} \tag{34b}$$

From equation (27), we see that it is possible also to construct the $t$-matrix in the wave representation (states $| \mathbf{k} \rangle$) in terms of operations on the finite matrix $\langle n | t | m \rangle$

$$\langle \mathbf{q} | t | \mathbf{k} \rangle = \frac{1}{D} \frac{\Omega}{(2\pi)^3} \sum_{n,m,j} e^{i(k \cdot R_m - q \cdot R_n)} \langle n | V | j \rangle \langle j | P | m \rangle \tag{35}$$

Equation (35) will not be the most useful expression for the $t$-matrix elements when the potential $V$ possesses interesting symmetry properties. For the case of the magnetic impurity in the spin-wave system, the perturbation potential has the symmetry of the point group. Since the potential cannot have matrix elements between functions which transform according to different irreducible representations, the introduction of such functions will simplify the evaluation of equation (35).

We will denote the representations by indices $\alpha, \beta$. Let us introduce symmetrized combinations of the functions $\phi$ (which are just symmetrized combinations of plane waves).

$$C_\beta^{(0)}(\mathbf{k}, R_m) = \sum_m{}' u(\beta, R_m) \phi^{(0)}(\mathbf{k}, R_m) \tag{36}$$

The coefficients $U(\beta, \mathbf{R}_m)$ are the elements of a unitary transformation. The index $\beta$ is understood to include designation of the row of a degenerate representation. The sum over $m$ in equation (36) includes all the different vectors $\mathbf{R}_m$ which can be found from any one of them by applying the operators of the group. (Thus, all the lattice vectors in equation (36) have the same length). Such a restricted sum is indicated by a prime on the summation sign. The coefficients $U$ obey the relations

$$\sideset{}{'}\sum_m U(\alpha, \mathbf{R}_m)U^+(\mathbf{R}_m, \beta) = \delta_{\alpha\beta} \tag{37a}$$

$$\sum_\beta U^+(\mathbf{R}_m, \beta)U(\beta, \mathbf{R}_n) = \delta_{m,n} \tag{37b}$$

The sum in equation (37b) includes all the rows of degenerate representations. Equation (37b) can be used to introduce the symmetrized functions into equation (35). Then we can define transformed matrices $V$, and $\mathcal{G}$, which are diagonal in the representation index:

$$\delta_{\alpha\beta} V^\alpha_{nm} = \sideset{}{'}\sum_{nm} U(\beta, \mathbf{R}_n) \langle n | V | m \rangle U^+(\mathbf{R}_m, \alpha) \tag{38}$$

A similar formula applies to $\mathcal{G}$. After the symmetrized functions are indroduced, we find that the determinant $D$ factors into a product of terms from the representations, (including rows of degenerate representations).

$$D = \prod_\beta D_\beta \tag{39}$$

Likewise, since after the transformation, the matrices $\mathcal{G}$ and $V$ are block diagonal. We can write

$$\langle \mathbf{q} | t | \mathbf{k} \rangle = \sum_\beta \langle \mathbf{q} | t_\beta | \mathbf{k} \rangle \tag{40}$$

In writing the "partial wave" $t$-matrices, it is convenient to indicate the rows of degenerate representations explicitly by an additional index: Then we interpret (40) as applying to representations only and sum over the index $\nu$ in the equation we will write for $t_\beta$:

$$\langle \mathbf{q} | t_\beta | \mathbf{k} \rangle = \frac{1}{D_\beta} \sum_{mjn} V^{(\beta)}_{mj} P^{(\beta)}_{nj} \sum_\nu C^+_{\beta,\nu}(\mathbf{q}, \mathbf{R}_m) C_{\beta,\nu}(\mathbf{k}, \mathbf{R}_n) \tag{41}$$

$$[1 - \mathcal{G}_\beta V_\beta]^{-1} = \frac{P_\beta}{D_\beta} \tag{42}$$

The submatrices $\mathcal{G}_\beta$, $V_\beta$ and $P_\beta$ as well as the determinant $D_\beta$ are the same for each row of a degenerate representation. The manner in which equation (41) reduces to the results of ordinary scattering theory in the limit of a spherically symmetric scatterer in free space will be discussed elsewhere. It suffices to say at the present that the angular dependence of the scattering amplitude is contained in the function $C_\beta$ and the phase shifts are contained in $D_\beta$.

We recall from the theory of matrices that a finite matrix always has an inverse unless the determinant of that matrix vanishes. Let us suppose that for some (real) energy $E$, one of the factors $D_\beta$ vanishes. This corresponds to a pole of the scattering amplitude, and means that the equation $\psi = \mathscr{G}V\psi$ has a solution for that energy (there is no "incident wave"). This then is a "bound state" of the system. This, we see that the energies of bound states may be determined by solving the equation

$$D_\beta(E) = 0 \tag{43}$$

for each representation $\beta$.

We now have enough of the general theory to return to the specific spin-wave problem. We will restrict our considerations to cubic lattices. As before, there are $Z$ nearest neighbors, and the nonzero portion of the matrix $V$ is $(Z + 1) \times (Z + 1)$. The matrix elements are determined from equation (14). It is convenient to introduce dimensionless quantities by dividing out a factor $-4JS$. Then the $V$ matrix can be expressed as follows, using the notation of Callaway and Boyd [4]:

$$V = -4JS \begin{pmatrix} \epsilon & \eta & \eta & \cdots & \eta \\ \eta & \rho & 0 & \cdots & \vdots \\ \vdots & 0 & \rho & 0 \cdot & \vdots \\ \eta & 0 & 0 & \cdots & \rho \end{pmatrix} \tag{44}$$

where

$$\epsilon = \frac{1}{2}Z\left(1 - \frac{J'}{J}\right)$$

$$\eta = 2\frac{[J'(SS')^{1/2} - JS]}{4JS} = \frac{1}{2}\left[\frac{J'}{J}\left(\frac{S'}{S}\right)^{1/2} - 1\right]$$

$$\rho = -2\frac{(J'S' - JS)}{4JS} = \frac{1}{2}\left(1 - \frac{J'S'}{JS}\right) \tag{45}$$

The elements of the Green's function matrix will just be

$$\langle n|(E - H_0)^{-1}|m\rangle = \mathscr{G}(\mathbf{R}_n - \mathbf{R}_m)$$

We now have to consider the irreducible representations. They work out as follows: Degeneracies and approximate symmetries are shown in parentheses.

| Simple cubic lattice | | | Bcc lattice | | | Fcc lattice | | |
|---|---|---|---|---|---|---|---|---|
| $\Gamma_1$ | (1) | $(s)$ | $\Gamma_1$ | (1) | $(s)$ | $\Gamma_1$ | (1) | $(s)$ |
| $\Gamma_{15}$ | (3) | $(p)$ | $\Gamma_{15}$ | (3) | $(p)$ | $\Gamma_{15}$ | (3) | $(p)$ |
| $\Gamma_{12}$ | (2) | $(d)$ | $\Gamma'_{25}$ | (3) | $(d)$ | $\Gamma_{12}$ | (2) | $(d)$ |
| | | | $\Gamma_2$ | (1) | $(f)$ | $\Gamma'_{25}$ | (3) | $(d)$ |
| | | | | | | $\Gamma_{25}$ | (3) | $(f)$ |

Now let us consider the construction of the subdeterminants $D_\beta$. It is not actually necessary to go explicitly through the introduction of the symmetrizing matrix $U$, since one can use directly the formulas obtained in energy band theory; that is, one can take the formulas which have been given for symmetrized combinations of plane waves (for $k = 0$), and proceed as in the computation of potential matrix elements. Alternately, one can use the coefficients appearing in the symmetrized combination to set up the unitary transformation. Expression (46) is the transformation for the simple cubic lattice:

$$U(\beta, \mathbf{R}_n) = \begin{pmatrix} 1 & 0 & 0 & 0 & 0 & 0 & 0 \\ 0 & 1/\sqrt{6} & 1/\sqrt{6} & 1/\sqrt{6} & 1/\sqrt{6} & 1/\sqrt{6} & 1/\sqrt{6} \\ 0 & 1/\sqrt{2} & -1/\sqrt{2} & 0 & 0 & 0 & 0 \\ 0 & 0 & 0 & 1/\sqrt{2} & -1/\sqrt{2} & 0 & 0 \\ 0 & 0 & 0 & 0 & 0 & 1/\sqrt{2} & -1/\sqrt{2} \\ 0 & 1/2 & 1/2 & -1/2 & -1/2 & 0 & 0 \\ 0 & 1/\sqrt{12} & 1/\sqrt{12} & 1/\sqrt{12} & 1/\sqrt{12} & -1/\sqrt{3} & -1/\sqrt{3} \end{pmatrix}$$

$$(46)$$

Thus we construct $UVU^+$, and obtain

$$UVU^+ = -4JS \begin{pmatrix} \epsilon & \sqrt{6}\,\eta & 0 & 0 & 0 & 0 & 0 \\ \sqrt{6}\,\eta & p & 0 & 0 & 0 & 0 & 0 \\ 0 & 0 & p & 0 & 0 & 0 & 0 \\ 0 & 0 & 0 & p & 0 & 0 & 0 \\ 0 & 0 & 0 & 0 & p & 0 & 0 \\ 0 & 0 & 0 & 0 & 0 & p & 0 \\ 0 & 0 & 0 & 0 & 0 & 0 & p \end{pmatrix} \qquad (47)$$

We have chosen the following convention for ordering the columns of $U$

$$\mathbf{R}_1 = (0, 0, 0), \qquad \mathbf{R}_2 = a(1, 0, 0), \quad \mathbf{R}_3 = a(-1, 0, 0), \quad \mathbf{R}_4 = a(0, 1, 0)$$
$$\mathbf{R}_5 = a(0, -1, 0), \quad \mathbf{R}_6 = a(0, 0, 1), \quad \mathbf{R}_7 = a(0, 0, -1)$$

The rows of $U$ are arranged as follows:

(1) $\Gamma_1$     (2) $\Gamma_1$     (3) $\Gamma_{15}, x$     (4) $\Gamma_{15}, y$     (5) $\Gamma_{15}, z$

(6) $\Gamma_{12}, x^2 - y^2$     (7) $\Gamma_{12}, x^2 + y^2 - 2z^2$

We treat the Green's function in the following way. Before transformation

$$R_m \rightarrow$$

$$\mathcal{G}(R_m - R_n) = \begin{array}{c} R_n \\ \downarrow \end{array} \begin{pmatrix} \mathcal{G}(0,0,0) & \mathcal{G}(1,0,0) & \mathcal{G}(-1,0,0) & \mathcal{G}(0,1,0) & \cdots \\ \mathcal{G}(-1,0,0) & \mathcal{G}(0,0,0) & \mathcal{G}(-2,0,0) & \mathcal{G}(-1,1,0) & \cdots \\ \mathcal{G}(1,0,0) & \mathcal{G}(2,0,0) & \mathcal{G}(0,0,0) & \mathcal{G}(1,1,0) & \cdots \\ \mathcal{G}(0,-1,0) & \mathcal{G}(1,-1,0) & \mathcal{G}(-1,-1,0) & \mathcal{G}(0,0,0) & \cdots \\ \vdots & \vdots & \vdots & \vdots & \cdots \end{pmatrix}$$

$$(48)$$

However, we observe from the formula (29) for $\mathcal{G}$ that it is invariant under any operation in the point group. Proof: Let $\alpha$ be such an operation:

$$\mathcal{G}(\alpha R) = \frac{\Omega}{8\pi^3} \int \frac{e^{ik \cdot \alpha R}}{E - E(\mathbf{k})} d^3k = \frac{\Omega}{8\pi^3} \int \frac{e^{-i\alpha^{-1}k \cdot R}}{E - E(\mathbf{k})} d^3k$$

Now put $\mathbf{k}' = \alpha^{-1}\mathbf{k}$, $\mathbf{k} = \alpha\mathbf{k}'$. Note that $E(\alpha\mathbf{k}') = E(\mathbf{k}')$, since the energy has full point group symmetry

$$\mathcal{G}(\alpha R) = \frac{\Omega}{8\pi^3} \int \frac{e^{-ik' \cdot R}}{E - E(\alpha\mathbf{k}')} d^3k' = \mathcal{G}(R) \tag{49}$$

This result reduces greatly the number of independent Green's functions we have to consider.

We can now apply the unitary transformation to the matrix $\mathcal{G}$. After doing so, we obtain a transformed matrix like equation (47).

$$U\mathcal{G}U^+ = \begin{pmatrix} \mathcal{G}(0,0,0) & \sqrt{Z}\mathcal{G}(1,0,0) & 0 & 0 & 0 & 0 & 0 \\ \sqrt{Z}\mathcal{G}(1,0,0) & \mathcal{G}_{s,1} & 0 & 0 & 0 & 0 & 0 \\ 0 & 0 & \mathcal{G}_p & 0 & 0 & 0 & 0 \\ 0 & 0 & 0 & \mathcal{G}_p & 0 & 0 & 0 \\ 0 & 0 & 0 & 0 & \mathcal{G}_p & 0 & 0 \\ 0 & 0 & 0 & 0 & 0 & \mathcal{G}_d & 0 \\ 0 & 0 & 0 & 0 & 0 & 0 & \mathcal{G}_d \end{pmatrix} \tag{50}$$

In the transformed matrix, the terms are as follows

$$\mathscr{G}_{s,1} = \mathscr{G}(0, 0, 0) + 4\mathscr{G}(1, 1, 0) + \mathscr{G}(2, 0, 0)$$
$$\mathscr{G}_{p} = \mathscr{G}(0, 0, 0) - \mathscr{G}(2, 0, 0)$$
$$\mathscr{G}_{d} = \mathscr{G}(0, 0, 0) - 2\mathscr{G}(1, 1, 0) + \mathscr{G}(2, 0, 0) \tag{51}$$

We have emphasized the simple cubic lattice for reasons which become important later: namely, it is possible in this case to reduce the three-dimensional integrals for the Green's functions to one-dimensional integrals which can readily be evaluated numerically. The symmetrization of the potential and Green's function matrices proceeds in exactly the same way for other crystal structures. In the simple cubic lattice, we can show that $\mathscr{G}$ can be expressed in terms of the Fourier transform of a product of three Bessel functions.

To do this, put $R = a(l, m, n)$, where $l, m, n$ are integers. From equation (17) we have that $E = 4JS(3 - \cos k_x a - \cos k_y a - \cos k_z a)$. Then after a little algebra

$$\mathscr{G}(R) = \left(\frac{a}{2\pi}\right)^3 \left(-\frac{1}{4JS}\right) \int d^3k \frac{e^{ia(k_x l + k_y m + k_z n)}}{\epsilon - i\eta - (\cos k_x a + \cos k_y a + \cos k_z a)} \tag{52}$$

where $\epsilon - i\eta = 1/4JS[3 - E - i\varepsilon]$ Next, we introduce an auxiliary variable $t$, define $q_x = k_x a$ etc. and write

$$\mathscr{G}(R) = -\frac{1}{4JS}\frac{1}{(2\pi)^3} \int_0^\infty dt\, e^{-i(\epsilon - i\eta)t} \int_{-\pi}^{\pi} dq_x \int_{-\pi}^{\pi} dq_y \int_{-\pi}^{\pi} dq_z\, e^{i(q_x l + q_y m + q_z n)}$$
$$\times\, e^{it(\cos q_x + \cos q_y + \cos q_z)}$$

Now we use the integral relation for Bessel functions

$$J_p(t) = \frac{i^{(-p)}}{2\pi} \int_{-\pi}^{\pi} \exp[i(px + t\cos x)]\, dx \tag{53}$$

Afterward we can let $\eta \to 0$.

Thus

$$\mathscr{G}(R) = -\frac{i^{(l+m+n+1)}}{4JS} \int_0^\infty dt\, e^{-i\epsilon t} J_l(t) J_m(t) J_n(t) \tag{54}$$

There are other representations: for instance if $\epsilon > 3$ ($E < 0$) a representation in term of Bessel functions of imaginary argument can be obtained.

Now we are in a position to consider the subdeterminants explicitly. To simplify the notation, let us write

$$\mathscr{G}(\mathbf{R}) = -\frac{1}{4JS} G_{lmn} \tag{55}$$

with

$$G_{lmn} = i^{(l+m+n+1)} \int e^{-l\epsilon t} \cdots dt$$

Then, for the $p$ states we have

$$D_p = D_{15} = 1 - \rho(G_{000} - G_{200}) \tag{56}$$

And for the $d$ states

$$D_d = D_{12} = 1 - \rho(G_{000} - 2G_{110} + G_{200}) \tag{57}$$

For the $S$ state, we have a $2 \times 2$ matrix to consider. In that matrix, there appear the Green's functions $\mathscr{G}(0, 0, 0)$, $\mathscr{G}(1, 0, 0)$, and the combination $\mathscr{G}_{s,1}$. These objects are related by certain identities. For instance, we can use the fact that all of the functions $\mathscr{G}(1, 0, 0)$ are equal by symmetry to write

$$
\begin{aligned}
\mathscr{G}(1, 0, 0) &= \frac{\Omega z^{-1}}{8\pi^3} \int \frac{\sum_{\Delta} e^{ik \cdot \Delta}}{E - E(\mathbf{k})} d^3 k \\
&= \left(1 - \frac{E}{2JSZ}\right)\mathscr{G}(0, 0, 0) + \frac{1}{2JSZ}\frac{\Omega}{8\pi^3} \int d^3 k \frac{E - E(\mathbf{k})}{E - E(\mathbf{k})} \\
&= \left(1 - \frac{E}{2JSZ}\right)\mathscr{G}(0, 0, 0) + \frac{1}{2JSZ}
\end{aligned}
\tag{58}
$$

Next, we observe that we can single out any nearest neighbor vector, $\Delta_0$, say, and write

$$
\begin{aligned}
\mathscr{G}_{s,1} &= \frac{\Omega}{8\pi^3} \int e^{ik \cdot \Delta_0} \frac{\sum_{\Delta} e^{ik \cdot \Delta}}{E - E(\mathbf{k})} d^3 k \\
&= \frac{\Omega}{8\pi^3} \int e^{ik \cdot \Delta_0}\left(\frac{Z - E/2JS}{E - E(\mathbf{k})} + \frac{1}{2JS}\right) d^3 k
\end{aligned}
$$

However

$$\int e^{ik \cdot \Delta_0} d^3 k = 0 \tag{59}$$

so we get

$$\mathscr{G}_{s,1} = \left(Z - \frac{E}{2JS}\right)\mathscr{G}(1, 0, 0)$$

Equations (58) and (59) are the desired identities. They have been derived in such a way that it is evident they can easily be generalized to any cubic system.

At this point, we have to face a rather tedious calculation of $D_s$ which I won't repeat here. After some algebra, we find

$$
\begin{aligned}
D_s &= 1 + \frac{Z}{2}(G_{1,0,0} - G_{0,0,0})\left(1 - \frac{J'}{J}\right) + \frac{EG_{1,0,0}}{4JS}\left(1 - \frac{J'S'}{JS}\right) \\
&= \frac{J'}{J} + \frac{E}{4JS}\left\{G_{1,0,0}\left(1 - \frac{J'S'}{JS}\right) - G_{0,0,0}\left(1 - \frac{J'}{J}\right)\right\}
\end{aligned}
\tag{60}
$$

From the second line of equation (60), we see that for small energies $D_s \approx J'/J$.

The energies of bound spin wave states of a particular type ($S$, $P$, or $d$) are found by equating the subdeterminants $D_s$, $D_p$, $D_d$ to zero, and solving for the energy. We have seen that for energies in the continuum, the Green's functions and hence the determinants are complex; thus for energies within the continuum, we do not expect to find solutions for real energies.

In order to facilitate calculation of the properties of the imperfect ferromagnet, it is convenient to define a phase shift for representation $\beta$ from the equation

$$
D_\beta = |D_\beta| e^{-i\delta_\beta}
\tag{61a}
$$

$$
\delta_\beta = \tan^{-1}\left[\frac{-\operatorname{Im} D_\beta}{\operatorname{Re} D_\beta}\right]
\tag{61b}
$$

The phase shifts defined in equation (61) are useful in two respects: (1) for low energies in a spherical band, they are the phase shifts which appear in the usual quantum-mechanical formula for the scattering amplitude, and (2) (see Callaway [2,3]) the change in the density of states due to scattering in representation $\beta$ is given by

$$
\Delta N_\beta = \frac{g_\beta}{\pi} \frac{d\delta_\beta}{dE}
\tag{62}
$$

The proof of equation (62) is relatively simple, it is given in Appendix A. Let us work out the phase shifts for the $\Gamma_1$ and $\Gamma_{15}$ representations. We will make throughout this discussion the "spherical band approximation," that is, we will replace the actual energy of the spin waves in the perfect crystal, equation (17), by the leading term in the expansion in power of $k$:

$$
E = \gamma \mathbf{k}^2 \quad \text{with} \quad \gamma = 2JSa^2
\tag{63}
$$

Now we require expressions for the imaginary part of the Green's functions. We have

$$\text{Im } \mathscr{G}(R) = -\frac{i\pi\Omega}{8\pi^3} \int d^3k \, e^{ik\cdot R}\delta[E - E(\mathbf{k})] \tag{64}$$

Define $q_0^2 = E/\gamma$. Then

$$\text{Im } \mathscr{G}(R) = \frac{-i\Omega}{8\pi^2\gamma} \int d^3k\delta(q_0^2 - k^2)e^{ik\cdot R} \tag{65}$$

$$= \frac{-i\Omega}{2\pi\gamma} \int k^2 dk \frac{\sin kR}{kR}\delta(q_0^2 - k^2) \tag{66}$$

$$= \frac{-i\Omega}{4\pi\gamma R} \sin q_0 R \tag{67}$$

Thus $\text{Im } \mathscr{G}(0) = -(\Omega/4\pi\gamma)q_0$, as a limit. Now consider $D_s$:

$$\text{Im } D_s = \frac{E}{4JS}\left\{\text{Im } G_{1,0,0}\left(1 - \frac{J'S'}{JS}\right) - \text{Im } G_{0,0,0}\left(1 - \frac{J'}{J}\right)\right\}$$

$$= -\frac{E(-\Omega)}{4\pi\gamma}\left[\frac{\sin q_0 a}{a}\left(1 - \frac{J'S'}{JS}\right) - q_0\left(1 - \frac{J'}{J}\right)\right]$$

Now we expand the sine, retaining only the first term, and

$$\text{Im } D_s = \frac{\gamma q_0^2 \Omega}{4\pi\gamma}q_0\left(1 - \frac{J'S'}{JS} - 1 + \frac{J'}{J}\right)$$

$$= \frac{q_0^3\Omega}{4\pi}\frac{J'}{J}\left(1 - \frac{S'}{S}\right) \tag{68}$$

Thus for $q_0$ small enough so that the tangent may be expanded and replaced by its leading term, we get

$$\delta_s = \frac{-\text{Im } D_s}{\text{Re } D_s} = \frac{-q_0^3\Omega}{4\pi}\frac{(J'/J)(1 - S'/S)}{J'/J}$$

$$= \frac{-q_0^3\Omega}{4\pi}\left(1 - \frac{S'}{S}\right) \tag{69}$$

Next, we turn to the $\Gamma_{15}$ representation. For small $E$, we use the value of $D_p$ at $E = 0$. From Wolfram and Callaway ([8]) we get $G_{000} - G_{200} = 0.42$ for $E = 0$. Thus $\text{Re } D_p = 1 - 0.42p$.

For the imaginary part, we have

$$\text{Im } D_p = -\rho[\text{Im } G_{000} - \text{Im } G_{200}]$$

$$= -\rho(-4JS)\frac{(-\Omega)}{4\pi\gamma}\left[q_0 - \frac{\sin 2q_0 a}{2a}\right]$$

$$= \frac{-\Omega\rho JS}{\pi\gamma}q_0\left(1 - \frac{\sin 2q_0 a}{2q_0 a}\right)$$

$$= \frac{-\Omega\rho JS}{\pi\gamma}q_0\left(1 - 1 + \frac{1}{6}\frac{(2q_0 a)^3}{2q_0 a} + \cdots\right)$$

$$= \frac{-\Omega\rho JS}{\pi\gamma}q_0 \cdot \frac{2}{3}(q_0 a)^2 = \frac{-q_0^3\Omega\rho}{3\pi} \tag{70}$$

Thus, again for $q_0$ small enough, we have

$$\delta_p = \frac{q_0^3\Omega\rho}{3\pi}\frac{1}{(1 - 0.42\rho)} \tag{71}$$

Now we note that both $\delta_s$ and $\delta_p$ are of order $q_0^3$. However, a similar analysis would show that $\delta_d$ is of order $q_0^5$, and to lowest order, can be neglected. We note that

$$\frac{d\delta_{s,p}}{dE} = \lambda_{s,p}\frac{dq_0^3}{dE} = \lambda_{s,p} \cdot 3q_0^2\frac{dq_0}{dE} = \frac{3\lambda_{s,p}}{2\gamma}q_0 \tag{72}$$

where $\lambda$ is a constant. The density of states for the perfect crystal is $\Omega E^{1/2}/4\pi^2\gamma^{3/2}$. To this we add the contribution from the scattering:

$$\frac{c}{\pi}\left(\frac{d\delta_s}{dE} + \frac{3d\delta_p}{dE}\right) = \frac{3cE^{1/2}\Omega}{8\pi^2\gamma^{3/2}}\left(-1 + \frac{S'}{S} + \frac{4\rho}{1 - 0.42\rho}\right) \tag{73}$$

where $C$ is the concentration of defect atoms. Of course, $C$ must be small for this theory to be valid. It is apparent that the density of states, including the impurities is still proportional to $E^{1/2}$, but with a different coefficient than in the perfect crystal. This suggests that we may define a changed effective mass, which we will denote by $\gamma_p$ by setting the density of states modified by the scattering equal to $\Omega E^{1/2}/4\pi^2\gamma_p^{3/2}$. Then we get

$$\frac{\Omega E^{1/2}}{4\pi^2\gamma_p^{3/2}} = \frac{\Omega E^{1/2}}{4\pi^2\gamma^{3/2}} + \frac{3cE^{1/2}\Omega}{8\pi^2\gamma^{3/2}}\left[-1 + \frac{S'}{S} + \frac{4\rho}{1 - 0.42\rho}\right]$$

or

$$\frac{1}{\gamma_p^{3/2}} = \frac{1}{\gamma^{3/2}}\left[1 - \frac{3c}{2}\left(1 - \frac{S'}{S} - \frac{4\rho}{1 - 0.42\rho}\right)\right]$$

To first order in $C$, we finally obtain

$$\gamma_p = \gamma\left[1 + C\left(1 - \frac{S'}{S} - \frac{4\rho}{1 - 0.42\rho}\right)\right] \tag{74}$$

This result is exact to first order in $C$ and agrees with results obtained by Murray [7] and Izyumov [6].

Now we will consider the possibility of resonant states within the continuum. Our previous treatment of scattering, involving the phase shifts and a low energy expansion, is not adequate in the cases of a resonance.

We have seen that localized states result when the equation $D_\beta(E) = 0$ has a solution for real $E$. There may also be solutions for complex $E = E_R + iE_i$. If $E_i$ is negative and small a scattering resonance will occur near $E_R$. If this definition seems somewhat vague, it cannot really be improved, since the basic concept of a scattering resonance is afflicted with the same disease—of course there are plenty of clearcut cases, but there are uncertain ones as well.

Since it is usually not convenient to compute Green's functions for complex values of the energy, it is usually not possible to solve the equation $D_\beta(E) = 0$ directly. Instead, we may take advantage of the fact that we are only interested in solutions which have a small imaginary part to proceed as follows: We look for solutions to the much simpler equation Re $D_\beta(E) = 0$ for real energies $E = E_0^{(\beta)}$, and then expand our functions in the vicinity of this point.

Thus for (complex) energies, close to $E_0^{(\beta)}$ we have

$$D_\beta(E) = (E - E_0^{(\beta)})D'_{\beta,R} + i[D_{\beta,i}(E_0^{(\beta)}) + (E - E_0^{(\beta)})D'_{\beta,i}(E_0^{(\beta)})] \tag{75}$$

Here we have written Re $D_\beta = D_{\beta,R} \ldots$, and a prime indicates differentiation with respect to energy. Let us now set $D_\beta = 0$ and solve for $E = E_R^{(\beta)} - i\Gamma_\beta/2$ by separating real and imaginary parts (We have chosen to write the imaginary part as $-\Gamma_\beta/2$ since this is in accord with the conventional definition of the width of a resonance). We obtain

$$E_R^{(\beta)} = E_0^{(\beta)} - \frac{D_{\beta,i}D'_{\beta,i}}{(D'_{\beta,R})^2 + (D'_{\beta,i})^2}$$

$$\approx E_0^{(\beta)} - \frac{D_{\beta,i}D'_{\beta,i}}{(D'_{\beta,R})^2} \tag{76a}$$

and

$$\Gamma_\beta = \frac{2D_{\beta,i}D'_{\beta,R}}{(D'_{\beta,R})^2 + (D'_{\beta,i})^2} \approx 2\frac{D_{\beta,i}}{D'_{\beta,R}} \tag{76b}$$

In the second step of equation (76), we have assumed that $|D'_{\beta,R}| \gg |D'_{\beta,i}|$ which is usually satisfied.

In the vicinity of a resonance the change in density of states has a characteristic resonance form. This may be derived as follows. Close to the resonance we may apply equation (75) to determine the phase shift, for representation $\beta$:

$$\tan \delta_\beta = \frac{-\Gamma_\beta}{2(E - E_0^{(\beta)})} \tag{77}$$

Equation (77) indicates that as $E$ passes through $E_0^{(\beta)}$ from below, the phase shift goes through $\pi/2$ (and is increasing for positive $\Gamma_\beta$). We may now apply equation (62) to determine $\Delta N_\beta$. On carrying out the differentiation, we get (after multiplying by the impurity concentration)

$$\Delta N_\beta = \frac{c g_\beta \Gamma_\beta}{2\pi} \frac{1}{(E - E_0^{(\beta)})^2 + \Gamma_\beta^2/4} \tag{78}$$

Equation (78) has the form expected for a scattering resonance. It appears in the density of states as well as in scattering cross sections; the connection occurring because a scattering cross section is proportional to the density of final states.

These results indicate that if the parameters of the magnetic impurity are suitable for a narrow resonance to develop at low energies, substantial alterations can be expected in the low-temperature thermodynamic properties of the system, such as the specific heat and magnetization. Wolfram and Callaway ([8]) showed that for reasonable values of the parameters $S'/S$ and $J'/J$, low energy $S$-wave resonances can occur. One would like to have the spin ratio $S'/S$ small and the exchange integral ratio $J'/J$ small as well—corresponding to a weakly coupled impurity. $P$-wave resonances occur near the top of the spin wave band and are probably less interesting.

We will conclude with a brief discussion of the spin wave scattering cross section. In the absence of resonances, the total cross section may be found immediately from the phase shifts given in equations (69) and (71). This approach is valid only at low energies, which is the only case we will consider.

From elementary scattering theory we have, if only $S$ and $P$ waves are included, the total cross section

$$\sigma_T = \frac{4\pi}{k^2}[\sin^2 \delta_s + 3 \sin^2 \delta_p] \tag{79}$$

since we are only considering small phase shifts, this reduces to

$$\sigma_T = \frac{4\pi}{k^2}[\delta_s^2 + 3\delta_p^2] \tag{80}$$

On substituting the phase shifts, this becomes

$$\sigma_T = \frac{a^2}{4\pi}k^4a^4\left[\left(1 - \frac{S'}{S}\right)^2 + \frac{16}{3}\frac{p^2}{(1 - 0.42p)^2}\right] \tag{81}$$

These results can be applied to determine the lifetime of a spin wave for small impurity concentrations through the formula

$$\frac{1}{\tau} = \frac{cv\sigma}{\Omega} \tag{82}$$

where $v$ is the spin-wave velocity $2\gamma k$. The lifetime is therefore proportional to $k^5$ and, for the simple cubic lattice, may be expressed as

$$\tau^{-1} = \frac{cE(k)}{2\pi}k^3a^3\left[\left(1 - \frac{S'}{S}\right)^2 + \frac{16}{3}\frac{p^2}{(1 - 0.42p)^2}\right] \tag{83}$$

To determine a mean free path appropriate for transport processes, we require however the so-called momentum-transfer cross section $\sigma_m$ which is given by

$$\sigma_m = 2\pi\int_0^\pi (1 - \cos\theta)\frac{d\sigma}{d\Omega}\sin\theta\, d\theta \tag{84}$$

in which $d\sigma/d\Omega$ is the differential cross section.

In our low-energy approximation, this is given by

$$\frac{d\sigma}{d\Omega} = \frac{1}{k^2}(\delta_s + 3\delta_p\cos\theta)^2 \tag{85}$$

Thus

$$\sigma_m = \frac{4\pi}{k^2}(\delta_s^2 - 2\delta_s\delta_p + 3\delta_p^2) \tag{86}$$

For the simple cubic lattice, we finally get

$$\sigma_m = \frac{a^2}{4\pi}k^4a^4\left[\left(1 - \frac{S'}{S}\right)^2 + \frac{8p}{3}\left(1 - \frac{S'}{S}\right)\frac{1}{1 - 0.42p} + \frac{16}{3}\frac{p^2}{(1 - 0.42p)^2}\right] \tag{87}$$

The mean free path for these purposes is given by the equation

$$l^{-1} = \frac{c\sigma_m}{\Omega} \tag{88}$$

An example of the effect of a resonance on the scattering cross section can be found in Callaway ([1]).

## APPENDIX A

Here we derive equation (62) relating the change in the density of states to the scattering phase shift.

We can define the density of states at energy $E$, $N(E)$ by the relation

$$N(E) = \frac{\Omega}{(2\pi)^3} \int d^3k \, \delta[E - E(\mathbf{k})] \qquad \text{(A-1)}$$

This formula is not particularly convenient for formal use. An equivalent formula is the following:

$$N(E) = -\frac{1}{\pi} \text{Im} \, Tr \frac{1}{E^+ - H} \qquad \text{(A-2)}$$

where $E^+ = E + i\epsilon$. The standard identiy

$$\frac{1}{x + i\epsilon} = P\left(\frac{1}{x}\right) - i\pi\delta(x)$$

leads from (A-2) to (A-1). We can now rewrite (A-2) as

$$N(E) = -\frac{1}{\pi} \text{Im} \left\{ Tr\left[\frac{d}{dE} \ln (E^+ - H)\right]\right\}$$

$$= -\frac{1}{\pi} \text{Im} \frac{d}{dE}\left[\ln \det (E^+ - H)\right] \qquad \text{(A-3)}$$

The equivalence of the two lines of equation (A-3) is most easily seen in the spcetral representation, but since the trace and the determinant are both invariant under unitary transformations, it holds in any representation. The contribution from the impurity state may be separated by writing

$$E^+ - H = (E^+ - H_0)\left(I - \frac{1}{E^+ - H_0} V\right)$$

The density of states of the crystal with the impurity replaced by a normal atom is denoted by $G(E)$, and can be written

$$G(E) = -\frac{1}{\pi} \text{Im} \left\{\frac{d}{dE}\left[\ln \det (E^+ - H_0)\right]\right\} \qquad \text{(A-4)}$$

Since the determinant of the product of two matrices is equal to the product of the determinants, the change in density of states due to the defect is

$$\Delta N = N(E) - G(E) = -\frac{1}{\pi} \, \text{Im} \left\{ \frac{d}{dE} \ln \det \left[ I - \frac{1}{E^+ - H_0} V \right] \right\} \quad \text{(A-5)}$$

The matrix $I - [1/(E^+ - H_0)]V = I - \mathcal{G}V$ is shown in block form in equation (32) of the main text. It can be seen by inspection of this form that the determinant of the entire matrix is equal to the determinant of the upper left-hand ($n \times n$) block. Thus we have

$$\Delta N = -\frac{1}{\pi} \, \text{Im} \, \frac{d}{dE} \ln D \quad \text{(A-6)}$$

where $D$ is given in equation (33a). Application of the unitary transformation which separates the irreducible representations causes the determinant to factor as described in equation (39). On substitution of equation (39) into (A-6) we find

$$\Delta N = \sum_\beta \Delta N_\beta \quad \text{(A-7a)}$$

where $\Delta N_\beta$ is the contribution of representation $\beta$ to the change in density of states, and is given by

$$\Delta N_\beta = -\frac{q_\beta}{\pi} \, \text{Im} \, \frac{dD_\beta}{dE} \quad \text{(A-7b)}$$

Finally we introduce the phase shift according to equation (61a)

$$D_\beta = |D_\beta| e^{-i\delta_\beta} \quad \text{(61a)}$$

Substitution into (A-7b) then yields

$$\Delta N_\beta = \frac{q_\beta}{\pi} \, \frac{d\delta_\beta}{dE} \quad \text{(62)}$$

which is just equation (62).

ACKNOWLEDGMENT

The work reported here has been supported by the U.S. Air Force Office of Scientific Research.

## REFERENCES

1.  J. Callaway, *Phys. Rev.* **132**: 2003 (1963).
2.  J. Callaway, *J. Math. Phys.* **5**: 783 (1964); *Phys. Rev.* **154**: 515 (1967).

3. J. Callaway, *Energy Band Theory*, Academic Press, New York, 1964, App. 2.
4. J. Callaway, and R. Boyd *Phys. Rev.* **134**: A1655 (1964).
5. D.L. Huber, *Phys. Rev.* **146**: 387, (1966).
6. Y. Izyumov, *Proc. Phys. Soc. (London)* **87**: 505, (1966).
7. G.A. Murray; *Proc. Phys. Soc. (London)* **89**: 87 (1966).
8. T. Wolfram and J. Callaway, *Phys. Rev.* **130**: 2207, (1963).

# Individual and Collective Excitations in an Electron Gas

## P. Resibois

*Institute of Physics*
*Université Libre*
*Brussels, Belgium*

---

## 1. INTRODUCTION ([1])

The simplest description of electrons in metals is given by the free particle model; the electron–electron and electron–ion interactions are completely neglected and the total energy of the system is simply given by

$$E^0 = \sum_{k,\sigma} \epsilon_k n^0_{k,\sigma} \tag{1.1}$$

where $\epsilon_k = \hbar^2 k^2/2m$ is the kinetic energy of an electron with wavenumber $k$ and mass $m$ and $n^0_{k,\sigma}$ is the number of electrons in state $k$, $\sigma$ (spin $\sigma = \pm\frac{1}{2}$). Electrons are fermions and obey thus, at thermal equilibrium with temperature $T$, the well-known Fermi–Dirac law:

$$n^0_{k,\sigma}(T) = \frac{1}{\exp[(\epsilon_k - \mu)\beta] + 1} \tag{1.2}$$

where $\mu$ is the chemical potential and $\beta = 1/KT$.

In particular, at zero temperature, $\mu$ becomes the Fermi energy:

$$\mu = \epsilon_F \tag{1.3}$$

and is such that all states with $\epsilon_k < \epsilon_F$ are filled while the states with $\epsilon_k > \epsilon_F$ are empty (see Fig. 1). It is easily shown that if we define the Fermi momentum $p_F$ by:

$$\epsilon_F = \frac{p_F^2}{2m} \tag{1.4}$$

one has*

---

*From now on, we generally set $\hbar = 1$.

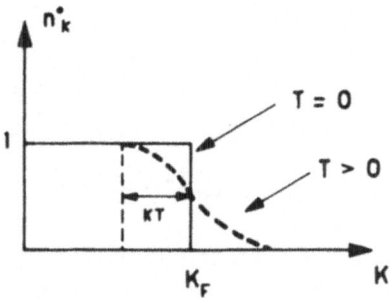

Figure 1

$$p_F = (3\pi^2 n)^{1/3} \tag{1.5}$$

where $n = N/\Omega$ ($N$: number of particles; $\Omega$; volume of the system). From these expressions, it is easy to calculate equilibrium properties; for instance, the number of excitations at $T \neq 0$ is proportional to $KT/\epsilon_F$ and the energy of each excitation is of the order of $KT$. We have

$$E(T) - E(0) \sim \frac{(KT)^2}{\epsilon_F}$$

and the specific heat is this proportional to $T$; more precisely, one has

$$\frac{C_v}{(NK)} = \frac{\pi^2}{2} \frac{KT}{\epsilon_F}$$

or from (1.4) and (1.5):

$$\frac{C_v}{\Omega} = \frac{m p_F K^2 T}{3} \tag{1.6}$$

Similarly, the magnetic susceptibility:

$$\frac{\chi}{N} = \frac{3\mu^2}{2\epsilon_F} \tag{1.7}$$

where $\mu$ is the Bohr magneton: $\mu = e\hbar/2mc$.

If we consider the transport properties of electrons, we clearly need to take into account some interaction. For instance, we may imagine that electrons are scattered by impurities; the transition probability for going from one state $k$ to another state $k'$ is given by the well-known "golden rule"

$$P(k \longrightarrow k') \sim 2\pi\delta(\epsilon_k - \epsilon_{k'})|\langle k \,|\, V \,|\, k'\rangle|^2 \tag{1.8}$$

where $\langle k|V|k'\rangle$ is the matrix element of the electron–impurity interaction potential $V$. From this result, we may then derive the well-known expression for the transport coefficients. For instance, one finds for the electrical conductivity at $T = 0$:

$$\sigma = -\frac{e^2}{\Omega}\sum_k k^2 \cdot \tau_k \cdot \frac{\partial n_k^0(0)}{\partial \epsilon_k} \tag{1.9}$$

where $n_k^0(0)$ is the equilibrium distribution (1.2) taken at $T = 0$ and $\tau_k$ is a relaxation time, closely related to the transition probability (1.8).

At equilibrium as well as out of equilibrium, the simple form of the result comes from the fact that all properties may be described as the sum of contributions coming from *individual excitations* (see equations (1.1) and (1.9)): these are simply here the various electrons in the system, characterized by their wavenumber $k$. Yet, when we want to consider electrons in a real metal, it is clear that these formulas cannot *a priori* be considered as valid. Indeed, at least two important aspects of the problem are neglected in the free-particle model:

1. The electron–ion interactions have been completely neglected; in particular, no account was taken of the periodic structure of the lattice.
2. The electron–electron interactions have also been discarded. The solution of the first difficulty is fairly well understood; one has to describe the electrons by Block wave functions:

$$\psi_{k,n}(r) = e^{ikr} \cdot u_{k,n}(r) \tag{1.10}$$

characterized by two quantum numbers: the first one $k$ corresponds to the pseudomomentum of the excitation ($|k| < \pi/a$, where $a$ is the lattice periodicity), while $n$ is a quantum number characterizing the different energy bands of the system; $u_{k,n}$ is a function having the lattice periodicity.

However, for most purposes, it may be shown that the main role of this new representation is simply to introduce an *effective mass* into the energy spectrum, i.e., the results of the free-particle theory are recovered provided one makes the replacement:

$$m \longrightarrow \tilde{m} \tag{1.11}$$

where $\tilde{m}$ is defined in terms of the spectrum of the energy bands. In particular, the linear dependence of the specific heat is still valid, as well as the temperature independence of magnetic susceptibility. As this aspect of the problem may be understood in terms of a one-particle model, we shall not discuss it any further here. More specifically we shall always discuss a very idealized model in which the lattice structure is completely negelected and ions are replaced by a continuous positive background, the only role of which is to ensure electroneutrality.

More puzzling are the electron–electron interactions; indeed the Coulombic interaction between two electrons:

$$V(r) = \frac{e^2}{r} \tag{1.12}$$

is a *long-range force* decreasing only very slowly with the interparticle separation $r$. As such, it should involve simultaneously many electrons and thus strongly affect their properties. Moreover, in metals, *Coulombic interactions are strong* and this leads *a priori* to a supplementary reason for obtaining strong departure from the free-particle model. Indeed it may be shown that, in the zero temperature limit, the only dimensionless parameter which measures the strength of the Coulombic interaction is the interelectronic mean distance in Bohr units, which is given by

$$r_s = \left(\frac{3}{4\pi}\right)^{1/3} \frac{me}{\hbar^2 n^{1/3}} \tag{1.13}$$

When typical numerical values are inserted, $r_s$ is not a small parameter ($2 < r_s < 5$).

Yet it is an experimental fact that the independent-particle model reproduces quite well the main features of electrons in metals, except for the numerical factors appearing in the physical laws. For instance, one finds

$$\frac{C_v}{\Omega} = \frac{m^* p_k K^2 T}{3} \tag{1.14}$$

except form small corrections; here $m^*$ is an "effective mass" which is different from the mass introduced in (1.11). Similar results hold for the magnetic susceptibility ($\mu^*$) or for transport properties ($\tau^*$). It is the aim of these lectures to indicate how, although the Coulomb interactions are strong and long range, the independent-particle model still furnishes a good approximation to the behavior of the electron gas. We shall show also that, for certain properties, this description is not valid and should be supplemented by a *collective description*, involving the so-called "plasmons."

In order to be as simple as possible, we shall, however, not solve this problem in its full generality; we shall rather consider separately the problem of long-range forces in a system where the interactions are weak [from (1.13) this corresponds to a high-density system] and, then, we shall discuss the problem of strong forces, assuming short-range effects. It would be a rather technical problem, but very difficult in practice, to incorporate the two features within a single model: we shall not discuss this question here.

## 2. DIELECTRIC FORMULATION OF THE ELECTRON GAS THEORY ([2])

Instead of considering the interactions between different electrons in the system, which is a difficult problem (although soluble) it is simpler to study the response of the system to an external weak test charge, with wave-number $q$ and frequency $\omega$:

$$e\eta_q(t, r) = e\eta_q^0 \exp [i(qr - \omega t)] \tag{2.1}$$

It can be shown that all the results obtained within this formalism are readily applicable to the system in the absence of this test charge. Let us stress that the motion of $\eta_q(r, t)$ is treated as *given*; the dynamical equations do not apply to the time evolution of $\eta_q(r, t)$. The interaction of the test charge with one given electron, located at $r_i$ is

$$V^{e_i}(r_i) = e^2 \int dr \frac{\eta_q(r)}{|r_i - r|} \tag{2.2}$$

According to well-known rules (see Landau and Lifshitz: *Quantum Mechanics*), we may go from a one-particle operator to its many-body equivalent in second quantization by the substitution (we neglect spin here):

$$f(r) \rightarrow \sum_{k,l} a_k^+ a_l \langle k|f|l \rangle \tag{2.3}$$

with $\langle k|f|l \rangle = \Omega^{-1} \int dr \exp(-ikr) f(r) \exp(ilr)$. Here $a_k^+$, $a_k$, respectively, are the well-known creation and destruction operators, obeying the anticommutation rules:

$$a_k^+ a_{k'} + a_{k'} a_k^+ = \delta_{kk'} \tag{2.4}$$

This leads to the following form for the total interaction in second quantization:

$$V^e = \sum_i V^{e_i}(r_i) = \sum_q \frac{4\pi e^2}{q} \hat{n}_q \eta_q^0 \exp (-i\omega t) \tag{2.5}$$

where we have introduced the local density operator

$$\hat{n}(r) = \sum_i \delta(r - r_i) = \frac{1}{\Omega} \sum_q \hat{n}_q \exp (iqr) \tag{2.6}$$

giving the number of electrons at point $r$ with

$$\hat{n}_q = \sum_k a_{k+q}^+ a_k \tag{2.7}$$

We shall now calculate the density fluctuation built up in the system in the presence of the test charge. For $t < 0$, before this test charge is introduced, the system is at equilibrium and characterized by the density matrix

$$\rho^{eq} = \exp\frac{[-\beta H]}{Z} \tag{2.8}$$

($\beta = 1/KT$); $Z =$ partition function), with the Hamiltonian

$$H = \sum_k \epsilon_k a_k^+ a_k + \frac{1}{2\Omega} \sum_{k,l,q} \frac{e^2}{q^2} a_{k+q}^+ a_{l-q}^+ a_l a_k \tag{2.9}$$

The density of the particle is then a constant:

$$n = \text{Tr } \hat{n}(r)\rho^{eq} = \frac{N}{\Omega}$$

Let us now look for the particle density for $t > 0$. For this, we write the Von Neuman equation for the density matrix:

$$i\frac{\partial \rho}{\partial t} = [H, \rho] + [V^e, \rho]$$

and decompose $\rho$ into two parts:

$$\rho = \rho^{eq} + \delta\rho$$

with

$$\delta\rho \propto \eta_q^0 \quad \text{(small)}$$

If we linearise the Von Neuman equation, we get then

$$i\frac{\partial \delta\rho}{\partial t} = [H, \delta\rho] + [V^e, \rho^{eq}] + O(\eta_q^{0^2}) \tag{2.11}$$

(N.B.: $[H, \rho^{eq}] = 0$).
The formal solution of (2.11) is

$$\rho(t) = \rho^{eq} + \frac{1}{i}\int_0^t \exp[-iH(t-t')][V^e, \rho^{eq}] \exp[iH(t-t')] \, dt' \tag{2.12}$$

The local density is now

$$n(r, t) = \text{Tr } \rho(t)\hat{n}(r) \tag{2.13}$$

and, because of the special form of the potential (2.6) it is readily seen that, in the long-term limit, the density fluctuation takes the form:

$$n(r, t) - n = \delta n(r, t) = \delta n_q(\omega) \exp[i(qr - \omega t)] \qquad (2.14)$$

After straightforward algebra, involving equations (2.13, 2.10, 2.12, 2.14, 2.5, and 2.6), we get for $t \to \infty$

$$\delta n_q(\omega) = \frac{4\pi e^2}{\Omega q^2} \eta_q^0 \frac{1}{i} \int_0^\infty \exp[i\omega\tau] \, \mathrm{Tr} \, \hat{n}_q(\tau)[\hat{n}_{-q}(0), \rho^{eq}] \, d\tau \qquad (2.15)$$

where

$$\hat{n}_q(\tau) = \exp(+iH\tau)\hat{n}_q \exp(-iH\tau) \text{ and } \tau = t - t'$$

This expression is closely related to a *"two-particle Green's function."* We shall however not discuss this question here and introduce now the *dielectric constant* $\epsilon_q(\omega)$.

We start form the macroscopic Maxwell equations. We have for the electric displacement

$$\mathbf{\nabla} \cdot \mathbf{D} = 4\pi n_{\text{true}} \qquad (2.16)$$

where $n_{\text{true}}$ is the "true" density of charge (here the external probe), while the electric field $\mathbf{E}$ obeys

$$\mathbf{\nabla} \cdot \mathbf{E} = 4\pi[n_{\text{true}} + n_{\text{ind}}] \qquad (2.17)$$

Here the induced charge $n_{\text{ind}}$ is the charge fluctuation of the electron gas, calculated in (2.15). In macroscopic theory, the dielectric constant $\epsilon$ is defined by

$$\mathbf{D} = \epsilon \mathbf{E} \qquad (2.18)$$

which is only valid for slowly varying phenomena (both in space and time).

We define now a generalized dielectric constant $\epsilon_q(\omega)$:

$$D_q(\omega) = \epsilon_q(\omega)E_q(\omega) \qquad (2.19)$$

where $D_q(\omega)$ and $E_q(\omega)$ are obtained by Fourier transform of equations (2.16, and 2.17) (*assumed to be valid for all q and $\omega$*). We get

$$iq \, D_q(\omega) = 4\pi e \eta_q^0 \exp(-i\omega t)$$

$$iq \, E_q(\omega) = 4\pi e[\eta_q^0 \exp(-i\omega t) + \delta n_q(\omega) \exp(-i\omega t)]$$

and we get then from (2.15)

$$\frac{1}{\epsilon_q(\omega)} = 1 + \frac{\delta n_q(\omega)}{\eta_q^0} = 1 + \frac{4\pi e^2}{\Omega q^2}\left[\frac{1}{i} \int_0^\infty d\tau \exp(i\omega\tau) \, \mathrm{Tr} \, \hat{n}_q(\tau)[\hat{n}_{-q}(0), \rho^{eq}]\right]$$

$$(2.20)$$

This quantity is independent of the test charge because we have a linear theory.

A more explicit form is obtained for $\epsilon_q(\omega)$ if we assume that the exact eigenstates and eigenvalues of the Hamiltonian $H$ are known:

$$H|\alpha\rangle = E_\alpha|\alpha\rangle$$

We can then evaluate the trace and the time integral in (2.20); for this a convergence factor $\gamma$ is introduced and we get

$$\frac{1}{\epsilon_q(\omega)} = 1 - \frac{4\pi e^2}{\Omega q^2} \sum_{\alpha\beta} \rho_\alpha(T) \left[ \frac{|\langle\beta|\hat{n}_q|\alpha\rangle|^2}{\omega + \omega_{\beta\alpha} + i\gamma} + \frac{|\langle\beta|\hat{n}_{-q}|\alpha\rangle|^2}{-\omega + \omega_{\beta\alpha} - i\gamma} \right] \quad (2.2\text{i})$$

where $\rho_\alpha(T) = \exp[-E_\alpha/KT]/Z$ and $\omega_{\beta\alpha} = E_\beta - E_\alpha$. In order to derive this formula, we have used the property

$$\langle\alpha|\hat{n}_{-q}|\beta\rangle = \langle\beta|\hat{n}_q|\alpha\rangle^* \quad (2.22)$$

a consequence of the reality condition $\hat{n}_{-q} = \hat{n}_q^*$.

This expression may be case into an alternative form, if we note that, by inversion symmetry, we have

$$|\langle\alpha|\hat{n}_q|\beta\rangle|^2 = |\langle\beta|\hat{n}_q|\alpha\rangle|^2 \quad (2.23)$$

We then interchange $\alpha$ and $\beta$ in the second term of the right-hand side of equation (2.21) and write.

$$\frac{1}{\epsilon_q(\omega)} = 1 - \frac{4\pi e^2}{\Omega q^2} \sum_{\alpha\beta} |\langle\alpha|\hat{n}_q|\beta\rangle|^2 \frac{\rho_\alpha(T) - \rho_\beta(T)}{(\omega + \omega_{\beta\alpha} + i\gamma)} \quad (2.24)$$

A number of exact theorems can be proved about the rigorous formulas (2.20) or (2.24) (Kramers–Kronig or fluctuation dissipation theorem: asymptotic forms for $\omega \rightarrow \infty$). In particular, it has been shown that the free energy of the system at temperature $T$ (and the ground state energy at $T = 0$) can be calculated exactly from $\epsilon_q(\omega)$.

However, as we are mainly interested in the nature of the excitations in the electron gas, we shall not discuss this problem here. We shall rather look for an approximate, but more explicit, form of $\epsilon_q(\omega)$, valid in a dense system and deduce from it a number of interesting consequences about the physical properties of the excitations.

Of course the simplest possible approximation amounts to neglecting completely the interactions between the electrons in the right-hand side of (2.9) (*Hartree–Fock approximation*):

$$H \simeq H_0, \quad E_\alpha \simeq E_\alpha^0 = \sum_k n_k \epsilon_k$$

and thus:

$$|\alpha\rangle \equiv |n_{k_1} \cdots n_{k_i} \cdots\rangle \quad \text{(free-particle state)}$$
$$\equiv |n\rangle \tag{2.25}$$

We define then the *exact polarizability*:

$$\frac{1}{\epsilon_q(\omega)} = 1 - 4\pi\alpha_q(\omega) \tag{2.26}$$

and use the notation $\tilde{\alpha}_q(\omega)$ for the *approximate result* coming from (2.25):

$$\tilde{\alpha}_q(\omega) = \frac{e^2}{\Omega q^2} \sum_{n,m} |\langle n|\hat{n}_q|m\rangle|^2 \frac{p_n(T) - p_m(T)}{(\omega + \omega_{mn} + i\gamma)} \tag{2.27}$$

But the only nonvanishing matrix elements $\langle n|\hat{n}_q|m\rangle$ correspond to (see 2.7)):

$$|m\rangle \equiv |n + 1_k - 1_{k+q}\rangle$$
$$E_m^0 = E_n^0 + \epsilon_k - \epsilon_{k+q}$$

Equation (2.27) becomes then

$$\tilde{\alpha}_q(\omega) = \frac{e^2}{\Omega q^2} \sum_k \sum_n \langle n|a_{k+q}^+ a_k a_k^+ a_{k+q}|n\rangle \frac{e^{-\beta E_n^0}}{Z^0} \frac{1 - \exp[-\beta(\epsilon_k - \epsilon_{k+q})]}{\omega - (\epsilon_{k+q} - \epsilon_k) + i\gamma}$$

$$= \frac{e^2}{\Omega q^2} \sum_k [\text{Tr } \hat{n}_{k+q}(1 - \hat{n}_k)\rho^{eq}] \frac{(1 - \exp[-\beta(\epsilon_k - \epsilon_{k+q})])}{\omega - (\epsilon_{k+q} - \epsilon_k) + i\gamma}$$

$$= \frac{e^2}{\Omega q^2} \sum_k n_{k+q}^0 (e - n_k^0) \frac{(1 - \exp[-\beta(\epsilon_k - \epsilon_{k+q})])}{\omega - (\epsilon_{k+q} - \epsilon_k) + i\gamma}$$

$$\tilde{\alpha}_q(\omega) = \frac{e^2}{\Omega q^2} \sum_k \frac{n_{k+q}^0 - n_k^0}{\omega - \omega_q(k) + i\gamma} \tag{2.28}$$

where we have introduced the Fermi–Dirac distribution (1.2) and the definition

$$\omega_q(k) = \epsilon_{k+q} - \epsilon_k$$

However, in developing this approximation, we have made an incorrect assumption: we have supposed that the electrons were only weakly responding to the displacement field of the test charge, which amounts to neglecting all the effects due to the polarization of the electron gas. Including these latter effects is a complicated problem, which involves discussion of the many-body aspects of the question. There is, however, a simple method which allows us to derive the correct result: we shall assume that the polarizability

$\tilde{\alpha}_q(\omega)$, as defined by (2.27), corresponds to the response of the system to the true local field:

$$\tilde{\alpha}_q(\omega) = \frac{P_q(\omega)}{E_q(\omega)} \qquad (2.29)$$

while, of course, the *exact* polarizability, equation (2.26), corresponds *by definition* to

$$\alpha_q(\omega) = \frac{P_q(\omega)}{D_q(\omega)} \qquad (2.30)$$

The justification of equation (2.29) stems from the assumption that, if it was not for the polarization effects, we could indeed consider the electrons as weakly responding: if we thus imagine a system without polarization of the medium but with an "external" field so adjusted that this polarization is included in its definition, our calculation for $\tilde{\alpha}_q(\omega)$ should be exact!

Let us suppose that the test charge is put on the plates of a condenser; it produces a displacement **D** and also induces a polarization charge $-4\pi\mathbf{P}$ on the surface of the medium. In order ot obtain the true electric field, we carve our a small sphere around the given electron; **E** is then given by the sum:

1. Of the displacement field **D**.
2. Of the surface polarization $-4\pi\mathbf{P}$
3. Of the Lorentz field on the surface of the sphere $\frac{4}{3}\pi\mathbf{P}$;
4. Of the field due to the molecules inside the sphere.

If *we neglect* (3) and (4), i.e., *all short-range effects*, we get

$$\mathbf{E} = \mathbf{D} - 4\pi\mathbf{P} \qquad (2.31)$$

Assuming this result to hold on the microscopic scale for frequency $\omega$ and wavenumber $q$, we get easily

$$\frac{1}{\epsilon_q(\omega)} = 1 - 4\pi\alpha_q(\omega) = \frac{E_q(\omega)}{D_q(\omega)} = \frac{E_q(\omega)}{E_q(\omega) + 4\pi P_q(\omega)} = \frac{1}{1 + 4\pi\tilde{\alpha}_q(\omega)} \qquad (2.32)$$

which indicates that the approximation $\tilde{\alpha}_q(\omega)$ for the exact polarizability is the first term in the infinite series development of $(1 + 4\pi\tilde{\alpha}_q(\omega))^{-1}$. Combining (2.28) and (2.32) we get

$$\epsilon_q(\omega) = 1 + \frac{4\pi e^2}{\Omega q^2} \sum_k \frac{n_{k+q}^0 - n_k^0}{\omega - \omega_q(k) + i\gamma} \qquad (2.29\text{-}bis)$$

It may be shown that this formula gives an exact result in the high-density, limit where the parameter (1.13) is small (*Random phase approximation*).

## CONSEQUENCES

### (1) Small Wavenumber–Low-Frequency Limit

Remember that one of the important questions to understand is the role of the long-range forces in an electron gas. As our external probe is also a Coulomb charge, it is pertinent, at least for a qualitative understanding, to look at the behavior of the dielectric constant $\epsilon_q(\omega)$ in the small wavenumber limit (which corresponds to large distances). If we are also interested in low-frequency phenomena, we have then the following result:

$$\lim_{\substack{q \to 0 \\ \omega \to 0}} \epsilon_q(\omega) = 1 + \frac{k_{\text{T.F.}}^2}{q^2} \qquad (2.30\text{-}bis)$$

with

$$k_{\text{T.F.}}^2 = -\frac{4\pi e^2}{\Omega} \sum_k \frac{\partial n_k^0(0)}{\partial \epsilon_k} = \frac{6\pi n e^2}{\epsilon_F} \qquad (2.31\text{-}bis)$$

This length $k_{\text{T.F.}}^{-1}$ is precisely the same as comes out of the more elementary Thomas-Fermi theory: it expresses the fact that for long distances (or $k \to 0$), the Coulomb force exerted by the test charge is completely screened: *due to polarization of the medium, the long range Coulomb force in an electron gas is only effective over distances of the order $k_{\text{T.F.}}^{-1}$ in the low-frequency limit:*

$$V_q \longrightarrow V_q^{\text{eff}}(\omega) = \frac{V_q}{\epsilon_q(\omega)} = \frac{4\pi e^2}{(q^2 + k_{\text{T.F.}}^2)} \qquad (2.32\text{-}bis)$$

This explains why Coulomb forces, although long-range, are rather ineffective in a dense electron gas. *Proof of (2.30-bis)*: from (2.29-bis), for small $q$ and $\omega \to 0$:

$$\epsilon_q(\omega) \simeq 1 + \frac{4\pi e^2}{\Omega q^2} \sum_k \frac{q \cdot \partial n^0/\partial k}{-kq/m}$$

$$\simeq 1 - \frac{4\pi e^2}{\Omega q^2} \sum_k \frac{\partial n^0}{\partial \epsilon_k}$$

The sum over k is easily performed because

$$\lim_{T \to 0} \frac{\partial n^0}{\partial \epsilon_k} = -\delta(\epsilon_k - \epsilon_F)$$

and leads to (2.31-*bis*).

## (2) Small Wavenumber–High-Frequency Limit Plasmon

However, it is not true even for $q \to 0$, that the high-frequency dielectric constant acts as a screening factor; indeed, we have

$$\lim_{\substack{q \to 0 \\ \omega \to \infty}} \epsilon_q(\omega) = 1 - \frac{\omega_{\text{pl.}}^2}{\omega^2} \tag{2.33}$$

with

$$\omega_{\text{pl}}^2 = \frac{4\pi e^2 n}{m} \tag{2.34}$$

*Proof of equation (2.33):*

$$\lim_{\substack{q \to 0 \\ \omega \to 0}} \epsilon_p(\omega) \simeq 1 + \frac{4\pi e^2}{\Omega q^2} \sum_k \frac{q \cdot \partial n_k^0/\partial k}{\omega(1 - kq/m\omega + \cdots)}$$

But $\sum_k \partial n_k^0/\partial k = 0$; thus:

$$\epsilon_q(\omega) \simeq 1 + \frac{4\pi e^2}{q^2 \omega^2} \frac{1}{\Omega} \sum_k q^2 \frac{k}{m} \cdot \frac{\partial n_k^0}{\partial k} = 1 - \frac{\omega_{\text{pl}}^2}{\omega^2}$$

We see also that $\omega \to \infty$ means $\omega \gg v_F q$.

Far from screening the Coulomb potential, the dielectric constant strongly enhances this interaction around $\omega = \omega_{\text{pl}}$. More precisely, we see from (2.20) that when $\epsilon_q(\omega) = 0$ a density fluctuation may be created *even in the absence of an external probe*: this corresponds to a new eigenstate of the system, called a *plasmon*, which describes a *collective oscillation* of the electron gas. Its interpretation is the following: if we create a charge disturbance in the system at the initial time, the electrons will move in order to screen this charge; they will, however, generally overshoot the equilibrium

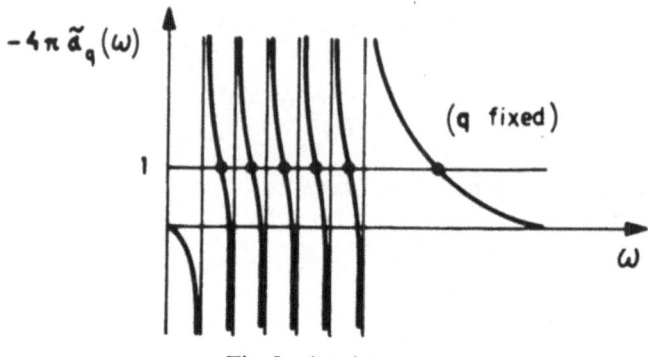

Fig. 2—$4\pi\tilde{a}_q(\omega) = 1$.

(neutrality) position and they will be pulled back again: these oscillations may then persist indefinitely.

In order to get an idea of the solution of

$$\epsilon_q(\omega) = 1 + 4\pi\tilde{a}_q(\omega) = 0 \tag{2.36}$$

for every $(q, \omega)$ let us neglect in (2.29-*bis*) the convergence factor $\gamma$ (we take principal parts) and let us work with a discrete (but dense) spectrum. The graphical solution of (2.36) is then sketched in Fig. 2 for $T = 0$. The vertical line represents the possible zeros of

$$\omega = \omega_q(k) = \frac{kq}{m} + \frac{q^2}{2m} \tag{2.35-bis}$$

where the denominator of $\alpha_q(\omega)$ vanishes.

If we consider positive frequencies only, this happens $M$ times [$M = 0(N)$] for points contained between the minimum value

$$\omega_q(k) = 0 \qquad \left(\frac{kq}{m} = -\frac{q^2}{2m}\right) \tag{2.36-bis}$$

up to the maximum value compatible with $k + q > k_F$, $k < k_F$; this occurs for $q$ parallel to $k$ and $k = k_F = mv_F$:

$$\omega_q^{\max} = qv_F + \frac{q^2}{2m} \tag{2.37}$$

We see from the graphical analysis that Re then $\epsilon_q(\omega)$ vanishes $(M - 1)$ times for values very close to the energy of a free particle–hole pair (up to order $1/N$); these solutions express the fact that the *Coulombic interactions* between the particles in the system [which in our analysis have been approximately accounted for by (2.32-*bis*)] *do not affect strongly the individual free-fermion excitations.* In the limit of a continuous spectrum ($\sum_k \rightarrow d^3k$ and the factor $i\gamma \neq 0$) these excitations are of course not generally stable because the electrons scatter each other: this leads to the imaginary part of the dielectric constant. *There is, however, one root which lies outside the continuum:* it corresponds to the plasmon (2.35). In the small-momentum case, we have indeed from (2.35) and (2.37):

$$\omega_{\text{pl}} > \omega_q^{\text{max}} \tag{2.38}$$

Even if we go to the limit of the continuous spectrum, this root remains stable because it lies outside of the continuum of particle–hole states:

$$\operatorname{Im} \epsilon_q(\omega_{\text{pl}}) = 0 \atop q_{\text{pl}} \to 0 \tag{2.39}$$

What happens now when the wavenumber increases? It may be shown that one has approximately:

$$\omega_q^{\text{pl}} \simeq \omega_{\text{pl}}\left[1 + \frac{q^2 v_F^2}{\omega_{\text{pl}}^2}\right] \tag{2.40}$$

while the maximum frequency (2.37) rises more rapidly. When a value of $q$ is reached such that

$$\omega_q^{\text{pl}} = \omega_q^{\text{max}} \tag{2.41}$$

the plasmon mode is no longer separated from the continuum; it merges into it and becomes unstable. This occurs, in first approximation, for

$$q^* v_F \simeq \omega_{\text{pl}}$$

$$q^* = \frac{\omega_{\text{pl}}}{v_F} \simeq k_{\text{T.F.}} \tag{2.42}$$

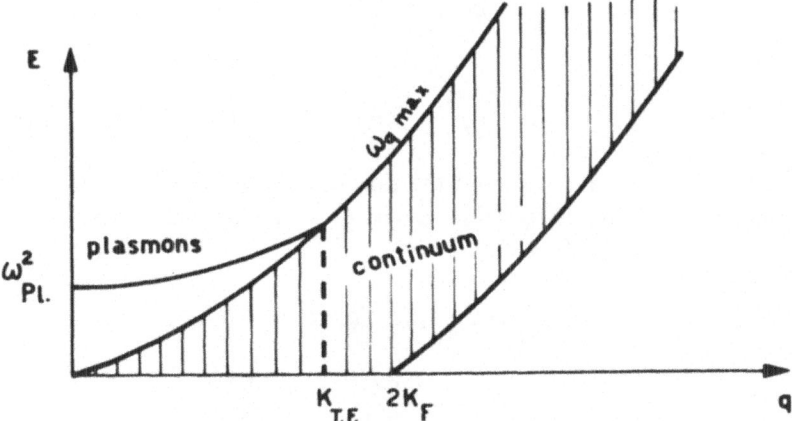

Fig. 3—Energy of excitations in a dense electron gas.

Stable collective oscillation only exists for small wavenumbers, smaller than the Thomas–Fermi length $k_{T.F.}^{-1}$. The situation is sketched in Fig. (3).

## 3. EXPERIMENTAL EVIDENCE FOR PLASMON

When typical metallic electron densities are inserted into (2.35), one finds

$$\hbar\omega_{pl} \simeq 12 \text{ eV}$$

which is a very large energy; there is thus no thermal excitation of plasmons. It can be shown that the plasmon leads to an important contribution to the ground state energy, but we do not consider this problem here.

The reason for this large energy is that one plasmon involves the correlated motion of many electrons, moving coherently; however, no single electron is strongly affected. Moreover, this is such a large energy with respect to the possible energy transfers from one single electron that individual and collective excitations remain largely independent.

How can one put plasmon effects in evidence experimentally? This can be done by energy loss experiments. Let us consider one fast electron moving in the medium. Its interaction with the system is

$$H_{\text{int}}(R) = \sum_j \frac{e^2}{|R - r_j|} \equiv \sum_q \frac{4\pi e^2}{q^2\Omega} \hat{n}_{-q}\, e^{-iqR} \tag{2.43}$$

as shown above in a similar case. It is a well-known result that the transition probability for the system to undergo a transition involving a momentum transfer $q$ and an energy transfer $\omega$ is

$$P(q, \omega) = 2\pi\left(\frac{4\pi e^2}{q^2}\right)^2 \sum_n |(\hat{n}_q)_{0n}|^2 \delta(\omega_{n0} - \omega) \tag{2.44}$$

where

$$\omega_{n0} = E_{\text{in}} - E_{\text{final}} = \frac{k^2}{2m} - \frac{(k - q)^2}{2m} = -\mathbf{q}\cdot\mathbf{v} + O(q^2) \tag{2.45}$$

In this formula we have assumed the plasmon to be initially in its ground state and we have denoted by $v$ the velocity of the incident electron.

The energy loss is then

$$\frac{dW(q, \omega)}{dt} = \omega\frac{dP(q, \omega)}{dt}$$

$$\equiv -\frac{8\pi e^2}{q^2\Omega} \omega \text{ Im} \frac{1}{\epsilon_q(\omega)}$$

because the imaginary part in (2.24) just invoves the same factors as in (2.45) when the limit $T \to 0$ is taken. Thus we get

$$\frac{dW}{dt} = -\frac{e^2}{\pi^2} \int \frac{d^3q}{q^2} \int_0^\infty d\omega \omega. \text{ Im } \left\{ \frac{1}{\epsilon_q(\omega)} \right\} \delta(\omega - \mathbf{q} \cdot \mathbf{v}) \qquad (2.46)$$

We get essentially two contributions to (2.46) for given $q$:

a) The continuous spectrum running from 0 to $\omega_q^{max}$ and corresponding to a screened interaction with individual electrons.

b) A strong peak at the plasmon frequency. Indeed we have

$$\text{Im } \frac{1}{\epsilon_q(\omega)} = \frac{-\text{Im } \epsilon_q(\omega)}{[\text{Im } \epsilon_q(\omega)]^2 + [\text{Re } \epsilon_q(\omega)]^2} \qquad (2.47)$$

As $\text{Im } \epsilon_q(\omega) \simeq 0$ for $\omega \simeq \omega_{pl}$ and [see (2.33)]

$$\text{Re } \epsilon_q(\omega) = 1 - \frac{\omega_{pl}^2}{\omega^2} \qquad (2.48)$$

using the well known formula

$$\lim_{\alpha \to 0} \frac{\alpha}{x^2 + \alpha^2} = \pi \delta(x) \qquad (2.49)$$

we get the important result

$$-\text{Im } \frac{1}{\epsilon_q(\omega)} = \pi \delta \left( 1 - \frac{\omega_{pl}^2}{\omega^2} \right) = \frac{\pi}{2} \omega_{pl} \delta(\omega - \omega_{pl}) \qquad (\omega \simeq \omega_{pl}) \quad (2.50)$$

We have sketched this result in Fig. (4).

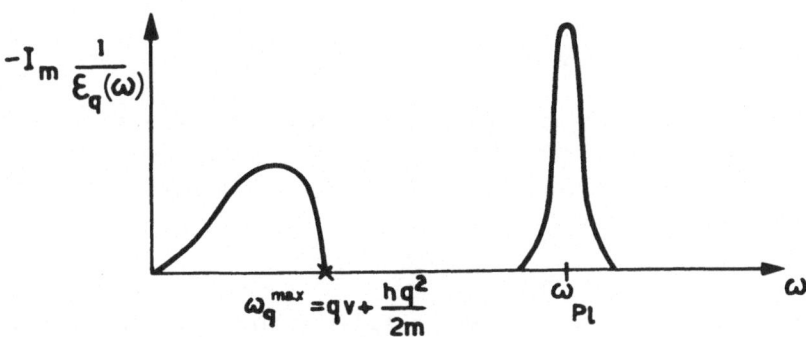

Fig. 4—Energy loss spectrum.

## 3. NORMAL FERMI LIQUID

Now that we have understood, at least qualitatively, the role of the long-range forces in a high-density electron gas, we would like also to have a knowledge of the effect of strong short forces on a many-fermion system.

### A. Landau Theory of Fermi Liquid ([3.1])

*a. The Model*

In 1956, Landau developed a beautiful theory, on a semi phenomenological level, that explains why normal interacting Fermion systems behave as a perfect gas. Consider a Fermi liquid (interacting normal Fermion system) with Hamiltonian:

$$H = H_0 + \lambda V \qquad (3.1)$$

When $\lambda = 0$, we have a perfect gas, entirely described by the occupation numbers $n_{p,\sigma}^0$. At $T = 0$.

$$n_{p,\sigma}^0 = \begin{cases} 1, p < p_F & p_F = (3\pi^2 n)^{1/3} \\ 0, p > p_F & \epsilon_p = \dfrac{p^2}{2m} \end{cases} \qquad (3.2)$$

In general, the interaction ($\lambda \neq 0$) will introduce transitions from one state to another and $\{n_{p,\sigma}^0\}$ will no longer be good a quantum number. However the basic postulate is: *the spectrum of elementary excitation of the Fermi liquid is the same as that of a perfect gas, at very low temperature.*

The momentum of each individual excitation is thus still a good quantum number. As a consequence: (1) the state of the Fermi liquid will be entirely described by giving the number of excitations (or quasi-particles) in each state of quantum number $(p, \sigma)$. (2) In particular, at $T = 0$, we shall have for the quasi-particle distribution

$$n_{p,\sigma}(T = 0) \equiv \bar{n}_{p,\sigma} = \begin{cases} 1, p < p_F \\ 0, p > p_F \end{cases} \qquad (3.3)$$

with the same Fermi momentum as for the noninteracting system. (3) The total number of quasi-particles is well defined (and equal to $N$); (4) The energy, which is of course no more defined by (1.1), is now a functional of the occupation numbers:

$$E = E(\{n_{p,\sigma}\}) \qquad (3.4)$$

(5) These excitations are stable for $p \sim p_F$ because their relaxation time tends to zero at $p = p_F$ ($T = 0$).

*Model Showing the Stability of the Excitations.* The interaction introduces scattering processes:

$$p_1 + p_2 \longrightarrow p_3 + p_4 \qquad (3.5)$$

Suppose we have one excitation $p_1 > p_F$; $p_2$ is of course inside the Fermi sphere $(p_2 < p_F)$, while $p_3$ and $p_4$ are accessible states $(p_3, p_4 > p_F)$. A real transition requires the conservation of energy:

$$\epsilon_{p_1} + \epsilon_{p_2} = \epsilon_{p_3} + \epsilon_{p_4} \qquad (3.5\text{-}bis)$$

which is impossible to satisfy when $p_1 \rightarrow p_F$. The excitations around the Fermi surfaces are thus stable.

However, we may formally develop the theory as if *all* excitations were stable because only those near the Fermi surface will contribute to physical properties. We may now define the energy of a quasi-particle as the energy increase when one quasi-particle with momentum $p$ and spin $\sigma$ is added.

Suppose a state of excitation $\{\bar{n}_{p,\sigma} + \delta n_{p,\sigma}\}$, where $\bar{n}_{p,\sigma}$ denotes the distribution of quasi-particles in the ground state. We have

$$\frac{E(\{n_{p,\sigma}\})}{\Omega} = \frac{E(\{\bar{n}_{p,\sigma} + \delta n_{p,\sigma}\})}{\Omega}$$

$$= \frac{E(\{\bar{n}_{p,\sigma}\})}{\Omega} + \frac{1}{\Omega} \sum_{p,\sigma} \frac{\partial E}{\partial n_{p,\sigma}}\Big|\{\bar{n}_{p,\sigma}\}\, \delta n_{p,\sigma} +$$

$$+ \frac{1}{2} \sum_{\substack{p,\sigma \\ p',\sigma'}} \frac{\partial^2 E}{\partial n_{p,\sigma}\partial n_{p',\sigma'}}\Big|\{\bar{n}_{p,\sigma}\}\, \delta n_{p,\sigma}\delta n_{p',\sigma'} + \cdots$$

which is written, by definition, as

$$= \frac{E(\{n_{p,\sigma}\})}{\Omega} + \frac{1}{\Omega} \sum_{p,\sigma} \bar{\epsilon}_p^0 \delta n_{p,\sigma} + \frac{1}{2\Omega^2} \sum_{p,\sigma,p',\sigma'} f_{p,\sigma,p'\sigma'} \delta n_{p,\sigma}\delta n_{p',\sigma'} + \cdots$$

where

$$f_{p\sigma,p'\sigma'} = \Omega \frac{\partial^2 E}{\partial n_{p,\sigma}\partial n_{p',\sigma'}}\Big|\{\bar{n}_{p\sigma}\}$$

and $\bar{\epsilon}_p^0$ is spin-independent. The energy of one quasi-particle is thus

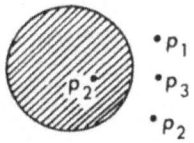

$$\tilde{\epsilon}_{p,\sigma} = \frac{\partial E}{\partial \bar{n}_{p,\sigma}}\bigg|_{\{\bar{n}_{p,\sigma}\}} \equiv \tilde{\epsilon}_p^0 + \frac{1}{\Omega}\sum_{p',\sigma'} f_{p\sigma,p'\sigma'}\delta n_{p',\sigma'} \qquad (3.7)$$

which is a functional of the state of the system.

Landau supposes that, at low but finite $T$,

$$n_{p,\sigma} = \frac{1}{\exp[+\beta(\tilde{\epsilon}_{p\sigma} - \epsilon_F)] + 1} \qquad (3.8)$$

### b. Consequences

When we have only one quasi-particle, we get from (3.7)

$$\tilde{\epsilon}_{p,\sigma} \equiv \tilde{\epsilon}_p^0$$

which may develop around $p_F$:

$$\tilde{\epsilon}_p^0 = \tilde{\epsilon}_{pF}^0 + v_F(p - p_F)$$

where

$$v_F = \frac{\partial \tilde{\epsilon}_p^0}{\partial p}\bigg|_{p_F} = \frac{p_F}{m^*} \qquad (3.9)$$

which defines the effective mass $m^*$.

By expressing the translation invariance of the system, Landau has established an important connection between $m^*$ and $f_{p\sigma,p'\sigma'}$ at the Fermi surface. If we write

$$f_{p\sigma,p'\sigma'}\big|_{\substack{|p|=p_F \\ |p'|=p_F}} = f_{\sigma,\sigma'}(\cos x)$$

$$\mathbf{p}\cdot\mathbf{p}' = |p_F|^2 \cos x$$

one has

$$\frac{1}{m^*} = \frac{1}{m} - \frac{p_F}{4\pi^2}\sum_\sigma \sum_{\sigma'} \int_{-1}^{+1} d(\cos x)\cdot\cos x\cdot f_{\sigma,\sigma'}(\cos x) \qquad (3.10)$$

We may also compute, for instance, the specific heat:

$$\frac{C_v}{\Omega} = \frac{\partial E/\Omega}{\partial T} \equiv \frac{\partial}{\partial T}E[n_{p,\sigma}(T)]$$

$$= \frac{\partial}{\partial T}E[\bar{n} + \delta n(T)]$$

$$= \frac{\partial}{\partial T}\left[E(\bar{n}) + \sum_{p,\sigma}\frac{\partial E}{\partial \bar{n}}\bigg|_{\bar{n}}\delta n_{p,\sigma}(T) + \cdots\right]$$

It may be shown that the higher-order terms are negligible. Moreover:

$$\frac{\partial}{\partial T}\delta n_{p,\sigma} = \frac{\partial}{\partial T}(n_{p,\sigma} - \bar{n}_{p,\sigma}) = \frac{\partial}{\partial T}n_{p,\sigma}$$

Thus

$$\frac{C_v}{\Omega} = \frac{\partial}{\partial T}\sum_{p,\sigma}\tilde{\epsilon}_p^0 n_{p,\sigma}\,(\tilde{\epsilon}_p^0; T) \qquad (3.11)$$

which is quite analogous to the free-particle expression and leads to

$$\frac{C_v}{\Omega} = \frac{m^* p_F K^2 T}{3}$$

Similarly, one shows the susceptibility

$$\frac{\chi}{\Omega} = \frac{3\mu^2}{2\epsilon_F}\gamma \qquad (3.12)$$

with:

$$\gamma = 1 + f(m^*, f_{\sigma,\sigma'}^{\text{exch.}})$$

Finally, for nonequilibrium properties, which are nondissipative at $T = 0$, the interaction between the quasi-particles is described by $f_{p,p'}$ instead of the forward scattering amplitude $\langle pp' | v | p'p \rangle$. This leads to "zero sound" with short-range forces and, again, to plasmons in the case of long-range forces.

## B. Microscopic Justification of the Landau Theory [1,2]

The most convenient tool for studying this problem at zero temperature is the one-particle Green's function. Although the detailed theory is difficult and would lead us far aside, it is rather easy to understand the main features of the theory in the equilibrium problem. At zero temperature, we define

$$G(x, x') = -i\langle T\{\psi(x)\psi^+(x')\}\rangle \qquad (3.13)$$

where we have neglected the spin variables and where

$$x = (r, t) \qquad x' = (r', t')$$
$$\psi(x) = \exp[iHt]\psi(r)\exp[-iHt]$$

(Heisenberg representation), and

$$\psi(r) = \sum_k a_k \frac{\exp[ikr]}{\Omega^{1/2}} \tag{3.14}$$

(destruction operator at point $r$). $T$ is the Dyson time-ordering operator such that:

$$T\{f(t)f(t')\} = f(t)f(t'), \qquad t > t'$$
$$= -f(t')f(t), \quad t' > t$$

$f$: Fermion operator and $\langle \ \rangle$ denotes average over the *exact* perturbed ground state.

Many general theorems are available for Green's functions but they will not be considered here. We just want to find the analogy between the one particle G.F. for the unperturbed system and for the perturbed Fermion system at zero temperature.

Consider first the free-particle case. For $t > t'$, we have

$$G^0(x, x') = -i < 0 | \exp[iH_0 t]\psi(r) \exp[-iH_0 t]$$
$$\times \exp[iH_0 t']\psi^+(r') \exp[-iH_0 t'] | 0\rangle$$

where $|0\rangle$ is the "vacuum of free particles, i.e., the state characterized by.

$$n_p^0 = 1, \quad p < p_F$$
$$= 0, \quad p > p_F$$

This expression is readily evaluated by considering the definition (3.14): $(t > t')$

$$G^0(x, x') = \frac{-1}{\Omega} \sum_k \sum_{k'} \langle 0 | a_k a_{k'}^+ | 0 \rangle \exp(ikr)$$
$$\times \exp(-ik'r') \exp(-i\epsilon_k t) \exp(i\epsilon_k t')$$
$$\equiv -\frac{1}{\Omega} \sum_k (1 - n_k^0) \exp[-i\epsilon_k(t - t')] \exp[ik(r - r')] \tag{3.15}$$

Similarly, for $t < t'$

$$G^0(x - x') \equiv -\frac{1}{\Omega} \sum_k (-n_k^0) \exp[-i\epsilon_k(t - t')] \exp[ik(r - r')] \tag{3.15-bis}$$

These expressions only depend on $\tau = t - t'$, $u = r - r'$. Defining the Fourier transform

$$G^0(p, \omega) = \int du \int_{-\infty}^{+\infty} d\tau G^0(u, \tau) \exp(i\omega\tau) \exp(-ipu) \tag{3.16}$$

and introducing the convergence factor $\epsilon$ for $t > 0$, $-\epsilon$ for $t < 0$, one gets easily

$$G^0(p, \omega) = \frac{1}{\omega - \epsilon_p + \epsilon \operatorname{sgn}(|p| - |p_F|)} \tag{3.17}$$

with

$$\operatorname{sgn}(x) = \begin{cases} +1, & x > 0 \\ -1, & x < 0 \end{cases}$$

Conversely, if (3.17) is given, we may find the nature of the excitation by Fourier inverstion with respect to $\omega$:

$$\tau > 0: G^0(p; \tau) = \int_{-\infty}^{+\infty} \frac{d\omega}{2\pi} \frac{\exp(-i\omega\tau)}{\omega - \epsilon_p + i\epsilon \operatorname{sgn}(|p| - |p_F|)}.$$

The integral is performed by closing the contour in the lower half-plane:

a) If $p < p_F$: pole at $\omega = \epsilon_p + i\epsilon$ in $S^+$ (upper half-plane)

$$G^0(p; t) = 0$$

b) If $p > p_F$: pole at $\omega = \epsilon_p - i\epsilon$ in $S^-$

Thus

$$\tau > 0: \quad G^0(p; \tau) = \exp(-i\epsilon_p - \epsilon)\tau\theta(p - p_F) \tag{3.19}$$

where $\theta(p - p_F)$ step function, which describes particle propagation. Similarly, for $\tau < 0$, we would obtain a hole propagation ($p < p_F$).

With this very elementary preparation, we are now able to understand the interacting system: it is a result derived by infinite order perturbation theory that

$$G(p, \omega) = \frac{1}{\omega - \epsilon_p - \sum_p(\omega) + i\epsilon \operatorname{sgn}(|p| - |p_F|)} \tag{3.20}$$

Thus, we obtain the same structure as for the free-particle case, except for the

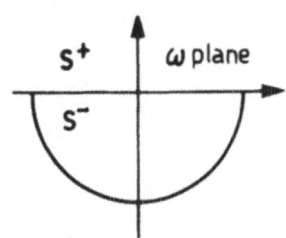

self-energy operator $\Sigma_p$ which describes the interaction of the particle of momentum $p$ with the medium. Now, in general

$$\Sigma_p(\omega) = \mathrm{Re}\, \Sigma_p(\omega) + i\Gamma_p(\omega) \qquad (3.20\text{-}bis)$$

and because of the imaginary part in $\Sigma_p(\omega)$, we may go directly to the limit $\epsilon \to 0$ in equation (2.20) (self-damping):

$$G(p, \omega) = \frac{1}{\omega - \epsilon_p - \Sigma_p(\omega)} \qquad (3.21)$$

If we define the chemical potential $\mu$ at absolute zero

$$\mu = E^0_{N+1} - E^0_N \qquad (3.22)$$

as the difference between the ground state energy of the system with $(N + 1)$ particles and the same quantity with $N$ particles ($E^0_{N+1}$ and $E^0_N$ are *a priori* unknown, but $\mu$ will later be identified with the Fermi energy), one can prove the exact theorem:

$$\Gamma_p(\omega) < 0, \quad \omega > \mu$$
$$\Gamma_p(\omega) > 0, \quad \omega < \mu$$

Because $\Gamma_p(\omega)$ is continuous around $\omega = \mu$ if perturbation calculus makes sense, this is equivalent to the statement (Van Hove's theorem)

$$\Gamma_p(\omega) = 0 \qquad \omega = \mu \qquad (3.23)$$

The time-dependent Green's function

$$G(p, t) = \int_{-\infty}^{+\infty} \frac{d\omega}{2\pi} G(p, \omega) \exp(-i\omega t) \qquad t > 0 \qquad (3.24)$$

will give us the required information about the nature of the excitation. For $t > 0$, we close the contour in the lower half-plane; the dominant behavior for long times of (3.24) will thus be determined from the poles of (3.21) closest to the real axis; let us denote by $\omega_1$ the *complex* solution of

$$\omega_1 = \epsilon^0_p - \Sigma_p(\omega_1) = 0 \qquad (3.25)$$

which is closest to the real axis.

We write $\omega_1$ as

$$\omega_1 = \epsilon_p + i\Gamma_p$$

and assume that the damping $\Gamma_p$ is small. Equation (3.25) gives then approximately

$$\tilde{\epsilon}_{p1} = \epsilon_p^0 + \text{Re} \sum_p(\tilde{\epsilon}_p) + 0(\Gamma_p) \tag{3.26a}$$

$$\Gamma_p = \text{Im} \sum_p(\tilde{\epsilon}_p) + 0(\Gamma_p^2) \tag{3.26b}$$

The reasoning is then in principle the following: (a) $\sum_p(\omega)$, being in principle known, determine the region $\omega \simeq \mu$ by imposing:

$$\text{Im} \sum_p(\mu) = 0 \tag{3.27}$$

thus $\mu$ is fixed. (b) The solution of (3.26a) being in principle determined and giving $\tilde{\epsilon}_p$ as a function of $p$,

$$\tilde{\epsilon}_p = f(p) \tag{3.28a}$$

determine the Fermi momentum $p_F$ by

$$\tilde{\epsilon}_{p_F} = f(p_F) \equiv \mu \tag{3.28b}$$

For Van Hove's theorem, we know that, if we take $p$ larger than and close to $p_F$, the condition that $\Gamma_p$ is small (and negative) will be satisfied. If we only retain the contribution coming from the pole closest to the real axis, we have

$$G(p, t) = \int_{-\infty}^{+\infty} \frac{d\omega}{2\pi} \frac{1}{\omega - \epsilon_p - \sum_p(\omega)} \exp(-i\omega t) \tag{3.29}$$

$$\simeq \int_{-\infty}^{+\infty} \frac{d\omega}{2\pi} \frac{1}{\omega - \epsilon_p - \sum_p(\tilde{\epsilon}_p) + \partial\epsilon_p/\partial\tilde{\epsilon}_p(\omega - \tilde{\epsilon}_p - i\Gamma_p)} e^{-i\omega t}$$

$$G(p, t) \simeq -iZ_p \exp[-i(\tilde{\epsilon}_p - i\Gamma_p)t]$$

with

$$Z_p = \left(1 - \frac{\partial\epsilon_p}{\partial\omega}\Big|_{\epsilon_p}\right)^{-1} \tag{3.30}$$

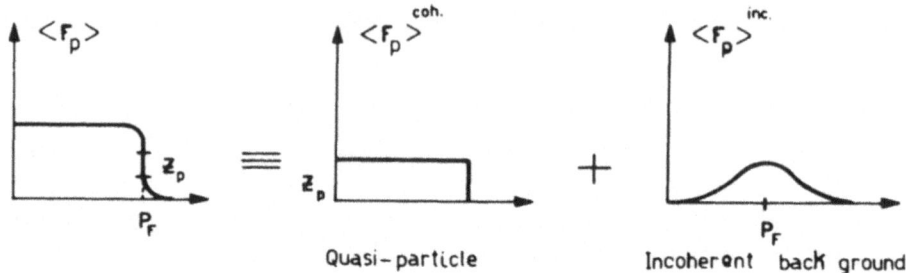

Fig. 5—Discontinuity at the Fermi suface.

We get thus: for momentum $p$ close to and above the Fermi momentum [determined in principle from (3.27)], the Green's function describes stable excitations (weak damping) similar to free particles with, however, a renormalized energy and a renormalized amplitude. Similarly for $p$ close to and below the Fermi momentum one obtains quasi-holes.

Another theorem which is rather easily proved is that the Fermi energy, defined by (3.22) corresponds to a sharp discontinuity in the one-particle momentum distribution function (d.f.) defined by

$$\langle F_p \rangle = \text{Tr } \hat{n}_p \rho^{eq}(T = 0)$$

In the interacting Fermi system, there is thus a "remembrance" of the unperturbed system: the Landau quasi-particle distribution function may also be shown to be just the "coherent," sharply discontinuous part of the momentum d.f. There exist many other theorems on Green's function which allow a more complete justification of the Landau theory; for instance:

1. The proof that the Fermi momentum $p_F$ is the same as in the free-particle case.
2. The mass equation.
3. The dynamical nondissipative properties (zero sound and plasmon).

We shall, however, not consider them here because the mathematical aspect becomes very involved.

## C. Quasi-Particle and Transport Properties

Until now, we have only considered equilibrium properties or dissipative properties which involve no "collisions" between the excitations. This is because the theory of Landau (as well as the dielectric approach) is only simple at zero temperature. As is well known, transport properties (involving entropy production) are only defined at finite temperature, except for some rather trivial problems like impurity scattering. Yet, precisely the same properties are observed as at equilibrium: the weak coupling approximation usually furnishes a very good description, although the interactions are both strong and long range (for instance $T^5$ law for electron–phonon resistence in metals). We are thus naturally lead to raise questions similar to those answered at equilibrium:

1. What is the role of the long-range forces?
2. What is the role of the strong forces?

The answer to the first question is rather well understood. For instance, it has been shown that the transport equation for a quantum plasma could be described by a weak coupling "Boltzmann equation" provided one uses a dynamical screened potential ([3.3]):

$$V_q = \frac{4\pi e^2}{q^2} \longrightarrow V_q(\omega) = \frac{V_q}{\epsilon_q(\omega)} \tag{3.31}$$

The answer to the second question is in a much less developed stage although preliminary results are encouraging.

Consider the weak coupling transport equation; as is well known, the momentum d.f. satisfies

$$\begin{aligned}
\partial_t \varphi_{p1} = \lambda^2 \sum |\langle p_1 p_2 | V | p_1' p_2' \rangle|^2 \delta(p_1 + p_2 - p_1' - p_2') 2\pi \\
\times \delta(\epsilon_{p_1} + \epsilon_{p_2} - \epsilon_{p_1'} - \epsilon_{p_2'}) \cdot [\varphi_{p_1} \varphi_{p_2} (1 - \varphi_{p_1'})(1 - \varphi_{p_2'}) \\
- \varphi_{p_1'} \varphi_{p_2'} (1 - \varphi_{p_1})(1 - \varphi_{p_2})] + \cdots \\
\equiv \lambda^2 C^{(2)}(\varphi)
\end{aligned} \tag{3.32}$$

Modern techniques in statistical mechanics allow writing down of equations to all orders [see, for instance, Prigogine, ([3.4])]:

$$\partial_t \varphi_{p_1} = \lambda^2 C^{(2)}(\varphi_1) + \lambda^4 C^{(4)}(\varphi_1) + \cdots \tag{3.33}$$

but no simple interpretation can be given to these higher-order terms (for instance, in some contribution, there is no energy conservation between initial and final states). However, it has been shown [Resibois, Dagonnier, Watabe ([3.5, 3.6])] that for fermions we may introduce a function defined by

$$n_p(t) = \sum_{p'} (\delta_{pp'}^{k\sigma} + \lambda^2 \gamma_{pp'}^{(2)}) \varphi_{p'}(t) \tag{3.34}$$

This new representation is such that at equilibrium $\varphi_p$ tends to the correct perturbed equilibrium momentum distribution, while $n_p$ tends to a Fermi–Dirac distribution with energies $\tilde{\epsilon}_p$ (quasi-particles). Equation (3.34) gives the same relationship as between Landau quasi-particle distribution and the momentum distribution. Then, one shows that (3.34) leads to

$$\partial_t n_{p_1} = \lambda^2 \tilde{C}^{(2)}(n) + \lambda^4 \tilde{C}^{(4)}(n) \qquad (3.35)$$

Here $\tilde{C}^{(2)}$ is a renormalized Born collision operator, constructed from (3.32) by the substitutions

$$V_{p_1 p_2, p_1' p_2'} \longrightarrow \sqrt{(Z_{p_1} Z_{p_2} Z_{p_1'} Z_{p_2'})}\, V_{p_1 p_2 p_1' p_2'}$$

$$\epsilon_{p_1} \longrightarrow \tilde{\epsilon}_{p_1}$$

while $\tilde{C}^{(4)}$ is a genuine fourth-order collision operator. Also, if we define the entropy of the interacting system by

$$S = -K \int n_p \ln n_p d_p - K \int (1 - n_p) \ln (1 - n_p)\, dp$$

it may be shown that this quantity goes to the correct equilibrium entropy (up to order $\lambda^2$).

Many similar problems have been considered (phonons; dense gas; polaron; electromagnetic interaction [see ($^{3.7}$)] but also only to lowest order ($\lambda^4$). It is a matter of conjecture whether these results may be generalized to all orders in $\lambda$ and to all temperatures. However, for low-temperature fermions, much hope exists that such a quasi-particle description of transport properties will be proved to be generally valid.

## REFERENCES

1. For general aspects of the free particle model, see for instance: R. Peierls, *Quantum Theory of Solids*, Clarendon Press, Oxford (1955).
2. For dielectric formulation of the electron gas theory, see for instance:
   1) D. Pines, *Elementary Excitations in Solids*, (1963) W.A. Benjamin & Co., N.Y.
   2) R. Brout and P. Carruthers, *Lectures on the Many-Electron Problem*, (1963) Interscience (John Wiley & Co.) N.Y.
3. For Normal Fermi Liquid, see for instance:
   1) Nozieres, *Problème à N corps*, (1963) Dunod.
   2) Abrikhosov, Gorkhov, Dzyaloshinski, *Methods of Quantum Field Theory in Statistical Physics*, (1963) Prentice Hall.
   3) R. Balescu, *Statistical Mechanics of Charged Particles*, (1963) Interscience (John Wiley & Co.), N.Y.
   4) Prigogine, *Non equilibrium Statistical Mechanics*, (1962) Interscience (John Wiley & Co.) N.Y.
   5) Resibois, *Phys. Rev.* **138 B**, 281 (1965); *Bull. Acad Sci. Belg.* **51**: 1288 (1964).
   6) M. Watabe and R. Dagonnier, *Phys. Rev.* **143**: 110 (1966).
   7) Henin, Prigogine, Resibois, and Watabe, *Phys. Letters* **7**: 253 (1965).

# Optical Investigation of Phonons and Plasmons

## E. Burstein

*Physics Department and Laboratory for the Research on the Structure of the Matter*
*University of Pennsylvania*
*Philadelphia, Pennsylvania*

## I. INVESTIGATION OF PHONONS BY RAMAN SCATTERING OF PHOTONS

In this lecture we will be concerned primarily with the Raman scattering by phonons. Scattering by other excitation modes such as scattering processes involving electronic transitions of paramagnetic atoms and scattering by spin waves and by plasmons have been discussed by Elliot and Loudon ([1]) and by Loudon ([2]). The special cases of scattering by plasmons in crystals lacking a center of inversion is discussed in Section III on "Optical Investigation of Plasmons."

The "mechanics" of the Raman scattering process is illustrated in Fig. 1. The incident radiation is characterized by $\mathbf{E}_0 = \hat{\mathbf{e}}_0 E_0 \exp i(\mathbf{k}_0 \cdot \mathbf{r} - \omega t)$ and the scattered radiation by $\mathbf{E}_s = \hat{\mathbf{e}}_s E_s \exp i(\mathbf{k}_s \cdot \mathbf{r} - \omega_s t)$, where $\hat{\mathbf{e}}_0$, $\omega_0$, $\mathbf{k}_0$ and $\hat{\mathbf{e}}_s$, $\omega_s$, $\mathbf{k}_s$ are the polarization vector, frequency, and wavevector of the incident and scattered radiation, respectively. The scattering process is visualized as follows: the $\mathbf{E}_0$ induces a dipole moment in the scattering center $\mathbf{M} = \alpha \hat{\mathbf{e}}_0 E_0$, where $\alpha$ is the electronic polarizability (second-rank tensor) of the scattering center. The induced electric dipole emits radiation having an electric field component in the direction of $\hat{\mathbf{e}}_s$ given by $\mathbf{M} \cdot \hat{\mathbf{e}}_s$. The intensity of the scattered radiation is accordingly proportional to $|\hat{\mathbf{e}}_s \cdot \alpha \cdot \hat{\mathbf{e}}_0 E_0|^2$.

In the case of scattering by phonons, the electronic polarizability $\alpha$ will be a function of the displacement amplitudes of the atoms and will therefore have a position and time dependence, corresponding to that of the phonon involved, $\exp i(\mathbf{q}_j \cdot \mathbf{r} - \omega_j t)$.

Quantum mechanically, the first-order scattering process can be visualized as involving a virtual electronic transition from an initial electronic state to an intermediate electronic state, induced by the incident photon of frequency $\omega_0$ and a return to the initial electronic state with the emission of a photon of frequency $\omega_s$ and the simultaneous creation or destruction of a phonon of frequency $\omega_j$ [3] (see Fig. 2). The requirements that energy and momentum be conserved are expressed by

$$\omega_0 = \omega_s \pm \omega_j \qquad \mathbf{k}_0 = \mathbf{k}_s \pm \mathbf{q}_j$$

where the $+$ sign corresponds to the creation of the phonon and the $-$ sign corresponds to the destruction of the phonon. The conservation of momentum relation is the familiar Bragas diffraction equation expressed in vector form.

In second-order scattering processes, two phonons take part, and these may both be created or destroyed or one may be created and the other destroyed. The corresponding energy and momentum conservation relation are given by

$$\omega_0 = \omega_s \pm \omega_j \pm \omega_j' \qquad \mathbf{k}_0 = \mathbf{k}_s \pm \mathbf{q}_j \pm \mathbf{q}_j'$$

Since $\mathbf{k}_0 \approx \mathbf{k}_s \approx 0$ for the photons, this means that in first-order (one-phonon) scattering the phonons involved will be such that $\mathbf{q}_j = 0$. In the case of second-order (two-phonon) scattering the phonons involved will be such that $|\mathbf{q}_j| = \pm |\mathbf{q}_j'|$.

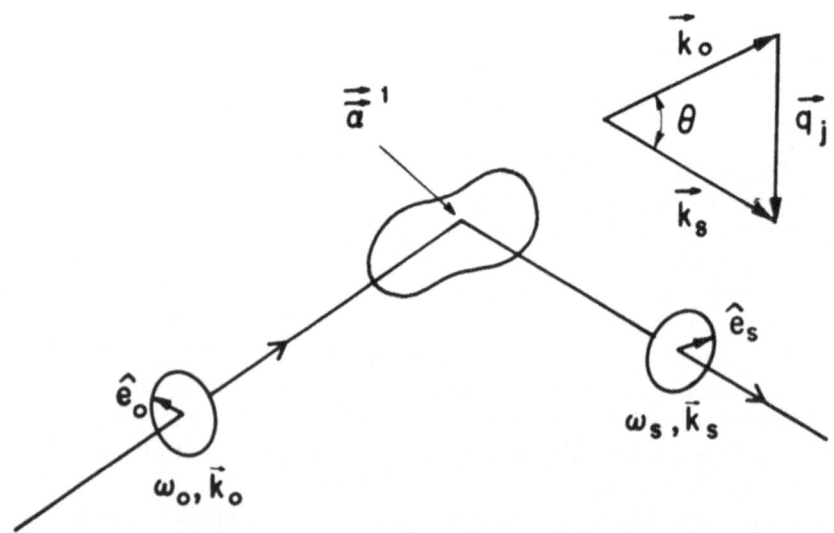

Fig. 1—Raman scattering geometry and corresponding wavevector diagram.

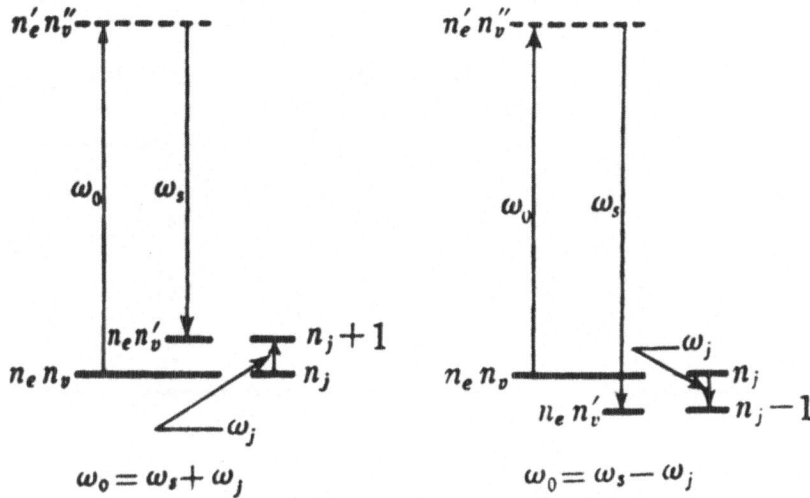

Fig. 2—The electronic and vibrational transitions which occur in Stokes ($\omega_0 = \omega_s + \omega_j$) and anti-Stokes ($\omega_0 = \omega_s - \omega_j$) Raman scattering.

The intensity of the scattered radiation will depend on the state of polarization of the incident and scattered radiation and on the electronic polarizability tensor as follows:

$$I_s(\omega_s, \hat{e}_s) \propto \omega_s^4 |\hat{e}_s \cdot d\alpha \cdot \hat{e}_0| E_0^2$$

where $d\alpha$ represents the change in the electronic polarizatility associated with the creation or destruction of the phonons. It can be expressed phenomenologically by a Taylor expansion in the relative displacement of the atoms

$$d\alpha = \alpha - \alpha_0 = \frac{\partial \alpha}{\partial u_j(q)} du_j(q) + \frac{\partial^2 \alpha}{\partial u_j(q_j) \partial u_j'(q_j)} du_j(q) du_j'(q') + \cdots$$

$$= \alpha_1 + \alpha_2 + \cdots$$

where $u_j(q)$, $u_j'(q')$, etc. represents the relative displacement of the atoms and the subscript indicates the vibration branch, i.e.,

$$u_j(\omega_j, q) = \frac{1}{\sqrt{NM}} d_j(q) Q_j(q) \exp i(q \cdot r - \omega_j t)$$

The first-order term $\alpha_1$ is responsible for first-order (or one-phonon) scattering, the second-order term $\alpha_2$ is responsible for second-order (or two-phonon) scattering, etc. It should be noted that first-order scattering involv-

## TABLE I

### The First-Order Polarizability Tensors for Cubic Crystals

(after Loudon, 1964)

| Cubic<br><br>Class | $\begin{pmatrix} a & & \\ & a & \\ & & a \end{pmatrix}$ | $\begin{pmatrix} b & & \\ & b & \\ & & -2b \end{pmatrix}$ | $\begin{pmatrix} -\sqrt{3b} & & \\ & -\sqrt{3b} & \\ & & 0 \end{pmatrix}$ | $\begin{pmatrix} & & \\ & & d \\ & d & \end{pmatrix}$ | $\begin{pmatrix} & & d \\ & & \\ d & & \end{pmatrix}$ | $\begin{pmatrix} & d & \\ d & & \\ & & \end{pmatrix}$ |
|---|---|---|---|---|---|---|
| 23   T | A | E | E | $F(x)$ | $F(y)$ | $F(z)$ |
| m3   $T_k$ | $A_g$ | $E_g$ | $E_g$ | $F$ | $F_g$ | $F$ |
| 432   O | $A_1$ | E | E | $F_2$ | $F_2$ | $F_2$ |
| 43m   $T_d$ | $A_1$ | E | E | $F_2(x)$ | $F_2(y)$ | $F_2(z)$ |
| m3m   $O_h$ | $A_{1g}$ | $E_g$ | $E_g$ | $F_{2g}$ | $F_{2g}$ | $F_{2g}$ |

ing acoustic phonons corresponds to Brillouin scattering, and $\partial\alpha/\partial u_j(q)$ corresponds to the photoelastic tensor [3].

The first-order polarizability coefficient $\partial\alpha/\partial u_j(q)$, which is responsible for first-order Raman scattering, is a third-rank tensor. The nonzero components of this tensor are determined by the symmetry of the phonons. They have been worked out for the various symmetries of the vibration modes that occur in the 32 crystal classes and are summarized in Loudon's 1964 review paper [2].

Table I contains the first-order polarizability tensors for the different symmetries of the vibration modes that occur in the cubic crystal classes. It should be noted that the tensor components are referred to the cubic crystallographic axes as reference axes. For purposes of compact notation, the $\mu\nu$ component of $\alpha_1$, due to atomic displacements along $\sigma$, is written as

$$(\alpha^1)_{\mu\nu} = \left(\frac{\partial\alpha}{\partial u_{j\sigma}}\right)_{\mu\nu} = \alpha_{\mu\nu,\sigma}$$

The intensity of the scattered radiation is accordingly expressed as

$$I_s(\omega_s, \hat{e}_s) \propto \omega_s^4 \sum_\sigma \sum_{\mu\nu} |\hat{e}_{s\mu} \cdot \alpha_{\mu\nu,\sigma} \cdot \hat{e}_{\partial\nu}| E_\partial^2$$

The $\alpha_{\mu\nu,\sigma}$ coefficients determine the polarization $\mu$ of the scattered radiation for a given polarization of the incident radiation $\nu$, corresponding to a given component $\sigma$ of the relative displacement of the atoms.

One must take into consideration the fact that the "scattering" phonon is coupled anharmonically to other phonons. $I_s(\omega_s, \hat{e}_s)$ will therefore include a factor of the form

$$\frac{\Gamma_j(\omega)}{\omega_j^2(q) + 2\omega_j(q)\Delta_j(\omega) - \omega^2 + 4\omega_j(q)\Gamma_j(\omega)}$$

where $\omega_j(q)$ is the resonant frequency in the absence of anharmonicity, $\Gamma_j(\omega)$ is the frequency-dependent lifetime of the phonon, and $\Delta_j(\omega)$ is the frequency shift due to the anharmonic coupling.

The frequency-dependent dampling constant $\Gamma_j(\omega)$ leads to additional structure, as well as to a broadening of the Raman lines. The additional structure may be considered as arising from a two-phonon scattering process with the "scattering phonon" acting as an intermediate state, in analogy with the anharmonic mechanisms for second-order infrared absorption where the infrared active phonon serves as the intermediate state. Finally, one must include the temperature-dependent factors for the Stokes and anti-Stokes lines. These differ from the corresponding factors for first-order infrared absorption, as shown in Table II. The difference in the temperature-dependence factors for infrared absorption and Raman scattering processes lies in the fact that both phonon absorption and emission processes contribute to the same frequency in infrared absorption, whereas they are separated into Stokes and anti-Stokes frequencies in Raman scattering.

## TABLE II

### Temperature-Dependent Factors for Infrared and Raman Processes

| Frequency | Infrared | Raman | |
|---|---|---|---|
| | | Stokes | Anti-Stokes |
| $\omega_j$ | $(\bar{n}+1) - (\bar{n}) = 1$ | $\bar{n}+1$ | $\bar{n}$ |
| $2\omega_j$ | $1 + 2\bar{n}_j$ | $1 + 2\bar{n}_j + \bar{n}_j^2$ | $\bar{n}_j^2$ |
| $\omega_j + \omega_j'$ | $1 + \bar{n}_j + \bar{n}_j'$ | $1 + \bar{n}_j\bar{n}_j' + \bar{n}_j + \bar{n}_j'$ | $\bar{n}_j\bar{n}_j'$ |

We list in Table III the infrared and Raman activity of the optical phonons in a number of simple cubic crystal structures.

Although the NaCl structure is Raman-inactive in first order it does exhibit a well-defined second-order Raman spectrum. The spectrum was first observed by Fermi and Rosetti (1931) and interpreted by them as due to the second-order terms in the electronic polarizability. The diamond and ZnS structures also exhibit second-order spectra. However, for these structures there are two mechanisms possible ([3]): (1) second-order polarizabillty mechanism; and (2) anharmonic mechanism. The anharmonic mechanism can only play a role when first-order scattering is allowed. Thus it cannot apply to the NaCl and CsCl types of crystals.

The structure of second-order Raman spectra is due primarily to the structure in the so called combined (two-phonon) density of phonon states.

## TABLE III

### First-Order Infrared and Raman Activity of Optical Phonons

| Crystal structure | Infrared | Raman | Remarks |
|---|---|---|---|
| NaCl | Yes | No | Atoms at centers of symmetry |
| CsCl | Yes | No | |
| ZnS | Yes | Yes | Structure lacks a center of symmetry |
| Diamond | No | Yes | Atoms not at center of symmetry, structure has center of symmetry |
| $CaF_2^a$ Optical mode (a) | Yes | No | Calcium sites at centers of symmetry |
| Optical mode (b) | No | Yes | Fluorine sites not at centers of symmetry |

[a]Mode (a) $\leftarrow\ominus\ \oplus\rightarrow\ \leftarrow\ominus$    mode (b) $\ominus\rightarrow\ \oplus\ \leftarrow\ominus$ ($\oplus$ Ca; $\ominus$ F)

In the early efforts to interpret the spectra, the peaks in the spectra were assumed to arise from phonon pairs at critical points in the Brillouin zone (B.Z.). Later the symmetries of the phonon at various symmetry points in the B.Z. were established, by group theoretical methods, for crystals having the diamond, zincblende and rocksalt structures ([5,6]). On the basis of these symmetries, it was possible to establish the selection rules for the pairs of phonons at the various symmetry points in the B. Z., which are active in second-order infrared absorption and Raman scattering processes. These selection rules were then combined with information about the two-phonon density of states obtained from theoretical or experimental $\omega$ vs. $q$ curves to establish phonon pair assignments for the peaks in the infrared absorption and Raman scattering spectra ([6-8]). The second-order infrared absorption and Raman spectra of the alkali halides are of particular interest in this connection, since the selection rules for two-phonon processes are quite restrictive for infrared absorption but are rather lenient for Raman scattering. This is illustrated in Tables IV and V for symmetry points at the zone boundary of the NaCl structure.

The absence of overtones in two-phonon infrared absorption is a general selection rule which is exhibited by all crystals having a center of inversion ([9]). The second-order selection rules for the NaCl structure account for the fact that there is little structure in the infrared spectra, whereas the Raman

## TABLE IV

### Infrared-Active Two-Phonon Combinations for NaCl([6])

| Symmetry point | Active combination |
|---|---|
| $\Gamma$ | None |
| $X$ | None |
| $L$ | TO + LA, TO + TA, LO + LA, LO + TA |
| $\left.\begin{array}{c}\Delta\\\Lambda\end{array}\right\}$ | All combinations |

## TABLE V

### Raman-Active Two-Phonon Overtones and Combinations for NaCl([6])

| Symmetry point | Active overtones and combinations |
|---|---|
| $\Gamma$ | 2LO, 2TO, LO + TO |
| $X$, $\Delta$ | 2LO, 2LA, LO + LA, ALL OTHER OVERTONES AND COMBINATIONS |
| $L$ | 2LO, 2TO, 2LA, 2TA, LO + TO, LA + TA |
| $\Lambda$ | ALL OVERTONES AND COMBINATIONS |

spectra have appreciable structure. This is illustrated by the infrared and Raman spectra of rocksalt shown in 3a Figs. and 3b. (The selection rules for infrared absorption are less restrictive for the zincblende structure due largely to the absence of a center of inversion.) The two-phonon assignments for the peaks in the Raman spectrum of NaCl are given in Table VI. Since the selection rules for Raman scattering are not restrictive in the case of crystals having the NaCl and CsCl structure, the structure in the Raman spectrum arises predominantly from the structure in the combined density of states. This was demonstrated for CsCl and CsBr by Karo, Hardy, and Morrison ([10]), who showed that the curves of the combined density of states derived from theoretical $\omega$ vs. $q$ curves exhibited structure quite similar to that of the experimental Raman spectra. It should be noted that peaks in the combined density of states occur when the two phonons both exhibit critical points at the same point in the B.Z., or when the two phonon branches involved have equal or opposite slopes at the same point in the B.Z. The results for CsBr are shown in Fig. 4.

The following is a brief summary of recent developments in the theory of Raman scattering.

a. Loudon ([11]) has carried out a theoretical calculation of the scatter-

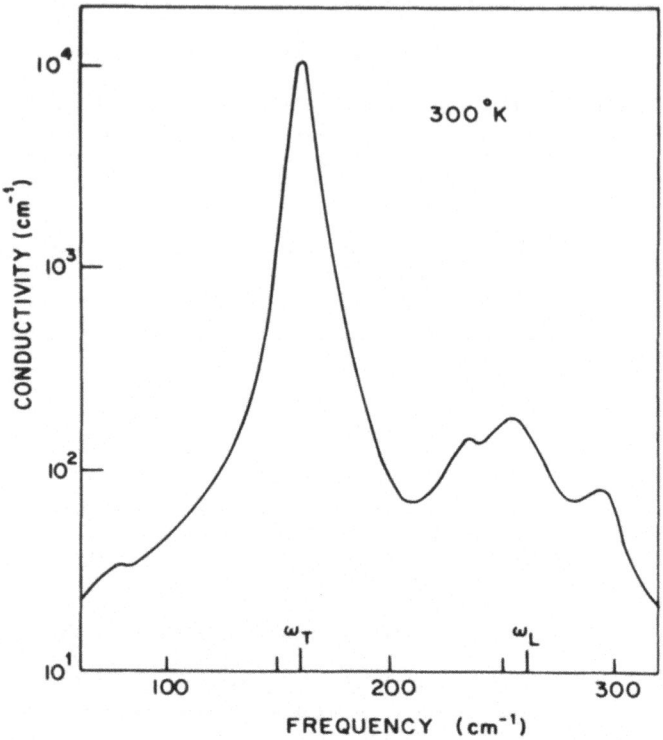

Fig. 3(a)—Room temperature infrared absorption spectrum of NaCl $\left[ \sigma(\omega) = \dfrac{\omega \epsilon_2(\omega)}{4\pi} \right]$ [After Smart, Willardson, Karo, and Hardy, *Proceedings International Conference on Lattice Dynamics*, Copenhagen 1963, New York, Pergamon Press, 1965; *J. Phys. Chem. Solids* **21**: (Suppl) 363 (1965).

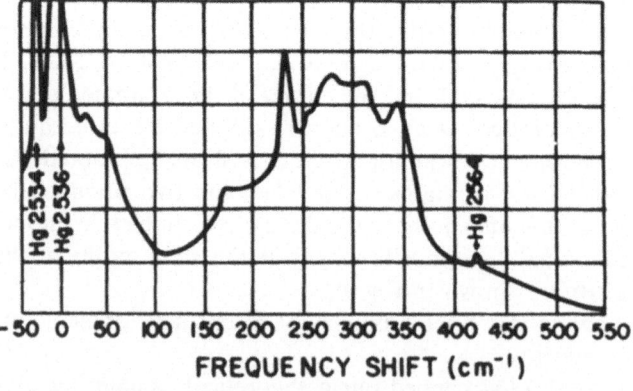

Fig. 3(b)—Room temperature Raman spectrum of NaCl. [From Welsh, Crawford, and Staple, *Nature* **164**: 737 (1949).

## TABLE VI

### Second-Order Raman Peaks in NaCl[6]

| Experimental[a] | Interpretation | Calculated for $X$ point |
|---|---|---|
| 31 | TO—LA($X$) | 30 |
| 55 | LA—TA($X$) | 56 |
| | LO—LA($\Delta$) | |
| | LA—TA($L$) | |
| 174 | 2TA($X$) | 174 |
| 234[b] | LA + TA($X$) | 230 |
| | LA + TA($\Delta$) | |
| | 2TA($L$) | |
| 256 | TO + TA($X$) | 260 |
| | 2TO($L$) | |
| 275 | LA + TA($L$) | |
| 285 | 2LA($X$) | 286 |
| 299 | LO + TA($X$) | 299 |
| | 2LA($\Delta$) | |
| 314 | TO + LA($X$) | 316 |
| | 2LA($L$) | |
| 346 | 2TO($X$) | 346 |
| | LO + LA($\Delta$) | |
| | LO + TO($L$) | |
| 415 (broad) | 2LO($X$) | 414 |
| | (among others) | |

[a]Peak positions in cm$^{-1}$
[b]Strongest peak.

ing efficiency, $I_s/I_0$, for diamond, based on a deformation potential mechanism for $\partial\alpha/\partial u$. His results indicate values of the order of $10^{-6}$ to $10^{-7}$ for the scattering efficiency in diamond.

   b. Cowley ([12]) has carried out a calculation of the second-order Raman spectrum for KBr and NaI based on the shell model, and only nearest neighbor contributions to the second-order electronic polarizability. His theoretical spectra exhibit the general features of the experimental spectrum. However, his curves indicate relatively strong peaks due to combinations (and overtones) of the $q \approx 0$ LO and TO modes which are not present in the experimental curves. The previous calculation of the second-order Raman spectrum of an alkali halide was that for NaCl by Born and Bradburn ([13]).

   c. Cowley ([14]) has also carried out a calculation of the second-order Raman spectrum for diamond taking into account anharmonic processes as well as the second-order charge in the electronic polarizability. His calculated spectrum is in rough agreement with the observed spectrum.

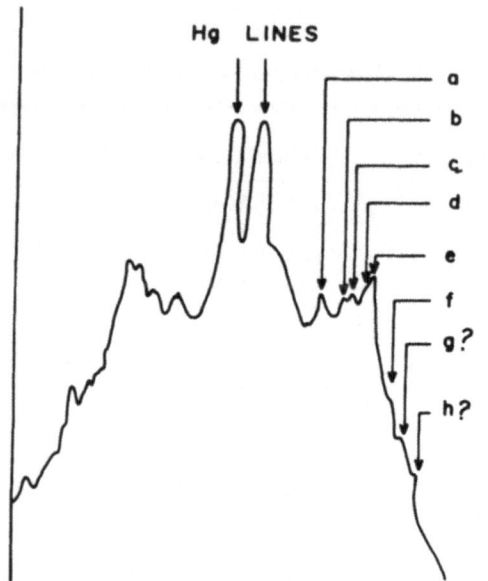

Figure 4–(a)

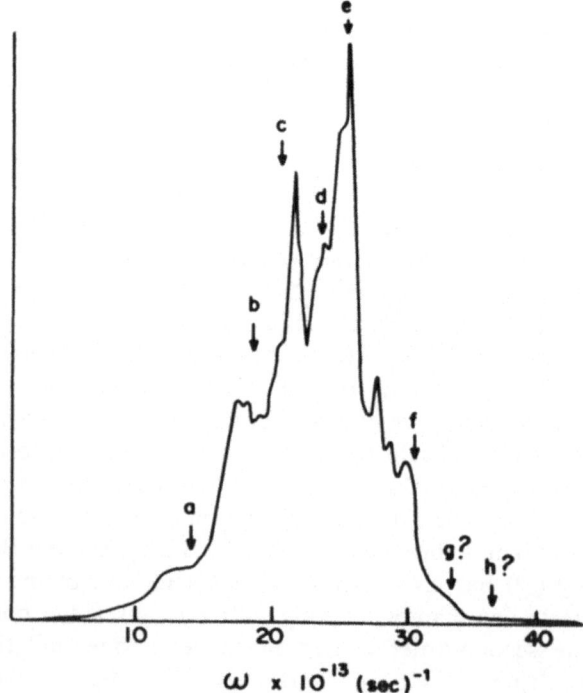

Figure 4–(b)

The previous calculation of the second-order spectrum of diamond was that by Smith ([15]), whose calculations were based on the second-order polarizability processes and did not include anharmonic processes.

A brief discussion is in order concerning the experimental "mechanics" of Raman scattering and about the relative advantages and disadvantages of Raman scattering and optical resonance absorption studies.

Early experiments were carried out using prisms or grating spectrometers and photographic techniques. The mercury are was used as an excitation source. This had the advantage, for the case of 2537 Å excitation, that

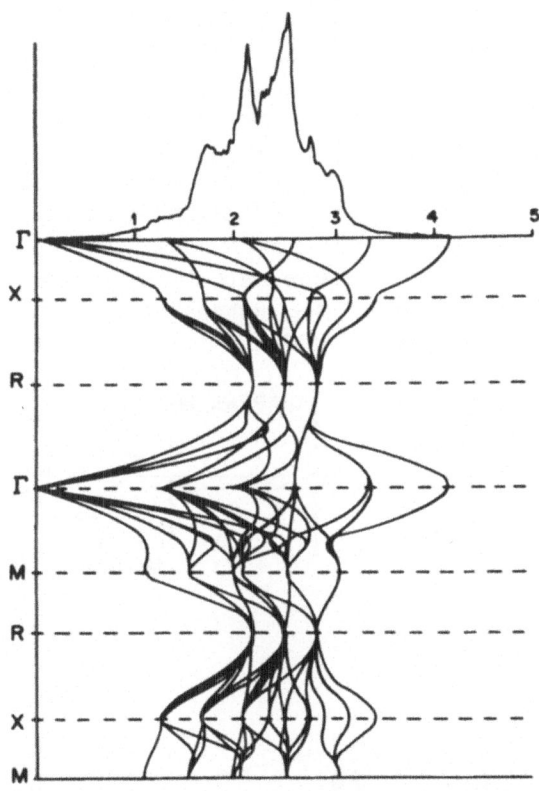

Figure 4–(c)

Fig. 4—Two-phonon dispersion curve (a), two-phonon density of states (b), and the observed Raman spectrum (c) of CsBr. The combination bands are labeled a,b, etc. Features marked with a query may be third-order lines. [After Karo, Hardy, and Morrison, *J. de Phys* **26**: 668 (1965).] The observed spectrum is that of Stekhanov, Korol'kov, and Eliashberg, *Sov. Phys. Solid State* **4**: 945 (1962).

one could pass the scattered radiation through mercury vapor and remove the 2537 Å radiation by "self-reversal." The photographic technique prevented high quantitative precision. Also, in order to obtain adequately intense spectra, the crystal was irradiated over a large solid angle and the scattered radiation was collected over a large solid angle.

Since the advent of the CW laser, there has been a resurgence of effort in the field ([16]). Highly monochromatic sources having intensities of the order of 50 mW (He–Ne gas with $\lambda = 6838$ Å) to 1 W (argon gas laser with $\lambda \approx 5000$ Å and yittrium aluminum garnet solid state laser with $\lambda = 1.0648$) are now available. These combined with photon counting techniques make possible quantitative measurements of scattering intensities and moreover allow one to define more accurately the directions of the incident and scattered beam, as well as to use polarized radiation for both incident and scattered beams. Typical experimental setups are shown in Figs. 5 and 6 for transparent and opaque crystals, respectively.

Raman scattering measurements are generally carried out at frequencies for which the media are transparent. In this respect Raman scattering experiments have an advantage over optical (resonance) absorption measurements where one encounters strong anomalous dispersion effects. One can make a detailed study of the structure of the Raman spectrum without having to resort to a Kronig–Kramers dispersion analysis of the data. In the case of first-order scattering in diamond it should be possible to investigate the de-

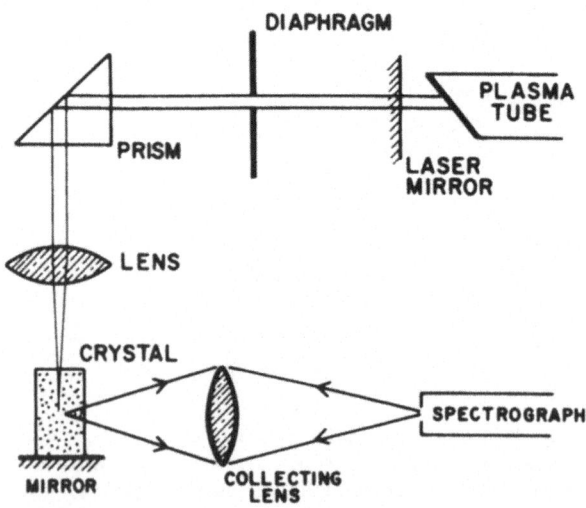

Fig. 5—Diagram of "input optics" for the measurement of the Raman scattering spectra of transparent crystals. [From J. P. Russel, *J. de Physique* **26**: 620 (1965).]

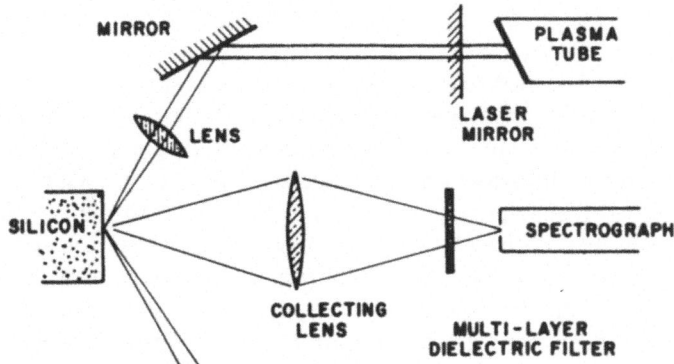

Fig. 6—Diagram of "input optics" for the measurement of the Raman scattering spectra of opaque crystals [From J.P. Russel *J. de Phys* **26**: 620 (1965).]

tailed shape of the Raman lines which are broadened by anharmonicity. For this purpose, it will be necessary to work at a higher resolution than that available with conventional grating or prism instruments $\Delta \nu \approx 1 \text{ cm}^{-1}$). This can be accomplished by combining the Raman spectrograph with a Fabry–Perot interferometer, to attain resolutions of 0.01 cm or better. The limitation in resolution under these circumstances will very likely be set by the decrease in the scattered intensity with increase in resolution.

It is possible to carry out Raman scattering measurements on opaque crystals. Thus Russell ([16]) at R.R.E. in England observed the first-order Raman Stokes line for silicon using a He–Ne CW laser as exiciting source and observing the "backward" scattered radiation (Fig. 6). The scattered intensity is considerably reduced since the effective scattering volume is limited to the region of the skin depth. The dependence of the scattering intensity on the frequency as $\omega_0$ approaches a resonance frequency has been discussed by Loudon ([17]), and Porto *et al.* ([18]) have recently reported the results of first-order Raman scattering experiments on CdS using different laser sources with frequencies smaller than and greater than that corresponding to the interband absorption edge.

It is also possible, in Raman scattering experiments, to use forward scattering, with appropriate filters to remove the excitation radiation and, thereby, to achieve $|\mathbf{k}_0 - \mathbf{k}_s|$ values much closer to zero. Experiments of this type have been carried out for GaP by Henry and Hopfield ([19]) and by Porto, Tell, and Damen ([20]), who were able, in this way, to observe Raman scattering by polaritons, i.e., by coupled photon–phonon modes. These experiments are discussed in Section 11 of the Lecture on the Interaction of the E.M. Radiation with Plasmons and Phonons.

Optical resonance absorption experiments and Raman scattering experiments differ in another fundamental way. The resonance absorption matrix elements involve the electric moment which is a first-rank tensor, whereas the Raman scattering matrix elements involve the electronic polarizability which is a second-rank tensor. The Raman scattering experiments accordingly provide more information about the symmetry of the vibration mode than is obtained from optical absorption experiments. In the case of media with cubic symmetry, the absorption is isotropic whereas the scattering is anisotropic. One can, by carrying out scattering measurements with polarized radiation, obtain information about the symmetry of scattering center. This applies both to lattice modes of vibration and to localized modes which are active in first order, as well as to combinations of modes which participate in higher-order scattering processes. The situation is, in some respects, analogous to the studies of polarized luminescence, which yield information about the symmetry of the fluorescent centers.

I would like to emphasize at this point the importance of carrying out Raman scattering experiments with polarized incident and scattered radiation for different orientations of the crystal [21]. Such experiments have been carried out in only a relatively small number of cases, notably ZnO [22] and CdS [23]. Previous experiments on CdS and ZnO by Poulet and Mathieu [24] were carried out without polarized radiation and only for a limited number of crystal orientations. Furthermore, their analysis was carried out essentially in terms of a molecular model in which the wavevector of the phonons is ignored. As one consequence of the limited orientations used, their spectra for CdS exhibited a second-order peak that was stronger than the first-order lines. As shown by Tell, Damen, and Porto [23], this was essentially due to the fact that the wavevectors of the incident and scattered beams did not correspond to appropriate wavevectors for the Raman-active phonons.

## II. ELECTRIC FIELD INDUCED ONE-PHONON INFRARED ABSORPTION AND ONE-PHONON RAMAN SCATTERING

Electric field induced infrared (I.R.) absorption and Raman scattering in crystals are manifestations of "morphic effects" which arise when the symmetry of the crystal is lowered by the application of an external force. The more interesting effects are those which involve (1) changes in the selection rules for photon absorption or photon scattering processes and (2) a splittings of the degeneracies of the normal modes of vibration. Similar effects may also be observed for impurity modes having appropriate symmetries [25, 26].

## 1. Infrared Absorption

In crystals having the diamond structure, the effective ionic charge of the atoms is zero and there is no first-order (one-phonon) resonance absorption of EM radiation by the fundamental ($q \approx 0$) optical vibration modes. As is well known, these crystals do exhibit well-defined, although weak, absorption bands which are due to higher-order electric moment effects [27]. As shown by Burstein and Ganesan [28] and by Szigetti [29] an externally applied static electric field $E^0$ induces an "effective" ionic charge on the atoms so that the $q \approx 0$ transverse optical (TO) and longitudinal optic (LO) vibration modes which are Raman-active also become I.R.-active. The electric field may also be expected to cause a splitting of the triply degenerate modes at $q \approx 0$ which will depend on-the direction of the applied field. The effect, which is essentially the same as that predicted by Condon [30] for homonuclear molecules and observed experimentally by Crawford and Dagy [31] for hydrogen molecules at fields of $10^5$ V/cm, may be visualized as follows: An applied electric field induces a dipole moment in each primative unit cell. The relative displacement of the atoms in the $q \approx 0$ optical vibration modes produces a change in the electronic polarizability of the atoms, and consequently, the induced dipole moment varies in amplitude and orientation at the frequency of the modes. The induced "effective" ionic charge

$$e_j^* = \left(\frac{\partial M}{\partial u_j}\right)_E$$

(where $M$ is the electronic moment per primitive unit cell and $u$ is the relative displacement of the two atoms in the unit cell) is proportional to the change in the electronic polarizability with the relative displacement of the atoms ($\partial\alpha/\partial u_j$). A measurement of the strength of the absorption band which is proportional to

$$e^{*2} = \left|\frac{\partial\alpha}{\partial u}E^0\right|^2$$

can, therefore, provide quantitative information about the first-order Raman matrix elements.

Experiments on the electric field induced absorption in diamond have been carried out at the University of Pennsylvania [32]. A weak band ($\Delta T = 0.25\%$ at 1332 cm$^{-1}$) was observed for an applied field of $1.2 \times 10^5$ V/cm.

From the magnitude of the absorption constant $A = 6 \times 10^{-7} E^2$/cm, one obtains a value of $4 \times 10^{-16}$ cm$^2$ for $\partial\alpha/\partial u$, which corresponds to a Raman scattering efficiency $I_s/I_0$ (per unit length of crystal per unit solid angle) of $3 \times 10^{-7}$. It is of interest to note that Crawford and MacDonald

($^{33}$) obtain from electric field induced absorption measurements a value of $3.8 \times 10^{-16}$ cm² for $(\partial\alpha/U)$ for the $\theta$ branch ($\Delta J = 0$) of molecular hydrogen.

## 2. Raman Scattering

An applied (static) field can also induce a first-order Raman spectrum in crystals having the NaCl or CsCl structure. In effect, the applied electric field lowers the symmetry of the crystal to that in which the $q \approx 0$ TO and LO modes are Raman active. The resulting symmetry and, correspondingly, the form of the electronic polarizability matrix element for Raman scattering depends on the direction of the applied field. For example, with $\mathbf{E}^0$ along a [111] direction, the crystal acquires a polar trigonal axis. The electric field may also be expected to produce a shift in the frequency of the $q \approx 0$ LO and TO modes, splitting the double degeneracy of the TO modes when the field is applied along directions other than [100] and [111]. Both TO and LO modes will participate in the field-induced Raman scattering and the scattered intensity will be proportional to $(E^0)^2$.

It is instructive to compare the magnitude of the electric-field-induced Raman scattering with second-order Raman scattering involving two $q \approx 0$ optical vibration modes. The latter can be viewed as phonon-induced Raman scattering in which one $q \approx 0$ mode produces a displacement of the atoms from their equilibrium center of symmetry positions and the other $q \approx 0$ mode produces the change in the electronic polarizability. The electric field induced contribution to the change in the electric polarizability can be written phenomenologically as

$$\frac{\partial^2\alpha}{\partial u_j(0)\partial E^0}E^0 \, du_j(0) = \frac{\partial\alpha}{\partial u_j} du_j(0) = \alpha_1(E)$$

where

$$\alpha(E) = \left(\frac{\partial\alpha}{\partial E^0}\right)E^0$$

and the subscripts denote a $q \approx 0$ TO or LO mode. This can also be expressed in terms of the electric field induced relative displacements of the positive and negative ions $\partial u/\partial E^0$ as follows

$$\alpha_1(E) = \left(\frac{\partial^2\alpha}{\partial u_j(0)\partial u^0}\right)\frac{du^0}{dE^0}E^0 \, du_j(0)$$

The coefficient $\partial^2\alpha/\partial u_j(0)\partial u^0$ is essentially the same as that appearing in the second-order electronic polarizability term $\partial^2\alpha/\partial u_j(0)\partial u_{j'}(0))$. The ratio of the scattering efficiencies of the electric-field-induced first-order scattering and the second-order Raman scattering involving two $q \approx 0$ modes will then depend on the relative magnitude of $du$ produced by the applied field and by the $q \approx 0$ mode. Experiments to observe a field-induced Raman scattering in NaCl and to observe the corresponding second-order Raman scattering due to two optical modes have been carried out by A. Pinzcuk and W. Taylor, at the University of Pennsylvania. However, no field-induced scattering has been observed for fields up to $5 \times 10$ V/cm.

## III. OPTICAL INVESTIGATION OF PLASMONS

### 1. Background

The first experimental evidence for plasmons in solids was obtained from studies of the energy loss of electrons which were backscattered from bulk solids ([34]). It was noted that electrons of moderate energy ($\sim 10$ keV) lost energy in discrete amounts. For a number of metals the loss spectra exhibited a series of sharp peaks approximately equally spaced. In 1951 Bohm and Pines ([35]) attributed such losses to the excitations of the electron conduction plasma, i.e., to plasmons. Energy loss peaks were also observed for insulators and these were similarly attributed to the excitation of the "collective" excitations involving bound electrons ([36]). Since the pioneering work of Rydberg, the techniques have been refined and the experiments ([37]) were carried out on thin films. The results obtained by Marton et al. ([38]) are shown in Fig. 7. The equally spaced loss peaks correspond to energy losses of $\Delta \mathscr{E} = \hbar \omega_p, 2\hbar \omega_p, \cdots$ to the electron plasma. There is also a continuum loss indicating other loss mechanisms. In 1957 Ritchie ([39]) showed that for metals there should be both a volume loss of energy with $\Delta \mathscr{E} = \hbar \omega_p$ and a surface loss of energy, with $\Delta \mathscr{E} = \hbar \omega_p / \sqrt{2}$ corresponding to an excitation of a "surface plasmon." This theory has been borne out by Powell and Swan ([40]) in experiments on clean unoxidized aluminum surfaces, in which they showed that there were both volume and a surface losses, and that the surface loss tended to disappear as the aluminum surface was oxidized.

In general, the identification of the electron energy losses as due to the volume or surface plasmons has generally been made on the basis of the match between the losses and the corresponding plasmon frequencies. There have been extensive efforts to obtain further independent verification. In 1958 Ferrel ([41]) predicted that the decay of plasma oscillations in thin films would take place with the emission of monochromatic electromagnetic (EM) radiation at an angle to the normal to the surface of the film. He showed that the probability of emission should depend on the film thickness, the

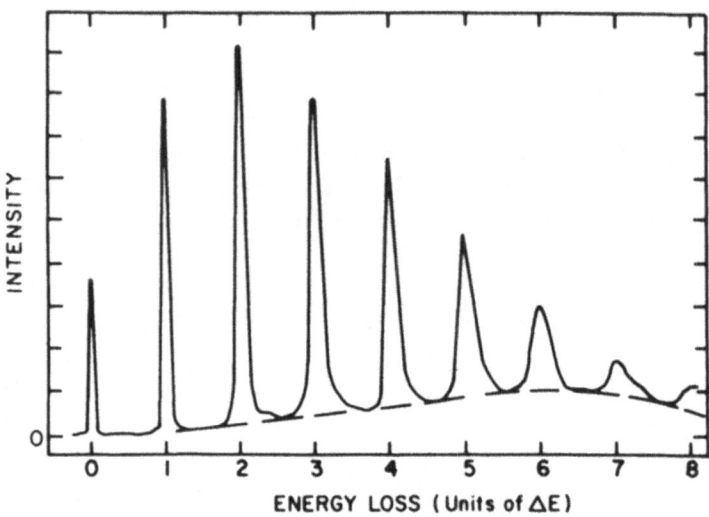

Fig. 7—Spectrum of energy losses for 20 keV electrons passing through a 2580 Å aluminum foil. [From Marton, Simpson, Fowler, and Swanson, *Phys. Rev.* **126**: 182 (1962)].

angle of observation, and the energy of the incident electrons which excite the plasmons. In experiments on Ag films by Steinman [42] and by Brown *et al.* [43] a peak in the emission spectra was found at 3300 Å corresponding to a plasmon energy of 3.8 eV.

It was then suggested by Silin and Fetisov [44] that the emitted EM radiation actually corresponded to "transition radiation," first postulated by Ginsburg and Frank in 1946 [45]. Transition radiation is emitted when a classical point-charged particle crosses a plane interface between two media of different dielectric constants. The radiation results from the change in the electric field surrounding the particle as it makes a transition, at constant velocity, from one medium into another. Transition radiation is in the same category as Cerenkov radiation, in that it involves the emission of radiation during the uniform motion of a charged particle and like Cerenkov radiation it is determined completely by the optical properties of the media, i.e., the radiation is a macroscopic phenomenon and can be calculated from Maxwell's equations and from the optical constants of the two media. The emission of radiation by plasmons excited by electrons is on this basis simply a part of the total emitted transition radiation whose presence is manifested as a peak in the emission spectra (Fig. 8). At present, studies of transition radiation and their interpretation in terms of the macroscopic optical constants are being carried out for a large number of materials including inorganic and organic solids (see review paper of Birchoff [46]).

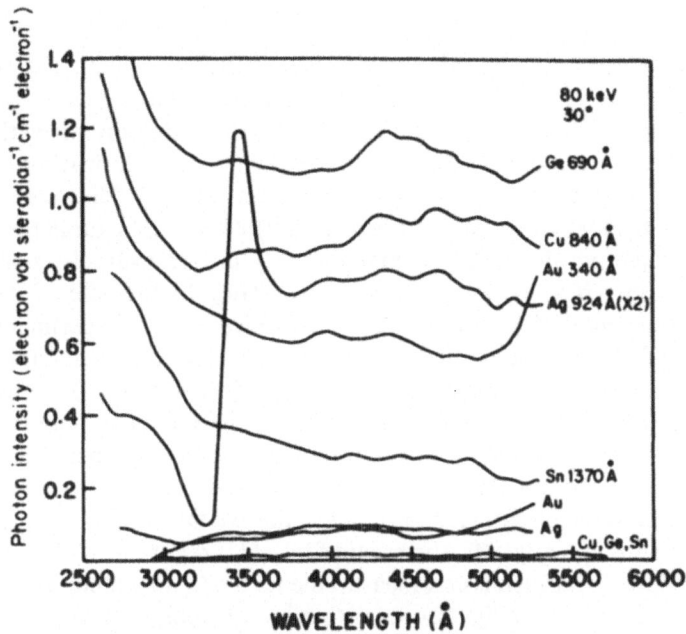

Fig. 8—Optical emission spectra from 80-keV electron irradiated foils of Ge, Au, Ag, and Sn at 30° to the electron beam direction [From R. D. Birkhoff, *International Symposium Physical Processes in Radiation Biology*, New York, Academic Press, 1964. 145.]

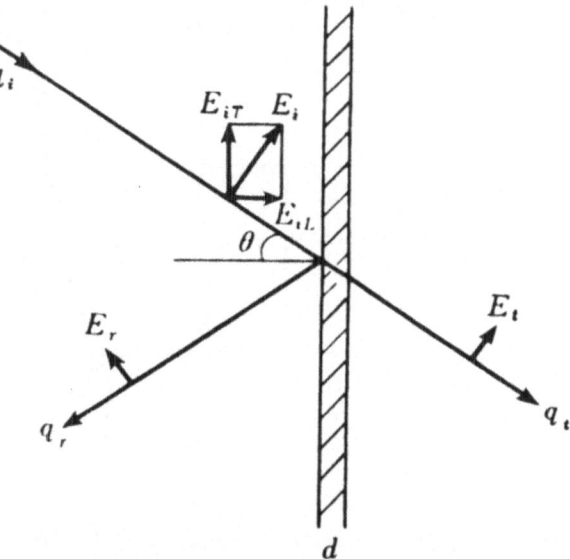

Fig. 9—Schematic diagram of oblique incidence transmission (and reflection) of thin films with plane of polarization in the plane of incidence.

Another approach to the optical investigations of plasmons was also proposed by Ferrel and Stern ([47]). They suggested that since plasmons can emit EM radiation, they should also absorb EM radiation when the latter is incident at an oblique angle with the electric field polarization vector in the plane of incidence. Under these conditions the EM field has a normal component relative to the film surface (Fig. 9) which can couple to the plasmons propagating along the normal to the film. Accordingly, one should observe minima in the transmission at $\mathscr{E} = \hbar\omega_p$. In the case of $q \approx 0$ LO phonons Berreman ([48]) showed both theoretically and experimentally (in thin films of alkali halides) that the oblique incidence transmission spectra exhibit minima at the frequencies of the $q \approx 0$ TO and LO phonons, $\omega_T$ and $\omega_L$, respectively, when the polarization of the radiation field was in the plane of incidence, and only a single minimum at $\omega = \omega_T$ when the polarization field was perpendicular to the plane of incidence. In these experiments, the excitation of the phonon is analogous to the excitation of the "longitudinal plasmon." Subsequently, McAlister and Stern ([49]) reported results of oblique incidence thin-film transmission measurements on Ag in which they observed a minimum at $\omega = \omega_p$ (Fig. 10). Investigation of oblique incidence thin-film transmission spectra have also been carried out for the $q \approx 0$ plasmons as

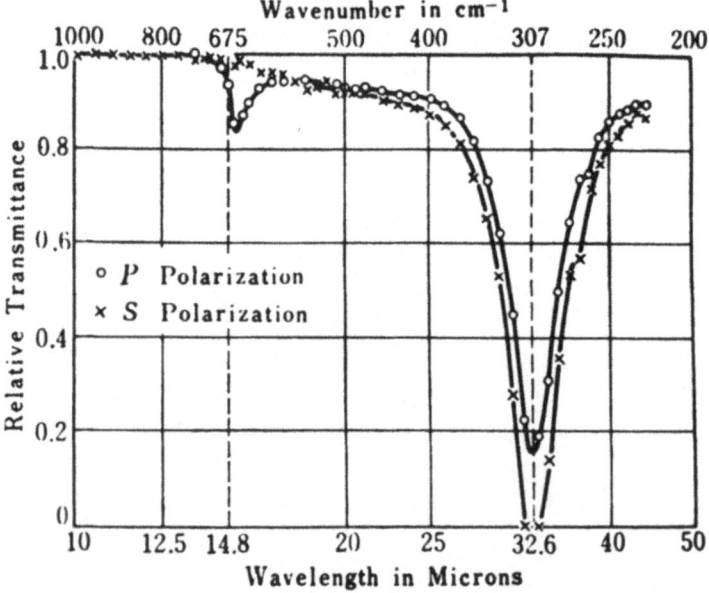

Fig. 10—Oblique incidence transmission spectra of LiF. The data indicated by $x$ are for the plane of polarization perpendicular, and that indicated by $o$ are for the plane of polarization parallel to the plane of incidence. [From D. W. Berreman, *Phys. Rev.* **130**: 2193 (1963)].

well as for the $q \approx 0$ LO phonons in semiconductors, and we may expect it to become a fairly routine tool for determining the frequencies of longitudinal excitations. (See also the discussion in Section IV on the Coupling of Plasmons with Other Longitudinal Excitation Modes.)

Before proceeding to a more detailed theoretical discussion of the interaction of plasmons with EM radiation, a few words may be in order regarding the nature of plasmons. First, there are the plasmons which involve all the electrons in the solid, both "free" and "bound." These exist in insulators as well as in metals. The plamsa frequency is given by

$$\omega_{pT}^2 = \frac{4\pi N e^2}{m}$$

where $N$ is the total density of electrons and $m$ is the mass of the electrons in free space. The plasmons energy $\hbar\omega_{pT}$ is greater than the binding energy of the electrons so the bound electrons behave as if they were "free." Then there are the plasmons associated with the "free" carriers. The plasma frequency is given by

$$\omega_p^2 = \frac{4\pi n e^2}{m^* \epsilon_0(\omega_p)}$$

where $n$ is the density of "free" electrons, $m^*$ is the effective mass of the electrons, and $\epsilon_0$ is the so-called high-frequency dielectric constant. In either case one must treat the coupling between the two longitudinal modes explicitly in order to determine both the frequency of the $q \approx 0$ plasmon-like mode and that of the $q \approx 0$ LO phonon-like mode. Coupled longitudinal excitations involving plasmons, such as coupled plasmon–phonon modes and coupled plasmon–exciton (or interband excitation) modes are discussed further in Section V.

## 2. The Interaction of Electromagnetic Radiation with Longitudinal (Collective) Excitations

It is of interest first to consider the general interaction of EM radiation with an assembly of electric dipole oscillators, since this can be readily specialized to plasmons, LO phonons, etc. For purposes of the present discussion, we will consider the EM radiation in the form of plane waves, $\mathbf{E} = \hat{\mathbf{e}} E_0 \exp i(\mathbf{k}, \mathbf{r} - \omega t)$ but neglect spatial dispersion, i.e., $\exp i\mathbf{k} \cdot \mathbf{r} \approx 1$. The medium will be considered to be in the form of a slab with lateral dimensions much greater than the wavelength. We also denote by the subscripts $T$ and $L$ transverse and longitudinal components relative to the slab normals. Thus for normal incidence the $\mathbf{E}$ field of the radiation is transverse. For oblique incidence with the plane of polarization in the plane of incidence the **E** field has both a transverse ($T$) and a longitudinal ($L$) component (Fig. 9).

We obtain the resonance frequencies of the transverse and longitudinal normal modes for an isotropic assembly of electric dispole oscillators and also the transverse and longitudinal dielectric constants, $\epsilon_T(\omega)$ and $\epsilon_L(\omega)$, by means of classical equations of motion. The boundary conditions are included in the equations of motion without having to use Maxwell's equations explicitly.

For our purposes we consider each electric dipole oscillator to consist of two atoms, with effective charges $\pm$ and masses $m_+$ and $m_-$. The equations of motion of the relative displacement, **u**, of the two charged masses, are given by

$$\bar{m}\ddot{u}_T + \bar{m}\gamma_T\dot{u}_T + \bar{m}\omega_0 u_T = e^* E_{T_{eff}}$$
$$\bar{m}\ddot{u}_L + \bar{m}\gamma_L\dot{u}_L + \bar{m}\omega_0 u_L = e^2 E_{L_{eff}}$$

where $\bar{m} = (m + m_-)/(m_+ + m_-)$ is the reduced mass; $\bar{m}\omega_0^2$ is the restoring free constant; $\gamma = 1/\tau$ is the damping constant; and $E_{T_{eff}}$ and $E_{L_{eff}}$ are the transverse and longitudinal components of the effective (or local) electric field. The effective fields, $E_{eff} = \hat{e}E_{0_{eff}} \exp(-i\omega t)$, are given by

$$\mathbf{E}_{eff} = \mathbf{E}_{ext} - L\mathbf{P} + \mathbf{E}^L$$

where $\mathbf{E}_{exp}$ is the external field; **P** is the macroscopic (average) polarization; $L$ is the depolarization factor which for the slab configuration is equal to zero for transverse fields, and $4\pi$ for longitudinal fields; and $\mathbf{E}^L$ is the Lorentz dipole field. For an isotropic medium $\mathbf{E}^L$ is given by

$$\mathbf{E}^L = \sum_{sphere}' E_d + \frac{4\pi}{3}\mathbf{P}$$

where $(4\pi/3)\mathbf{P}$ is the contribution to $\mathbf{E}^L$ from the polarization charge on a sphere centered at the oscillator under consideration whose diameter is large compared with the distance between oscillators and the sum is taken over the dipole contributions to the field at the oscillator due to the induced dipoles at the other oscillator within the sphere. For a random or a cubic array of oscillators

$$\sum_{sphere}' E_d = 0 \qquad \mathbf{E}^L = \frac{4\pi}{3}\mathbf{P}$$

$$\mathbf{E}_{eff} = \mathbf{E}_{ext} - L\mathbf{P} + \frac{4\pi}{3}\mathbf{P}$$

The transverse and longitudinal components of $\mathbf{E}_{\text{eff}}$ are accordingly given by

$$\mathbf{E}_{T_{\text{eff}}} = \mathbf{E}_{T_{\text{ext}}} + \frac{4\pi}{3}\mathbf{P}_T$$

$$\mathbf{E}_{L_{\text{eff}}} = \mathbf{E}_{L_{\text{ext}}} - 4\pi\mathbf{P}_L + \frac{4\pi}{3}\mathbf{P} = \mathbf{E}_{L_{\text{ext}}} - \frac{8\pi}{3}\mathbf{P}_L$$

The corresponding components of the polarization of the medium are given by

$$\mathbf{P}_T = ne^*u_T + \chi_{0T}E_{T_{\text{eff}}}$$
$$= ne^*u_T + n\alpha_0 E_{T_{\text{ext}}}$$
$$\mathbf{P}_L = ne^*u_L + \chi_{0L}(E_{L_{\text{ext}}} - 4\pi P_L)$$
$$= \frac{ne^*u + n\alpha_0 E_{L_{\text{ext}}}}{\epsilon_0}$$

where $n$ is the density of the oscillators, $\alpha_0 = \alpha_+ + \alpha_-$ is the "electronic" polarizability of the oscillators, $\epsilon_0 = 1 + 4\pi n\alpha_0$, and, for simplicity, we have assumed that the electronic wavefunctions are extended so that there is no Lorentz dipole contribution to the effective field acting on the electrons.

Upon introducing an $\exp[-i\omega t]$ time dependence for $\mathbf{u}$, $\mathbf{P}$, and $\mathbf{E}$, we obtain the following transverse and longitudinal equations of motion:

$$\bar{m}\ddot{u}_T + \bar{m}\gamma_T u_T + \bar{m}\omega_T^2 u_T = e^*\mathbf{E}_{T_{\text{ext}}}$$
$$\bar{m}\ddot{u}_L + \bar{m}\gamma_L u_L + \bar{m}\omega_L^2 u_L = e^*\mathbf{E}_{L_{\text{ext}}}$$

where

$$\omega_T^2 = \omega_0^2 - \frac{4\pi}{3}\frac{ne_T^{*2}}{\bar{m}}$$

$$\omega_L^2 = \omega_T^2 + \frac{4\pi ne_T^{*2}}{\bar{m}\epsilon_0} = \omega_T^2 + \Omega^2$$

We note that $\Omega^2 = 4\pi ne^*e_T^{*2}/\bar{m}\epsilon_0$ represents the effective "plasma frequency" of the oscillators. Thus we see that the assembly of oscillators exhibits transverse and longitudinal resonance frequencies, $\omega_T$ and $\omega_L$, respectively. The fact that $\omega_L > \omega_T$ is due to the added contribution to the restoring force from the longitudinal "depolarization" field, $\mathbf{E}_L = -4\pi\mathbf{P}_L$.

The transverse resonance frequency is itself lower than $\omega_0$ because of the reduction in the restoring force arising from the Lorentz field $\mathbf{E}_T^L = (4\pi/3)\mathbf{P}$. It should be pointed out that a longitudinal polarization field $\mathbf{E}_L = -4\pi\mathbf{P}_L$ exists whenever $\operatorname{div}\mathbf{P} \neq 0$. The latter occurs either when there are surface polarization charges, as in the case of a uniformly polarized slab, or when the wavevector is not equal to zero.

The above equations of motion and resonance frequencies apply directly to $q \approx 0$ TO and LO phonons in polar solids. We can also apply the equation of motion to the case of an assembly of electrons in a uniformly charged positive background which exhibits plasma oscillations. This is accomplished by setting $\omega_0 = 0$ and taking into account the fact that the effective field seen by electrons is just the macroscopic avarge field $E$. The corresponding equation of motion for the displacements of the electrons relative to the uniform positive charged background is given by

$$m^* \ddot{u}_T + m^* \gamma u_T = e E_{T_{ext}}$$

$$m^* \ddot{u}_L + m^* \gamma u_L + \frac{4\pi n e^2}{\epsilon_0} = \frac{e E_{L_{ext}}}{\epsilon_0}$$

where $m^*$ is the effective mass and $e$ is the electronic charge. The corresponding transverse and longitudinal resonance frequencies are given by $\omega_T = 0$ and $\omega_L^2 = 4\pi n e^2 / m^* \epsilon_0 = \omega_p^2$.

The optical response of the medium is generally characterized by the complex frequency-dependent transverse dielectric constant This can be obtained from the relation

$$P_T = \chi_T E_T = \left( \frac{\epsilon_T - 1}{4\pi} \right) E_T$$

We obtain directly

$$\epsilon_T(\omega) = \epsilon_{T_1}(\omega) + i\epsilon_{T_2}(\omega) = \epsilon_0 + \frac{\Omega^2}{\omega_T^2 - \omega^2 - i\gamma_T \omega}$$

$$= \epsilon_0 + \frac{\Omega^2(\omega_T^2 - \omega^2)}{(\omega_T^2 \omega^2)^2 + \gamma_T^2 \omega^2} + \frac{\Omega^2 \gamma_T \omega}{(\omega_T^2 - \omega^2)^2 + \gamma_T^2 \omega^2}$$

It should also be noted that the corresponding longitudinal dielectric constant $\epsilon_L(\omega)$ is identical to $\epsilon_T(\omega)$. This is a well-known result which holds for $q \approx 0$. The terms $\epsilon_T(\omega)$ and $\epsilon_L(\omega)$ will, however, differ appreciably when $q \neq 0$. A plot of the real and imaginary parts of $\epsilon_T(\omega)$ is shown in Fig. 11 and 12 for electric dipole excitations and for plasmons. We note that there is a region above $\omega_T$, in which $\epsilon_{T_1}(\omega)$ is negative. Also $\epsilon_{T_1}(\omega)$ exhibits a dielectric anomaly, i.e., $\epsilon_{T_1}(\omega) = 0$, at a frequency which corresponds to the longitudinal resonance frequency, $\omega_L$, of the system. That this is so follows from the fact that $\epsilon_L(\omega) = \epsilon_T(\omega)$ for $q \approx 0$, so that when $\epsilon_T(\omega)$ goes to zero, $\epsilon_L(\omega)$ goes to zero. The longitudinal resonance frequency corresponds to the condition $D_L(\omega) = \epsilon_L(\omega)E_L(\omega) = 0$, which for finite $E_L(\omega)$ occurs when $\epsilon_1(\omega) = 0$.

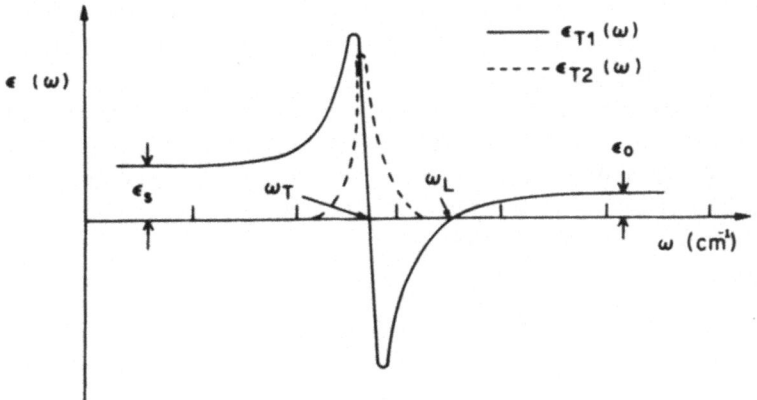

Fig. 11—Curves of $\epsilon_{T_1}(\omega)$ and $\epsilon_{T_2}(\omega)$ *vs.* $\omega$ for an assembly of electric dipole oscillators.

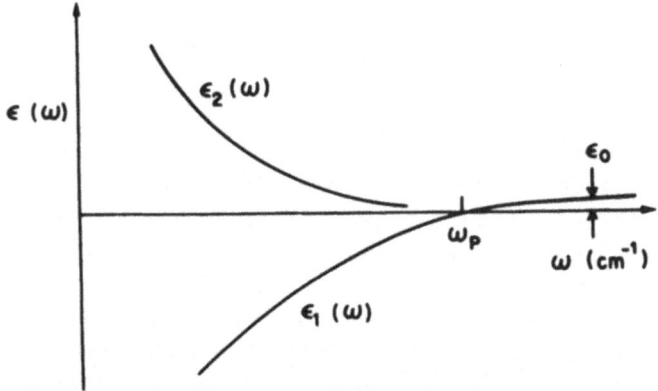

Fig. 12—Curves of $\epsilon_1(\omega)$ and $\epsilon_2(\omega)$ *vs.* $\omega$ for an electron plasma.

When $\omega$ is set equal to zero, we obtain the expression for the static dielectric constant

$$\epsilon_T(\omega = 0) = \epsilon_s = \epsilon_0 + \frac{\Omega^2}{\omega_T^2}$$

The relation between $\omega_L^2$ and $\omega_T^2$ can be experssed in terms of $\epsilon_s$ and $\epsilon_0$

$$\frac{\omega_L^2}{\omega_L^2} = 1 + \frac{\Omega^2}{\omega_T^2} = \frac{\epsilon_s}{\epsilon_0}$$

This relation, which is known as the Lyddane–Sachs–Teller relation, is a macroscopic relationship which does not depend on the microscopic model.

Although derived for $q \approx 0$ TO and LO phonons it applies equally well to the $q \approx 0$ transverse and longitudinal excitation modes of any assembly of electric dipole oscillators.

In the frequency region where $\epsilon_1(\omega)$ is negative and $\epsilon_2(\omega)$ is small, the wavevector is predominantly imaginary. Accordingly, in this frequency region the EM radiation does not propagate through the medium but rather is specularly reflected at the front surface of a thick slab.

The reflectivity $R(\omega)$ is given by

$$R(\omega) = \frac{[n(\omega) - 1]^2 + \kappa(\omega)^2}{[n(\omega) + 1]^2 + \kappa(\omega)^2}$$

where

$$\epsilon(\omega) = n^2(\omega) - \kappa(\omega)^2$$
$$\epsilon_2(\omega) = 2n(\omega)\kappa(\omega)$$
$$\eta(\omega) = n(\omega) + i\kappa(\omega)$$

is the complex refractive index. Curves of $R(\omega)$ vs. wave are shown (for zero damping) in Figs. 13 are 14 electric depole oscillators and plasmons, respectively. The minimum in reflection corresponds to $\epsilon_1(\omega) = 1$. By carrying out a Kronig–Kramers dispersion analysis of $R(\omega)$, one can obtain the optical constants, $n(\omega)$ and $\kappa(\omega)$ and $\epsilon_1(\omega)$ and $\epsilon_2(\omega)$. The longitudinal resonance frequency ($\omega_L$ or $\omega_p$) is given by the frequency at which $\epsilon_{T_1}(\omega) = 0$ and the transverse resonance frequency ($\omega_T$) is given by the frequency at which $\epsilon_{T_2}(\omega)$ is a maximum.

One can determine $\omega_T$ directly by carrying out normal incidence transmission measurements on thin films whose thickness is very much smaller than the wavelength in the medium, i.e., less than the "skin depth." For this case the large reflection at the front surface is canceled by the corresponding reflection from the back surface except at frequencies near $\omega_T$, where absorption occurs. The films accordingly exhibit a transmission minimum and a reflection peak at $\omega = \omega_T$.

As discussed earlier it is possible to obtain a coupling of the EM radiation with the longitudinal as well as the transverse excitation modes for *oblique* incidence of the EM radiation with the $E$ field polarized in the plane of incidence (Fig. 9). A Fabry–Perot analysis of the oblique incidence transmission of EM radiation along with Maxwell's equations, boundary conditions, and the front and back surface yields the following expressions for the transmission of EM radiation in the limit $\delta = 2\pi d/\lambda_0 \ll 1$ [48]

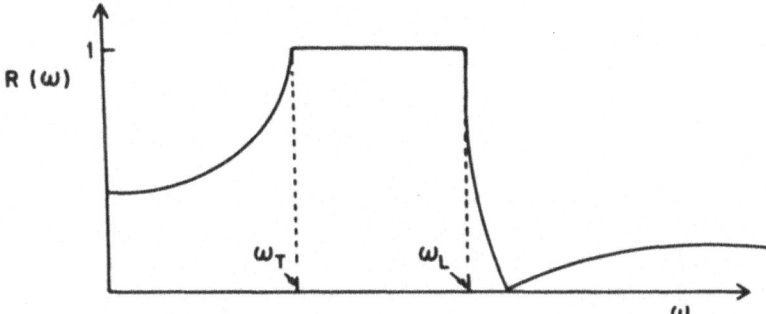

Fig. 13—Reflectance curve for an assembly of electric dipole oscillators.

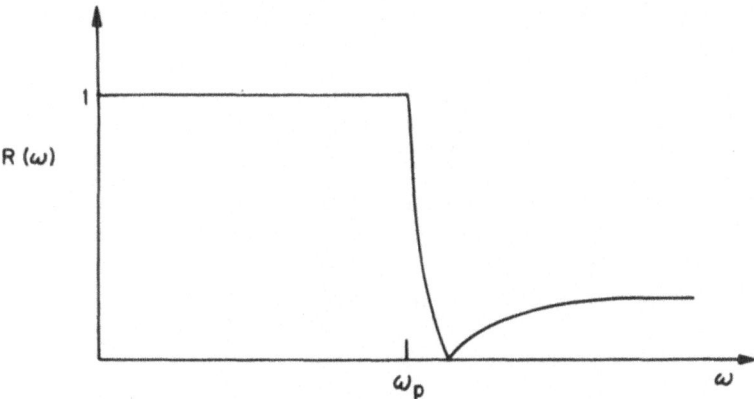

Fig. 14—Reflectance curve for an electron plasma.

$$T_{\perp}(\omega) = 1 - \frac{\delta}{\cos\theta}\epsilon_2(\omega)$$

$$T_{//}(\omega) = 1 - \frac{\delta}{\cos\theta}\left[\epsilon_2(\omega)\cos^2\theta + \frac{\epsilon_2(\omega)}{\epsilon_1(\omega)^2 + \epsilon_2(\omega)}\sin^2\theta\right]$$

where $\perp$ and $//$ designate polarization perpendicular and parallel to the plane of incidence, respectively, and $\theta$ is the angle of incidence. The factor $\delta/\cos\theta$ represents the "effective" thickness of the film. The factors $\cos^2\theta$ and $\sin^2\theta$ arise from the absolute squares of the electric field components $E_T^2 = E_{//}^2\cos^2\theta$ and $E_L^2 = E_{//}\sin^2\theta$. In the expression for $T_{//}(\omega)$ the term

$$\frac{\delta}{\cos\theta}[\epsilon_2(\omega)\cos^2\theta]$$

is due to the coupling of the radiation to the transverse modes of excitation (absent from plasmas) and the term

$$\frac{\delta}{\cos\theta}\left[\frac{\epsilon_2(\omega)\sin^2\epsilon}{\epsilon_1(\omega)^2 + \epsilon_2(\omega)^2}\right]$$

is due to the coupling of the radiation to the longitudinal mode of excitation The factor $\epsilon_2(\omega)/[\epsilon_1(\omega)^2 + \epsilon_2(\omega)^2]$ is equal to $-\text{Im}\,[1/\epsilon(\omega)]$. In the excitation of plasmons by electrons this factor measures the coupling of the longitudinal field of the electrons with the plasmons ([50]). As shown by Burstein *et al.* ([51]) it is convenient to express the coupling of the EM radiation to the transverse and longitudinal excitation modes in terms of a dielectric response function $\epsilon_T^r(\omega) = 1 + 4\pi\chi_T^r(\omega)$ and $\epsilon_L^r(\omega) = 1 + 4\pi\chi_L^r(\omega)$ where $\chi_L^r(\omega)$ and $\chi_L^r(\omega)$ are the corresponding dielectric susceptibility response functions, defined by

$$\mathbf{P}_T = \chi_T^{(r)}\mathbf{E}_{T_{ext}}$$
$$\mathbf{P}_L = \chi_L^{(r)}\mathbf{E}_{L_{ext}}$$

(See also discussion of these quantities in the Lectures of Resibois.)

The dielectric response functions are analogs of the conductivity functions introduced by Dresselhaus, *et al.* ([52]) in their treatment of magnetoplasma oscillations at microwave frequencies. Since

$$\mathbf{E}_{T_{ext}} = \mathbf{E}_T$$

is follows that $\chi_T^r(\omega) = \chi_T(\omega)$ and $\epsilon_T^r(\omega) = \epsilon_T(\omega)$ On the other hand, $\mathbf{E}_{L_{ext}} \neq \mathbf{E}_L$, and one finds that $\epsilon_L(\omega) = 2 - 1/\epsilon_L(\omega)$ and, in particular, that

$$\epsilon_{L_2}(\omega) = -I_m\left[\frac{1}{\epsilon_L(\omega)}\right] = \frac{\epsilon_2(\omega)}{\epsilon_1(\omega)^2 + \epsilon_2(\omega)^2}$$

One can then rewrite the expressions for $T_{//}(\omega)$ and $T_{\perp}(\omega)$ in the simple form

$$T_{//}(\omega) = 1 - \frac{\delta}{\cos\theta}[\epsilon_{T_2}^r(\omega)\cos^2\theta + \epsilon_{L_2}^r(\omega)\sin^2\theta]$$
$$T_{\perp}(\omega) = 1 - \delta\cos\theta\,\epsilon_{T_2}^r(\omega)$$

The derivation of $\epsilon_L^r(\omega)$ proceeds simply from the equation of motion by solving for $\mathbf{u}_L(\omega)$ in terms of $\mathbf{E}_{L_{ext}}$. One obtains directly

$$\mathbf{u}_L(\omega) = \frac{(e^*/m)\mathbf{E}_{L_{ext}}}{\epsilon_0(\omega_L^2 - \omega^2 - i\omega\gamma_L)}$$

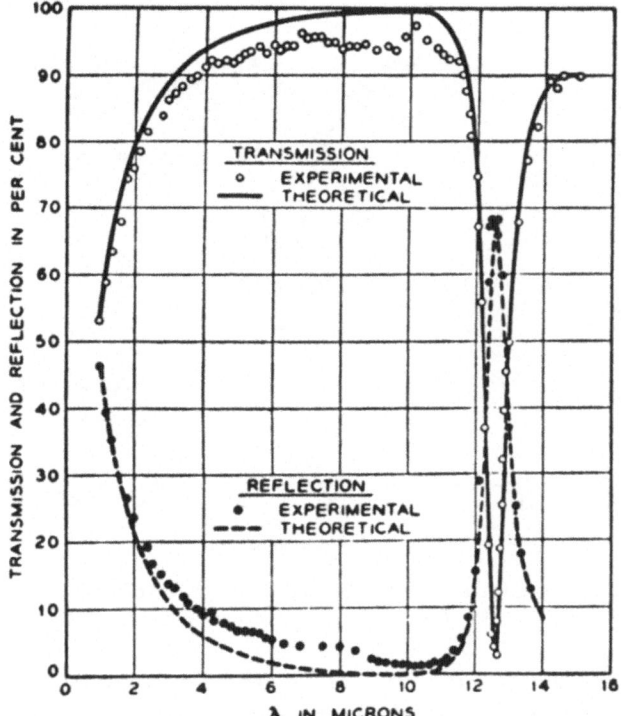

Fig. 15—Transmission and reflection curves for a thin film of SiC, 0.06 $\mu$ in thickness. The deviation of the curves from 100% at shorter wavelengths is due to electronic transitions. [From Spitzer, Kleinman, and Frosch, *Phys. Rev.* **113**: 133 (1959)].

Introducing this into the expression for $P_L(\omega)$ one obtains

$$P_L(\omega) = \frac{1}{\epsilon_0^2} \frac{(ne^{*2}/m)E_{L_{ext}}}{\omega_L^2 - \omega_T^2 - i\gamma\omega} + \frac{\chi_0 E_{L_{ext}}}{\epsilon_0} = \chi_L^r(\omega)E_{L_{ext}}$$

$$\epsilon_L^r(\omega) = 1 + 4\pi\chi_L(\omega) = 1 + \frac{\epsilon - 1}{\epsilon_0} + \frac{\Omega^2}{\epsilon_0^2(\omega_L^2 - \omega^2 - i\omega\gamma_L)}$$

In the case of alkali halides, Berreman used evaporated films having a thickness of the order of 0.1 $\mu$ and was able to obtain transmission spectra uncomplicated by reflection effects (Fig. 10) (For $\delta \ll 1$, the reflection goes as $\delta^2$ whereas the transmission goes as $\delta$). Oblique incidence transmission data for semiconductor films several microns thick also reveal the minima at $\omega_T$ and $\omega_L$ (Fig. 15 are 16) but cannot be used directly to obtain quantitative information about $\epsilon_{L_2}^r(\omega)$, i.e., one must obtain the reflection data as well. The oblique incidence transmission spectra obtained by McAlister and

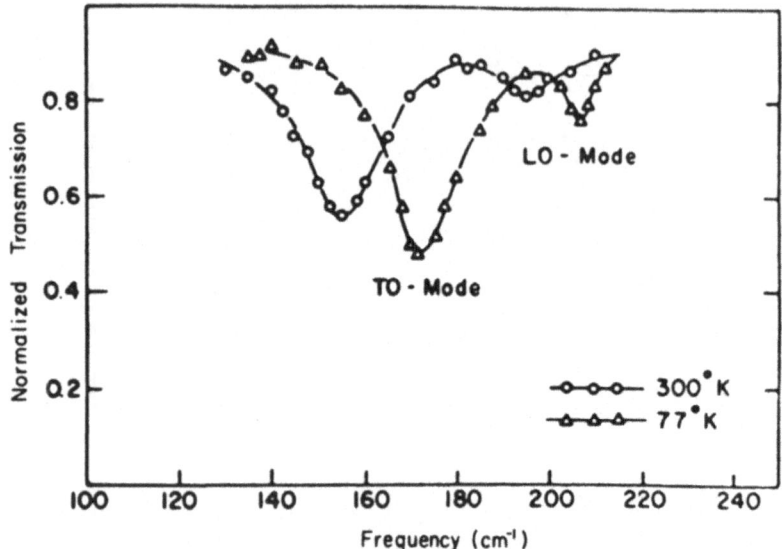

Fig. 16—Oblique incidence (45°) transmission spectra for epitaxial 1-$\mu$-thick CuCl grown on silicon (S. Iwasa, Thesis, University of Pennsylvania, 1965).

Stern ([49]) for evaporated thin films of Ag exhibit a transmission minimum at $\hbar\omega = 3.8$ eV in agreement with the plasmon energy $\hbar\omega_p$ determined from electron energy loss studies.

## IV. POLARITONS

As a result of the strong interaction between the EM radiation (photons) and the electric dipole excitations (TO phonons) the propagating state can be no longer be described as a photon or an electric dipole oscillation, but as a mixture of the two "elementary" excitation known as a polariton. In the case of TO phonons, the $\omega$ vs. $q$ curves for the photons and the TO phonons near $q = 0$ are shown in Fig. 17.

In the absence of coupling, the dispersion curves are straight lines which cross at $\mathbf{k} = q$. The photon–phonon interaction results in the $\omega$ vs. $q$ dispersion curves given by the solid lines ([50]). The slope of the $\omega$ vs. $q$ curve for $\omega < \omega_T$ corresponds to velocity of the photons $v = c/\epsilon_s^{1/2}$, whereas that for $\omega > \omega_L$ corresponds to a velocity $v = c/\epsilon_0^{1/2}$. For $\omega < \omega_T$, the polariton is predominantly photon (eletromagnetic) in character whereas for $\omega \approx \omega_T$ the polariton is predominantly phonon in character. At frequencies such that $\omega_T < \omega < \omega_L$, the propagation vector is predominantly imaginary. For completeness the curves for the interaction of photons with (electron)

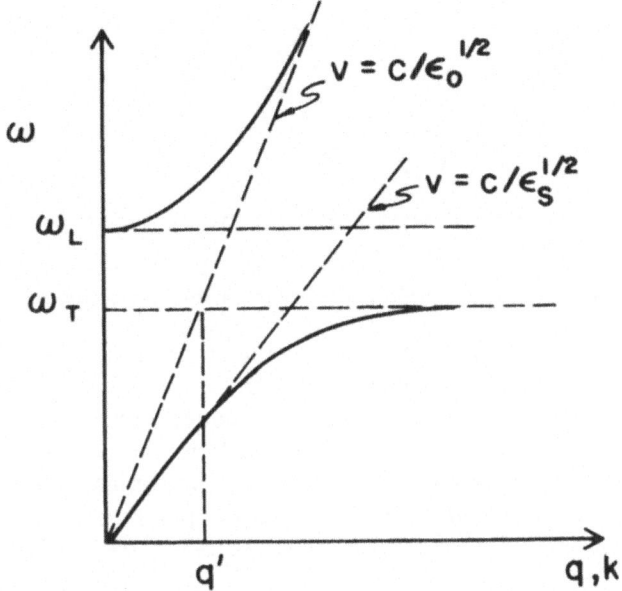

Fig. 17—The dispersion curves for polaritons (coupled photon–electric dipole excitation modes).

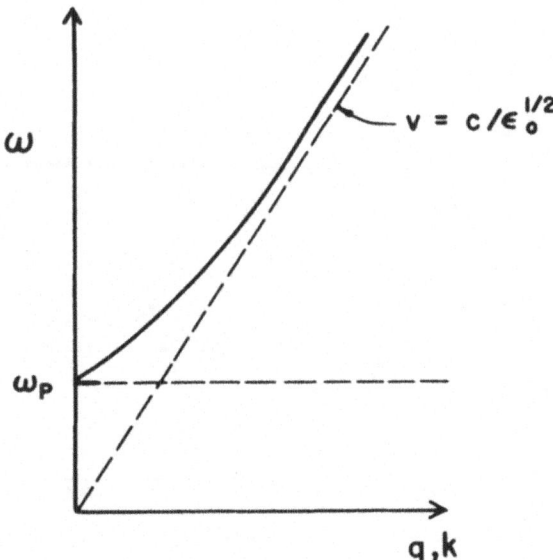

Fig. 18—The dispersion curve for the polariton modes of an electron plasma.

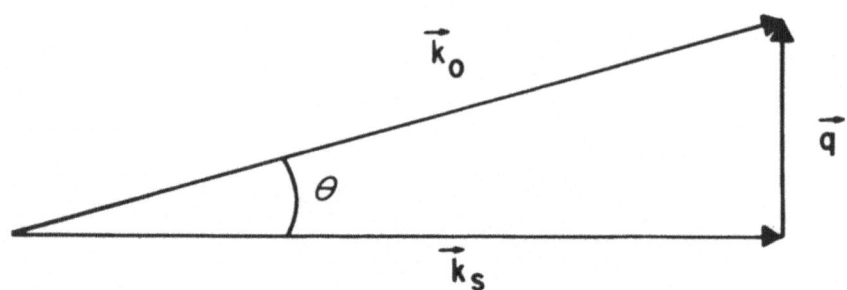

Fig. 19—The wavevector diagram for small-angle scattering.

plasmons is shown in Fig. 18. In this case the propagation vector is predominantly imaginary for all frequencies below $\omega_p$.

The existence of polaritons consisting of coupled photon–phonon modes has recently been demonstrated by Henry and Hopfield ([54]) in Raman scattering experiments on GaP carried out at small scattering angles. GaP has a zincblende structure and the $q \approx 0$ TO and LO phonons are Raman-active. The scattering geometry used in the experiments is shown in Fig. 19.

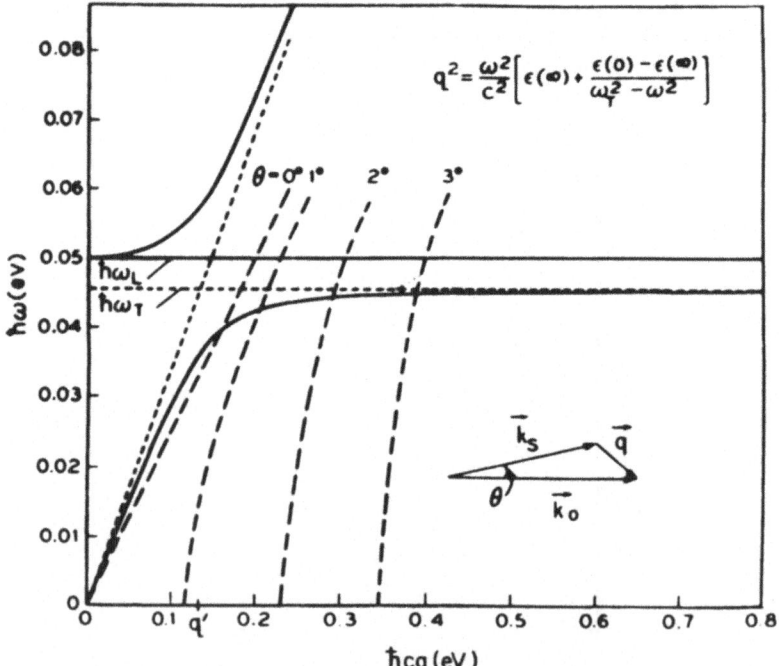

Fig. 20—The polariton dispersion curves for GaP. The dashed curves for 0°, 1°, 2°, and 3° represent the momentum and energy conservation conditions. [From C. H. Henry and J.J. Hopfield, *Phys. Rev. Letters* **15**: 964 (1965).]

For $\theta < 3°$ the frequency shift of the Stokes phonon emission line for the TO phonons is decreased to a value below $\omega_T$. The energy and wavevector of the polaritons taking part in the scattering are determined by

$$\hbar\omega_p = \hbar\omega_0 - \hbar\omega_s$$
$$q = (k_0^2 + k_s^2 - 2k_0k_s\cos\theta)^{1/2}$$

To a good approximation the wavevector can also be expressed as ($^{54}$)

$$q = \left[\left(\frac{\partial k}{\partial\omega}\right)_{\omega=\omega_0}^2 \omega_q^2 + k_0k_s\theta^2\right]^{1/2}$$

where $k_0 = n(\omega_0)\omega_0/c$, $k_s = (\omega_s)\omega_s/c$, and $n(\omega)$ is the frequency-dependent refractive index. The allowed values for $\omega$ and $q$, which occur for different values of $\theta$ given by the above expression for $q$, are indicated by the dashed lines in Fig. 20. For $\theta = 0$ the curve is a straight line with the same slope as that for the photons at $\omega = \omega_0[$i.e., $\omega_q/q = (\partial\omega/\partial q)\omega_0]$ which is smaller than the slope of the $\omega$ vs. $q$ dispersion curve for $\omega < \omega_T$. The crossing of these

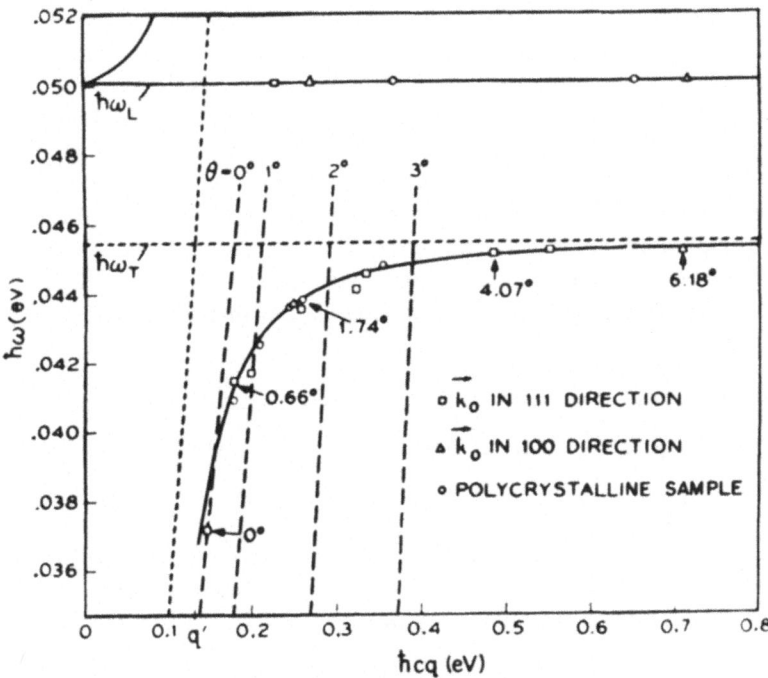

Fig. 21—The observed $\omega$, $q$ points for small-angle Raman scattering by polaritions in GaP. [From C. H. Henry and J.J. Hopfield, *Phys. Rev. Letters* **15**: 964 (1965).]

lines with the polariton dispersion curves determines the polaritons which are "emitted" in the scattering process. For $\theta = 0$ the frequency of the emitted polariton is 20% below $\omega_T$. The Raman scattering experiments were carried out using a He–Ne laser operating at 6328 Å. The experimental results for the Stokes frequency shifts obtained for different values (Fig. 21) are found to be in good agreement with the theoretical dispersion curves based on the expression for $\epsilon_T(\omega)$ derived from infrared measurements by Kleinman and Spitzer ([55]).

Scattering by polaritons in an optically anisotropic crystal has been observed by Porto, Tell, and Damen ([56]), who carried out nearly forward scattering in ZnO (wurtzite structure). For the incident beam polarized perpendicular to the $C$-axis and the scattered beam polarized parallel to the $C$-axis, the frequency of the emitted polariton can be shifted from 407 to 140 cm$^{-1}$ ($\theta = 0.63°$).

## V. COUPLING OF PLASMONS WITH OTHER LONGITUDINAL EXCITATION MODES

The simplest case to consider is that of coupled plasmon–LO phonon modes which occur in polar semiconductors containing free electrons. An indication of such coupling is found in the fact that for $\omega_p > \omega_L$ the plasmon frequency is given by $\omega_p^2 = 4\pi n e^2 / m^* \epsilon_0$ whereas for $\omega_p < \omega_L$ it is given by $\omega_p^2 = 4\pi n e^2 / m^* \epsilon_s$. The plasmons and the LO phonons are coupled by the self-consistent macroscopic polarization field, $E_L = -4\pi P_L$, which is set up by the displacement of the atoms and the electrons ([57]). The coupled plasmon–LO phonon modes can be determined in a straight forward manner from the equations of motion for the two types of longitudinal excitation. Omitting the damping term and ignoring spatial dispersion effects for simplicity the equations are ([58])

$$m_1 \ddot{u}_1 + m_1 \omega_T^2 u_1 = e_1 (E_{\text{ext}} - 4\pi P) \qquad \text{(Atoms)}$$
$$m_2 \ddot{u}_2 = e_2 (E_{\text{ext}} - 4\pi P) \qquad \text{(Electrons)}$$

where

$$P = n_1 e_1 u_1 + n_2 e_2 u_2 + \chi_0 (E_{\text{ext}} - 4\pi P)$$

is the longitudinal polarization. Upon introducing the $\exp(-i\omega t)$ time-dependence for the variables we obtain

$$(-\omega^2 + \omega_L^2) u_1 + \frac{m_2}{m_1} \omega_p^2 u_2 = \frac{e_1}{m_1} E_{\text{ext}}$$

$$\frac{m_1}{m_2} (\omega_L^2 - \omega_T^2) n_1 + (-\omega^2 + \omega_p^2) u_2 = \frac{e_2}{m_2} E_{\text{ext}}$$

The normal mode frequencies are obtained by setting the determinant of the coefficients of $u_1$ and $u_2$ equal to zero, i.e., $E_{ext} = 0$. One obtains a quadratic equation in $\omega^2$ which has two real roots $\Omega_+ = \omega_-^2$ and $\Omega_- = \omega_+^2$ corresponding to two coupled plasmon–LO phonon normal modes for a given $\omega_p$. A plot of the dependence of the high $(\omega_1)$ and low $(\omega_2)$ normal mode frequencies on $n$, the free carrier concentration (or on $\omega_p^2$ which is proportional to $n$) is given in Fig. 22.

At low carrier concentration $(\omega_p < \omega_L)$ the $\Omega_+$ mode is mainly an LO phonon and the $\Omega_-$ mode is mainly a plasmon screened by the low-frequency (static) dielectric constant. At high concentrations $(\omega_p > \omega_L)$ the $\Omega_+$ mode is mainly a plasmon screened by $\epsilon_0$, whereas the $\Omega_-$ mode corresponds to an LO phonon screened essentially by the dielectric constant of the free carriers,

$$\omega_2^2 = \omega_T^2 + \frac{4\pi n_1 e_1^2}{m_1 \epsilon(\omega)} \approx \omega_T^2$$

where

$$\epsilon(\omega) = \epsilon_0 - \frac{\omega_p^2}{\omega_2^2} \approx -\frac{\omega_p^2}{\omega_2^2}$$

In the high-frequency $(\Omega_+)$ mode the displacement of the atoms and the electrons relative to the positively charged background are in phase, whereas

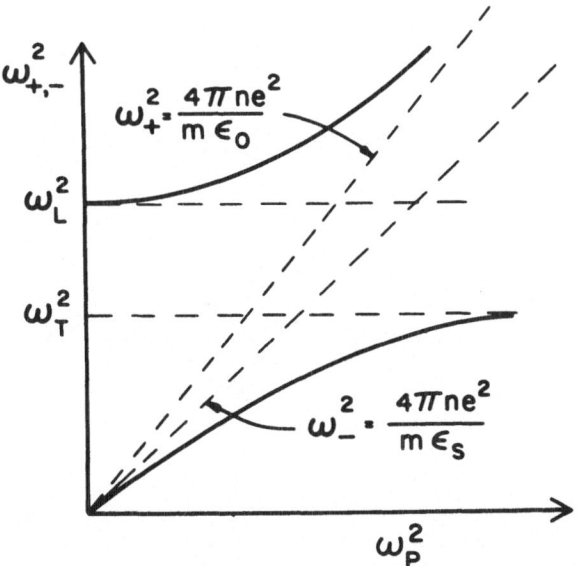

Fig. 22—The dispersion curve for coupled plasmon–LO phonon modes.

in the low-frequency ($\Omega_-$) mode, the two displacements are 180° out of phase.

One can derive an expression for $\epsilon_T(\omega)$ by solving for $u_{T_1}$ and $u_{T_2}$ in terms of $E_T(\omega)$ and introducing the resulting experssion for $u_{T_1}$ and $u_{T_2}$ into the equation for $P_T(\omega)$. One obtains an expression for $\epsilon_T(\omega)$ which turns out to be the sum of the independent contribution from interband transitions, lattice vibrations, and free carriers, i.e.,

$$\epsilon_T(\omega) = \epsilon_0\left(1 - \frac{\omega_p^2}{\omega^2} + \frac{\Omega_L^2}{\omega_T^2 - \omega^2}\right)$$

The frequency dependence of $\epsilon_T(\omega)$ is shown in Fig. 23 for the two cases of $\omega_p > \omega_L$ and $\omega_p < \omega_L$. We note that there are two transverse dielectric anomalies: $\epsilon_T(\omega) = 0$. For isotropic media the frequencies at which the dielectric anomalies occur correspond to the high- and low-frequency coupled plasmon–phonon modes [i.e., the frequencies of the longitudinal coupled modes are given by the roots of $\epsilon_T(\omega) = \epsilon_L(\omega)$]. As in the case of individual uncoupled longitudinal modes, one can observe the excitation of the modes by EM waves by carrying out oblique incidence transmission experiments on

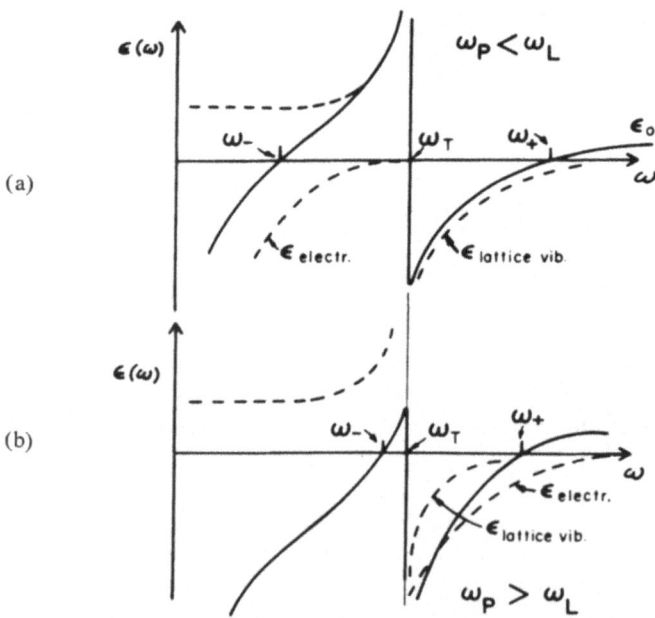

Fig. 23—Curves of $\epsilon_1(\omega)$ *vs.* $\omega$ for a diatomic cubic polar crystal containing an electron plasma for (a) $\omega_P < \omega_L$ and (b) $\omega_P > \omega_L$.

thin films. Such experiments have been carried out by Iwasa ([58]) for InSb and GaAs containing $10^{18}$ electrons/cm³. For these carrier densities, $\omega_p \gg \omega_L$, and it was possible to observe well-defined transmission minima corresponding to $\omega_i^2 \approx 4\pi n e^2/m^*\epsilon_0$. Since $\omega_2$ lies close to $\omega_T$ the transmission minimum associated with the $\omega_2$ mode was masked by that for the $q \approx 0$ TO mode.

It is of interest to consider the frequencies of the propagating EM and plasmon–LO phonon modes in the region of small $\mathbf{q}$. These are given by $\omega^2 = c^2q^2/\epsilon_T(\omega)$ and a plot of the dispersion curves is shown in Fig. 24 for $\omega_p > \omega_L$ and $\omega_p < \omega_L$. It should be pointed out that for $\omega_p > \omega_L$ the electron screens the effective plasma frequency of the atoms.

In their inelastic experiments on PbTe containing $10^{19}$ free carriers/cm³, Cowley and Dolling ([59]) found that the frequency of the LO mode increased

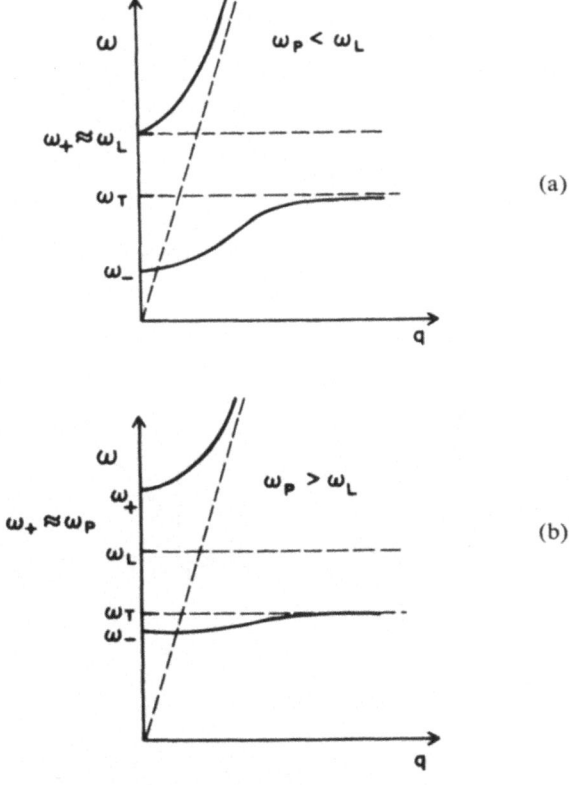

Fig. 24—The polariton dispersion curves of a polar cubic diatomic crystal containing an electron plasma for (a) $\omega_P < \omega_L$ and (b) for $\omega_P > \omega_L$.

with increasing $q$. They interpreted the results in terms of the decrease in screening of the LO phonon by the free electron as $q$ is increased.

For finite $q$ values, one must include the $q$ dependence of the frequencies of the plasmon and LO phonons as well as the $q$ dependence of the dielectric constant. Furthermore $\epsilon_L(\omega, q)$ is not equal to $\epsilon_T(\omega, q)$. In the case of semiconductors containing an appreciable carrier density, i.e., $n \approx 10^{18}/\text{cm}^3$ the $\mathbf{r}_s$ values are small (i.e., $\mathbf{r}_s = 0.34$ for GaAs containing $10^{18}$ electrons/cm$^3$) and to a good approximation, it is possible to use the following expressions for $\epsilon_L(\omega, q)$ and for the wavevector-dependent plasmon frequency $\omega_p(q)$

$$\epsilon(\omega, q) = \epsilon_0\left[1 - \frac{\omega_p^L}{\omega^2}\left(1 + \frac{q^2 v_F^2}{\omega_p^2}\right)\right]$$

$$\omega_p^2 = \omega_p^2\left(1 + \frac{q^2}{k_{TF}^2 \omega_p^2}\right)$$

shere $k_{TF}$ is the Thomas-Fermi wavevector. (See discussion of $\omega_p(q)$ and $\epsilon(q, \omega)$ in the Lectures by Resibois.) At carrier densities much greater than

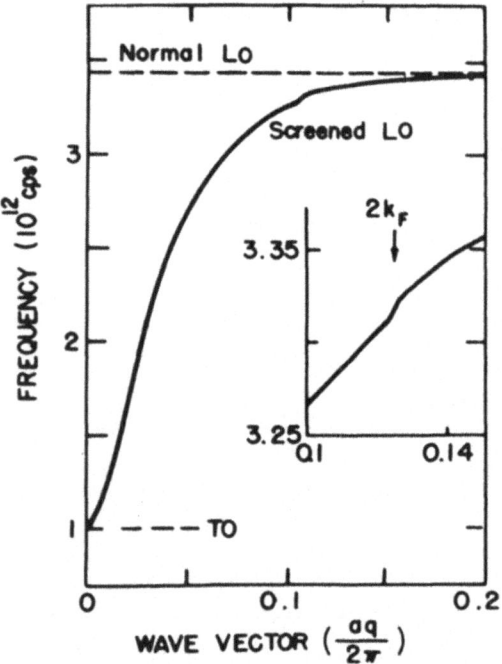

Fig. 25—The calculated LO dispersion curve in doped PbTe. [From R.A. Cowley and G.J. Dolling, *Phys. Rev. Letters* 14: 549 (1965).]

$10^{18}/cm^3$, $\omega_p(q)$ is generally much larger than $\omega_L(q)$ and the frequencies of the two coupled modes are given by ($^{59}$)

$$\Omega_+^2(q) = \omega_p^2 \left( 1 + \frac{q^2}{k_{TF}^2 \omega_p^2} \right)$$

and

$$\Omega_-^2(q) = \omega_T^2(q) + \frac{\Omega^2(q)\epsilon_0}{\epsilon(\omega, q)}$$

where

$$\epsilon(\omega, q) = \epsilon_0 \left( 1 + \frac{k_{TF}^2}{q^2} \right)$$

At values of $q$ such that $q \gg k_{TF}$, $\epsilon(\omega, q) \approx \epsilon_0$, so that

$$\omega_-^2(q) = \omega_T^2(q) + \Omega^2(q) \approx \omega_L^2(q)$$

Free carrier plasmas will also couple with longitudinal interband electron transitions. Such coupled modes manifest themselves very clearly in the optical properties of Ag ($^{60}$). In the case of Ag, the "free" carrier plasmon frequency would, in the absence of interband transitions, normally correspond to an energy of $\sim 9$ eV. The reflection spectrum (Fig. 26a) and the derived spectrum of the energy loss function, $-I_m[1/\epsilon(\omega)] = \epsilon_2(\omega)/\epsilon_1(\omega)^2 + \epsilon_2(\omega)^2$ (Fig. 26b) clearly show the existence of a longitudinal excitation mode at $\hbar\omega_p = 3.8$ eV. (The existence of this excitation mode has also been demonstrated in an electron energy loss, transition radiation, and oblique incidence transmission studies described earlier in this lecture. The curves of $\epsilon_1(\omega)$ and $\epsilon_2(\omega)$ (Fig. 26c) show that $\epsilon_1(\omega)$ goes to zero and that $\epsilon_2(\omega)$ is relatively small at this energy. The excitation mode corresponds to a plasmon coupled with electron excitation of the electron in the $d$ band. The resonance may also be considered to be a plasmon screened by the frequency-dependent dielectric constant due to interband transitions.

The frequency of the excitation mode corresponds to the frequency at which $\epsilon_1(\omega) = 0$ where

$$\epsilon_1(\omega) = 1 + \epsilon_1^{IB}(\omega) - \frac{\omega_p^2}{\omega^2}$$

and $\epsilon_1^{IB}(\omega)$ is the contribution to $\epsilon_1(\omega)$ from interband transitions. On setting $\epsilon_1(\omega) = 0$ one obtains

$$\omega_{p'}^2 = \frac{\omega_p^2}{1 + \epsilon^{IB}(\omega)}$$

In the case of Ag, $\omega_{p'}$ falls below the absorption edge for interband transitions so that $\epsilon^{IB}(\omega)$ is predominantly real. It should be noted that the energy loss

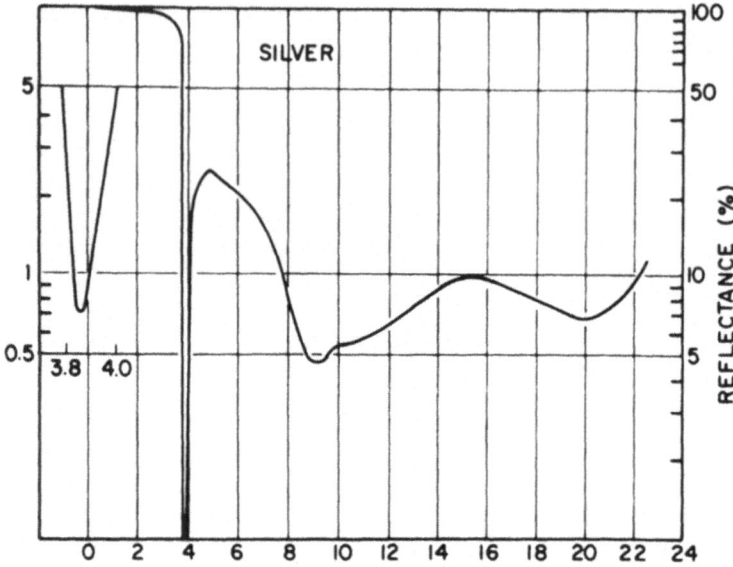

Figure 26–(a)

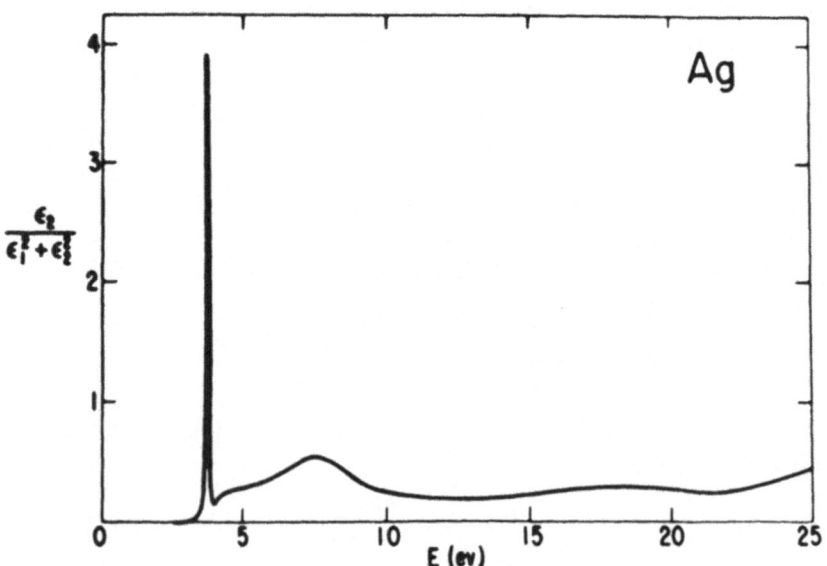

Figure 26–(b)

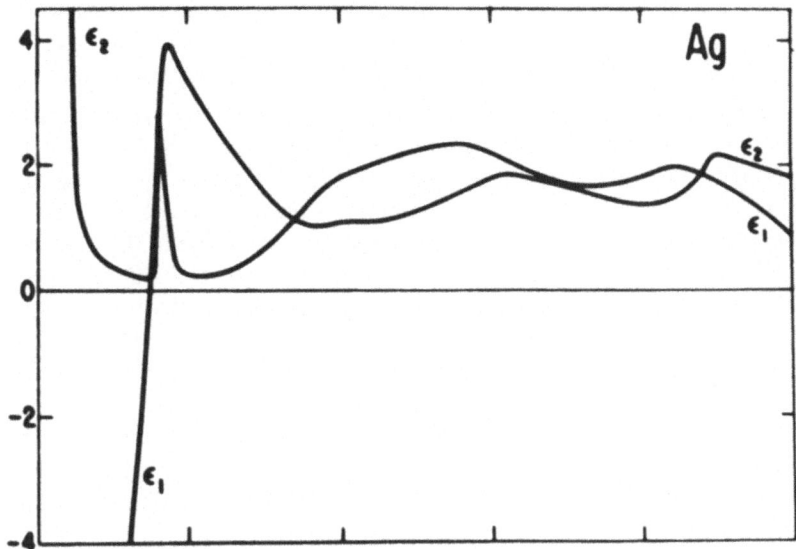

Figure 26–(c)

Fig. 26—The optical properties of silver: (a) the reflectance spectrum (b) the
$\epsilon_2/(\epsilon_1^2 + \epsilon_2^2) = |-\text{Im}\,(1/\epsilon)|$  *vs.* $\hbar\omega$  spectrum (c) the $\epsilon_1(\omega)$ and $\epsilon_2(\omega)$
spectra. [From H. Ehrenreich and H.R. Philipp, *Phys. Rev.* **128**: 1622 (1962).]

function also peaks at 7.5 eV near the free electron plasmon frequency. How-
ever, in this region neither $\epsilon_1(\omega)$ nor $\epsilon_2(\omega)$ vanish although both are small.
This peak corresponds to the plasmon frequency of the free electrons, which
is shifted to lower energies by the interaction with interband transitions. In
the case of Cu only the high-frequency plasmalike excitation mode is observed
The low-energy resonance does not appear since the negative free-electron
contribution to $\epsilon_1(\omega)$ is not sufficiently compensated by the positive contribution from interband transition of the $d$ electrons to cause $\epsilon_1(\omega)$ to go to
zero, i.e., the peak in $\epsilon_1^{\text{IB}}(\omega)$ occurs at a lower frequency where $\omega_p^2/\omega^2$ is
appreciable, so that there is no cancellation. As pointed out by Philips
and Ehrenreich the existence of the longitudinal excitation at 3.8 eV in Ag
is a somewhat accidental consequence of the position and strength of the
peak in $\epsilon_1^{\text{IB}}(\omega)$.

## VI. RAMAN SCATTERING BY PLASMONS IN ZINCBLENDE (ZnS) TYPE CRYSTALS

There has been considerable interest in recent years in Raman scattering of photons by plasmons in semiconductors.

Theoretical estimates of scattering cross sections have been obtained
for intraband scattering ([61]). Also the case of interband scattering in the
presence of an external magnetic field has been treated by Wolf ([62]). For the
case of plasmons in crystals lacking a center of symmetry, the scattering
mechanism should be similar to that for LO phonons ([63]). Mooradian and
Wright ([64]) have recently reported well-defined Raman scattering spectra
for GaAs (using a 1.0648 $\mu$ yttrium aluminum garnet CW laser, where GaAs
is transparent). The spectra (Fig. 27) clearly show lines due to the two coupled
plasmon–LO phonon modes as well as a line due to the TO phonon. The
position of the coupled mode peaks are in good agreement with the expected
positions of the coupled mode (Fig. 28). The magnitude of the scattering
matrix elements for the coupled plasmon–LO phonon modes is attributed
to the "phonon strength" of the mode.

An expression for the phonon strength $S$ has been obtained by Varga
([57]). His results for the phonon strength of the two coupled plasmon–LO
phonon modes in GaAs plotted as a function of carrier concentration are
shown in Fig. 29.

At low concentrations ($\omega_p < \omega_L$), the phonon strength is close to unity
for the high-frequency ($\omega_1$) plasmonlike mode and it is essentially zero for
the low-frequency ($\omega_L$) phononlike mode. At high concentration, the phonon
strength is close to zero for the high-frequency plasmonlike mode and close
to unity for the low-frequency phononlike mode. Varga also finds that the
phonon strength obeys the sum rule $\sum \omega_j S_j = \omega_L$, where the subscript $j =$
1, 2 denotes the particular coupled mode.

There is actually an important additional contribution to the scattering
matrix elements which arises from the "electro-optic" effect of the longitudinal electric field associated with each of the coupled modes ([65]). Thus for
high carrier densities such that $\omega_p > \omega_L$ in crystals which exhibit an appreciable electro-optic effect, the Raman scattering by the high-frequency plasmonlike mode, for which the phonon strength is close to zero, will be due
almost entirely to the electro-optic effect. The magnitude of the scattering
matrix element for a given mode is determined by

$$\alpha_1 = \frac{\partial \alpha}{\partial Q_j} dQ_j$$

where $Q$ is the normal coordinate of the mode. In the case of the coupled
plasmon–LO phonon modes, $\partial \alpha / \partial Q_j$ can be expanded as a sum of three
terms:

$$\left(\frac{\partial \alpha}{\partial Q_j}\right) = \left(\frac{\partial \alpha}{\partial Q_1}\right)_{U_2, E}\frac{du_1}{\partial Q_j} + \left(\frac{\partial \alpha}{\partial u_2}\right)_{U_1, E}\frac{du_2}{\partial Q_j} + \left(\frac{\partial \alpha}{\partial E_j}\right)_{U_1, U_2}\frac{dE_j}{\partial Q_j}$$

where $(\partial \alpha / \partial u_1)_{U_2, E} du_1$ is the first-order change in the electromagnetic

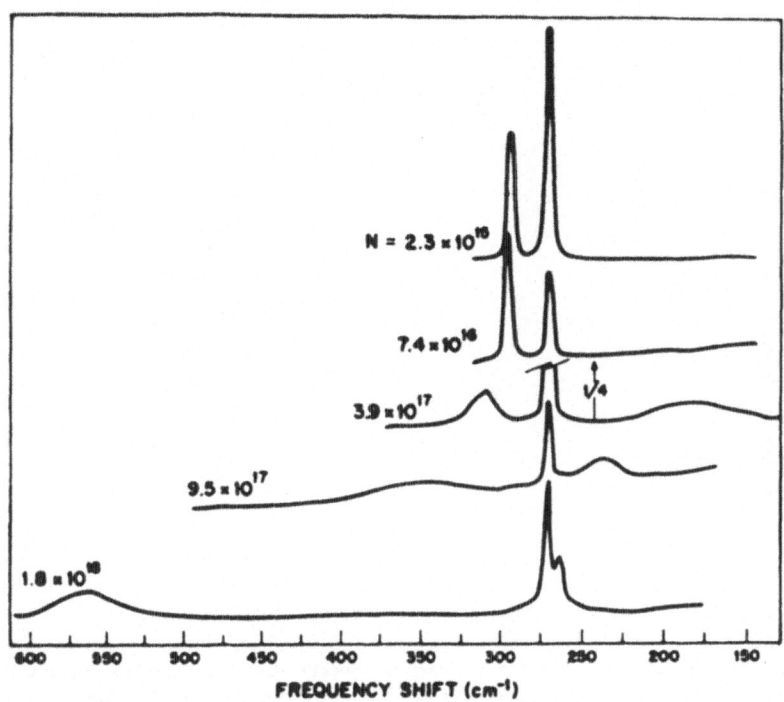

Fig. 27—The Raman scattering spectra of *n*-type GaAs crystals containing various concentrations of electrons [From A.A. Mooradian and G.B. Wright, *Phys. Rev. Letters* **16**: 999 (1967).]

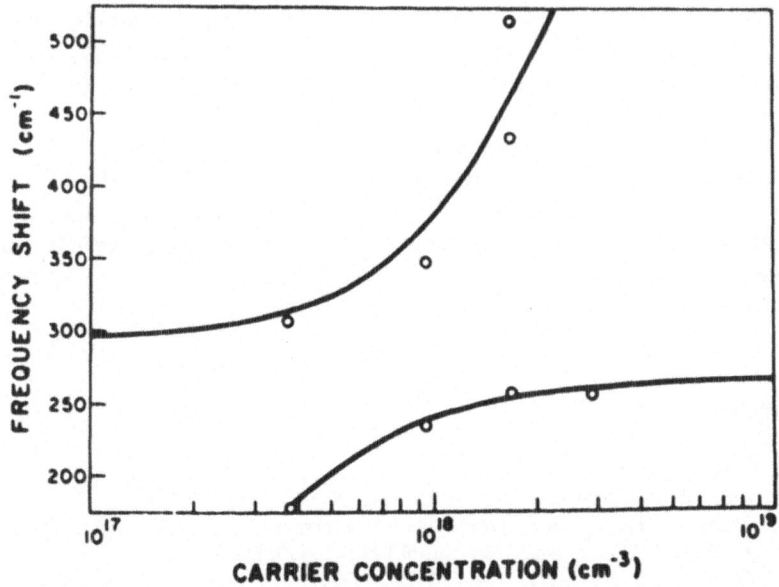

Fig. 28—The observed frequency shifts for the coupled plasmon–LO phonon modes in *n*-type GaAs. [From A.A. Mooradian and G.B. Wright, *Phys. Rev. Letters* **16**: 999 (1967).]

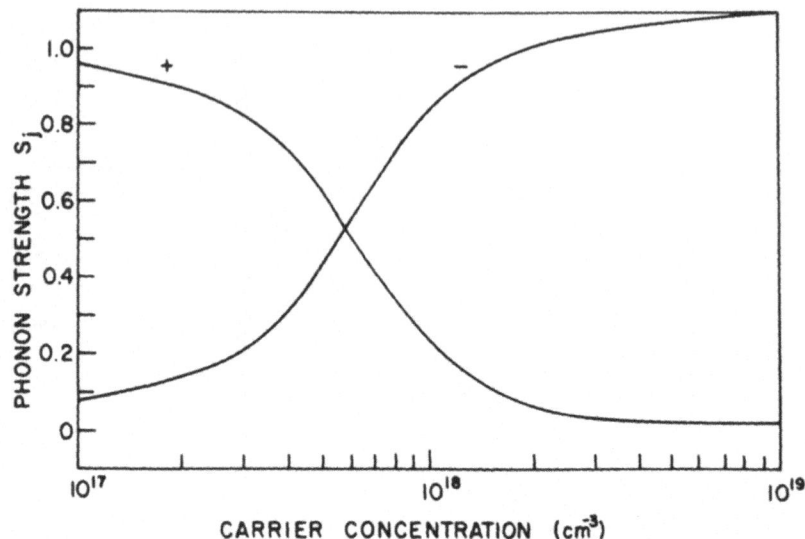

Fig. 29—The phonon strength *vs.* carrier concentration for coupled plasmon-LO phonon modes in *n*-type GaAs. [From A.A. Mooradian and G.B. Wright, *Phys. Rev. Letters* **16**: 999 (1967).]

polarizability induced by the displacement of the atoms: $(\partial\alpha/\partial u_2)_{U_1, E}du_2$ is the corresponding first-order change in the electron polarizability induced by the displacement of the electrons relative to positive charged background, and $(\partial\alpha/\partial E_j)dE_j$ is the change in polarizability induced by the longitudinal electric field associated with the mode.

In the first-order Raman scattering experiments on ZnS carried out by Couture-Mathiev *et al.* ([66]), it was found that the intensity of the LO line is greater than that of the TO line. Poulet ([67]) has attributed this difference in intensity to the existence of an electro-optic contribution to $\partial\alpha/\partial Q$ and to the fact that $E_T = 0$ for the TO phonons, whereas $E_L = -4\pi neu_L$ for the LO plasmons. [See also discussion by Loudon ([68]).]

# REFERENCES

1. R.J. Elliott and R. Loudon, *Phys. Rev. Letters* **3**: 189 (1963).
2. R. Loudon, *Advances in Physics (Phil. Mag. Suppl.)* **13**: 423 (1964).
3. E. Burstein, *Proceedings International Conference Lattice Dynamics* (Pergamon Press 1964) *J. Phys.Chem Solids* (Suppl.) **21**: 315 (1965).
4. E. Fermi and F. Rasetti, *Z. Phys.* **71**: 689 (1931).
5. J. Birman *Phys. Rev.* **131**: 1489 (1963).
6. E. Burstein, F.A. Johnson, and R. Loudon, *Phys. Rev.* **139**: 1239 (1965).
7. S. Ganesan, E. Burstein, A.M. Karo, and J.R. Hardy *J. de Phys.* **26**: 639 (1965).

8. R. Loudon and F.A. Johnson, *Proc. Roy. Soc.* **281**: 224 (1964).
9. R. Loudon *Proc. Phys. Soc. London* **82**: 393 (1963A).
10. A.M. Karo, J.R. Hardy, and I Morrison, *J. de Phys.* **26**: 668 (1965).
11. R. Loudon, *Proc. Roy Soc.* **A275**: 218 (1963).
12. R.A. Cowley, *Proc. Phys. Soc. London* **84**: 281 (1964).
13. M. Born and M. Bradburn, *Proc. Roy. Soc.* **A188**: 161 (1947).
14. R.A. Cowley, *J. de Phys.* **26**: 659 (1965).
15. H.M. J. Smith, *Phil. Trans.* **A241**: 105 (1948).
16. J.P. Russel, *J. de Phys.* **26**: 620 (1965).
17. R. Loudon, *J. de Phys.* **26**: 677 (1965).
18. S.P.S. Porto, B. Tell, and T.C. Damen, *Phys. Rev. Letters* **144**: 771 (1966).
19. C.H. Henry and J. J. Hopfield, *Phys. Rev. Letters* **15**: 964 (1965).
20. S.P.S. Porto, B. Tell, and T.C. Damen, *Phys. Rev. Letters* **16**: 450 (1965).
21. L. Kleinman, *Solid State Communication*, **3**: 47 (1965).
22. T.C. Damen, S.P.S. Porto, and B. Tell, *Phys. Rev.* **142**: 570 (1966).
23. B. Tell, T.C. Damen, and S.P.S. Porto, *Phys. Rev.* **144**: 771 (1966).
24. H. Poulet and J.P. Mathieu, *Am. Phys. (Paris)* **9**: 549 (1964).
25. W. Hayes, H.F. MacDonald, and R.J. Elliott, *Phys. Rev. Letters* **15**: 961 (1965).
26. A. Maradudin, S. Ganesan, and E. Burstein, *Phys. Rev.* **163**: 882 (1967).
27. M. Lax and E. Burstein, *Phys. Rev.* **97**: 39 (1955).
28. E. Burstein and S. Ganesan, *J. de Phys.* **26**: 637 (1965).
29. B. Szigeti, *Proceedings International Conference Lattice Dynamics*, Copenhagen 1963; *J. Phys. Chem. Solids* (Suppl.) **21**: 405 (1964).
30. E. Condon, *Phys. Rev.* **41**: 759 (1939).
31. M.F. Crawford and I.R. Dagg, *Phys. Rev.* **91**: 1569 (1963).
32. E. Anastassakis, S. Iwasa, and E Burstein, *Phys. Rev. Letters* **17**: 1051 (1966).
33. M.F. Crawford and R.E. MacDonald, *Can. J. Phys.* **36**: 1022 (1958).
34. E. Rydberg, *Proc. Roy. Soc. (London) Ser. A* **127**: 111 (1930).
35. D. Bohm and D. Pines, *Phys. Rev.* **87**: 625 (1951).
36. D. Pines, *Rev. Mod. Phys.* **28**: 184 (1956); also *Elementary Excitation in Solids*, W.A. Benjamin Inc., New York, 1963.
37. G. Ruthermann, *Naturw.* **29**: 648 (1951).
38. L. Marton, J.A. Simpson, H.A. Fowler, and H. Swanson, *Phys. Rev.* **126**: 182 (1962).
39. R.H. Ritchie, *Phys. Rev.* **106**: 874 (1957); also R.H. Ritchie and H.B. Eldridge, *Phys. Rev.* **126**: 1935 (1962).
40. C.J. Powell and J.B. Swann, *Phys. Rev.* **118**: 640 (1960).
41. R.A. Ferrel, *Phys. Rev.* **111**: 1214 (1958).
42. S. Steinmann, *Phys. Rev. Letters* **5**: 470 (1960); *Z. Physik* **163**: 92 (1961).
43. R.W. Brown, P. Wessel, and E.P. Trounson, *Phys. Rev. Letters* **5**: 472 (1960); also A.F. Frank, E.T. Arakawa, and D. Birchoff *Phys. Rev.* **126**: 1947 (1962).
44. V.P. Silin and E.P. Fetiso, *Phys. Rev. Letters* **7**: 374 (1961); also E.A. Stern, *Phys. Rev. Letters* **8**: 7 (1962).
45. V.I. Ginsberg and I.M. Frank, JETP **16**: 1 (1946); also I.M. Frank and V.I. Ginsberg, *J. Phys.* (USSR) **9**: 353 (1945).
46. R.D. Birchoff *Physical Processes in Radiation Biology*, (Academic Press, New York, 1964) p. 145 (Review paper); and E.T. Arakawa, R.W. Hamm, W.F. Hansom and Tom Jelinek, *Proceedings Conference on Optical Properties of Metals Paris*, 1965.
47. R.A. Ferrel and E.A. Stern, *Am. J. Phys.* **30**: 810 (1962).
48. D.W. Berreman, *Phys. Rev.* **130**: 2193 (1963).
49. A.J. McAlister and E.A. Stern, *Phys. Rev.* **132**: 1599 (1963); also S. Yamaguchi, *J. Phys. Soc. Jap.* **17**: 1172 (1962); and M Hattori, K. Yamada, and H. Suzuki, *J. Phys. Soc. Jap.* **18**: 203 (1963).

50. H. Frohlich and H. Pelzer, *Proc. Phys. Soc. (London)* **68**: 525 (1955).
51. E. Burstein, S. Iwasa, and Y. Sawada, *Proceedings E. Fermi International School of Physics* 1966. *Course XXXIV Optical Properties of Solids*, Academic Press, New York, 1966, p. 421.
52. G. Dresselhouse, A. Kip, and C. Kittel, *Phys. Rev.* **100**: 618 (1955).
53. M. Born and K. Huang, *Dynamical Theory of Crystal Lattices*, Clarendon Press Oxford, 1954.
54. C.H. Henry and J.J. Hopfield, *Phys. Rev. Letters* **15**: 964 (1966).
55. D.A. Kleinman and W.G. Spitzer, *Phys. Rev.* **118**: 110 (1960).
56. S.P.S. Porto, B. Tell, and T.C. Damen, *Phys. Rev. Letters* **16**: 450 (1966).
57. B.J. Varga, *Phys. Rev.* **137**: A1896 (1965).
58. S. Iwasa, Y. Sawada, E. Burstein, and E. Palik, Proceedings International Conference on Physics of Semiconductors, Kyoto, 1966, *J. Phys. Soc. Jap.* (Suppl.) **21**: 742 (1966); and S. Iwasa (Thesis, Physics Department, Pennsylvania, 1965).
59. R.A. Cowley and Dolling, *Phys. Rev. Letters* **14**: 549 (1965).
60. H. Ehrenreich and H. R. Philipp, *Phys. Rev.* **128**: 1622 (1962); see also H. Ehrenreich, *Proceedings E. Fermi, International School of Physics Course* **34** "*The Optical Properties of Solids*," (Academic Press, New York 1964).
61. A.L. McWhorter, *Proceedings International Conference on Physics of Quantum Electronics*, McGraw Hill New York, 1965, p. A-10.
62. D.A. Wolff, *Phys. Rev. Letters* **16**: 225 (1965).
63. E. Burstein, *J. de Phys.* **11**: 688 (1965) (Discussion).
64. A. Mooradian and G.B. Wright, *Phys. Rev. Letters* **16**: 999 (1966).
65. E. Burstein, A. Pinzcuk, and S. Iwasa, *Phys. Rev.* **157**: 611 (1967).
66. J. Couture-Mathieu, H. Poulet, and J.P. Mathieu, *C.R. Acad. Sciences* **234**: 1761 (1952).
67. H. Poulet, *C.R. Acad. Sciences* **238**: 70 (1954); also *Ann. de Phys.* (*Paris*) **10**: 908 (1955).
68. R. Loudon, *Proc. Roy. Soc.* **A275**: 218 (1963).

# Free and Bound Excitons

## J. J. Hopfield

*Department of Physics*
*Princeton University*
*Princeton, New Jersey*

## I. EXCITONS AS ELECTRONIC STATES

An exciton is an electronic excitation characterized by an energy and wavevector. There are two extreme points of view toward exciton states. In molecular crystals (e.g., anthracene) the low-lying excitations are in lowest approximation the excited states of the individual molecules. Let $\varphi_i$ denote the state of the entire crystal in which atom $i$ is excited and the other atoms are in the ground state. $N$ atoms are present in the crystal. The $N$ degenerate states $\varphi_i$ will have matrix elements of the Hamiltonian connecting them. For the simple case of one molecule per unit cell, the exciton wavefunctions

$$\Psi_k = \sum_{i=1}^{N} |\varphi_i\rangle e^{i\mathbf{k}\cdot\mathbf{R}_i}$$

will diagonalize the matrix elements connecting these $N$ states. The $N$ excitons $\Psi_k$ will have, as a result, wavevector-dependent energies. This point of view was first described by Frenkel.

In insulators of high electron density, energy band structure will dominate the excitation spectrum. Given a wavevector $\mathbf{k}$, however, there are many excitations of electron and hole pairs having a total wavevector $\mathbf{k}$. If $|\Psi_0\rangle$ is the ground state Hartree–Fock wavefunction of the crystal, $a_{ck}^+$ a Fermion creation operator for an electron in the conduction band of wavevector $\mathbf{k}$, and $a_{vk}$ is the Fermion annihilation operator for the valence band, then any state of the form

$$a_{vk'+k} a_{ck'}^+ | \Psi_0\rangle$$

is a state having one pair excited and wavevector $\mathbf{k}$. The lowest such excited uncorrelated pair state has the energy gap for its excitation energy. Electron–hole Coulomb interactions introduce correlations between electron

and hole, so the actual energy eigenstates of the crystal might better be taken
to have the form

$$\Psi_{exc} = \sum_k g_k a_{vk'+k} a_{ck'}^+ | \Psi_0 \rangle$$

where $g_k$ is a set of coefficients to be determined by trying to solve the many-
boyd Schroedinger equation. This point of view was first described by
Wannier and by Mott, and greatly extended by Elliott and by Dresselhaus.
When the exciton binding energy is small, the many-body equation reduces
in effective mass approximation to a problem of two particles attracting each
other by a (screened) Coulomb interaction, and, in the simplest isotropic
case, will produce a series of exciton bands displaced from the band gap by
scaled hydrogen atom energy levels. The exciton bands are generated through
the energy associated with center of mass motion.

Both models of an exciton describe a correlated electron–hole pair which
moves through the crystal as a unit. This unit need not be stable. Excitons
at energies above the lowest band gap are well known, and excitons in some
metals can also be expected to occur. Spin waves are a special example
of excitons. Although the wavefunctions $\Psi_k$ and $\Psi_{exc}$ given above seem
reasonable, they are not appropriate for direct use. In each case, the use of
such a wavefunction omits the screening of the Coulomb interactions, by
the "appropriate" dielectric constant. Thus these wavefunctions cannot be
used in a variational calculation for exciton energies. This is a major stum-
bling block in trying to improve exciton calculations.

Real excitons in most systems have an energy level structure complicated
by valence and conduction band structure, by spin and by electron–phonon
couplings. As elementary excitations, however, they can be classified as belonging to irreducible representations of the symmetry group of the crystal.
In optical transitions taken in the dipole approximation, only those transitions belonging at $\mathbf{k} = 0$ (long-wavelength limit) to the *vector* representations
will be coupled to light in a one-photon process. In a rigid lattice, such excitons would produce optical absorption lines.

The experimental evidence for excitons is of many varieties. Perhaps the
first evidence for Frenkel excitons (as distinct from local excitations) was
observed in molecular crystals with two molecules per unit cell. Davydov
noted that in such circumstances there would be two exciton branches having
different energies at $\mathbf{k} = 0$. These Davydov splittings have been observed in
many materials. The excitons in alkali halides and solid rare gases have also
been unambiguously observed. These cases will be covered in lectures by
Baldini.

Weakly bound excitons treatable in the effective mass approximation
are common in semiconuctors. Experimental observations by Gross and

co-workers, extended by Nikitine, showed the existence of a series of "hydrogenic" exciton absorption lines at energies given by the expected

$$E_b - E_B\left(\frac{1}{n^2}\right)$$

in $Cu_2O$. These absorption lines also obey the prediction by Elliott that the absorption strength corresponding to a (forbidden) series with the $n = 1$ line missing should go as $(n^2 - 1)/n^5$. The quantitative agreement between theory and experiment for about five lines of the series is striking. Apparently $Cu_2O$ is the only known example of a direct band gap material with simple, isotropic bands. All other semiconductors have complications of band structure which make detailed calculations depend on many experimental band parameters, and such calculations are not feasible for more than a few levels.

In CdS and CdSe the exciton levels have been examined in detail in the presence of a magnetic field. Since the electron and hole rotate in the same direction about their common center of mass, while they have opposite charges, they contribute with opposite sign to the orbital magnetic moment of an exciton. Hexagonal CdS and CdSe have optical selection rules which permit exciton states having a $p$-like relative angular momentum of electron and hole to be observed in some modes of polarization. (In cubic crystals such as CdTe, the $S$-like excitons are observed. In $Cu_2O$ the $p$-like excitons were also observed, but linewidth problems prevent in $Cu_2O$ the detailed Zeeman effect studies which have been carried out in CdS and CdSe.) Zeeman and energy level studies on $n = 1$; $n = 2$; $n = 3$ states in these crystals have yielded the anisotropic effective masses and $g$-values of electrons and holes in these materials from the observed exciton lines. The adequate agreement between these optically determined parameters and the same parameters evaluated in other fashions might be regarded as a demonstration of the existence of excitons.

Exciton motion has been demonstrated by Thomas in Cds through a magneto-Stark effect. The excitons with which light directly (but let us assume weakly) interacts are those having the same wavevectors as the light, and are therefore moving through the crystal with a velocity

$$v = \frac{\hbar c}{m} \cdot \frac{2\pi}{\lambda} \cdot n$$

where $m$ is the exciton mass, $n$ the index of refraction, and $\lambda$ the vacuum wavelength. In a magnetic field, such excitons see an equivalent electric field $(v/c) \times H$ in addition to the magnetic field. This effective electric field due to exciton motion must be added to any electric field present for calculation of Stark splittings. Thomas was able to measure the effective field $(v/c) \times H$ through the asymmetry it introduced in the Stark effect in the presence of a magnetic field. As the effective field is directly caused by the exciton motion, it is a clear demonstration of an exciton velocity.

In indirect band gap materials such as Ge; Si; C; GaP; SiC; AgCl; AgBr to name a few, the lowest energy electron–hole pair excitation energy and the lowest energy excitons have wavevectors of the order of a zone boundary wavevector. Such excitons and electron–hole pairs cannot be directly made by light. Electron–phonon coupling provides a mechanism for seeing such excitons, as the state

$$(\text{Phonon}_{-k}) + (\text{Exciton}_k)$$

has zero wavevector and can be reached in an optical process. The matrix element depends only weakly on $\mathbf{k}$. The energy of this total state as a function of $\mathbf{k}$ changes from its value at the minimum chiefly because of the $E(k)$ of the exciton. The phonon has such small curvature in its dispersion relation that its energy can be regarded as constant near the threshold for indirect processes. Thus the absorption at zero temperature should behave as

$$C\sqrt{E - E_{\text{exc.}} - E_{\text{ph. at min}}}$$

the energy dependence simply coming from the $E^{1/2}$ density of states of the exciton. The observation by McFarlane *et al.* of such absorption edges in indirect band gap semiconductors [in which the free electron–hole pair creation absorption coefficient in Born approximation is proportional to $(E - E_{\text{threshold}})^2$] is strong evidence for excitons in indirect band gap materials.

The interaction of excitons and phonons broadens the exciton resonances observed in optical absorption. At the same time, it produces a low-energy tail of the lowest energy exciton which is then observed as the "shape of the fundamental absorption edge." The approximate form of this shape is given at least at high temperatures by Urbach's rule

$$\alpha(E) = C \exp\left[-\gamma \frac{(E - E_0)}{KT}\right]$$

where $\gamma$ $C$ and $E_0$ are constants. $\gamma$ decreases with increasing electron–phonon coupling, running from about 0.8 for alkali halides to values of 2-4 for typical semiconductors.

Many speculations have been made on the origin of Urbach's rule in strong coupling systems such as alkali halides. While theories have been constructed which fit experiments, few predictions have been made, and particular models for particular situations seem to be called for. The weaker electron–phonon coupling characteristics of semiconductors permit a more detailed experimental investigation of the experimental shape of the funda-

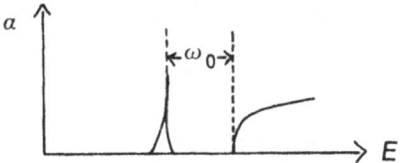

mental absorption edge, while permitting first-principle calculations to be made. We shall limit out comments to the following simple case.

At zero temperature, a single exciton band should, in conjunction with coupling to phonons of a single frequency $\omega_0$, produce an absorption spectrum of the form consisting of a zero phonon line, followed by a continuum due to transitions whose final state is exciton $\mathbf{k}$ + phonon ($-\mathbf{k}$). This continuum begins at an energy $\hbar\omega_0$ above the zero phonon line. At a finite temperature, the zero phonon line gains width due to collisions with phonons. In addition, transitions in which a phonon can be absorbed, beginning at an energy $\hbar\omega_0$ below the zero phonon line, are now possible. Their rate is proportional to the phonon occupation number

$$\bar{n} = \frac{1}{1 - \exp\left[-\hbar\omega_0/kT\right]}$$

As higher numbers of phonons are taken into account, the absorption edge will consist of a series of thresholds at the energies $E_{\text{exc.}} - n\hbar\omega_0$. Such thresholds, associated with the strong electron–optical phonon coupling in semiconductors, produce observable thresholds in the fundamental edge of most II—IV compounds.

Segall has recently done a thorough job of calculating the one- and two-phonon parts of the optical absorption edge. The problem in the calculation is the perturbation summation over intermediate states, very crudely approximated by earlier calculations. His calculations show that quantitative agreement between experiment and theory can be obtained with no adjustable parameters in CdTe (a crystal carefully studied over the appropriate energy range by Marple) over a temperature range from 0°K up to about 100°K, at which point the experimental data are also well represented by Urbach's rule.

In this case, Urbach's rule appears to be understandable from first principles. The model is, however, inappropriate for large electron–phonon couplings. Probably the greatest mystery associated with Urbach's rule is the wide variety of exciton–phonon coupling Hamiltonians which apparently (i.e., experimentally) yield this simple rule in the high-temperature limit.

When exciton lines have finite widths, whether through phonon or through being intrinsically metastable exciton line shapes are not necessarily simple. In general, the exciton line will lie on a continuous background of other optical processes to other states. Since the Hamiltonian also couples the exciton to these other states, there are in general interference effects between the "exciton transition" and the background on which it lies. The general theory of this interference effect can be expressed in terms of the Breit–Wigner formula, and has been treated in detail by Fano for the case of atomic excitations lying in the continuum. Experimental examples in which this kind of interference occur in exciton states can clearly be seen in the asymmetry of the $n = 2$ and higher lines in the yellow exciton series of $Cu_2O$ and in some of the higher band excitons of alkali halides and rare gas crystals.

## II. ASPECTS OF POLARITONS

The strong interactions betwen light and matter are through the interaction of the electric field of the light wave with the polarization field of the matter. Since weak interactions can be later added as a perturbation, it is the strong interaction problem which should be solved first. The set of linearized equations which describes the optics of a hypothetical homogeneous medium having this electric interactions is

   a)  Maxwell's equations

   b)  $\mathbf{B} = \mathbf{H}$                                                    (1)

   c)  $\mathbf{P}(\mathbf{r}, t) = \iint d^3r' dt' \, \underset{\sim}{\mathbf{K}}(r - r', t - t') \mathbf{E}(r', t')$

where the tensor kernel $\underset{\sim}{\mathbf{K}}$ is a function of only the two variables. Static magnetic fields enter the problem through the form of $\underset{\sim}{\mathbf{K}}$. The Fourier transformed relation equivalent to $c$

$$\mathbf{P}(\mathbf{k}, \omega) = \underset{\sim}{\alpha}(\mathbf{k}, \omega) \mathbf{E}(\mathbf{k}, \omega) \tag{2}$$

is more useful, where $\underset{\sim}{\alpha}(\mathbf{k}, \omega)$ is the polarizability tensor. A form such as 1 or 2 is valid only for relatively long wavelengths compared to lattice dimensions and for frequencies low compared to X-ray Bragg reflection frequencies.

Given a particular $k$, the real and imaginary parts of any component of the polarizability tensor are connected by a Kramers–Kronig relation because the electric field causes the polarization. As a result, any component of $\underset{\sim}{\alpha}$ can be written in the form

$$\alpha_{ij}(k, \omega) = \sum_n \frac{\beta_n(k)}{\omega_n(k) - \omega} \tag{3}$$

where $\omega_n$ is a complex resonant frequency whose imaginary part is positive, and $\beta_n$ and $\omega_n$ are functions of $k$. $\beta_n$ and $\alpha_n$ also depend on the components $i$ and $j$. The sum of $n$ includes the possibility of integration. In the most interesting cases, levels are sharp, principal axes can be chosen, and $\alpha_{ii}$ approximated over a narrow frequency range by the "one level" approximation

$$\alpha_{ii}(k, \omega) \simeq \frac{\beta(k)}{\omega^2(k) - \omega^2} + \gamma(k) \tag{4}$$

The extreme long-wavelength limit is obtained by setting $k$ equal to zero in classical optics. The polarizability, the conventional $E$ and $H$ boundary conditions, and Maxwell's equations in this case describe completely the solution to the linearized problem.

If the imaginary part of $\omega_n$ is small compared to its real part, $\omega_n$ might be thought of as "a resonant frequency of the solid." As such, it is to be expected that it will result from a Schroedinger equation calculation of the energy levels. Some care must be taken in making this assertion, for the interaction of radiation with a solid is not as simple as the interaction with atoms.

In the case of atoms, the calculation of the atomic polarizability is straightforward. First, the Coulomb gauge is used to separate the longitudinal Coulomb interactions from the transverse electromagnetic field. In the Coulomb gauge, Coulomb interactions are not retarded. The Schroedinger equation is then solved for the atomic energy levels. Radiative corrections to the real part of the energy level positions are of the order of the cube of the fine-structure constant and can be safely neglected except as a cause of linewith. In the case of a solid a similar procedure must be used, and the energy levels of the system first calculated in the Coulomb gauge. Represent the polarization response to an applied external vector potential $A_0 e^{ik \cdot r} e^{i\omega t}$ in the long-wavelength limit by

$$\mathbf{P} = \frac{-i\omega}{C}\alpha'A_0 \quad \text{or} \quad \mathbf{P} = \underset{\sim}{\alpha}'\mathbf{E}_{ext} \tag{5}$$

The polarizability which must be defined for use in Maxwell's equations is the response to the total field. The total field contains an additional term from the electric field of the charge density $\nabla \cdot \mathbf{P}$,

$$\mathbf{E}_{total} = E_{ext} - 4\pi k(k \cdot \mathbf{P}) \tag{6}$$

and the polarizability tensor $\underset{\sim}{\alpha}$ for use in Maxwell's equations defined by the relation between polarization and total field is simply

$$\mathbf{P} = \underset{\sim}{\alpha}(\omega, k)\mathbf{E}_{\text{total}}$$

$$\underset{\sim}{\alpha} = \left[\underset{\sim}{\alpha}'(\omega, k)^{-1} - \frac{4\pi}{k^2}\begin{pmatrix} k_{xx} & k_{xy} & k_{xz} \\ k_{yx} & k_{yy} & k_{yz} \\ k_{zx} & k_{zy} & k_{zz} \end{pmatrix}\right]^{-1} \tag{7}$$

This peculiar looking form of $\underset{\sim}{\alpha}$ in terms of $\underset{\sim}{\alpha}'$ is a consequence of the fact that the computation of $\underset{\sim}{\alpha}'$ contained some (but not all) electric fields in the original crystal Hamiltonian. A special example in which the meaning of this expression is readily apparent is the long-wavelength limit in a cubic crystal. In such a case the solutions to Maxwell's equations can be chosen purely longitudinal or purely transverse; $\underset{\sim}{\alpha}$ is a scalar times the unit matrix. Two effective polarizabilities are found,

Transverse waves

$$\alpha = \alpha'_{\text{trans}} \tag{8a}$$

Longitudinal waves

$$\alpha = \frac{\alpha'_{\text{long}}}{1 - 4\pi\alpha'_{\text{long}}} \tag{8b}$$

The physical condition that, in the long-wavelength limit, these two polarizabilities (8a) and (8b) must be the same in a cubic crystal leads to the relation

$$\alpha_{\text{long}} = \frac{\alpha'_{\text{trans}}}{1 + 4\pi\alpha'_{\text{trans}}} \tag{9}$$

One immediate consequence of (9) is that the poles of the longitudinal response function, the longitudinal frequencies, are the zeroes of the transverse dielectric function. This relation generates the plasma frequency in the case of free electrons and the Lyddane–Sachs–Teller relation for optical phonons. The effect of the coupling between the solid and the radiation field is rather different from that between a gas of atoms and the radiation field. When the number of atoms per cubic wavelength becomes large, it becomes necessary to use extended excitation states, or excitation waves, to describe the states of the solid. The coupling to the radiation field has *important* effects for a *few* such states, rather than as in the case of tenuous media, a *weak* effect for *all* the atomic states.

Consider a simple dielectric model with a polarizability independent of wavevector. For electromagnetic waves in a cubic crystal or in a high symmetry direction in a uniaxial crystal (e.g., $k$ in the basal plane) the polarizability (for each polarization in the uniaxial case) is simply a function of frequency, and the index of refraction is

$$n^2 = 1 + 4\pi\alpha(\omega) \tag{10}$$

Since

$$n^2 = \frac{c^2 k^2}{\omega^2}$$

for propagating waves, the dispersion relation

$$\frac{c^2 k^2}{\omega^2} = 1 + 4\pi\alpha(\omega) \tag{11}$$

defines $k$ as a function of $\omega$ for the actual propagating waves in crystals. The interpretation of this dispersion relation is particularly trivial in the case of an $\alpha(\omega)$ representing an isolated lossless oscillator:

$$\alpha(\omega) = \frac{\beta}{1 - \omega^2/\omega_0^2} \tag{12}$$

Figure 1 shows the dispersion curve. For any given $\omega$, two states ($\neq k$) of the same frequency exist, corresponding to waves propagating in the two directions.

The two values of $\omega$ for each given **k** correspond to the two states "light"

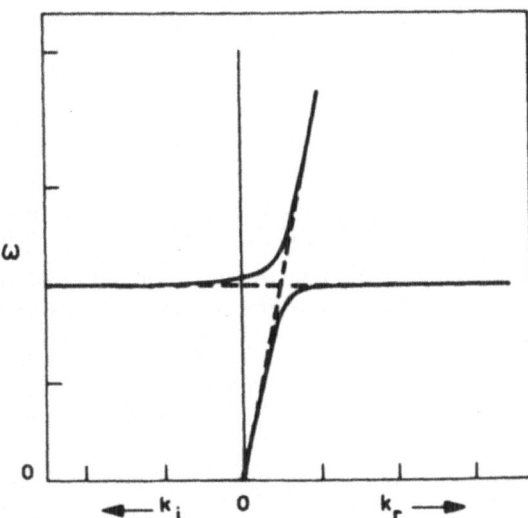

Fig. 1—The frequency as a function of wavelength for the transverse polaritons of a simple oscillator. Real wavevectors are plotted to the right and imaginary wavevectors to the left. The dotted lines show the dispersion relations in the uncoupled system.

and "polarization oscillator" of the uncoupled problem. The actual normal modes of the coupled problem have been called polaritons. Depending on $\omega$ and $k$, the polariton ranges from "lightlike" to "polarization oscillator like." These polariton dispersion curves are simply one means of describing classical optics. They describe the actual normal modes of the coupled crystal and radiation field.

If the polaritons were simply an alternative method of describing index of refraction phenomena, they would not be a particularly useful concept. The physical situations in which they are useful constructs are generally describable as weak-level crossings, When an external "force" of arbitrary nature having the form

$$F(\mathbf{r}, t) = F_0 e^{i\mathbf{k}\cdot\mathbf{r}} e^{i\omega t}$$

is *weakly* coupled to the electronic transitions in a crystal, the crystal will resonantly absorb energy when the relation between the applied $k$ and $\omega$ matches the dispersion relations of the crystal normal modes. If in this weak perturbation $k$ were held fixed and $\omega$ varied, an absorption line would be seen at the $\omega(k)$ corresponding to the crystal normal modes. In the crystal Hamiltonian all strong couplings *including the radiation field* should be present. The normal modes of the crystal referring to polarization oscillators must, therefore, be taken as polaritons. For large wavevectors the distinction between polaritons and the uncoupled polarization is unimportant at optical frequencies, but for small wavevector ($ck \approx \omega$) it is the polariton dispersion relation which will be measured.

The polariton dispersion relation has been observed in two kinds of experiments. The simpler experiment in concept was performed in ZnO and CdS by D.G. Thomas. The microscopic origin of the polarization oscillators in these experiments are the exciton states. In these hexagonal crystals, light propagating perpendicular to the $c$-axis experiences different polarizabilities in the two directions of polarization. For light polarized parallel to the $c$-axis, no dispersion effects were important, and this light can be taken to have the dispersion curve

$$\omega = \frac{ck}{n_{//}}$$

This light was used as the probing signal of wavevector $\mathbf{k}$ and frequency $\omega$. In ZnO and CdS the lowest energy excitations couple to light polarized perpendicular to the hexagonal axis. In the vicinity of such an exciton resonance, polaritons can be constructed which will have a dispersion curve like that in Fig. 2.

These polaritons are normally not coupled at all to light polarized parallel to the $c$-axis. When the absorption of light polarized parallel to the $c$-

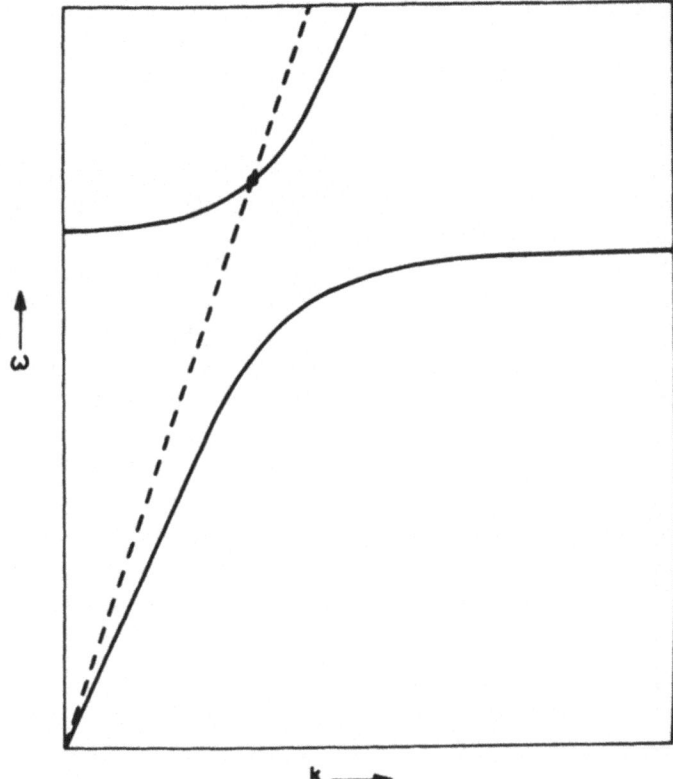

Fig. 2—The polariton dispersion relations for $k \perp c$ in a typical uniaxial crystal near an oscillator polarized perpendicular to the axis. The dashed curve is the dispersion relation for polaritons polarized parallel to the axis, and the solid curve for polaritons polarized perpendicular to the axis.

axis is measured, the polariton crossing ($\odot$ in Fig. 2) is normally not observed as an absorption line because of this vanishing coupling.

A weak magnetic field parallel to the direction of propagation provides a coupling between these polaritons and "light" (another polariton) polarized parallel to the $c$-axis, and results in absorption lines in this polarization at the intersection of the probe $\omega(k)$ curve with the polariton dispersion curves. The equality of $\omega$ and $k$ can be simply expressed as $n_{//} = n_{\perp}$, an isotropic point of the crystal.

This effect is explicitly contained in classical optics if the magneto-optic problem is solved in a uniaxial crystal in this geometry. Polaritons are, however, a useful point of view because in terms of polaritons one can qualitatively understand the origin of these absorption lines on a physical

basis. These weak absorption lines do not correspond to electronic state
energies calculable from the Schroedinger equation with Coulomb interactions alone. It should be noted that the physical manifestation of the level
crossing depends to some extent on the polariton damping and the magnitude of the magnetic field.

Henry and Hopfield used Raman scattering to measure a polariton
dispersion curve in GaP. The polarization oscillator in this case was the
optical phonon. In this cubic crystal which lacks inversion symmetry, the
optical phonons are also Raman-active. In a Raman experiment in which
one photon $\omega$, $\mathbf{k}$, is Raman-scattered to a final state $\omega'$, $\mathbf{k}'$, the normal mode
of the crystal excited is the normal mode at $\omega-\omega'$, $\mathbf{k}-\mathbf{k}'$. In terms of a weakly
coupled driving force, the Raman scattering simply provides a nonlinear
mechanism for driving the infrared polaritons at the photon difference frequency and difference wavevector. The measurement of Raman scattering
at fixed scattering angle defines a geometry in which $\Delta\mathbf{k}$ can be calculated

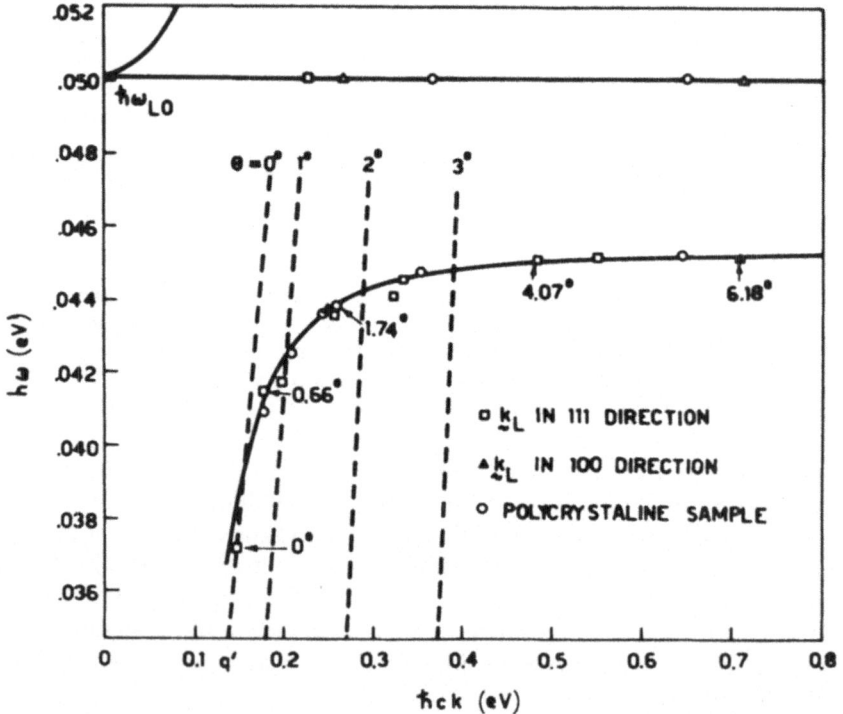

Fig. 3—$\Delta\omega$ $vs.$ $\Delta k$ (dashed lines) for several angles of incidence. The experimentally
measured $\Delta\omega$ and $\Delta k$ determined through such curves the polariton dispersion
relation (points shown). The solid lines are the calculated dispersion curve (after
Henry and Hopfield).

from $\Delta\omega$. In this geometry, when a Raman line is observed at a given $\Delta\omega$, $\Delta\mathbf{k}$ can be calculated and one point of the polariton dispersion curve determined. The experimental results are shown in Fig. 3. The calculated polariton curves (solid lines) are based on infrared reflectivity. Similar results have been obtained by Porto for ZnO. By utilizing the geometric flexibility possible in uniaxial crystals, Porto was able to extend the measured curve much lower in frequency.

At $\mathbf{k} = 0$ the polariton dispersion curve and corresponding longitudinal branch have the same frequency. This coincidence is a necessary consequence of the inability of a wave locally to be able to distinguish longitudinal from transverse in a time shorter than $\lambda/c$. Thus for $\omega\lambda/v \gg 1$, the polariton normal mode frequency must be independent of the direction of $\mathbf{k}$ for a given polarization.

## IV. EFFECTS OF SPATIAL DISPERSION

Consider an electromagnetic wave propagating in a crystal in a direction of high symmetry such that a particular transverse polarization of the electromagnetic wave can be guaranteed by symmetry considerations even for finite wavevectors. In this restricted geometry it suffices to define a scalar polarizability response $\alpha(k, \omega)$ to an applied transverse field $E(k, \omega)$. The solution to Maxwell's equations then yields the polariton dispersion relation

$$\frac{c^2 k^2}{\omega^2} = 1 + 4\pi\alpha(k, \omega) \equiv \epsilon(k, \omega) \tag{13}$$

If $\alpha$ is independent of $k$, normal optics results, and this equation has for any given $\omega$, two solutions corresponding to right and left running (possibly attenuated) waves. In a lossless medium there can be stop bands (when $\epsilon$ is negative) having purely imaginary wavevectors (surface waves). When $\alpha$ depends on $\mathbf{k}$, there will in general be several right or left running solutions to the dispersion relation for any given $\omega$, and possibly additional surface waves. The primary origin of these additional waves is the fact that if a resonant frequency of a crystal is wavevector-dependent, additional mechanisms then exist for the propagation of energy even in the absence of transverse electromagnetic fields. The coupling of electromagnetic fields to these propagating energy states preserves this multiplicity of modes which can propagate energy at a given frequency. Polariton dispersion relations are shown in Fig. 4 for the case of positive mass spatial dispersion (Fig. 4a) and negative mass spatial dispersion (Fig. 4b). These curves should be contrasted with Fig. 1. Because physical effects which might be observable depend on the sign of the mass and Fig 4a is the usual case for exciton resonances in direct

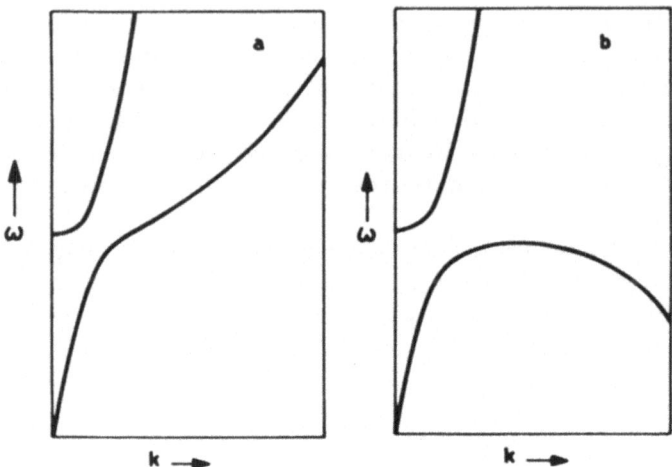

Fig. 4—The dispersion relation for real wavevectors for polaritons
in a positive mass (4a) and negative mass (4b) case. The "mass" is a
measure of the $k = 0$ curvature of the uncoupled polarization wave.

band gap semiconductors, only the positive mass case will be treated. (Spatial
dispersion effects associated with optical lattice vibrations have not been
observed).

While the solutions to the dispersion relation determine the propagating
modes within the crystal, the solution to the physical problem of the reflectivity or the transmission of light through a crystal is dependent on a boundary condition which will establish how much of a given mode is present
when light is externally incident on a crystal. From an experimental point
of view there are two major effects of spatial dispersion on the reflectivity
of semiconductors in the vicinity of exciton transition. First, under circumstances in which the exciton level width would result in a reflectivity
very near 1.0 if classical optics were valid, the observed reflectivities remain
well below this level. (Independent evidence of the sharpness of the excitons
is simultaneously observable in sharp structure in the reflectance). This
qualitative effect is an immediate result of the polariton dispersion relation
shown in Fig. 4a. Since for any $\omega$ there is now at least one propagating (real
wavevector) polariton, no "stop band" will exist, and virtually any boundary
condition will lead to a reflectivity which does not approach 1.0 even in the
limit of zero damping. The second experimental effect often observed is a
sharp spike in the reflectivity at the frequency $\omega_l$ (see Fig. 5) at which the
dispersion relation crosses $k^2 = 0$. A particular boundary condition was
constructed which produced such a spike. It is well known in the theory of
the scattering of electrons from atoms that at the energy of onset of a simple

*inelastic* scattering threshold the elastic scattering cross section necessarily shows a singular behavior and exhibits as a function of energy a vertical tangent at the threshold enengy $E_t$. The possible forms of the elastic scattering cross section behaves as $\sqrt{|E - E_t|}$ times a constant on each side of $E_t$. The possible forms of the elastic scattering cross section in the vicinity of the threshold energy are shown in Fig. 6. Reflectivity can be interpreted as an elastic scattering cross section of photons by a crystal. The energy $\omega_l$ is the threshold frequency above which an additional channel exists for propagation inside the crystal. From the point of view of reflected photons this is an inelastic channel. As a result, the possible forms of the reflectivity at $\omega_l$ are precisely those of Fig. 6. (The elastic scattering radial wave equation is also effectively one-dimensional).

It therefore appears that any general choice of a lossless boundary condition will, with spatial dispersion, lead to vertical tangents and the possibility of reflectivity spikes at $\omega_l$ as experimentally observed in several substances. No supposition that the surface is intrinsic to the semiconductor is necessary.

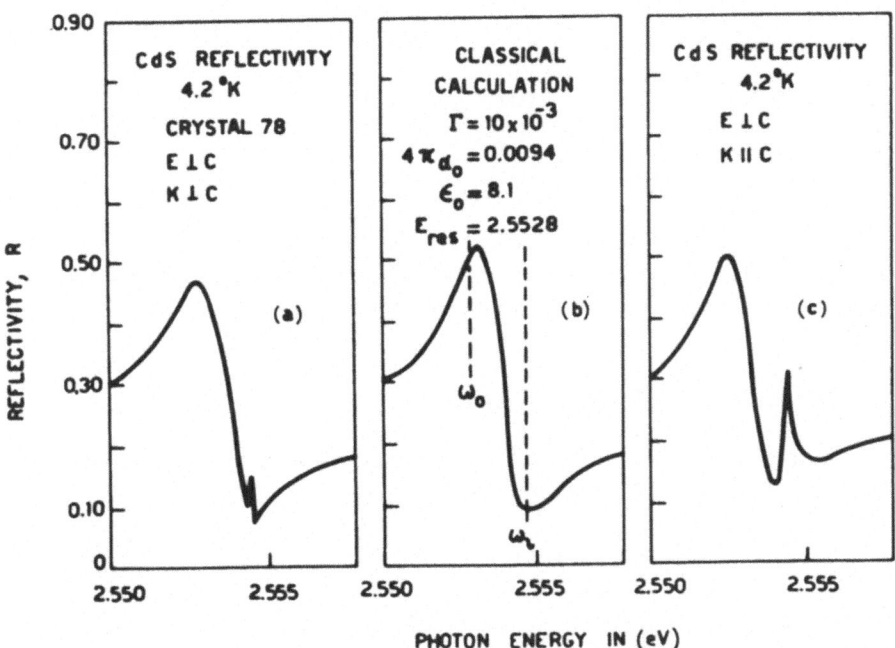

Fig. 5—The normal incidence reflectivity of CdS in the vicinity of the lowest energy exciton peak. Two different but classically equivalent geometries are shown. A classical fit it given for comparison. The additional sharp peak is at 2.5545 eV (after Hopfield and Thomas).

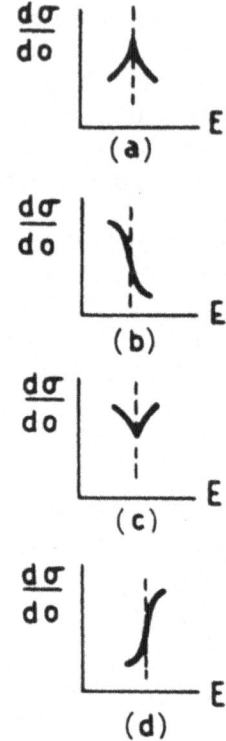

Fig. 6—The possible forms of the reflectivity as a function of frequency near the frequency $\omega_l$ (where $\epsilon(\omega, 0)$ vanishes) for negligible damping (after Landau and Lifshitz). These curves are also the form of the elastic scattering at the threshhold of an inelastic channel.

Only a few special boundary conditions lack such spikes. The boundary condition used by Hopfield and Thomas to produce the reflectivity spikes at $\omega_l$ is a special example demonstrating this general effect.

## V. FLUORESCENCE OF POLARITONS

Simple excitons in a direct band gap crystal without coupling to light can be characterized by a dispersion relation

$$E(k) = E_0 + \frac{h^2 k^2}{2m_{ex}}$$

When exitons are created by generating electron–hole pairs, they might be hoped to come to a kinetic energy equilibrium with the lattice temperature before recombination by whatever mechanism is operative. For low exciton concentration the kinetic energy distribution would be Maxwellian, like that of any particle with a parabolic $E(k)$.

The polariton dispersion relation (Fig. 4a) is not parabolic. The most important difference is that it does not have a lowest energy state near $E_0$, as the pure exciton does, but instead tails off toward zero energy as $k$ approaches zero. If at low temperature an exciton is generated and it truly comes to equilibrium among the polariton states before recombination, the probable polariton state before annihilation would lie far down this tail, and the probability of finding a polariton of energy around $E_0$ at annihilation would be neglibibly small. When at low temperatures ($KT \ll E_0$) fluorescence is in fact observed near $E_0$ for polaritons, the polariton kinetic energy distribution is necessarily nonthermal! Polariton fluorescence is observed through collisions of the polaritons with the crystal boundary. At any such collision, there is a probability that the polariton is reflected back into the crystal and a probability that the polariton disappears and a photon propagates away outside the crystal. These probabilities can be calculated from the boundary conditions of Section IV. In fact, for frequencies below the longitudinal frequency, the probability that a polariton is reflected (rather than transmitted) in a lossless system is the same as the externally measured optical reflectivity in the same geometry.

There are many physical factors entering the calculation of the polariton distribution observed in fluorescence. The generation of energetic electron–hole pairs in restricted spatial volume leads to an initially nonthermal polariton distribution. The distribution which is sampled in fluorescence is then determined by the history in space and in momentum distribution of this population. The energy loss processes which control the energy thermalization would need to be rapid compared to polariton loss mechanisms (collisions with the surface, or with absorbing impurities) to obtain a thermal distribution of energies. The spatial extent over which the polaritons are generated would need to be long compared to appropriate polariton mean free paths in order that the fluorescence be characteristic of a volume independent of energy.

It is known experimentally that when a crystal with a direct band gap has electron–hole pairs generated in it by some mechanism, intrinsic emission "lines" are often seen in the exciton energies $E_0$. Because the energy relaxation time for hot particles in semiconductors is generally faster than for cool particles, the kinetic energy distribution of polaritons at energies well above $E_0$ is perhaps characterized by the lattice temperature. A large uniformly excited crystal would at such high energies show a fluorescence spectrum (when viewed at normal incidence) of the form

$$[1 - R(E)]e^{-E/KT} \qquad E/KT \ll 1 \qquad (21)$$

where $E$ is the photon energy and $R(E)$ is the (externally) measured normal incidence reflectivity. This form can be obtained by detailed balance. The two effects responsible for cutting off this spectrum at lower energies are the failure of the polaritons to survive long enough to reach low kinetic energies and the polariton mean free path becoming comparable to the excitation depth or the sample size (i.e., the sample becomes transparent). At higher energies, above the longitudinal frequency, one of the polariton branches can again have mean-free-path considerations entering the observed fluorescence. Indeed, it is not certain in typical photoluminescence experiments whether there is any energy range over which the shape (21) independent of mechanism can be expected to hold.

The fluorescence of excitons is apparently a complicated matter, depending on the details of experiment and on the strengths of polariton–phonon and polariton–impurity interactions. In summary of the present state of understanding, in spite of the fact that polariton fluorescence has been frequently observed, the systematic studies necessary to its quantitative understanding have not been made.

## III. BOUND EXCITONS

Impurities or defects which bind electrons or holes are well known in insulating crystals. Among cases of simple donor centers having an excess nuclear charge of one are substitutional P in Si, Cl in CdS, and the F-center in alkali halides. Let us represent the excess positive charge of such a center by $\oplus$, the electron by $-$, and the neutral donor therefore by $\oplus -$. The $H^-$ ion is a stable ion. In the effective mass approximation the complex $\oplus =$, having two electrons bound to the donor, should also be stable. The $F'$ center is such a center.

This negatively charged center $\oplus =$, can necessarily bind a hole $+$ through long-range forces. The bound exciton state $\oplus = +$ will therefore exist as an excited state of the crystal containing neutral donors and is an analog of the $H_2$ molecule. The $H_2^+$ molecular analog $\oplus - +$ should also be stable under some circumstances.

Uncharged defects, such as C substitutional for Si in silicon, N for P in GaP, O for Te in ZeTe, and I for Cl in KCl can also act as electron or hole traps. If such a substitution traps an electron or hole, long-range Coulomb forces should then bind a hole or electron and a bound exciton state should be possible here also. Such a state might be represented by $\bigcirc - +$.

Bound exciton states are optically observed as emission and absorption lines lying below the fundamental absorption edge of a solid. When the chemistry of the crystal is under control, the origin of the lines can be identified directly, as was done with the bound exciton states associated with F-centers in alkali halides. (These states are seen in absorption as the $\alpha$ and $\beta$ bands.) The centers $\oplus = +$ and $\ominus + -$ were chemically identified in silicon.

Zeeman studies have also proved a fruitful means of classifying bound exciton states in systems in which the crystal perfection is not under perfect control. The energy band structure of a typical cubic crystal has a fourfold-degenerate valence band. A simple bound hole state will split into four sublevels in a magnetic field. A typical bound electron has two magnetic spin levels. This the excited and unexcited states of an exciton bound to a donor have the structure

$$\oplus = + \ \equiv$$
$$\oplus - \quad =$$

in a magnetic field. The two electrons on the bound exciton will be in the spin singlet state and do not contribute to the magnetic moment. Of the eight possible optical transitions, two are forbidden, and the other six allowed, resulting in a characteristic six-line pattern. The $g$-values of electron and hole can be obtained. Low-temperature observation of the thermalization in the initial state permit the determination of the $g$-value and multiplicity of the initial state. Of course a sharp zero phonon line is almost a necessity for such experiments. This method of study has been widely used in semiconductors.

Bound exciton optical transitions are often accompanied by sidebands due to phonon cooperation. The primary (but *not* the only) mechanism of phonon cooperation is usually due to the difference in equilibrium position of the atoms in the crystal when the bound exciton is present and when it is absent. In an optical transition, this difference is dissipated in a statistical emission (at finite temperature, or absorption) of phonons. When excitons are weakly bound (i.e., the size of this bound exciton is large compared to a lattice constant), the phonon coupling in the bound state is simplified, as the properties of the bound exciton state and its phonon couplings can usually then be calculated from the properties of electrons and holes near the band extrema and their phonon couplings.

Bound exciton transitions provide a useful method of studying the electronic and lattice vibrational properties of defects. The primary distinction between transitions being described here as bound excition transitions and optical transitions in something like a rare earth impurity is that (1) the electronic states are derived from the bands and the electronic properties of the bound exciton are thus related to other electronic properties of the crystal, and (2) the typical bound exciton is produced by a simple defect in the lattice whose binding resembles that of the host crystal. The perturbation of such a defect on the phonon spectrum is relatively simple and a useful probe of lattice dynamics.

For the purpose of demonstrating the nature and wealth of information obtainable from bound exciton, the rest of this lecture will describe various aspects of the system N substituted for P in GaP. This system has been studied by D. G. Thomas.

The ground state of this system has no particles bound to the nitrogen atoms. The nitrogen atom represents a short range but deep potential well probably attractive to electrons and deep enough to bind an exciton. (This is not knowm *a priori*, but is an experimentel observation.) The binding energy of the exciton to an N atom is about 0.010 eV, and the binding energy of the exciton itself about .010 eV also.

Already there is an interesting prediction available from these optical observations. Because the N potential well is deep and short-range, the scattering cross section of low-kinetic-energy conduction band electrons can be calculated from the binding energy of an electron to the N atom. (The problem is analogous to the calculation of the low-energy $S$-wave contribution to the neutron—proton scattering cross section from the binding energy of the deuteron.) The low binding energy of the electrons ($\sim 0.01$ eV.) leads to a nearly maximal cross section for scattering of electrons having energies of the order of 0.01 eV and greater. (The maximum $S$-wave cross section is $4\pi/k^2$, where $k$ is the wavevector of the electron.) This cross section should reduce the nitrogen temperature mobility of the electrons (but not of holes) in GaP by an order of magnitude at the maximum nitrogen concentration ($\sim 10^{18}$) Thomas has been able to achieve.

The electron and hole "spin–spin" interaction splits the excited state into an upper $J = 1$ component to which an optical dipole process is allowed and a lower $J = 2$ state to which dipole processes are forbidden. Emission spectra at low temperatures exhibit both lines, as the $J = 2$ state makes up for its weak oscillator strength by a large Boltzmann factor in its favor.

The emission spectrum at 4.2°K is shown in Fig. 7 (after Thomas and Hopfield). The zero phonon $J = 1$ and $J = 2$ lines are about equal in intensity. Several critical points in the phonon spectrum are clearly observable in the phonon wings. The $\mathbf{k} = 0$ longitudinal phonon is also a pronounced feature. This longitudinal optical phonon coupling is due to long-range electric fields, and is of symmetry $\Gamma_1$. The LO and TO phonons are therefore not of equal strength in their phonon cooperation. There appears to be a gap in the phonon spectrum between the optical and acoustic branches.

The substitution of a light N atom for the heavier P is expected to result in a local mode. The N–Ga and N–P bonds should not have radically different force constants. The single local mode does not cooperate strongly with the optical transition for symmetry reasons (it has symmetry $\Gamma_{15}$).

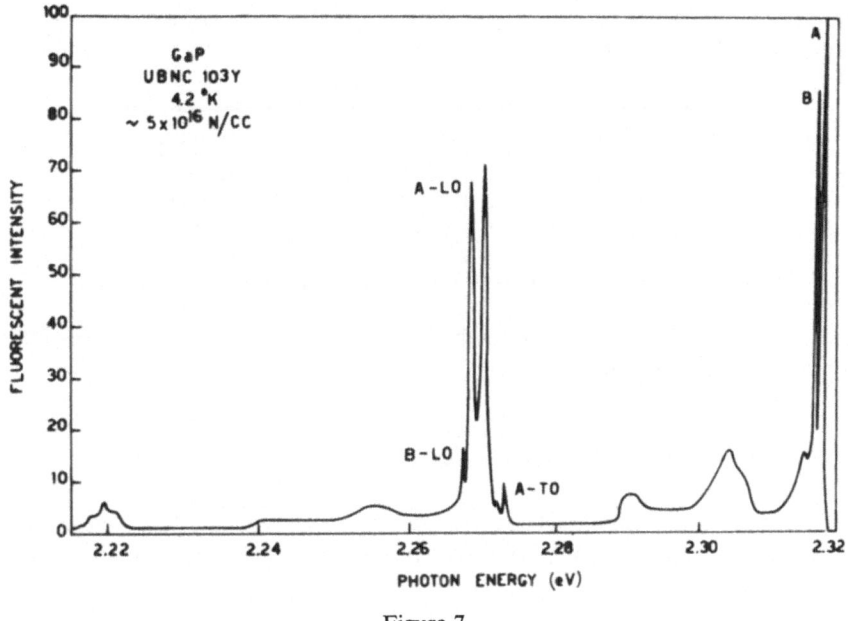

Figure 7

A transition in which the local mode is *doubly* excited can produce an excita-tion of symmetry $\Gamma_1$, and can result in a line in the emission spectrum (unfor-tunately, just off scale to the left in Fig. 7.) Bismuth substituted for N also binds as an exciton. Thomas has observed its spectrum, which appears to have a low frequency resonance in the continuum and perhaps gap local modes.

Optical emission and absorption lines produced by N atoms on nearest-neighbor P sites, on second neighbor sites, and out to about ninth neighbors can be seen in crystals having large N concentrations. Lines associated with pairs have absorptions strengths which go as the square of the N concentra-tion. That a particular set of optical transitions is associated with, say fourth nearest neighbors can be established in several ways. First, the $J = 1$ and $J = 2$ lines split into more lines in the lowered symmetry of NN pairs, and some estimate of the symmetry can be obtained from this increased fine structure. Second, the binding energy increases with decreasing pair separa-tion. Third, the Zeeman splittings of pairs depend in a characteristic manner on the orientation of the pair axis relative to the magnetic field. Finally, the number of $n$th neighbor sites is a rather erratic function of $n$, and the pair emission and absorption patterns are both proportional to the number of pair sites.

The local mode in which the two N atoms vibrate along the line joining them and 180° out of phase has the correct symmetry to couple to the optical transitions. The pair transitions show the following local made frequencies.

| Neighbor position | $N_{14}$ | $N_{13}$ |
| --- | --- | --- |
| First | 0.0610 eV | 0.0589 |
| Second | 0.0580 eV | 0.0562 |
| Third | 0.0613 eV | 0.0595 |
| Fourth | 0.0614 eV | 0.0596 |

Thomas has measured these lines both for $N_{14}$ and $N_{15}$. The isotope effect is given in perturbation theory by

$$\frac{\Delta\omega}{\omega} = \frac{1}{2}\frac{\Delta m}{m} \times \left(\begin{array}{l}\text{Fraction of kinetic energy of mode on}\\ \text{mass substition site}\end{array}\right)$$

The above measurements of the isotope effect shows this last factor to be about 0.9; i.e., the energy of the load mode is 90% localized in the motion of the N atom.

The separation dependence of the local mode frequency should be a stringent test of the understanding of the local mode in GaP.

The observations so far discussed represent emission spectra, for which in the final bound state there are phonons present but no electrons and holes. The absorption spectrum of N in GaP has been studied by Thomas and Dean, and is shown in Fig. 8. At high nitrogen concentrations such as this one, the

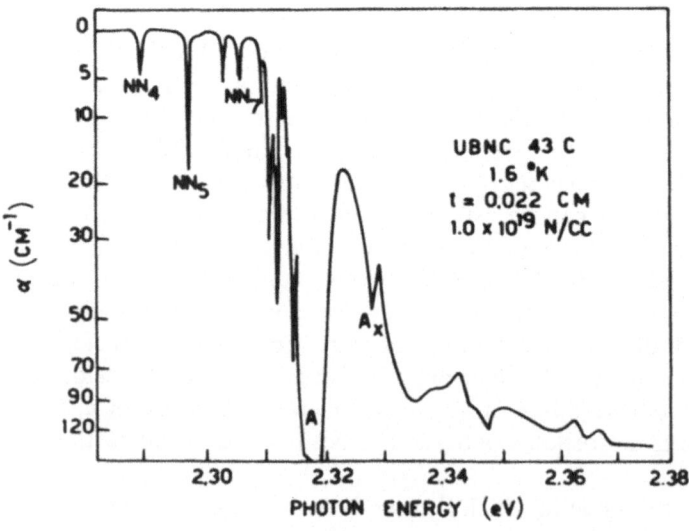

Figure 8

intrinsic optical absorption is essentially negligible near the absorption edge compared to the absorption introduced by the nitrogen.

The zero phonon line seen in emission is clearly present, and a few pair absorption lines can be seen. The strong continuous absorption above the exciton energy gap provides another method for studying the electronic effects for the N substitution. The difference in phonon cooperation between absorption and emission is extreme. In absorption, the optical phonon cooperation produces a *decrease* in the net absorption. This is a result of the overlapping between the two coupled absorption mechanisms, and is a classic example of the nonadditivity of absorption strengths. In the present case, the two "final states" which overlap are: (a) bound exciton and phonon, and (b) free exciton and no phonon.

The examples quoted so far are representative of the nature of information available from studies of bound excitons. Particularly in systems where zero phonon lines can be observed, bound exciton spectra provide a simple path to much detailed and quantitive information.

## REFERENCES

M. Born and K. Huang, *Dynamical Theory of Crystal Lattices*, Oxford University Press, London, 1954, p. 82 *ff.*

U. Fano, *Phys. Rev.* **103**: 1202 (1956).

J.J. Hopfield, *Phys. Rev.* **112**: 1555(1958).

R.S. Knox, *Theory of Excitons*, Academic Press, New York, 1963, p. 103 *ff.*

J.J. Hopfield and D.G. Thomas, *Phys. Rev. Letters* **15**: 22(1965).

C.H. Henry and J.J. Hopfield, *Phys. Rev. Letters* **15**: 964(1965).

R. Loudon, *Proc. Phys. Soc. (London)* **82**: 393(1963).

S.P.S. Porto, *Phys. Rev. Letters* **16**: 450(1966).

U.M. Agranovich and V.L. Ginzburg, *Usp. Fiz. Nauk* **76**: 643 (1962); translation—*Soviet Physics—Uspekhi* **5**: 323(1962) .

S.I. Pekar, *Fiz. Tverd. Tela* **4**: 1301 (1962); translation–*Soviet Phys.—Solid State* **4**: 953 (1962).

J.J. Hopfield and D.G. Thomas, *Phys. Rev.* **132**: 563(1963).

L.D. Landau and E.M. Lifshitz, *Quantum Mechanics—Non-Relativistic Theory*, Pergamon Press, Oxford, 1958, pp. 565–571 (2nd edition).

J.J. Hopfield and D.G. Thomas, *Phys. Rev.* **122**: 35 (1962).

M.A. Lampert, *Phys. Rev. Letters* **1**: 450 (1958).

J.R. Haynes, *Phys. Rev. Letters* **4**: 505 (1960).

Choyke, Hamilton, and Patrick, *Phys. Rev.* **133**: A1163 (1964).

J.J. Hopfield, *International Conference on the Physics of Semiconductors*, Paris, 1964, p. 725 (Dunod).

D.G. Thomas and J.J. Hopfield, *Phys. Rev.* **128**: 2135 (1960).

D.G. Thomas and J.J. Hopfield, *Phys. Rev.* **150**: 2 (1966).

Trumbore, Gershenzon, and Thomas, *Appl. Phys. Letters* **9**: 4 (1966).

Hopfield, Thomas, and Lynch, *Phys. Rev. Letters* **17**: 312 (1966)

Hopfield, Dean, and Thomas, "Interference between intermediate states in the optical properties of nitrogen doped GaP", *Phys. Rev.* **158**: 748 (1967).

J.J. Hopfield, "Elastic Scattering at Inelastic Thresholds,"*Lectures of the Tokyo Summer Institute for Theoretical Physics* (in press).

R.S. Knox, *"Theory of Excitons"*, Supplement 5.

F. Seitz and D. Turnbull, Ed. *Solid State Physics*, Academic Press, New York, 1963.

J.C. Phillips, "The Fundamental Optical Spectra of Solids," Seitz and Turnbull,*op. cit.*, **18** (1966).

B. Segal, *Phys. Rev.* **150**: 734 (1966).

D.T.F. Marple, *Phys. Rev.* **150**: 728 (1966).

G.D. Mahan, *Phys. Rev.* **145**: 602 (1966).

B.I. Halperin, *Phys. Rev.* **139**: A104 (1965).

## References added in proof

W.C. Tait, *Phys. Rev.* **166**: 769 (1968).

D.G. Thomas and J.J. Hopfield, *Phys. Rev.* **175**: 1021 (1968).

R.A. Faulkner and J.J. Hopfield in *Localized Excitations in Solids*, (R.F. Wallis, Ed.), Plenum Press, New York (1968).

J.C. Phillips, *Phys. Rev. Lett.* **22**: 285 (1969).

Y. Toyozawa and J. Hermanson, *Phys. Rev. Lett.* **21**: 1637 (1968).

W.C. Walker, D.M. Roessler, and Eugene Loh, *Phys. Rev. Lett.* **20**: 847 (1968).

R.S. Meltzer et al., *Phys. Rev. Lett.* **21**: 913 (1968).

S. Freeman and J.J. Hopfield, *Phys. Rev. Lett.* **21**: 910 (1968).

# Experimental Studies of Excitons in the Rare Gas Solids and the Alkali Halides

G. Baldini

*Istituto di Fisica*
*Università Degli Studi*
*Milan, Italy*

## INTRODUCTION

It is well known that the optical excitation spectra of the rare gas solids (RGS) and the alkali halides (AH) consist of several peaks which occur in the vacuum ultraviolet. The AH have been investigated over several years both experimentally and theoretically but we cannot say that satisfactory interpretations of the numerous structures which appear in their spectra have been given.

As for the RGS, we know perhaps less but the close analogy between the optical properties of the RGS and those of the AH will be of considerable help in the analyses we intend to make in these lectures. Here we will give a summary of the most significant experimental results which have been published in the last few years together with recent unpublished optical measurements. Evidence, then, will be brought forward that a few types of excitons are necessary in order to interpret the low-energy optical transitions whereas other controversial exciton models, recently indicated as responsible for some of the peaks, appear not to be required.

From the theoretical side, the band structure calculations seem to yield results which should allow a more complete interpretation of the data. There is also evidence that the character of the atomic states is somewhat reflected in the band structure of the AH and therefore it will be possible to understand better the optical properties of these solids and, because of the similarity, also the properties of the RGS.

## A. OPTICAL CONSTANTS

It is well known that a great deal of information about the electronic structure of solids is contained in their optical constants. This point can be illustrated schematically as follows. In the one electron approximation a wavefunction is constructed which is essentially a linear combination of atomic functions and which obeys to the Bloch theorem. This function is labeled by the crystal momentum $\mathbf{k}$ and by a band index $j$. Transitions between filled states $(j, \mathbf{k})$ and empty states $(l, \mathbf{k})$ intervene in the dielectric function which can be split, in general, into free electron, $\epsilon_f(\omega)$, and bound electron, $\epsilon_b(\omega)$, contributions according to [1]

$$\epsilon_f = 1 - \frac{4\pi n e^2}{m\omega^2(1 + 1/\omega\tau)} = 1 - \frac{\omega_p^2}{\omega^2(1 + 1/\omega\tau)} \tag{1}$$

and

$$\epsilon_b = -\frac{e^2}{\pi^2 m} \sum_{jl} f_{jl}(\mathbf{k}) \left(\omega_{jl} \pm \frac{i}{\tau} + \omega\right)^{-1} \left(\omega_{jl} + \frac{i}{\tau} - \omega\right)^{-1} \tag{2}$$

where $f_{jl}(\mathbf{k})$ is the well-known oscillator strength for the $j \rightarrow l$ transition at $\mathbf{k}$.

Since in these lectures we consider insulators, we can omit the $\epsilon_f$ contribution and since the most interesting part of the dielectric function is the imaginary term, with suitable manipulation and neglecting the interaction with the lattice vibrations we obtain

$$\epsilon_{b2} = \frac{e^2 h^2}{m} \sum_{jl} \frac{1}{\Omega} \int \frac{f_{jl}(k)dS_{\mathbf{k}}}{E_{jl}(\nabla_{\mathbf{k}}E_j - \nabla_{\mathbf{k}}E_l)} \tag{3}$$

and

$$\epsilon_{b2} \simeq \frac{2\pi h e^2}{m\omega} \sum_{jl} \bar{f}_{jl} \left(\frac{dN}{dE_{jl}}\right) \hbar\omega \tag{4}$$

We also have

$$\frac{dN}{dE_{jl}} = \frac{1}{\Omega} \int \frac{dS_{\mathbf{k}}}{\nabla_{\mathbf{k}}(E_j - E_l)} \tag{5}$$

and $S$ a surface of constant energy given by $\hbar\omega = E_l(\mathbf{k}) - E_j(\mathbf{k})$. It is evident that $\epsilon_2(\omega)$ will reflect the behavior of $\nabla_{\mathbf{k}}E_{jl}$ and singularities will occur at frequencies which are solutions of

$$\nabla_{\mathbf{k}}E_j(\mathbf{k}) - \nabla_{\mathbf{k}}E_l(\mathbf{k}) = 0 \tag{6}$$

i.e., where the joint density of states $dN/dE_{jl}$ has a singularity. This point has been discussed by several authors and a convenient reference is the article by Phillips [1].

We know that $\epsilon_1$ and $\epsilon_2$ are related to each other by the Kramers–Kronig relations and, therefore, from the knowledge of, say, $\epsilon_2$ over the entire spectrum (here by entire spectrum we mean the whole spectral region where the valence electrons are active) it is easy to compute $\epsilon_1$:

$$\epsilon_1(\omega_0) = 1 + \frac{1}{\pi} \int_{-\infty}^{+\infty} \frac{\epsilon_2(\omega)}{\omega - \omega_0} d\omega \qquad (7)$$

Experimentally, however, we cannot obtain directly the dielectric response but we rather get either the absorption constant or the reflection coefficient. These in general are given in terms of the refractive index which is related to the dielectric constant by the well-known expressions

$$\epsilon_1 = n^2 - k^2$$

$$\epsilon_2 = 2nk$$

If one measures the absorption, properly corrected for reflection losses, $k$ is obtained or the absorption constant $\alpha = 4\pi k/\lambda$. On the other hand, when the reflectivity is measured, $R = rr^*$, at normal incidence both $n$ and $k$ appear together:

$$r = \frac{n + ik - 1}{n + ik + 1} = \rho \exp{(i\vartheta)} \qquad (8)$$

with the phase $\vartheta$ given by the Kramers–Kronig expressions [2]: Then it is possible to determine $n(\omega)$ and $k(\omega)$ or $\epsilon_1(\omega)$ and $\epsilon_2(\omega)$.

There are however several ways of avoiding the Kramers–Kronig analysis of a wide spectrum and they essentially consist of a few, at least two, measurements at the same frequency. These could be two reflectivity measures at different angles, a reflectivity measure and an absorption measure, two reflectivity measures on thin samples of different thickness or transmission measures on two samples. Many of these techniques are described in the literature [3]. A method which we have employed with success on thin AH film is the following. The reflectivity is measured at the free surface of the film $R_1$ and also at the rear surface $R_2$ and at the interface between film and substrate of index $n_0$. From the two expressions, where $n_0$ is known,

$$R_1 = \frac{(n - 1)^2 + k^2}{(n + 1)^2 + k^2} \qquad (9)$$

$$R_2 = \frac{(n - n_0)^2 + k^2}{(n + n_0)^2 + k^2} \qquad (10)$$

one easily gets $n$ and $k$ or $\epsilon_1$ and $\epsilon_2$.

This method appears to be very good for $k$ not too small ($\geq 0.3$); when $k$ is small, one can measure also the transmission of the sample. Many studies have been made on the AH by employing films. The proper procedure for measuring $\alpha$ correctly is that which takes into account reflection losses, i.e., the measured transmission is given, in the absence of interference effects, by

$$T = \frac{(1 - R_1)(1 - R_2)}{1 - R_1 R_2 e^{-2\alpha t}} e^{-\alpha t} \tag{11}$$

where $R_1$ and $R_2$ have been given above. The common expression $T = e^{-\alpha t}$ ceases to be valid when $R_1$ or $R_2$, are large and this happens, usually, at frequencies where $\alpha$ also is large. Measurements of $T$ with two samples, provided that the thickness $t$ is known, will give the correct $\alpha$. If instead one measures the reflectivity from two films the proper expression is

$$R = R_1 + R_2 \frac{(1 - R_1)^2 e^{-2\alpha t}}{1 - R_1 R_2 e^{-2\alpha t}} \tag{12}$$

which approaches $R_1$ for $\alpha t \gg 1$. If the reflectivity is measured at two different angles, or at the same angle but with polarizations $s$ and $p$, the Fresnel formulas give $n$ and $k$.

Among the most widely used techniques we should recall also the ellipticity measurements which are again based on the use of the Fresnel formulas and polarized light. Variations of the above-sketched methods are found in the literature to which the reader is referred ([4]).

At the beginning of this section we stressed the point that the important optical constant is $\epsilon_2$. However, we should note that both $R$ and $\alpha$ have peaks or other peculiar features at about the same frequency where $\epsilon_2$ has them. The important thing to remember is that the peaks in $R$ or $\alpha$ might be slightly shifted when compared with $\epsilon_2$ and therefore the comparison with quantitative calculations might not be too satisfactory. On the other hand, we should emphasize the point that the theoretical calculations are not free from approximations and so the comparison with the direct measurements can be sufficient in most cases.

## B. EXPERIMENTAL TECHNIQUES FOR THE STUDY OF OPTICAL CONSTANTS

The exciton and interband transitions for AH and RGS begin at $\hbar\omega > 5$ eV and therefore proper monochromators or spectrophotometers must be used. Vacuum UV monochromators consist essentially of a concave grating, blazed for the proper wavelength, and two slits. Light sources for this region are the hydrogen discharge which can be employed down to 900 Å and the rare gas discharge which is useful down to 500 Å. These light sources are relatively weak and they emit both continuous and line spectra. It has been realized that in order to do high-resolution spectroscopy in the VUV, especially below 1000 Å, the synchrotron radiation has to be used. Transmission measurements have been made on thin films of AH ([5]) and RGS ([6]) deposited on suitable substrates, using the method illustrated in the above section. Transmission measurements are limited by the scattered light which is always present with the monochromatic beam. In the VUV it is fairly difficult to measure transmissions which are much less than 1%.

Reflectivity measurements on single crystals of AH ([7,8]) seem to be a better approach since their reflectivity very seldom goes below 1% and the samples are in the form of single crystals, whereas the thin films used for transmission experiments do not generally have very good crystalline structure. A recent study of AH at 55°K has shown that some peaks are better resolved in reflectivity from single crystals, ([9]) than in absorption from thin films. We have employed a double beam continuous recording technique for the region 5–14 eV with very satisfactory results. The monochromatic beam at the exit slit of the monochromator had been split off by means of a LiF plate at 45°. The transmitted light was incident on the sample under an angle of 22° and, after reflection, on a photomultiplier. The LiF reflected light was fed into a second photomultiplier. The outputs of the two dc amplifiers which followed the photomultipliers were fed into logarithmic converters and their difference taken thus obtaining directly on the recorder the log of the reflectivity. Corrections for the spectral dependence of the splitting ratio of LiF were easily made.

## C. ENERGY LEVELS OF RARE GAS ATOMS ([10]) AND ALKALI ATOMS ([11])

As it will be shown in the following section, tightly bound excitons characterize the low-energy spectra of both the AH and the RGS and therefore it is reasonable to expect that atomic like transitions play an important role in determining their optical constants. Because both the AH and the RGS have closed-shell configurations the similarities in their spectra are not unexpected. We can consider now the energy levels of a rare gas atom. The spectra of rare gas atoms are usually interpreted in terms of the $jl$ coupling scheme in which the total angular momentum of the outer $p$ shell is coupled to the orbital angular momentum of the excited electron and the resulting states then coupled to the spin of the electron ([10]).

We will focus our attention on the $4p^55s$ and $4p^54d$ excited configurations of Kr. Two optically allowed transitions can occur between the $p^6$ ground state and the lowest $4p^55s$ excited state because of spin–orbit interaction of the hole in the $p$ shell (spin–orbit doublet). For similar reasons three transitions are allowed from the $p^6$ shell to the $4p^54d$ excited state. Similar behavior is displayed by Ar and Xe atoms but the important differences should be pointed out. The spin–orbit doublet separation is large for Xe, 1.2 eV; small for Ar, 0.2 eV; and intermediate for Kr, 0.6 eV. The relative energy of the $s$ and $d$ excited states is dependent upon the electrostatic interaction of the excited electron with the core electrons. The minimum energy separation between $s$ and $d$ states is 0.2 eV in Xe and $\sim 2$ eV in Ar whereas for Kr we get 1.35 eV.

In the case of the AH crystals two kinds of atoms are involved and therefore both the atomic energy levels of positive and negative ions are expected to contribute to the optical properties. The valence electrons can be described essentially by the halogen $p^6 \rightarrow p^5$ transitions whereas the excited electron is expected to belong to a combination of both the halogen and alkali ion electron states. Splittings similar to those of the rare gases are to be expected for the lowest doublet of the AH. The energy of the higher excited states which involve also the $d$ states of the alkali atoms should depend on the cation in the lattice.

## D. COMPARISON OF CRYSTAL SPECTRA AND ATOMIC LEVELS

Since the ions of the AH are isoelectronic with the rare gas atoms we expect both the AH and the RGS to show similar behavior as far as their optical properties are concerned. In our comparison, the chlorides (Fig. 1) are matched with solid argon (Fig. 2), the bromides (Fig. 3) with solid krypton (Fig. 4), and the iodides (Fig. 5) with solid xenon (Fig. 6). We see in Fig. 1 that the absorption doublet of Ar $(A_1, B_1)$ at $\sim 12$ eV occurs close to the atomic resonant doublet $p^6 - p^54s$ with structure at higher energies ($\sim 14$ eV) in the region of the $3d$ excited states. A comparison with one of the chlorides (RbCl in Fig. 2) shows a doublet $A_1, B_1$ (at energies lower than those of argon since we deal with the excited electron of the Cl$^-$ ion, and a broad peak at $\sim 9$ eV which could be due to the $d$ excited states of Cl or Rb.

Similar results are obtained for KCl. In NaCl, on the other hand, the peak at higher energy is missing. The chlorides show some weak structure after the minimum in reflectivity. These states do not seem to correspond to allowed atomic transitions. This structure appears only in crystals and not in thin films.

The comparison between Kr and the bromides shows again the doublet $p^6 \rightarrow p^5 5s$ of Kr at 10.2 and 10.8 eV similar to the doublet of KBr at 6.8 and 7.3 eV. Again the structure at 12–13 eV in Kr and 8–9 eV in RbBr is attributed to the excited $d$ states. Approximately the same spectrum is obtained also for KBr.

For the two remaining spectra, those of Xe and RbI, the situation appears different at first. The analogy is again satisfactory if the spin–orbit doublet $A_1$, $B_1$ of Xe at 8.3 and 9.5 eV is compared with the doublet $A_1$, $B_1$ of RbI at 5.7 and 7.0 eV. The $d$ states above 10 eV in Xe seem to fall between the components of the spin–orbit doublet of the iodides. The weak lines at

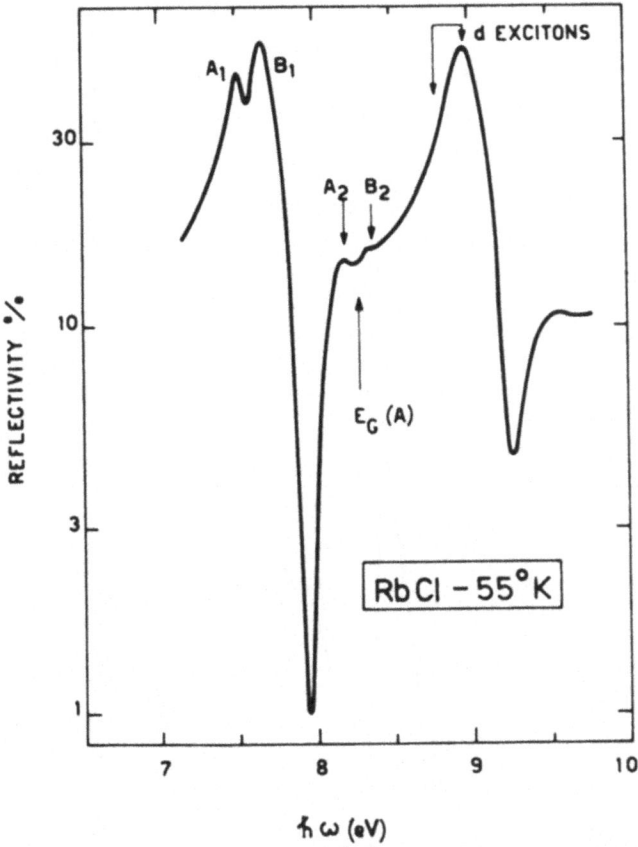

Fig. 1—Reflectivity spectrum of a RbCl single crystal. $A_1$, $A_2$ and $B_1$, $B_2$ are supposed to be Wannier exciton lines. $E_G(A)$ is the band gap associated with the $A$ series which is characterized by the excited electron in an $s$-like state. The $d$ excitons are assumed to arise from the energy gap at $X$, see Fig. 7.

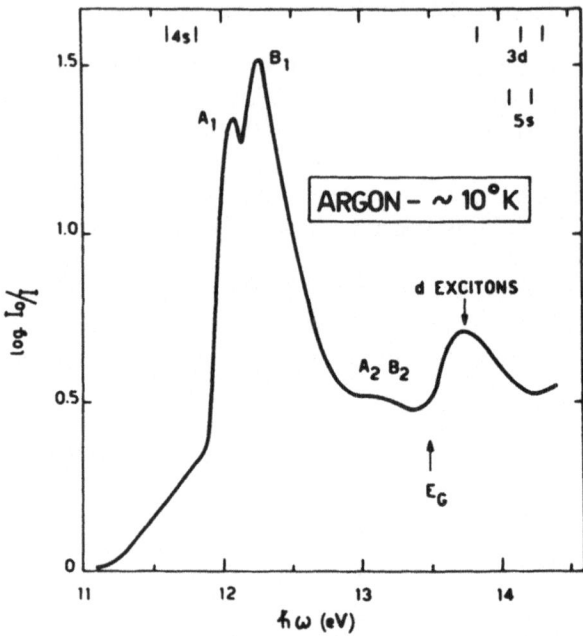

Fig. 2—Absorption spectrum of Ar. Atomic levels are indicated by the vertical segments. See also Fig. 1.

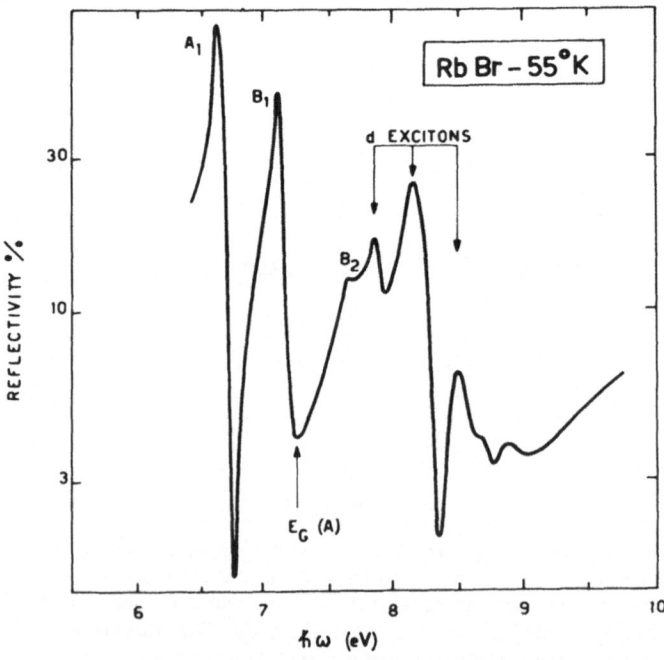

Fig. 3—Reflectivity spectrum of an RbBr single crystal. See also Fig. 1.

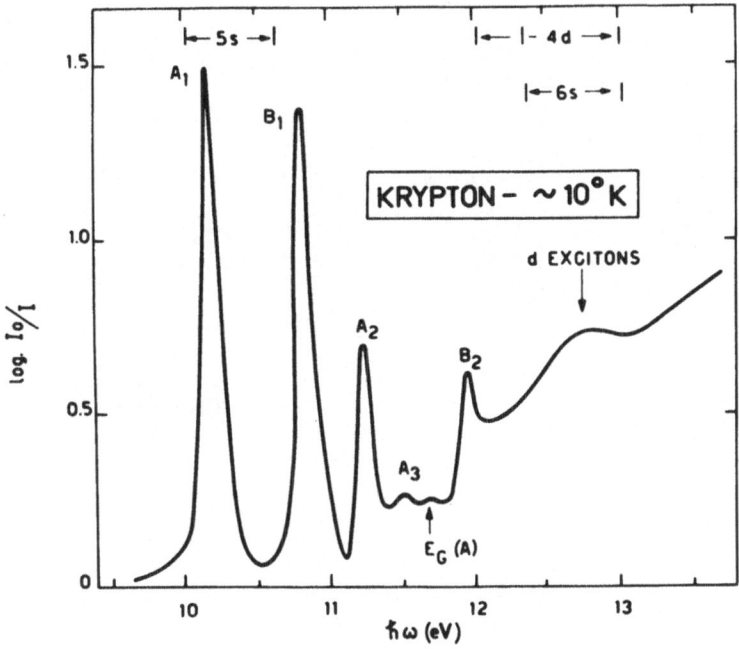

Fig. 4—Absorption spectrum of annealed Kr. See also Figs. 1 and 2.

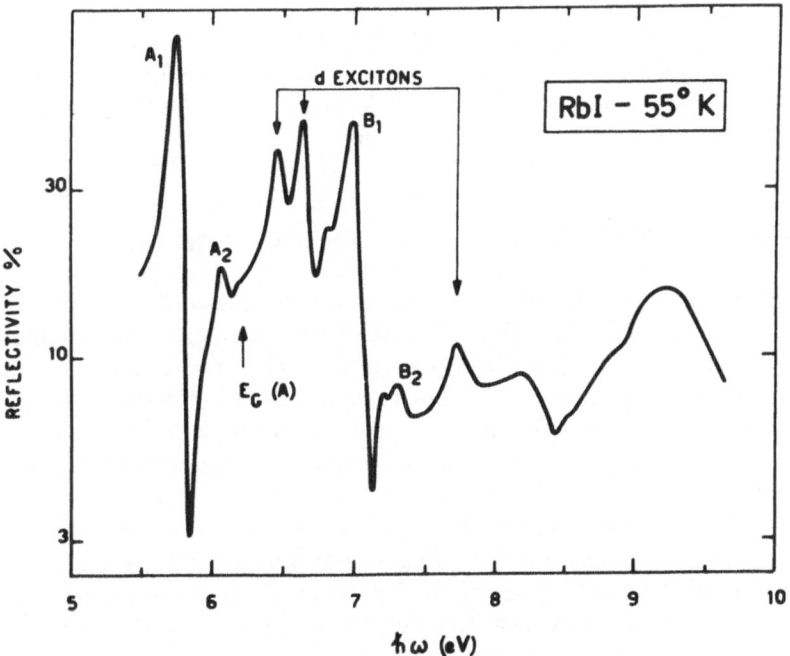

Fig. 5—Reflectivity spectrum of an RbI single crystal. See also Fig. 1.

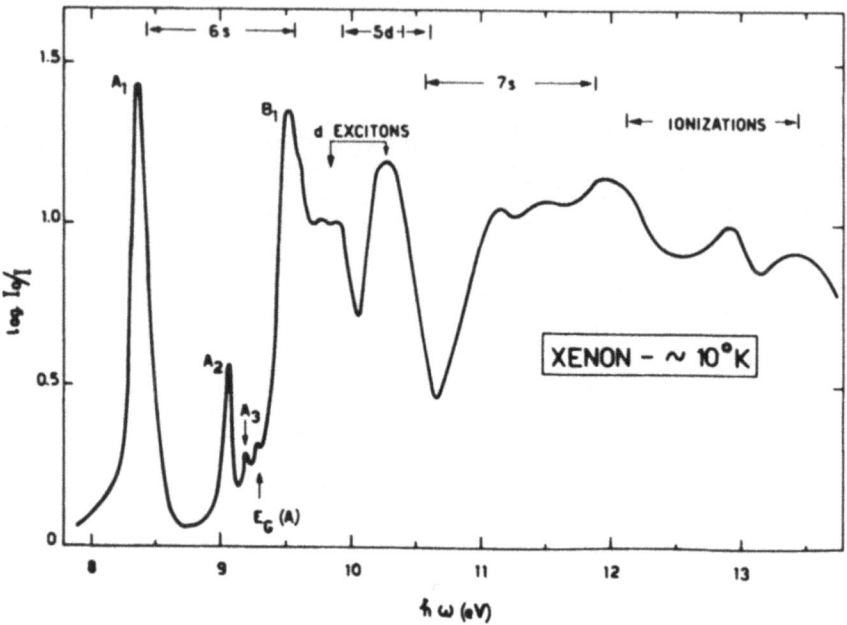

Fig. 6—Absorption spectrum of annealed Xe. See also Figs. 1 and 2.

9.1 eV in Xe and at 6.1 eV in RbI ($A_2$) are not attributable to atomlike transitions and we will see that they derive from the intrinsic nature of the crystals. Similar considerations apply to KI whereas in NaI some of the structures, at ∼6.5 eV in RbI, are missing.

## E. EXCITON MODELS FOR THE RGS AND THE AH

There is no doubt that most of the low-lying peaks characteristic of the optical properties of the AH and the RGS are due to excitons. Singularities of the type given by condition (6) do not in general give peaks but rather broad structure with a more or less steep rise or fall. Furthermore, photo-conductivity measurements indicate that optical stimulation in the first peak does not give an increment to the photocurrent. We must accept the idea that the peaks are due to excitons also because the rather small dielectric constant of both kinds of solids is certainly responsible for a large electron–hole inter-action which is equivalent to saying responsible for strong exciton lines. Several models have been proposed for the excitons in these solids and we refer the reader to the review by Knox ([12]) confining ourselves only to the enumeration of the various models.

The Frenkel model is a description of the exciton as a tightly bound electron–hole pair with both the electron and the hole located at the same lattice point at a given time. There is also momentum associated with the exciton so that it can move through the lattice. Here the exciton is described in terms of atomiclike wavefunctions and its energy is approximately equal to the atomic energy. Perhaps this model applies to the spin–orbit doublets of all the spectra shown here but certainly the best examples are the molecular crystals.

If the electron and the hole are allowed to take large separations—several lattice spacings—then the pair is weakly bound and the assumption of a potential between the pair of the type $r^{-1}$ is fairly reasonable. This is essentially the Wannier model which should be observed as a series of lines of decreasing intensity and conveging to a fixed energy if the line widths are smaller than the spacing between two peaks. Among the best known examples we have the visible spectrum of $Cu_2O$ where two series of Wannier excitons are observed. This model because of its simplicity is particularly useful since, when applicable, it gives a large amount of information about the electronic structure of the crystal. In fact, the absorption peaks occur at energies given by $\hbar\omega_n = E_G - G/n^2$ with $n = 1, 2 \ldots$ and both the band gap $E_G$ and the binding energy $G$ are readily calculated directly from the spectrum. Simple expressions are also given for the exciton radii and the binding energy in terms of the reduced mass of the exciton and the dielectric constant of the crystal. Care must be taken in choosing the proper dielectric constant and in order to clarify this point let us consider different cases[12]. When the electron is close to the hole the kinetic energy of the pair is high and we can assume that the rotational frequency of the pair is greater than $E_G/\hbar$, the resonant frequency of the valence electrons. In such a case the electron–hole pair will not polarize the lattice, i.e., $\epsilon$ is equal to unity since the valence electrons cannot follow the internal motion of the exciton. So we write

$$\omega_{\text{exc}} = \frac{\hbar}{\mu d^2} > \frac{E_G}{\hbar}.$$

For $E_G = 2 \text{ eV}$ and $\mu = \frac{1}{2}m_e$ (reduced mass of the exciton we find $d \simeq 2.5$ Å. It has to be noted at this point that for such a small value of $d$ the Wannier model might collapse because the potential is probably not Coulomb-like. When $d^2$ becomes larger than $\hbar^2/\mu E_G$ then the exciton will be able to polarize the valence electrons and the Coulomb interaction will be reduced by the factor $\epsilon_\infty^{-2} = n^{-4}$. When the exciton rotational frequency becomes so small that the atoms or ions in the lattice will follow its motion, we have

$$\omega = \frac{\hbar}{\mu d^2} \sim \omega_0$$

Where $\omega_0$ is the optical vibrational frequency. If

$$d^2 \geq \hbar/\mu\omega_0 = d_L^2 \sim 30 \text{ Å}$$

(for $\mu = \frac{1}{2}$ and $\omega_0 = 3 \times 10^{13}$ sec$^{-1}$) then the static dielectric constant $\epsilon_0$ should be employed.

Intermediate between the above models are the transfer model and the excitation model. In the transfer model applied to AH, the exciton energy is calculated from the following process: a charge is removed from a negative ion and added to one of the positive neighbors. The calculation by Hilsch and Pohl gives a reasonable value of the energy for the AH doublets:

$$E = E_A - E_I + \frac{\alpha_M e^2}{a}$$

in spite of the oversimplification of the model. The wavefunctions of the $s$ and $d$ unfilled alkali orbitals apparently play an important role since there are marked differences between Na and K halides at energies above those of the doublets. The energy difference between the lowest unoccupied $s$ and $d$ levels is 1 eV larger in Na than it is in K. At this point it can be noted that the six alkali neighbors of a halogen ion allow the formation of six linear combinations of their wavefunctions. One has $s$ symmetry, two transform into each other like $d$-functions and three have $p$-like symmetry. Transitions from the $p$ halogen valence states are allowed to the $s$- and $d$-like states, forbidden to the other states.

Finally there are so called excitation models ([12]) which consist essentially of setting up appropriate wavefunctions for the electron which is moving in the potential well generated by the hole. However, the existence of states higher than the lowest excited state has not been considered and a serious test of the model is not possible.

## F. INTERPRETATION OF THE LOW LYING PEAKS IN TERMS OF THE WANNIER EXCITON

We see in the spectra of the RGS that a few lines ($A_1$, $A_1$) correspond to atomic transitions but, on the other hand, other structures ($A_2$, $B_2$) do not seem to be closely related to the nature of the constituent atoms. The lines which occur close to the atomic transitions are expected to arise from a localized or Frenkel type exciton whereas for the others another model should be more appropriate. We may expect the same to occur in the AH. The Wannier model has been applied to both the AH and the RGS by Fischer and Hilsch ([13]) and by the author ([6]). The considerations of the preceding section should be kept in mind especially for the lowest line the $A_1$ exciton since for it the dielectric constant could be as small as unity. The Wannier formula, written here as

## TABLE I

| | Reference | $\epsilon_\infty$ | $\mu$(me) | $G$(eV) | $E_G$(eV) | $r_1$(Å) | $a$(Å) |
|---|---|---|---|---|---|---|---|
| NaI | (13) | 3.16 | 0.23 | 0.31 | 5.96 | 7.28 | 2.2 |
| KI | (13) | 2.82 | 0.26 | 0.45 | 6.32 | 5.75 | 2.2 |
| RbI | (13) | 2.72 | 0.28 | 0.50 | 6.24 | 5.15 | 2.2 |
| Kbr | (9) | 2.44 | 0.24 | 0.55 | 7.4 | 5.3 | 2.0 |
| NaCl | (13) | 2.39 | 0.28 | 0.67 | 8.72 | 4.52 | 1.8 |
| KCl | (13) | 2.23 | 0.27 | 0.73 | 8.68 | 4.38 | 1.8 |
| Xe | (6) | 2.20 | 0.31 | 0.86 | 9.3 | 3.8 | 2.2 |
| Kr | (6) | 1.80 | 0.4 | 1.7 | 11.7 | 2.4 | 2.0 |
| Ar | (6) | 1.67 | 0.3 | 1.3 | 13.5 | 2.0 | 1.8 |

$$\hbar(\omega_m - \omega_n) = G(m^{-2} - n^{-2})$$

has been applied to the $m = 1$ and $n = 2$ lines of the AH and Ar spectra, without an internal check on another line such as $A_3$, therefore the values that we give in Table I, especially $\mu$, might be affected by a nonnegligible error. The data for Kr and Xe in Table I have been obtained by employing $A_2$ and $A_3$ and probably they are more reliable since the Wannier model rests on the assumption of large orbits.

According to Elliott ([14]) a check on the validity of the Wannier model is based on the oscillator strength of the lines in the series. Since it is proportional to $n^{-3}$, a ratio of 8 : 1 between the strength of $A_1$ and $A_2$ should be found. This value is found only for the ratio $A_1/A_2$ of solid Xe but already in solid Kr we obtain 4 : 1 and, apparently, smaller ratios seem to correspond to the AH. A model intermediate between those of Frenkel and Wannier is probably the best description of the $A_1$ peak.

Unfortunately, the strong electron–phonon interaction present in the AH does not allow the separation of peaks with $n = 3, 4, \ldots$ from the continuous absorption or reflectance. Higher members of the series would give more solid information on the band structure. From photoconductivity measurements ([15]), however, it is found that at $\hbar\omega \simeq E_G$ the photocurrent increases rapidly and therefore the Wannier model accounts satisfactorily for the lowest lying transitions of both the AH and the RGS except, perhaps, the first peak. We note, in passing, that in general the photocunductivity measurements on the AH are not usually corrected for reflectivity $R$, and so the quantum yield should be divided by $(1 - R)$ in order to get correct values.

## G. BAND STRUCTURE AND "$d$" EXCITONS

Considerable attention has been given in recent years to the band structure of the AH ([16]) and the RGS ([17]). The typical band structure, notably that for KI Onodera et al. ([16]), is shown in Fig. 7. There are good reasons to

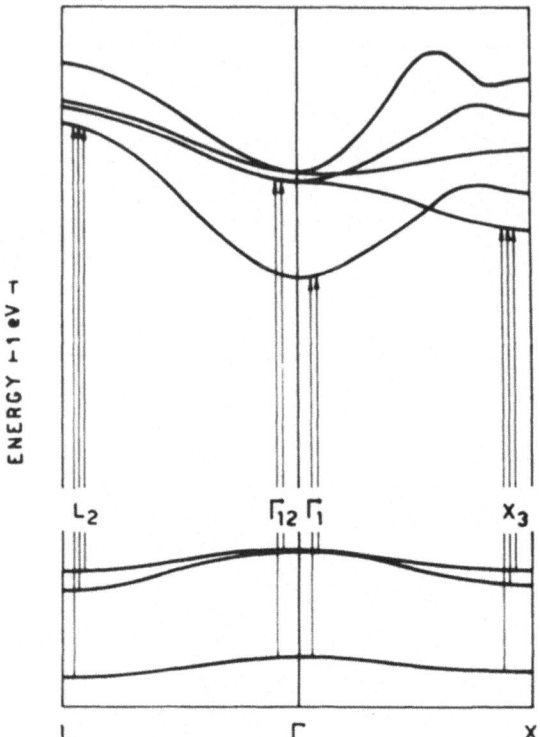

Fig. 7—Schematic band structure of an alkali
halide crystal. See Ref. 16.

expect similar behavior for the other AH and also for the RGS. According
to this picture and comparing solid Xe and the K and Rb iodides (Figs. 5
and 6), we can easily attribute the $A_1$ and $A_2$ peaks to excitons created from
the upper $\Gamma_{15}$ holes and the $\Gamma_1$ electrons. The spin–orbit splitting of the valence band accounts for $B_1$, ~1–2 eV shifted from $A_1$. It is hard to locate
$B_2$ but we can probably assume that the line is shifted from $A_2$ as much as
is $B_1$ from $A_1$, i.e., by the spin–orbit splitting of the valence band at $\Gamma_{15}$. The
analysis of the spectra of Xe and the iodides in terms of the band structure of
Fig. 7 brings into the picture the excitons at X. These states have been shown
to be related to the $d$ states of the alkali metal by Onodera et al. ($^{16}$) The
minimum in the conduction band at X guarantees the existence of excitons.
Then, since this minimum is fairly low it appears that these excitons, labeled
$d$ excitons, maintain a strong oscillator strength because they fall at energies
where the density of states of the conduction band is low, i.e., autoionization
effects become apparently negligible.

The higher energy peaks can probably be ascribed both to exciton states
and interband edges of energy higher than that of the $d$ excitons. No definite

attribution can evidently be given at present. It seems, however, that the introduction of saddle-point excitons ([18]) is not necessary for the interpretation of the optical properties at low energies. Experiments on the liquid phase of Xe ([19]) seem, on the other hand, to support the fact that the structure at 9.6–10.6 eV is due to compact excitons of $d$ character weakly influenced by the change solid-to-liquid phase. On the contrary, the existence of loose Wannier states, $n \geq 2$, has been shown to be strongly dependent upon the crystalline order ([6]). This analysis of the iodides and Xe suggests also that Xenon has a band structure similar to that of KI. Against this interpretation might appear the spectrum of NaI ([5]) which is missing the $d$ excitons. We can easily find the reason for the mismatch by noting that Na has a very high $d$ level and therefore the lowest conduction band at X (Fig. 7) will be pushed up in case of NaI. The contrary of course will happen for CsI ([5]) essentially for the fact that the $d$ level in Cs is very low. The different symmetry will account for extra shift of the $d$ excitons. It is difficult at present to locate interband transitions at energies higher that those of the $\Gamma$ and X edges. It is likely that $L$ interband edges will occur at $\hbar\omega > 9$ eV for RbI and $\hbar\omega > 11$ eV for Xe. In the AH, from photoelectric emission measurements ([20]), it is possible to locate the vacuum level which appears to be $\sim 2$ eV higher than $\Gamma_1$ and in agreement with photoconductivity data. In Xe, however, the photoelectric threshold is at 9.7 eV ([21]), only $\sim 0.4$ eV higher than $\Gamma_1$.

The above discussion can be immediately applied to the other AH and RGS. If RbBr (Fig. 3) and Kr (Fig. 4) are compared, the excitons at $\Gamma$ are easily recognized as the peaks $A_1, A_2, A_3 \ldots, B_1, B_2 \ldots$. The theoretical calculations for Kr ([17]) indicate that the $\Gamma$ exciton parameters ([6]) are consistent with the band structure. The excited $d$ states of atomic Kr occur at $\hbar\omega > 12$ eV, and it is likely that the $d$ excitons will occur above the $B_2$ line. In RbBr (Fig. 3) whose spectrum is similar to that of KBr, the $d$ excitons have been assigned at $\sim 8$ eV. The assumption is confirmed by the disappearance of the peaks in NaBr ([5]), in agreement with the above argument, and by the low-lying doublet at 7.1 eV in CsBr ([5]) located between the main spin–orbit doublet.

We are left now with the chlorides and argon, and we can make the following remarks. The Wannier states $A_1$ and $A_2$ with their split-off partners $B_1$ and $B_2$ are easily recognized in the spectra of RbCl (Fig. 1) and of Ar (Fig. 2) for the $A_1$ and $B_1$ lines only. Again $d$ excitons which we have attributed to the X point in the Brillouin zone are apparent at 9 eV in RbCl and 13.7 eV in Ar. The behavior of other halides agrees with our suggestions and we note that KCl has essentially the same spectrum as RbCl. Again the $A_2$ and $B_2$ lines are shown in reflectivity data. NaCl does not possess a peak equivalent to that of RbCl at 9 eV whereas CsCl ([5]) both in the fcc and sc structure shows the $d$ peaks very low in energy.

The comparison of the band structure calculations available, those of

KCl and KI, together with the analysis of the spectra of the AH, and perhaps of the RGS, suggests some of the features of the band structure of the other alkali halides. The Wannier series should, in principle, give the curvature of the conduction band at $\Gamma_1$, and if an analogous series were found for the $d$ excitons the same could be done at $X$. However, we must remember the considerations made in Section E on the dielectric constant. The calculation of the exciton mass, essentially equal to the electron mass, since the hole mass is large ($m_h \sim 3$ to 4 times the free electron mass) gives us the proper band parameters only if the same dielectric constant is employed for all the lines. From the results presented here it seems that the $A_1$ line corresponds to a rather compact state, with dielectric constant not known, and therefore the conduction band curvature, when computed with this $A_1$ line may be affected by some error. A more favorable situation exists in the case of mixed AH and RGS ([22]) where several lines of the Wannier series are observed.

The $d$ excitons have been attributed here to the $X$ point in the Brillouin zone, but we must remember that other $d$-like states exist at $\Gamma$ ($\Gamma_{12}$ and $\Gamma_{25'}$) These states seem in general to be higher than the $X$ states but might be responsible together with the $X$ states, of the more complex structure of the $d$ excitons in Cs halides.

## CONCLUSIONS

From the analysis of the optical data on AH and RGS we have found that:

1. Tightly bound excitons exist and they correspond to the lowest excited $s$ and $d$ levels of the electron. It is very likely that one can make theoretical predictions of the exciton energies for these states using the Frenkel model,

2. There is no doubt that Wannier excitons are displayed by these solids and this has been confirmed also by two photon absorptions in KI and CsI. ([23])

3. Saddle-point excitons do not seem to play a role, as previously assumed, ([18]) at least at low energies, since they should depend upon the crystalline order.

4. Interband transitions occur from energies of the order of $\sim 1$ eV higher than that of the first exciton but a detailed comparison with the calculated band structure is not yet available.

5. The $d$-like excitons occur above the first gap, at $X$, but autoionization effects do not seem to reduce their lifetimes. It is apparent that these resonant states along with the $A_1$ and $B_1$ excitons are fairly localized and the interaction with the continuum is negligible.

6. The exciton-like absorption spectrum possesses most of the oscillator strength in a region of a few eV.

This last point can be justified by noting that the electron–hole interaction is not screened by a large dielectric constant as it occurs in semiconductors where interband effects are predominant. For both the AH and the RGS the small $\epsilon$ favors exciton effects over interband transitions. Also one sees that, excluding the confluence of two saddle points as a cause of all the lines, exciton effects are more evident in halides of light alkalis since these have larger $\epsilon$ and smaller lattice constants. Besides, since the $n = 1$ exciton is more tightly bound than the Wannier exciton its radius is smaller and the oscillator strength larger. If the optical data are extended to higher energies, as was done by Phillipp and Ehrenreich ([8]), it is seen by Kramers–Kronig analysis of the data measured at room temperature, that in the AH plasma resonances can occur at energies of the order of 11 to 14 eV, from KI to KCl. At this frequency the oscillator strength corresponding to the valence band has been exhausted and it can be assumed that $\epsilon(\omega) = 1 - \omega_p^2/\omega^2$ for this region. About six electrons for each molecule contribute to $\omega_p$. There will be strong damping of plasma oscillation since interband transition occurs at nearby energies. Evidence for the excitation of exciton states in AH with fast electrons, replacing photons, has been recently given by Creutzburg ([24]). The energy-loss spectra are very similar to the optical spectra in the exciton region. We must remember, however, that photon excitation produces transverse excitons whereas fast electrons generate longitudinal excitons.

ACKNOWLEDGMENT

I wish to thank B. Bosacchi for helpful discussions during the preparation of these notes.

## REFERENCES

1. See, for example, J.C. Phillips, *Solid State Physics*, Vol. 18, Academic Press, 1966, and references quoted therein.
2. F.C. Jahoda, *Phys. Rev.* **107**: 1261 (1957).
3. O.S. Heavens, *Physics of Thin Films*, vol. 2, ed. by G. Hass and R.E Thun, Academic Press, New York, 1964.
4. M.P. Givens, *Solid State Physics Vol. 6*, Academic Press, New York, 1958.
5. J. Eby, K. Teegarden, and D. Dutton, *Phys. Rev.* **116**: (1959) 1099.
   K. Teegarden and G. Baldini, *Phys. Rev.* **155**: 896 (1967).
6. G. Baldini, *Phys. Rev.* **128**: 1562 (1962).
7. P. Hartman, J. Nelson, and J. Siegfried, *Phys. Rev.* **105**: 2016 (1963).

8. H. Phillipp and H. Ehrenreich, *Phys. Rev.* **131**: 2016 (1963).
9. G. Baldini and B. Bosacchi, *Phys. Rev.* **166**: 863 (1968).
10. G.A. Cook, editor, *Argon, Helium, and the Rare Gases*, Interscience, New York 1961.
11. National Bureau of Standards, U.S. Circular No. 467, (1952).
12. R.S. Knox, *Solid State Physics*, Suppl. 5, Academic Press, New York, 1963.
13. F. Fischer and R. Hilsch, *Nachrichten Akademischer Wissenschaften, Goettingen* **8**: 241 (1959).
14. R.J. Elliott, *Phys. Rev.* **108**: 1384 (1957).
15. G.R. Huggett and K. Teegarden, *Phys. Rev.* **141**: 797 (1966).
16. Y. Onodera, M. Okazaki and T. Inui, *J. Phys. Soc. Japan* **21**: 2229 (1966); S. Oyama and T. Miyakawa, *J. Phys. Soc. Jap.* **21**: 868 (1966).
17. R. Knox and F. Bassani, *Phys. Rev.* **124**: 652 (1961); Mattheiss, *Phys. Rev.* **133**: 1399 (1964).
    W.B. Fowler, *Phys. Rev.* **132**: 1591 (1963).
18. J.C. Phillips, *Phys. Rev.* **136**: 1714 (1964).
19. D. Beaglehole, *Phys. Rev. Letters* **15**: 551 (1965).
20. E.A. Taft and H.R. Philipp, *J. Phys. Chem. Sol.* **3**: 1 (1957).
21. J.O 'Brien and K. Teegarden, *Phys. Rev. Letters* **17**: 919 (1966).
22. G. Baldini, *Phys. Rev.* **137**: A508 (1965); G. Baldini and K. Teegarden, *J. Phys. Chem. Solids*, **27**: 943 (1966).
23. J.J. Hopfield and J.M. Worlock, *Phys. Rev.* **137**: A 1455 (1965).
24. M. Creutzburg, *Zeits. f. Physik* **196**: 433 (1966).

# Localized Modes, Resonance Modes, and Correlation Functions*

## A. A. Maradudin

*Department of Physics*
*University of California*
*Irvine, California*

My mandate this morning is to discuss the theoretical aspects of localized and resonance vibration modes associated with impurity atoms in crystals. However, rather than giving a purely theoretical talk, that is, a purely formal talk, I prefer instead to introduce localized and resonance modes, and to discuss some of their properties, in the context of experimental methods for studying such exceptional vibration modes.

The types of experiments I should like to consider are infrared lattice vibration absorption measurements, Raman and neutron scattering experiments, the resonant absorption or $\gamma$-rays by nuclei bound in a crystal, spin-lattice relaxation time measurements, and specific heat measurements. I could also include in this list the optical analogs of the Mössbauer effect, *viz.*, the absorption of light by impurity centers which then undergo an electronic transition, which may or may not be phonon-assisted. However, such experiments will be described in detail at this conference by Professor Nardelli and I will accordingly not discuss them.

Although a first glance it may appear that there is very little similarity between these several kinds of experiments, in fact, from the theoretical point of view they have the common feature that the function of interest which enters into the interpretation of the experimental results is in each case conveniently expressed in the form of a Fourier transform with respect to time of a displacement–displacement or displacement–momentum correlation function. This does not mean that the kind of information about the

*This talk was presented at the Conference on Localized Excitations, Milan, Italy, July 25–26, 1966. The research was supported by the Air Force Office of Scientific Reserach, Office of Aerospace Research, United States Air Force, under AFOSR Grant Number 1080–66.

dynamical properties of perturbed crystals that one obtains in each kind of experiment is the same. Nevertheless, as we will see, there is a close similarity among the kinds of information one obtains from several of the experiments I will describe.

That the Fourier transform of a correlation function should arise in the theoretical interpretation of the experiments I have described above should not be too surprising. In almost every one of them what is being measured is a transition probability, which takes the form of an absorption coefficient or a scattering cross section for the interaction of whatever external probe we are using with the atomic vibrations of the perturbed crystal.

It is a well-known result of time-dependent perturbation theory that if a system is subjected to a time-dependent perturbation with a sinusoidal time dependence $O(r_1, \cdots, r_n) \sin \omega t$, where the $\{r_i\}$ are the coordinates appearing in the problem, the probability per unit time of the system making a transition from a given initial state $|m\rangle$ to some final state $|n\rangle$ with the absorption of energy $\hbar\omega$ from the external perturbation is ([1])

$$W_{m \to n} = \frac{\pi}{2\hbar}|\langle m|O|n\rangle|^2 \delta[\hbar\omega - (E_n - E_m)] \tag{1}$$

where $|m\rangle$ is an eigenstate of the Hamiltonian in the absence of the perturbation and $E_m$ is the corresponding energy eigenvalue. If we use a trick which was apparently used for the first time in the context of such problems by Lamb ([2]) and introduce the representation of the $\delta$-function given by

$$2\pi\delta(x) = \int_{-\infty}^{\infty} dt \, e^{itx} \tag{2}$$

we can rewrite the expression for $W_{m \to n}$ in the form

$$W_{m \to n} = \frac{1}{4\hbar^2} \int_{-\infty}^{\infty} dt \, e^{it\omega} \langle m|O(t)|n\rangle\langle n|O(0)|m\rangle \tag{3}$$

where $O(t)$ is the operator $O$ in the Heisenberg representation:

$$O(t) = e^{i(t/\hbar)H} O e^{-i(t/\hbar)H} \tag{4}$$

and $H$ is the Hamiltonian of the system in the absence of the perturbation. In most physical problems we are not interested in the transition of our system from a given initial state to a given final state, so that we must sum over all possible final states $n$. Moreover, we ordinarily do not know the precise initial state of the system when the perturbation acts on it, and the best that we can often do is to assume that the initial state is one of the possible states of the system with a probability which is given by a canonical distribution

$$\rho = \frac{e^{-\beta H}}{\text{Tr } e^{-\beta H}} \qquad \beta = \frac{1}{k_B T} \qquad\qquad (5)$$

where $k_B$ is Boltzmann's constant.

With these assumptions we obtain the result that the canoncically averaged total transition rate is given by

$$\sum_{mn} \frac{e^{-\beta E_m}}{\text{Tr } e^{-\beta H}} W_{m \to n} = \frac{1}{4\hbar^2} \int_{-\infty}^{\infty} dt \, e^{it\omega} \langle O(t)O(0)\rangle \qquad\qquad (6)$$

where the angular brackets mean

$$\langle A \rangle = \text{Tr } \rho A \qquad\qquad (7)$$

for any operator $A$.

Since absorption or scattering cross sections differ from the type of transition probability which I have described only by certain multiplicative factors related to the flux of energy or particles in the external perturbation, we see in this simple example how the Fourier transform with respect to time can arise in expressions for absorption coefficients or scattering cross sections.

With this introduction, I will now write down without proof the expressions for the functions of physical interest in each of the types of experiment I have considered above.

## A. Infrared Lattice Vibration Absorption

The $\mu\nu$-component of the imaginary part of the dielectric response tensor of a crystal is given by ([3])

$$\epsilon_{\mu\nu}^{(2)}(\omega) = \frac{\Lambda^2}{n(\omega)} \frac{2\pi}{\hbar V} \int_{-\infty}^{\infty} dt \, e^{-i\omega t} \langle M_\nu(t)M_\mu(0)\rangle \qquad\qquad (8)$$

In this expression $\Lambda$ is a factor which takes account of the fact that the macroscopic electric field inside the crystal may not be the same as the externally applied field $E(t)$, $V$ is the crystal volume, and $n(\omega) = [\exp \beta\hbar\omega - 1]^{-1}$. $M$ is the crystal dipole moment operator. The operator $O$ in this case is $-M \cdot E(t)$.

## B. Raman Scattering

The intensity of Raman scattering per unit solid angle can be expressed as ([4])

$$I(\Omega) = \frac{\omega_0^4}{2\pi c^3} \sum_{\alpha\gamma\beta\lambda} n_\alpha n_\beta i_{\alpha\gamma,\,\beta\lambda}(\Omega) E_\gamma^- E_\lambda^+ \tag{9a}$$

$$i_{\alpha\gamma,\,\beta\lambda}(\Omega) = \frac{1}{2\pi} \int_{-\infty}^{\infty} dt\, e^{-i\Omega t} \langle P_{\beta\lambda}(t)^* P_{\alpha\gamma}(0) \rangle \tag{9b}$$

$\omega_0$ is the frequency of the incident radiation, $\Omega = \omega - \omega_0$ is the shift in the frequency of the scattered light, $\mathbf{n}$ is a unit vector which is perpendicular to the direction of observation and describes the polarization of the scattered light, $\mathbf{E}^\pm$ are the positive and negative frequency components of the incident light, and $P_{\alpha\beta}$ is the $\alpha\beta$ component of the electronic polarizability tensor of the crystal. The interaction operator $O$ in this example is again $-\mathbf{M}\cdot\mathbf{E}(t)$, but in this case $\mathbf{M}$ is the contribution to the crystal dipole moment which is induced by the incident light acting through the electronic polarizability, $M_\alpha = \sum_\gamma P_{\alpha\gamma} E_\gamma$.

## C. Neutron Scattering by Bravais Crystals

The coherent and incoherent scattering cross sections in the harmonic approximation can be written compactly in the following forms [5]:

$$\left(\frac{d^2\sigma}{d\Omega d\epsilon_s}\right)_{\text{coh}} = \langle a \rangle^2 \frac{e^{-2M}}{2\pi\hbar} \frac{k_2}{k_1} \sum_{ll'} e^{-i\mathbf{\kappa}\cdot[\mathbf{x}(l)-\mathbf{x}(l')]} \int_{-\infty}^{\infty} dt\, e^{i\omega t} e^{\langle \mathbf{k}\cdot\mathbf{u}(lt)\,\mathbf{k}\cdot\mathbf{u}(l'0)\rangle} \tag{10a}$$

$$\left(\frac{d^2\sigma}{d\Omega d\epsilon_s}\right)_{\text{inc}} = [\langle a^2 \rangle - \langle a \rangle^2] \frac{e^{-2M}}{2\pi\hbar} \frac{k_2}{k_1} \sum_l \int_{-\infty}^{\infty} dt\, e^{i\omega t} e^{\langle \mathbf{k}\cdot\mathbf{u}(lt)\,\mathbf{k}\cdot\mathbf{u}(l0)\rangle} \tag{10b}$$

In these expressions

$$e^{-2M} = e^{-\langle [\mathbf{k}\cdot\mathbf{u}(l)]^2 \rangle}$$

is the Debye–Waller factor, $\hbar\mathbf{\kappa}$ is the momentum transfer from neutron to crystal: $\mathbf{k}_2 = \mathbf{k}_1 - \mathbf{\kappa}$, where $\mathbf{k}_1$ and $\mathbf{k}_2$ are the wavevectors of the incident and scattered neutrons, respectively; $\hbar\omega$ is the energy transfer from neutron to crystal: $\hbar\omega = \hbar^2(k_1^2 - k_2^2)/2m$; $\mathbf{x}(l)$ is the position vector of the equilibrium position of the $l$th atom, $\mathbf{u}(l)$ is the displacement of the $l$th atom from its equilibrium position, and $a_l$ is the nuclear scattering length associated with the $l$th nucleus. The interaction between a neutron and the crystal is through the Fermi pseudopotential, $O = \sum_l a_l \delta[\mathbf{r} - \mathbf{x}(l) - \mathbf{u}(l)]$.

## D. The Mössbauer Effect

The cross section for absorption of $\gamma$-rays by a nucleus bound in a crystal is [6]

$$\sigma_a(\omega) = \frac{1}{2}\sigma_0 \gamma e^{-2M} \int_{-\infty}^{\infty} dt\, e^{-i\omega t} e^{\langle \mathbf{k} \cdot \mathbf{u}(lt)\mathbf{k} \cdot \mathbf{u}(l0)\rangle} \tag{11}$$

where $l$ is the site of the absorbing nucleus; $\hbar\kappa$ is the momentum of the incident $\gamma$-ray; $\hbar\omega = E - E_0$, where $E$ is the energy of the incident $\gamma$-ray and $E_0$ is the difference between the energy of the lowest excited state of the nucleus and the ground state; $\gamma = \Gamma/2\hbar$ is the width of the excited nuclear state; and $\sigma_0$ is the resonance absorption cross section for the absorbing nucleus.

## E. Spin Lattice Relaxation

If we excite a nonequilibrium number of spin $\frac{1}{2}$ paramagnetic impurities into the higher-energy spin state, the spin lattice relaxation time $\tau$ gives the decay time for the distribution to relax to equilibrium ([7]):

$$\frac{1}{\tau} = [1 + e^{-\beta\hbar\omega_z}]\frac{1}{\hbar^2} \int_{-\infty}^{\infty} dt\, e^{i\omega_z t}\langle V(t)V(0)\rangle \tag{12}$$

In this expression $\hbar\omega_z$ is the energy difference between the two spin states of a spin $\frac{1}{2}$ paramagnetic impurity ion in some host crystal. The operator $V = \langle + | H_I | - \rangle$ is the matrix element of the spin lattice interaction Hamiltonian betewen the eigenfunctions of the spin Hamiltonian corresponding to the higher ($+$) and lower ($-$) energy spin states. The interaction is through the vibrational modulation of the crystalline electric field which acts on the orbital motion of the paramagnetic electrons to induce spin transitions.

## F. Specific Heat

The specific heat of an assembly of $3N$ harmonic oscillators is given by

$$C_v(T) = k_B \sum_s \frac{(\hbar\omega_s/2k_BT)^2}{\sinh^2(\hbar\omega_s/2k_BT)} = 3Nk_B \int_0^{\omega_L} g(\omega)\frac{(\hbar\omega/2k_BT)^2}{\sinh^2(\hbar\omega/2k_BT)}d\omega \tag{13}$$

where $\omega_s$ is the frequency of the $s$th oscillator.

In the second expression $g(\omega)$ is the frequency spectrum of the crystal and can be expressed equivalently in the forms

$$g(\omega) = \frac{2\omega}{3N} \sum_s \delta(\omega^2 - \omega_s^2) = 2\omega G(\omega^2) \tag{14}$$

$$3NG(\omega^2) = \frac{i}{2\pi\hbar|\omega|n(\omega)} \sum_{l\alpha} \int_{-\infty}^{\infty} dt\, e^{-i\omega t}\langle u_\alpha(lt)p_\alpha(l0)\rangle \tag{15}$$

The elements of both the crystal dipole moment and electronic polariz-ability operators $M_\mu$ and $P_{\mu\nu}$ can be expanded in powers of the atomic displacements, and for crystals containing impurities the terms linear in the displacements will not vanish even if they are zero in the perfect crystal ([8]):

$$M_\mu = \sum_{l\kappa\alpha} M_{\mu,\alpha}(l\kappa)u_\alpha(l\kappa) + \tfrac{1}{2}\sum_{l\kappa\alpha}\sum_{l'\kappa'\beta} M_{\mu,\alpha\beta}(l\kappa; l'\kappa')u_\alpha(l\kappa)u_\beta(l'\kappa') + \cdots$$

(16a)

$$P_{\mu\nu} = P_{\mu\nu}^{(0)} + \sum_{l\kappa\alpha} P_{\mu\nu,\alpha}(l\kappa)u_\alpha(l k) + \tfrac{1}{2}\sum_{l\kappa\alpha}\sum_{l'\kappa'\beta} P_{\mu\nu,\alpha\beta}(l\kappa; l'\kappa')u_\alpha(l\kappa)u_\beta(l'\kappa') + \cdots$$

(16b)

Similarly, the interaction Hamiltonian in the calculation of the spin-lattice relaxation time can be expanded in the same kind of series:

$$V = \sum_{l\kappa\alpha} V_\alpha(l\kappa)u_\alpha(l\kappa) + \tfrac{1}{2}\sum_{l\kappa\alpha}\sum_{l'\kappa'\beta} V_{\alpha\beta}(l\kappa; l'\kappa')u_\alpha(l\kappa)u_\beta(l'\kappa') + \cdots \quad (16c)$$

The structure of the coefficients in this expansion is such that it is really an expansion in the relative displacements of the impurity atom and the atoms of the host crystal with which it interacts. This is because only such displace-ments play a role in modulating the crystal field at the impurity site.

Thus in each case listed above we have a Fourier transform of a displace-ment–displacement or a displacement–momentum correlation function to calculate.

Since these correlation functions appear in the interpretation of experi-mental results and at the same time provide a convenient way of introducing localized and resonance modes into the dynamical theory of perturbed crys-tals, we now turn to a discussion of how these functions can be computed.

The time-independent equations of motion of a perfect Bravais crystal can be written in the form

$$\sum_{l'\beta} [M\omega^2\delta_{ll'}\delta_{\alpha\beta} - \Phi_{\alpha\beta}^{(0)}(ll')]u_\beta(l') = 0 \quad (17)$$

where $M$ is the atomic mass, and the $\{\Phi_{\alpha\beta}(ll')\}$ are the atomic force constants, or, more compactly, in matrix form as $\mathbf{Lu} = 0$. The corresponding equations for a perturbed crystal are

$$\sum_{l'\beta} [M_l\omega^2\delta_{ll'}\delta_{\alpha\beta} - \Phi_{\alpha\beta}(ll')]u_\beta(l') = 0 \quad (18a)$$

or

$$(\mathbf{L} - \delta\mathbf{L})\mathbf{u} = 0 \quad (18b)$$

If we assume that $u_\alpha(l) = B_\alpha(l)/(M_l)^{1/2}$, the latter equations can be rewritten as

$$\sum_{l'\beta} D_{\alpha\beta}(ll')B_{\beta}^{(s)}(l') = \omega_s^2 B_{\alpha}^{(s)}(l) \qquad (19)$$

where the matrix $\mathbf{D}$ is the dynamical matrix, whose elements are given by

$$D_{\alpha\beta}(ll') = \frac{\Phi_{\alpha\beta}(ll')}{(M_l M_{l'})^{1/2}} \qquad (20)$$

and we have indexed by $s\,(=1, 2, 3, \ldots, 3N)$ the $3N$ solutions of this eigenvalue equation. The eigenvectors $\{B_{\alpha}^{(s)}(l)\}$ can be chosen to be orthonormal and complete:

$$\sum_{l\alpha} B_{\alpha}^{(s)}(l)B_{\alpha}^{(s')}(l) = \delta_{ss'}$$

$$\sum_{s} B_{\alpha}^{(s)}(l)B_{\beta}^{(s)}(l') = \delta_{ll'}\delta_{\alpha\beta} \qquad (21)$$

and provide the basis for a normal coordinate transformation to phonon creation and destruction operators:

$$u_{\alpha}(lt) = \left(\frac{\hbar}{2M_l}\right)^{1/2} \sum_s B_{\alpha}^{(s)}(l)\frac{1}{\omega_s^{1/2}}(b_s e^{-i\omega_s t} + b_s^{\dagger}e^{i\omega_s t}) \qquad (22a)$$

$$p_{\alpha}(lt) = \frac{1}{i}\left(\frac{\hbar M_l}{2}\right)^{1/2} \sum_s B_{\alpha}^{(s)}(l)\omega_s^{1/2}(b_s e^{-i\omega_s t} - b_s^{\dagger}e^{i\omega_s t}) \qquad (22b)$$

With the aid of these expansions the following correlation functions are readily obtained:

$$\langle u_{\alpha}(lt)u_{\beta}(l'0)\rangle = \frac{\hbar}{2(M_l M_{l'})^{1/2}} \sum_s \frac{B_{\alpha}^{(s)}(l)B_{\beta}^{(s)}(l')}{\omega_s}\{n(\omega_s)e^{i\omega_s t} + [n(\omega_s) + 1]e^{-i\omega_s t}\} \qquad (23a)$$

$$\langle u_{\alpha}(lt)p_{\beta}(l'0)\rangle = \frac{\hbar}{2i}\left(\frac{M_{l'}}{M_l}\right)^{1/2} \sum_s B_{\alpha}^{(s)}(l)B_{\beta}^{(s)}(l')\{n(\omega_s)e^{i\omega_s t} - [n(\omega_s) + 1]e^{-i\omega_s t}\} \qquad (23b)$$

with Fourier transforms

$$\int_{-\infty}^{\infty} dt\, e^{-i\omega t}\langle u_{\alpha}(lt)u_{\beta}(l'0)\rangle = \frac{2\pi\hbar n(\omega)}{(M_l M_{l'})^{1/2}} \operatorname{sgn} \omega \sum_s B_{\alpha}^{(s)}(l)B_{\beta}^{(s)}(l')\delta(\omega^2 - \omega_s^2)$$

$$= \operatorname{Im} 2\hbar n(\omega)\operatorname{sgn} \omega U_{\alpha\beta}(ll'; \omega^2 - i0) \qquad (24a)$$

$$\int_{-\infty}^{\infty} dt\, e^{-i\omega t}\langle u_{\alpha}(lt)p_{\beta}(l'0)\rangle = \frac{1}{i}2\pi\hbar\,|\omega|\,n(\omega)\left(\frac{M_{l'}}{M_l}\right)^{1/2}$$

$$\times \sum_s B_{\alpha}^{(s)}(l)B_{\beta}^{(s)}(l')\delta(\omega^2 - \omega_s^2)$$

$$= \frac{1}{i}\operatorname{Im} 2\hbar\,|\omega|\,n(\omega)M_{l'}U_{\alpha\beta}(ll'; \omega^2 - i0) \qquad (24b)$$

where we have defined

$$U_{\alpha\beta}(ll'; \omega^2) = \frac{1}{(M_l M_{l'})^{1/2}} \sum_s \frac{B_\alpha^{(s)}(l) B_\beta^{(s)}(l')}{\omega^2 - \omega_s^2} \qquad (25)$$

A little reflection shows that in fact,

$$U_{\alpha\beta}(ll'; \omega^2) = (\mathbf{L} - \delta\mathbf{L})_{\substack{ll' \\ \alpha\beta}}^{-1} \qquad (26)$$

If we define the inverse matrix

$$G_{\alpha\beta}(ll', \omega^2) = (\mathbf{L}^{-1})_{\substack{ll' \\ \alpha\beta}} \qquad (27)$$

we can write a matrix equation for the matrix $\mathbf{U}$:

$$\mathbf{U} = \mathbf{G} + \mathbf{G}\delta\mathbf{L}\mathbf{U} \qquad (28)$$

Inasmuch as the matrix $\mathbf{G}$ refers to the unperturbed crystal, we can regard it as known, at least in principle. In fact, a common representation for $G_{\alpha\beta}(ll', \omega^2)$ is $(^9)$

$$G_{\alpha\beta}(ll'; \omega^2) = \frac{1}{NM} \sum_{\mathbf{k}j} \frac{e_\alpha(\mathbf{k}j) e_\beta(\mathbf{k}j)}{\omega^2 - \omega_j^2(\mathbf{k})} \cos \mathbf{k} \cdot [\mathbf{x}(l) - \mathbf{x}(l')] \qquad (29)$$

where the sum on $\mathbf{k}$ runs over the $N$ allowed values this variable can assume inside the first Brillouin zone of the crystal, and which are specified by the cyclic boundary conditions on the atomic displacements. The index $j (= 1, 2, 3)$ labels the three branches of the phonon spectrum associated with each value of $\mathbf{k}$, and $\omega_j(\mathbf{k})$ and $e_\alpha(\mathbf{k}j)$ are the frequency of the normal mode $(\mathbf{k}j)$ and the corresponding unit polarization vector, respectively.

Let us determine the form of $U_{\alpha\beta}(ll'; \omega^2)$ in a case which is simple, but which has some practical interest. This is the case of a cubic Bravais crystal containing an isotopic impurity of mass $M'$ at the lattice site which marks the origin of our coordinate system. In this case the perturbation matrix $\delta\mathbf{L}$ has the form

$$\delta L_{\alpha\beta}(ll'; \omega^2) = \epsilon M \omega^2 \delta_{l0} \delta_{l'0} \delta_{\alpha\beta} \qquad \epsilon = 1 - \frac{M'}{M} \qquad (30)$$

and the equation for $U_{\alpha\beta}(ll'; \omega^2)$ takes the form

$$U_{\alpha\beta}(ll'; \omega^2) = G_{\alpha\beta}(ll'; \omega^2) + \epsilon M \omega^2 \sum_\gamma G_{\alpha\gamma}(l0; \omega^2) U_{\gamma\beta}(0l'; \omega^2) \qquad (31)$$

and we see that if we focus our attention on the particular element $U_{\alpha\beta}(00; \omega^2)$ then because the cubic symmetry of the host crystal forces $G_{\alpha\beta}(00; \omega^2)$ to be isotropic, $G_{\alpha\beta}(00; \omega^2) = \delta_{\alpha\beta} G(\omega^2)$, we find that

$$U_{\alpha\beta}(00;\omega^2) = \delta_{\alpha\beta}\frac{G(\omega^2)}{1 - \epsilon M\omega^2 G(\omega^2)} \tag{32}$$

From the representation of $U_{\alpha\beta}(00;\omega^2)$ in the form

$$U_{\alpha\beta}(00;\omega^2) = \frac{1}{M_0}\sum_s \frac{B_\alpha^{(s)}(0)B_\beta^{(s)}(0)}{\omega^2 - \omega_s^2} \tag{33}$$

we see that as a function of $\omega$ this function has a simple pole at each of the normal-mode frequencies of the perturbed crystal. This means that the normal-mode frequencies of the perturbed crystal are given by the poles of the function $G(\omega^2)$ and by the zeroes of the function $1 - \epsilon M\omega^2 G(\omega^2)$. Now, the function $G(\omega^2)$ in this case is given by

$$G(\omega^2) = \frac{1}{3NM}\sum_{kj}\frac{1}{\omega^2 - \omega_j^2(\mathbf{k})} \tag{34}$$

where we have used the cubic symmetry of the host crystal, and we see that this function has a simple pole at each of the unperturbed normal-mode frequencies. It might appear therefore the $U_{\alpha\beta}(00;\omega^2)$ has more poles than $G(\omega^2)$ because the equation

$$1 = \epsilon M\omega^2 G(\omega^2) \tag{35}$$

certainly has solutions too. In fact, however, it can be shown by a slightly different kind of calculation that the latter equation gives the frequencies of only those modes which have been perturbed by the introduction of the impurity atom into the crystal [10]. Many of the modes of the perturbed crystals have the same frequencies as they did in the unperturbed crystal. In particular, all modes of the perfect crystal which have a node at the impurity site will not feel the change in the mass of the atom at this site because it is not vibrating. The frequencies of these modes of the perturbed crystal are given by the poles of $G(\omega^2)$, but these poles now have a smaller residue in the quotient for $U_{\alpha\beta}(00;\omega^2)$ than in the expression for $G(\omega^2)$ itself because of the denominator, so that the total number of poles of $U_{\alpha\beta}(00;\omega^2)$ does not in fact exceed the number of degrees of freedom in the perturbed crystal.

If we rewrite the last equation as

$$\frac{1}{\epsilon} = \frac{\omega^2}{3N}\sum_{kj}\frac{1}{\omega^2 - \omega_j^2(\mathbf{k})} \tag{36}$$

we can solve it graphically for the frequencies of the perturbed modes and this solution is shown in Fig. 1. In general, for $\epsilon > 0$, which corresponds to a light impurity, the frequencies of the perturbed modes are shifted to slightly

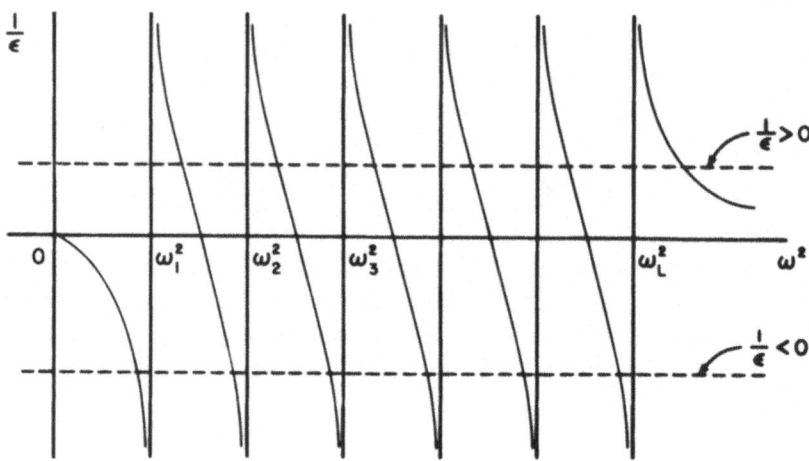

Fig. 1—The graphical solution of equation (36).

higher values compared against their values in the unperturbed crystal, but by no more than the distance to the next unperturbed frequency. When the impurity is heavy, the frequencies are shifted down, but again by no more than the distance to the next lower unperturbed frequencies. If $\epsilon$ is positive and sufficiently large (but less than 1) a mode can occur whose frequency lies above $\omega_L$, the maximum frequency of the unperturbed crystal. Its frequency can be as high as you like if $\epsilon$ is made sufficiently close to unity because there is no unperturbed mode with a frequency lying above $\omega_L$ by definition. Such a mode is called a localized mode because the fact that its frequency lies in a forbidden or stop band for the unperturbed crystal means that it is a non-propagating mode. In fact, the displacements of the atoms in this mode die off faster than exponentially with increasing distance from the impurity site.

For the Fourier transform of the correlation function $\langle u_\alpha(0; t)u_\beta(0; 0)\rangle$ we need Im $U_{\alpha\beta}(00; \omega^2 - i0)$. If we define two functions

$$G_0(\omega^2) = \frac{1}{3N} \sum_{\mathbf{k}j} \delta[\omega^2 - \omega_j^2(\mathbf{k})] \tag{37a}$$

$$\tilde{G}_0(\omega^2) = \frac{1}{3N} \sum_{\mathbf{k}j} \frac{1}{[\omega^2 - \omega_j^2(\mathbf{k})]_p} \tag{37b}$$

which are the imaginary and real parts of $MG(\omega^2 - i0)$, respectively, then in terms of these functions we have that

$$\text{Im } U_{\alpha\beta}(00; \omega^2 - i0) = \frac{\delta_{\alpha\beta}}{M} \frac{\pi G_0(\omega^2)}{[1 - \epsilon\omega^2\tilde{G}_0(\omega^2)]^2 + \pi^2\epsilon^2\omega^4 G_0^2(\omega^2)} \tag{38}$$

for all frequencies for which $G_0(\omega^2)$ does not vanish. However, since $G_0(\omega^2)$ is a nonnegative function of $\omega^2$, and in fact is recognized to be the distribution function of the squared frequencies of the unperturbed host lattice, we can say that the last equation applies whenever $\omega$ is in the band of frequencies allowed to the normal modes of the unperturbed host crystal.

If we have a localized vibration mode with frequency $\omega_0$, for which $1 - \epsilon M \omega_0^2 G(\omega_0^2) = 0$, then we obtain an additional contribution to $\operatorname{Im} U_{\alpha\beta}(00; \omega^2 - i0)$, viz.,

$$\operatorname{Im} U_{\alpha\beta}(00; \omega^2 - i0) = \frac{\delta_{\alpha\beta}}{M} \frac{\pi \delta(\omega^2 - \omega_0^2)}{\epsilon^2 \omega_0^2 B(\omega_0^2)} \tag{39a}$$

$$B(\omega_0^2) = \frac{1}{3N} \sum_{kj} \frac{\omega_j^2(\mathbf{k})}{[\omega_0^2 - \omega_j^2(\mathbf{k})]^2} \tag{39b}$$

Thus a localized mode contributes a $\delta$-function peak, centered at its frequency, to the Fourier transform of the correlation function $\langle u_\alpha(0; t) u_\beta(0; 0) \rangle$ or $\langle u_\alpha(0; t) p_\beta(0; 0) \rangle$.

However, reference to the form $U_{\alpha\beta}(00; \omega^2)$ takes when $\omega^2$ lies in an allowed band for the unperturbed frequencies shows that if the equation

$$1 = \epsilon \omega_r^2 \tilde{G}_0(\omega_r^2) \tag{40}$$

is satisfied for some frequency $\omega_r$, then $\operatorname{Im} U_{\alpha\beta}(00; \omega^2 - i0)$ can have a resonance character in the vicinity of this frequency, viz.,

$$\operatorname{Im} U_{\alpha\beta}(00; \omega^2 - i0) = \frac{\delta_{\alpha\beta}}{M} \frac{1}{2\epsilon^2 \omega_r^3 \tilde{B}(\omega_r^2)} \frac{\tfrac{1}{2}\Gamma}{(\omega - \omega_r)^2 + \tfrac{1}{4}\Gamma^2} \tag{41}$$

where

$$\Gamma = \frac{\pi \omega_r G_0(\omega_r^2)}{\tilde{B}(\omega_r^2)}$$

$$\tilde{B}(\omega_r^2) = \frac{1}{3N} \sum_{kj} \frac{\omega_j^2(\mathbf{k})}{[\omega_r^2 - \omega_j^2(\mathbf{k})]_p^2} \tag{42}$$

Of course, in order that there be a real resonance we must have $\Gamma \ll \omega_L$, otherwise the peak is indistinguishable from the background. The frequency $\omega_r$ at which $1 = \epsilon \omega_r^2 \tilde{G}_0(\omega_r^2)$ and at which $\operatorname{Im} U_{\alpha\beta}(00; \omega^2 - i0)$ has a resonance-type peak is called the frequency of a resonance mode. However, it should be emphasized that unlike a localized mode, which is an exact eigenstate of the perturbed crystal in the harmonic approximation, the resonance mode is not an exact eigenstate of the crystal Hamiltonian.

In Fig. 2 we have plotted $\operatorname{Im} U_{\alpha\beta}(00; \omega^2 - i0)$ for $\epsilon > 0$, $\epsilon = 0$, and

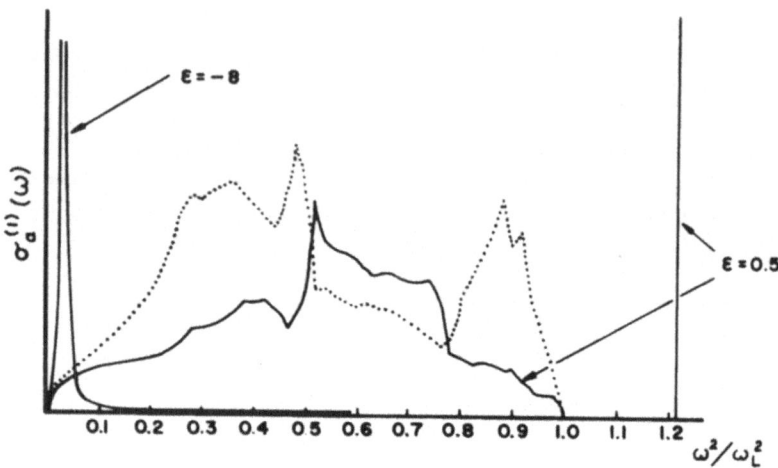

Fig. 2—A plot of the function $G_0(\omega^2)/\{[1 - \epsilon\omega^2\tilde{G}_0(\omega^2)]^2 + \pi^2\epsilon^2\omega^4 G_0^2(\omega^2)\}$ for three different values of the impurity mass relative to that of the atoms in the host crystal. The curves are plotted on the basis of a nearest neighbor, central-force model of a face-centered cubic crystal for the host crystal [A.A. Maradudin, *Rev. Mod. Phys.* **36**: 417 (1964).]

$\epsilon > 0$. In the first case the impurity gives rise to a localized mode; in the last case it gives rise to a low frequency resonance mode.

The preceding example, although simple, is not without physical interest. In a first approximation it allows us to calculate the one-phonon infrared absorption coefficient for a crystal of solid argon containing xenon or krypton impurities, which has recently been measured by Jones and Woodfine [11]. At the same time, knowledge of the function Im $U_{\alpha\beta}(ll'; \omega^2 - i0)$ with minor modifications enables us to calculate the first order Raman spectrum of a crystal containing impurity atoms, when the impurities occupy sites possessing cubic symmetry, but lacking inversion symmetry, for example, substitutional impurities in crystals of the diamond structure [12]. This function will also give us the cross section for the absorption of $\gamma$-rays by a nucleus bound in a cubic crystal when the nucleus recoils and excites or de-excites one phonon [13]. In a somewhat cruder approximation the function Im $U_{\alpha\beta}(ll'; \omega^2 - i0)$ can also be used to give the cross section for the incoherent scattering of neutrons by a cubic Bravais crystal containing substitutional impurities [14]. In Fig. 3 we have displayed the experimental results of Rubin *et al.* [15] for the energy spectrum of neutrons scattered incoherently from a crystal of vanadium containing 4 at. % hydrogen. The portion of the spectrum below $\sim 38$ MeV reflects the frequency spectrum of pure vanadium, as would be suggested by equation (38), and the peak at $\sim 100$ MeV is attributed to scattering by localized vibration modes associated with the light hydrogen impurities. The latter conclusion is implied by equation (39). The large width of the latter peak may be attributed to the diffusive motion of the interstitial hydrogen atoms through the vanadium lattice. Finally, this result can be used to compute the change in the specific heat of a cubic crystal due to the introduction of substitutional impurities into it. In the first three cases which I have just mentioned the corresponding absorption coefficient or scattering cross section therefore is seen to be of the form of the distribution function for the squares of the normal-mode frequencies of the unperturbed host crystal, $G_0(\omega^2)$, weighted by a function which can be shown to give the frequency dependence of the mean-square amplitude of vibration of the substitutional defect atom ([16]).

These results illustrate two types of effects which defects in crystals can have on the vibrational properties of the crystal and which are reflected in the experimental results obtained from the types of experiments I have just listed. The first of these is the impurity atoms can give rise ot exceptional vibration modes called localized and resonance modes which contribute δ-function peaks and Lorentzian resonance peaks (in the harmonic approximation), respectively, to the absorption or scattering cross sections being investigated. The second is that if the perturbation of the lattice vibrations by the impurity atom is weak, in the sense that the latter introduces no resonance modes into the vibration spectrum, then the observed absorption spectrum or scattering cross section will be a reflection of the distribution of squared frequencies of the unperturbed host crystal. At the least, the various Van Hove singularities ([17]) present in the latter function should also be ob-

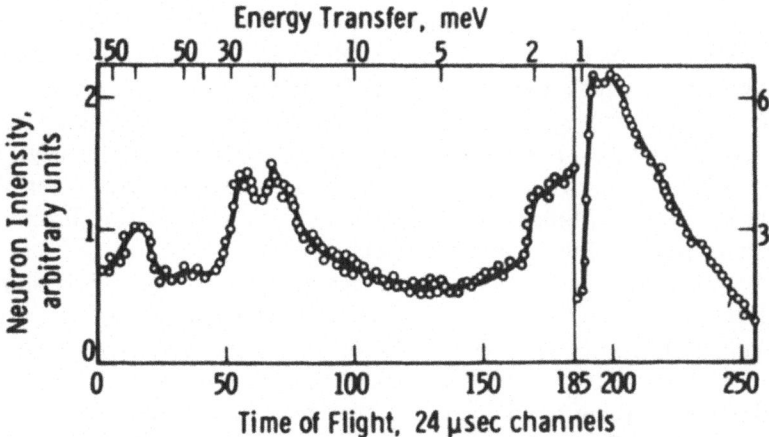

Fig. 3—The energy spectrum of neutrons scattered elastically and inelastically by vanadium containing 4 at .% hydrogen at 150°C. The background has been subtracted in this figure. Note the different scale for the elastic processes. [R. Rubin, J. Peretti, G. Verdan, and W. Kley, *Phys. Letters* **14**: 100 (1965).]

served in the former spectra. Thus the impurity atom can be used as a probe to give information about the dynamical properties of the perfect host crystal, in addition to providing us with an experimental means of studying localized excitations in solids.

I have not yet said anything about the relation of the function $U_{\alpha\beta}(ll';$ $\omega^2 - i0)$ to the one-phonon cross section for the coherent scattering of neutrons by the lattice vibrations of a perturbed crystal. The connection here is more complicated than in the cases considered before. A calculation by Elliott and Maradudin [18] in which finite concentrations of impurities were considered yielded the result that the coherent cross section as a function of energy loss from neutron to crystal for a fixed momentum transfer has the form

$$\sum_j \frac{[\boldsymbol{\kappa}\cdot\mathbf{e}(\boldsymbol{\kappa}j)]^2}{\omega_j(\boldsymbol{\kappa})} \frac{\Gamma_j(\boldsymbol{\kappa};\omega)}{[\omega^2 - \Omega_j^2(\boldsymbol{\kappa};\omega)]^2 + \omega_j^2(\boldsymbol{\kappa})\Gamma_j^2(\boldsymbol{\kappa};\omega)} \tag{43}$$

where

$$\Omega_j^2(\boldsymbol{\kappa};\omega) = \omega_j^2(\boldsymbol{\kappa}) + \epsilon p\omega_j^2(\boldsymbol{\kappa}) \, \mathrm{Re} \, \frac{U_{\alpha\alpha}(00;\omega^2 - i0)}{G(\omega^2 - i0)} \tag{44a}$$

$$\Gamma_j(\boldsymbol{\kappa};\omega) = \epsilon p\omega_j(\boldsymbol{\kappa})M\omega^2 \, \mathrm{Im} \, U_{\alpha\alpha}(00;\omega^2 - i0) \tag{44b}$$

and $p$ is the concentration of impurity atoms. This expression reflects the fact that the $\delta$-functions of $\omega^2 - \omega_j^2(\mathbf{k})$ which appear in the corresponding expressions for pure crystals in the harmonic approximation are broadened into finite peaks by the presence of impurities. From the known behavior of $\mathrm{Im}\, U_{\alpha\beta}(00;\omega^2 - i0)$ as a function of $\omega^2$, we can immediately predict that the widths of the neutron peaks will show a sudden increase as the energy transfer equals the value of the frequency of a resonance mode. At the same time, the frequency shift $\Omega_j^2(\mathbf{k},\omega) - \omega_j^2(\mathbf{k})$ should display an antiresonance behavior in the vicinity of a resonance-mode frequency. These behaviors of the width and shift are indeed observed in the experimental results of Bjerrum-Møller and Mackintosh [19] on a system of chromium in tungsten shown in Figs. 4 and 5, respectively.

In the preceding examples, the harmonic approximation is sufficient for the calculation of the effects of defects on a particular dynamical property of a crystal. In contrast, in the calculation of the temperature dependence of the spin-lattice relaxation time for a spin $\frac{1}{2}$ paramagnetic impurity in a crystal, the most striking consequence of the fact that the paramagnetic ion is an impurity and, in particular, that it gives rise to localized vibration modes, can only be calculated if the anharmonicity of the lattice vibrations is taken into account. Moreover, in the preceding examples we were concerned pri-

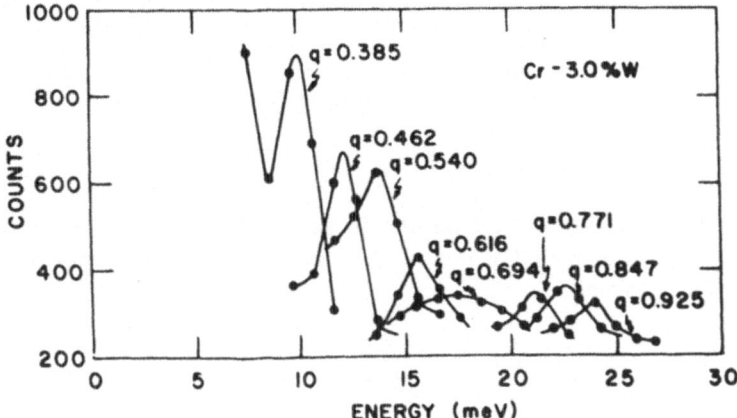

Fig. 4—Neutron groups in Cr containing 3 at .% W at room temperature. The phonon wavevector is in the [110] direction and is measured in reciprocal angstrom units. [H. Bjerrum–Møller and A. R. Mackintosh, *Phys. Rev. Letters* **15**: 623 (1965).]

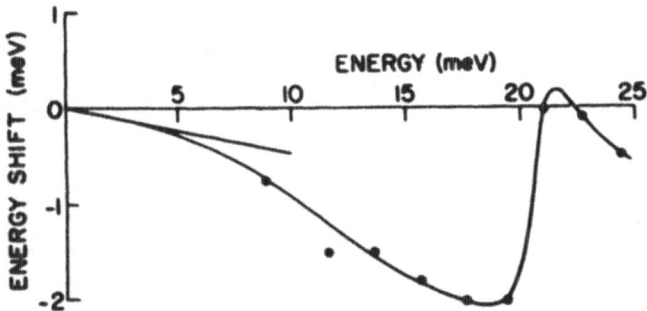

Fig. 5—Phonon energy shifts in Cr containing 3 at .% W as a function of the phonon energy in pure Cr. [H. Bjerrum–Møller and A.R. Mackintosh, *Phys. Rev. Letters* **15**: 623 (1965).]

marily with interactions between an external probe and lattice vibrations which can be described collectively as one-phonon processes, that is, processes in which the external probe excites or de-excites one quantum of vibrational energy. In discussing the spinlattice relaxation time, however, we focus our attention on the two-phonon processes, because it is on these processes that the localized modes introduced by the impurity make their greatest mark. The two-phonon processes are those which are contributed to the correlation function (12) by the second term on the right-hand side of equation (16c):

$$\left(\frac{1}{\tau}\right)_{2\text{-phonon}} = \frac{1}{4}[1 + e^{-\beta\hbar\omega_z}]\frac{1}{\hbar^2}\sum_{l_1\kappa_1\alpha_1}\cdots\sum_{l_4\kappa_4\alpha_4}V_{\alpha_1\alpha_2}(l_1\kappa_1; l_2\kappa_2)$$

$$\times V_{\alpha_3\alpha_4}(l_3\kappa_3; l_4\kappa_4)\int_{-\infty}^{\infty}dt\,e^{it\omega_z}\langle u_{\alpha_1}(l_1\kappa_1; t)u_{\alpha_2}(l_2\kappa_2; t)$$

$$\times u_{\alpha_3}(l_3\kappa_3; 0)u_{\alpha_4}(l_4\kappa_4; 0)\rangle \tag{45}$$

In the harmonic approximation the indicated Fourier transform can be evaluated formally with the aid of the normal coordinate transformation (22) to yield the result

$$\left(\frac{1}{\tau}\right)_{2\text{-phonon}} = \frac{\pi}{4}[1 + e^{-\beta\hbar\omega_z}]\sum_{s_1 s_2}\frac{V_{s_1 s_2}^2}{\omega_{s_1}\omega_{s_2}}$$

$$\times [(n_{s_1} + 1)(n_{s_2} + 1)\delta(\omega_z - \omega_{s_1} - \omega_{s_2}) + n_{s_1}n_{s_2}\delta(\omega_z + \omega_{s_1} + \omega_{s_2})$$

$$+ (n_{s_1} + 1)n_{s_2}\delta(\omega_z - \omega_{s_1} + \omega_{s_2}) + n_{s_1}(n_{s_2} + 1)\delta(\omega_z + \omega_{s_1} - \omega_{s_2})] \tag{46a}$$

where

$$n_s \equiv n(\omega_s)$$

and

$$V_{s_1 s_2} = \sum_{l_1\kappa_1\alpha_1}\sum_{l_2\kappa_2\alpha_2}B_{\alpha_1}^{(s_1)}(l_1\kappa_1)\frac{V_{\alpha_1\alpha_2}(l_1\kappa_1; l_2\kappa_2)}{(M_{\kappa_1}M_{\kappa_2})^{1/2}}B_{\alpha_2}^{(s_2)}(l_2\kappa_2) \tag{46b}$$

The sums over $s_1$ and $s_2$ on the right-hand side of equation (46a) run over all vibration modes of the crystal, including localized modes, if they occur. However, we see from equation (46a) that localized modes make no contribution to the two-phonon contribution to the spin-lattice relaxation time in the harmonic approximation. For if $s_1$ and $s_2$ both label a localized mode whose frequency is $\omega_0$, the argument of no $\delta$-function can vanish. The picture is changed if we take account of the anharmonic terms in the crystal potential energy, as was first pointed out by Klemens [20]. In the presence of anharmonicity, the interactions among the normal modes and the consequent uncertainty regarding the energy of a phonon, have the effect, crudely speaking, of relaxing the condition of strict conservation of energy which gives rise to the $\delta$-functions on the right-hand side of equation (46a). Localized vibration modes are therefore able to contribute to the spin-lattice relaxation time, primarily through the third and fourth terms on the right-hand side of equation (46a). The localized mode contribution to the spin-lattice relaxation time is found to have the form [7]

$$\left(\frac{1}{\tau}\right)_{\substack{2\text{-phonon}\\\text{local mode}}} = (\text{const.})\frac{\gamma_0}{\omega_z^2 + 4\gamma_0^2}e^{-\Theta_0/T} \tag{47}$$

where $\Theta_0 = \hbar\omega_0/k_B$, $\omega_0$ is the frequency of the localized mode, and $\gamma_0$ is the width of the localized mode, i.e., the uncertainty in its energy due to the possibility of giving up part of its energy to other modes through the anharmonic terms in the crystal potential energy. The term $\gamma_0^{-1}$ is often called the lifetime of the localized mode. The exponential dependence on temperature of the localized-mode contribution to the spin-lattice relaxation time distinguishes it from the contributions associated with the band modes and the one-phonon processes, which have a power-law dependence on temperature.

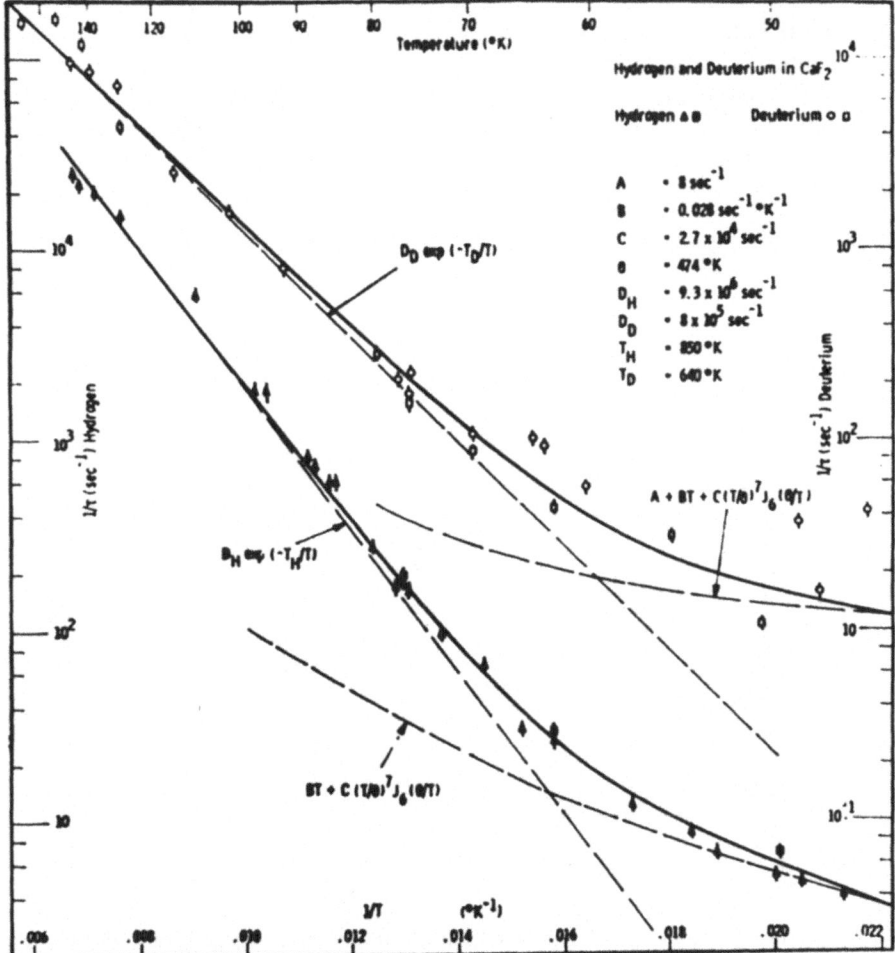

Fig. 6—The temperature dependence of the inverse relaxation time for atomic hydrogen and deuterium in interstitial sites in CaF$_2$. Note that the vertical scale for deuterium is shifted one decade from that for hydrogen. [D.W. Feldman, J.G. Castle, Jr., and J. Murphy, *Phys. Rev.* **138**: A1208 (1965).]

In Fig. 6 are shown the experimental data of Feldman, *et al.* ([21]) for the temperature dependence of the spin-lattice relaxation times associated with interstitial hydrogen and deuterium atoms in $CaF_2$. In each case a contribution of the form given by Eq. (47) is obtained. In the case of the hydrogen impurity, from the value of $\Theta_0 = 850 \pm 60°K$, a localized-mode frequency of $\omega_0 = 590.5 \pm 417 \text{ cm}^{-1}$ is obtained. This is rather higher than the maximum frequency of pure $CaF_2$, which is estimated to be $444.5 \text{ cm}^{-1}$ ([22]). The frequency $\omega_0$ obtained for the deuterium impurity lies very close to the maximum frequency of pure $CaF_2$, and suggests that this impurity may give rise to a resonance mode in the continuous spectrum, rather than to a localized mode. The contribution of resonance modes to the spin-lattice relaxation time of a paramagnetic impurity has recently been studied by Mills ([23]).

We come finally to the effect of defects on the specific heat of a crystal. It should be clear from the outset that localized vibration modes have less interesting effects on the specific heat than do resonance modes. The former, being high-frequency modes, are excited appreciably only at high tempera-

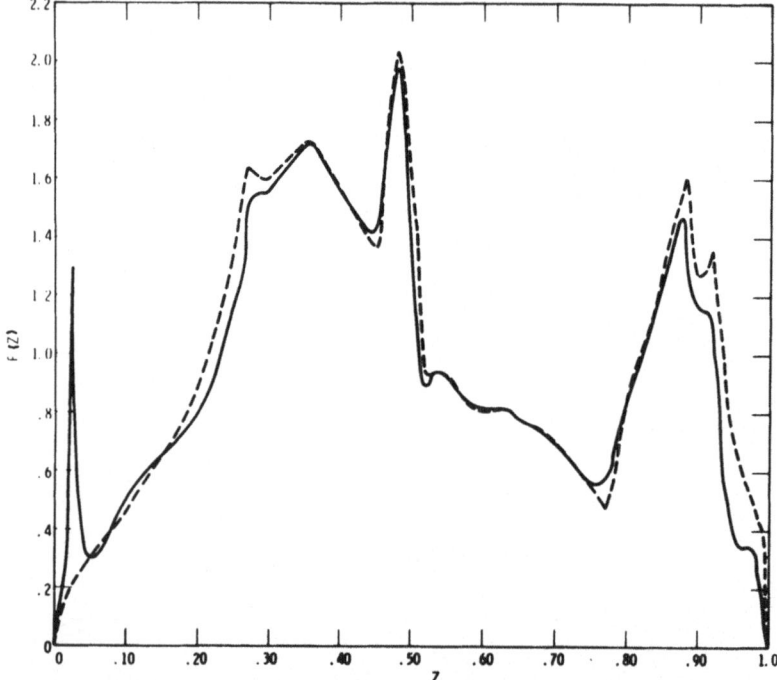

Fig. 7—The spectrum of squared frequencies for an isotopically disordered nearest neighbor, central-force model of a face-centered cubic crystal. The impurities are nine times as heavy as the atoms they replace, and their concentration is 2%. [B. Mozer and A.A. Maradudin, *Bull. Am. Phys. Soc.* [II] **8**: 193 (1963).]

tures, at which the specific heat has reached the classical or Dulong–Petit value. If impurities which give rise to low-frequency resonance modes are present in a crystal, we see from equations (15), (24), and (14) that they contribute a sharp spike to the low-frequency end of the frequency spectrum of the crystal. This is shown in Fig. 7 for heavy, substitutional, isotopic impurities. As was pointed out by Lehman and DeWames ([24]), and in greater detail by Kagan and Iosilevskii ([25]), such a spike in the frequency spectrum leads to an enhancement of the low-temperature specific heat of the impure crystal in the temperature range in which the resonance modes begin to be excited appreciably. This effect has recently been observed experimentally by Panova and Samoilov ([26]) for Mg containing 2.8 at .% Pb as an impurity, and by Cape et al. ([28]) for Mg containing 1 at .% Pb or Cd as impurities. In Fig. 8 is plotted the difference between the specific heat of Mg containing Pb impurities and that of pure Mg as determined by Cape et al. The excess specific heat is attributed to the excitation of low-frequency resonance modes associated with the heavy impurity atoms.

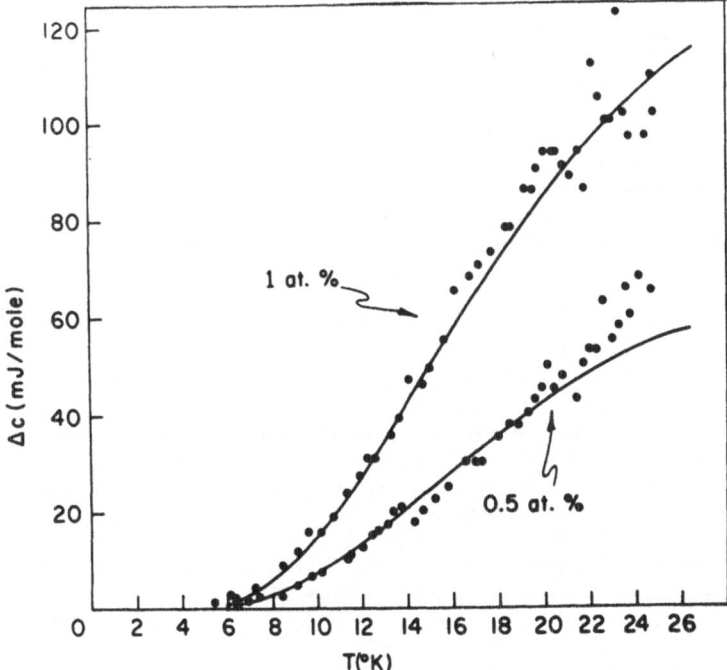

Fig. 8—The excess specific heat associated with low-frequency resonance modes due to Pb impurities in Mg. The data points were obtained by measuring the change in heat capacity relative to pure Mg. The solid curves are theoretical predictions. [J.A. Cape, G.W. Lehman, W.V. Johnston, and R.E. De Wames, *Phys. Rev. Letters* **16**: 892 (1966).]

Thus we have seen that the study of correlation functions of atomic displacements and momenta not only provides a means for understanding a variety of interaction processes between external probes and the atomic vibrations in crystals, but also gives us a convenient way of introducing the concepts of localized and resonance modes into the dynamical theory of imperfect crystals.

ACKNOWLEDGMENT

I should like to thank Dr. G. F. Nardelli for his invitation to participate in this conference, and for the hospitality extended to me in Milan.

## REFERENCES

1. L.I. Schiff, *Quantum Mechanics*, McGraw-Hill Book Co., Inc., New York, 1949, p. 195.
2. W.E. Lamb, Jr., *Phys. Rev.* **55**: 190 (1939).
3. This result follows, for example, from equation (8.14) of the article by A. A. Maradudin in *Astrophysics and the Many-Body Problem* (W. A. Benjamin, Inc., New York, 1963) p. 107, if the dielectric tensor is related to the dielectric susceptibility tensor by $\epsilon_{\mu\nu}(\omega) = \delta_{\mu\nu} + 4\pi\chi_{\mu\nu}(\omega)$.
4. M. Born and K. Huang, *Dynamical Theory of Crystal Lattices*, Oxford University Press, Oxford, 1954, p. 368.
5. L. Van Hove, *Phys. Rev.* **95**: 249 (1954).
6. K.S. Singwi and A. Sjölander, *Phys. Rev.* **120**: 1093 (1960).
7. A.A. Maradudin, *Solid State Physics* **19**: 1 (1966).
8. Reference 4, p. 219.
9. A.A. Maradudin, E.W. Montroll, and G.H. Weiss, *Theory of Lattice Dynamics in the Harmonic Approximation*, Academic Press, Inc., New York, 1963, p. 148.
10. A.A. Maradudin, in *Phonons and Phonon Interactions* edited by T.A. Bak, W.A. Benjamin, Inc., New York, 1964, p. 451.
11. G.O. Jones and J.M. Woodfine, *Proceedings IXth International Conference on Low Temperature Physics*, Plenum Press, New York, 1965, p. 1089.
12. Nguyen Xuan Xinh, Westinghouse Research Laboratories Scientific Paper 75–9F5–442–P8, Sept. 3, 1965 (unpublished).
13. A.A. Maradudin, *Rev. Mod. Phys.* **36**: 417 (1964).
14. A.A. Maradudin, *Solid State Physics* **18**: 273 (1966).
15. R. Rubin, J. Peretti, G. Verdan, and W. Kley, *Phys. Letters* **14**: 100 (1965).
16. Reference 10, p. 447.
17. L. Van Hove, *Phys. Rev.* **89**: 1189 (1953).
18. R.J. Elliott and A.A. Maradudin, in *Inelastic Scattering of Neutrons*, International Atomic Energy Agency, Vienna, 1965.
19. H. Bjerrum-Moller and A.R. Mackintosh, *Phys. Rev. Letters* **15**: 623 (1965).
20. P.G. Klemens, *Phys. Rev.* **125**: 1795 (1962).
21. D.W. Feldman, J.G. Castle, Jr., and J. Murphy, *Phys. Rev.* **138**: A1208 (1965).
22. S. Ganesan and R. Srinivasan, *Can. J. Phys.* **40**: 74 (1962).

23. D.L. Mills, *Phys. Rev.* **146**: 336 (1966).
24. G. Lehman and R.E. DeWames, *Phys. Rev.* **131**: 1008 (1963).
25. Yu. M. Kagan and Ya. A. Iosilevskii, *Zhur. Eksperim. i Teor. Fiz.* **45**: 819 (1963) [English translation: *Soviet Physics-JETP* **18**: 562 (1964)].
26. G. Kh. Panova and B.N. Samoilov, *Zh. Eksperim. i Teor. Fiz.* **49**: 456 (1965) [English translation: *Soviet Physics-JETP* **22**: 320 (1966)].
27. J.A. Cape, G.W. Lehman, W.V. Johnston, and R.E. DeWames, *Phys. Rev. Letters* **16**: 892 (1966).

# Surface Spin Waves[*]

## I. P. Ipatova and A. A. Klochikhin

*A.F. Ioffe Physico-Technical Institute*
*Leningrad, USSR*

and

## A. A. Maradudin and R. F. Wallis

*University of California*
*Irvine, California*

It is well known from the theory of elasticity ([1]) that a semi-infinite crystal possessing at least one stress-free boundary surface can have exceptional vibration modes which are localized in the vicinity of the boundary surface. These modes are the so-called Rayleigh waves in which the displacement pattern, while wavelike in directions parallel to the free surface, decays exponentially in amplitude with increasing distance into the crystal from the surface.

Because of the formal similarities between the theory of spin waves in ferromagnetic crystals and the theory of lattice dynamics in the harmonic approximation, when the spin operators of the former theory and the displacement amplitudes of the latter are expanded in terms of phonon creation and destruction operators ([2]), it is natural to ask if surface spin waves which are the analogs of the Rayleigh surface waves of elasticity theory can exist in semi-infinite ferromagnetic crystals. In the present note we give an affirmative answer to this question.

The method of demonstration employed here is one which has been used previously ([3]) for the study of the dynamical properties of atoms in the surface layers of a crystal. We assume a ferromagnetic simple cubic crystal of lattice parameter $a_0$. The position vector of the $l$th atom in the crystal is given by

---

[*]This paper was presented at the Conference on Localized Excitations in Solids, Milan, July 25–26, 1966. The research at the University of California was supported by the Air Force Office of Scientific Research, Office of Aerospace Research, United States Air Force, under AFOSR Grant Number 1080–66.

$$\mathbf{x}(l) = l_1 \mathbf{a}_1 + l_2 \mathbf{a}_2 + l_3 \mathbf{a}_3 \tag{1}$$

where the three primitive noncoplanar translation vectors $\mathbf{a}_1, \mathbf{a}_2, \mathbf{a}_3$ are given explicitly by

$$\mathbf{a}_1 = a_0(1,0,0) \qquad \mathbf{a}_2 = a_0(0,1,0) \qquad \mathbf{a}_3 = a_0(0,0,1) \tag{2}$$

and where $l_1, l_2, l_3$ are three integers, which can be positive, negative, or zero, and to which we refer collectively as $l$.

The Hamiltonian for the crystal is given by

$$H_0 = -Hg\beta \sum_l S_z(l) - \tfrac{1}{2} \sum_{ll'}{}' 2J(ll')\mathbf{S}(l)\cdot\mathbf{S}(l') \tag{3}$$

In equation (3), $\mathbf{S}(l)$ is the spin angular momentum operator of the atom at $\mathbf{x}(l)$, $H$ is the strength of an externally applied magnetic field, which has been assumed to be directed along the $z$-axis, $g$ is the spectroscopic splitting factor, and $\beta$ is the Bohr magneton. $J(ll')$ is the exchange integral between the atoms at the sites $l$ and $l'$. It will be assumed to depend on $l$ and $l'$ only through the magnitude $|\mathbf{x}(l) - \mathbf{x}(l')|$ of the separation between these sites, and to be non-zero only when $l$ and $l'$ are nearest or next-nearest neighbor sites. Finally, the prime on the second sum on the right-hand side of equation (3) means that the terms with $l = l'$ are to be omitted.

We assume further that the spin angular momentum operator $\mathbf{S}(l)$ associated with the $l$th atom obeys the cyclic boundary condition in a macro-crystal which has the shape of a cube whose edges have a length $La_0$, and which can be taken to be the crystal of physical interest. In the present context, the cyclic boundary condition takes the form

$$\mathbf{S}(l_1 + n_1 L, l_2 + n_2 L, l_3 + n_3 L) = \mathbf{S}(l_1, l_2, l_3) \tag{4}$$

where $n_1, n_2, n_3$ are arbitrary integers. Thus, the summation variables $l_1$, $l_2, l_3$, and $l'_1, l'_2, l'_3$ in equation (3) can each be assumed to take on the values $-L/2 + 1, \ldots, L/2$. The total number of atoms in the crystal, $L^3$, will be denoted by $N$.

We now carry out the Holstein–Primakoff [2] transformation from the spin angular momentum operators $\{\mathbf{S}(l)\}$ to phononlike creation and destruction operators $\{b_l^+\}$ and $\{b_l\}$ according to

$$
\begin{aligned}
S_x(l) &= (S/2)^{1/2}(b_l + b_l^+) \\
S_y(l) &= (1/i)(S/2)^{1/2}(b_l - b_l^+) \\
S_z(l) &= S - b_l^+ b_l
\end{aligned}
\tag{5}
$$

where $S$ is the magnitude of the spin on each atom in units of $\hbar$, and the operators $b_l^+$ and $b_l$ obey the commutation rules

$$[b_l, b_{l'}^+] = \delta_{ll'} \qquad [b_l, b_{l'}] = [b_l^+, b_{l'}^+] = 0 \qquad (6)$$

Substitution of equation (5) into equation (3), and the neglect of all terms of higher order in $b_l^+$ and $b_l$ than quadratic, yields the form of the Hamiltonian for a Heisenberg ferromagnet which provides the starting point for our subsequent discussion:

$$H_0 = E_0 + Hg\beta \sum_l b_l^+ b_l + S \sum_{ll'}{}' J(ll')(b_l^+ - b_{l'}^+)(b_l - b_{l'}) \qquad (7a)$$

$$E_0 = -g\beta SNH - S^2 \sum_{ll'}{}' J(ll') \qquad (7b)$$

The energy $E_0$ is the energy when all of the atomic magnets point in the direction of the external magnetic field $\mathbf{H} = (0, 0, H)$. In what follows we regard the Hamiltonian $H_0$ as the unperturbed Hamiltonian for the crystal.

We now create a pair of adjacent free surfaces in the crystal by setting equal to zero the exhange interactions between the spins on atoms which are on opposite sides of the plane $z = a_0/2$, where $a_0$ is the lattice parameter. We do this mathematically by subtracting such terms from the right-hand side of equation (7a). Thus, the Hamiltonian which describes a semi-infinite ferromagnetic crystal which is bounded by a pair of planes normal to the $z$-axis can be written as

$$H = H_0 - V \qquad (8a)$$

where

$$V = 2S \sum_{ll'}{}' \delta_{l_30}\delta_{l_3'1}J(ll')(b_l^+ - b_{l'}^+)(b_l - b_{l'}) \qquad (8b)$$

If we keep in mind the fact that the operators $\{\mathbf{S}(l)\}$, and therefore the operators $\{b_l^+\}$ and $\{b_l\}$, satisfy the cyclic boundary condition, we see that by cutting the crystal in the way we have, we have created a platelet of thickness $La_0$ in the $z$-direction and of infinite extent in the $x$- and $y$-directions. In each of these two directions the spin pattern is still periodic with periods $La_0$.

It is now convenient to take advantage of the fact that the cut crystal is still invariant against rigid body displacements of the type $l_1\mathbf{a}_1 + l_2\mathbf{a}_2$ (that is, parallel to the free surface) by re-expressing the Hamiltonian $H$ in terms of new operators $b_\mathbf{k}^+$ and $b_\mathbf{k}$ which are defined by

$$b_l = \frac{1}{N^{1/2}} \sum_\mathbf{k} e^{-i\mathbf{k}\cdot\mathbf{x}(l)} b_\mathbf{k}$$

$$b_l^+ = \frac{1}{N^{1/2}} \sum_\mathbf{k} e^{i\mathbf{k}\cdot\mathbf{x}(l)} b_\mathbf{k}^+ \qquad (9)$$

Equations (6) and (9) together imply that the operators $b_k^+$ and $b_k$ obey the following commutation rules:

$$[b_k, b_{k'}^+] = \Delta(k - k') \qquad [b_k^+, b_{k'}^+] = [b_k, b_{k'}] = 0 \tag{10}$$

where $\Delta(k)$ equals unity if $k$ equals zero, modulo $|2\pi$ times a reciprocal lattice vector, and vanishes otherwise. The wavevector $k$ appearing in these expressions takes on $N$ discrete values which are specified by the periodic boundary condition and which are uniformly distributed throughout the first Brillouin zone of the crystal with a density $\Omega/(2\pi)^3$, where $\Omega$ is the crystal volume.

In terms of these new operators the Hamiltonian $H$ becomes

$$H = E_0 + \sum_k \hbar\omega(k) b_k^+ b_k - \hbar \sum_{kk'} \delta(k_x - k_x')\delta(k_y - k_y') \sum_l V_l^*(k) V_l(k') b_k^+ b_{k'} \tag{11a}$$

where

$$\hbar\omega(k) = Hg\beta + 2S \sum_l{}' J(l)\,[1 - \cos k \cdot x(l)] \tag{11b}$$

and

$$V_l(k) = \delta_{l_3, -1}\left(\frac{2SJ(\bar{l})}{\hbar L}\right)^{1/2}\{1 - \exp[ik \cdot x(\bar{l})]\}$$
$$= V_l^*(-k) \tag{11c}$$

The symbol $\delta(k_x)$ equals unity if $k_x$ equals zero, modulo $2\pi$ times the $x$-component of a reciprocal lattice vector, and vanishes otherwise. In the present context, since both $k$ and $k'$ are restricted to lie inside the first Brillouin zone of the crystal, $\delta(k_x - k_x')$ is equivalent to the Kronecker symbol on $k_x$ and $k_x'$. The presence of the factor $\delta(k_x - k_x')\,\delta(k_y - k_y')$ in the third term on the right-hand side of equation (11a) is a reflection of the periodicity of the cut crystal in the $x$- and $y$-directions. The prime on the sum on the right-hand side of equation (11b) means that the term with $l = 0$ is to be excluded. The index $\bar{l}$ in (11a) and (11c) labels the relative coordinates of an atom in the plane $z = a_0(l_3 = 1)$ and its nearest neighbor and its four next-nearest neighbors in the plane $z = 0$ ($l_3 = 0$). These five coordinates will be given the following conventional lables in all that follows:

$$
\begin{aligned}
x(\bar{l}) = 1: &\quad a_0(00\bar{1}) \\
2: &\quad a_0(\bar{1}0\bar{1}) \\
3: &\quad a_0(0\bar{1}\bar{1}) \\
4: &\quad a_0(10\bar{1}) \\
5: &\quad a_0(01\bar{1})
\end{aligned} \tag{12}
$$

To obtain the excitation spectrum of the system described by the Hamiltonian (11a), we study the equation of motion of the operator $b_{\mathbf{k}}^+$. This is given by

$$i\hbar \dot{b}_{\mathbf{k}}^+ = [b_{\mathbf{k}}^+, H]$$
$$= -\hbar\omega(\mathbf{k})b_{\mathbf{k}}^+ + \hbar \sum_{\mathbf{h}} \delta(h_x - k_x)\delta(h_y - k_y) \sum_{l} V_l^*(\mathbf{h})V_l(\mathbf{k})b_{\mathbf{h}}^+ \qquad (13)$$

If we assume an harmonic time-dependence for $b_{\mathbf{k}}^+$, i.e.,

$$\dot{b}_{\mathbf{k}}^+ = i\omega b_{\mathbf{k}}^+ \qquad (14)$$

[the choice of the sign of the right-hand side being dictated by the fact that $b_{\mathbf{k}}^+$ is a creation operator], equation (13) can be written as

$$\omega b_{\mathbf{k}}^+ = \omega(\mathbf{k})b_{\mathbf{k}}^+ - \sum_{\mathbf{h}} \delta(h_x - k_x)\delta(h_y - k_y) \sum_{l} V_l^*(\mathbf{h})V_l(\mathbf{k})b_{\mathbf{h}}^+ \qquad (15)$$

Introducing an operator $d_l^+(k_x k_y)$ by the relation

$$d_l^+(k_x k_y) = \sum_{\mathbf{h}} \delta(h_x - k_x)\delta(h_y - k_y)V_l^*(\mathbf{h})b_{\mathbf{h}}^+ \qquad (16)$$

we can rewrite equation (15) further as

$$b_{\mathbf{k}}^+ = -\frac{1}{\omega - \omega(\mathbf{k})} \sum_{l} V_l(\mathbf{k})d_l^+(k_x k_y) \qquad (17)$$

From this equation and (16) it follows that $d_l^+(k_x k_y)$ satisfies the homogeneous equation

$$d_l^+(k_x k_y) = -\sum_{l'} M_{ll'}(k_x k_y; \omega) \, d_{l'}^+(k_x k_y) \qquad (18)$$

where $\mathbf{M}(k_x k_y; \omega)$ is a $5 \times 5$ matrix whose elements are given by

$$M_{ll'}(k_x k_y; \omega) = \sum_{k_z} [V_l^*(\mathbf{k})V_{l'}(\mathbf{k})]/[\omega - \omega(\mathbf{k})] \qquad (19)$$

If we substitute into Eq. (19) the expression for $V_l(\mathbf{k})$ given by Eq. (11c), we find that $M_{ll'}(k_x k_y; \omega)$ takes the form

$$M_{ll'}(k_x k_y; \omega) = \exp\left[-\frac{i}{2}a_0(k_x \bar{l}_1 + k_y \bar{l}_2)\right]$$
$$\times m_{ll'}(k_x k_y; \omega) \exp\left[\frac{i}{2}a_0(k_x \bar{l}_1' + k_y \bar{l}_2')\right] \qquad (20)$$

where

$$m_{ll'}(k_x k_y; \omega) = \frac{8S}{\hbar L}[J(\bar{l})J(\bar{l}')]^{1/2} \sum_{\mathbf{k}} [\sin \tfrac{1}{2}\mathbf{k}\cdot\mathbf{x}(\bar{l}) \sin \tfrac{1}{2}\mathbf{k}\cdot\mathbf{x}(\bar{l}')]/[\omega - \omega(\mathbf{k})] \quad (21)$$

In contrast with the matrix $\mathbf{M}(k_x k_y; \omega)$, which is merely Hermitian, the matrix $\mathbf{m}(k_x k_y; \omega)$ is real and symmetric in $\bar{l}$ and $\bar{l}'$. Equation (18) can therefore be rewritten in the form

$$\left\{ \exp\left[ \frac{i}{2} a_0(k_x \bar{l}_1 + k_y \bar{l}_2) \right] d_{\bar{l}}^+(k_x k_y) \right\} = -\sum_{\bar{l}'} m_{\bar{l}\bar{l}'}(k_x k_y; \omega)$$
$$\times \left\{ \exp\left[ \frac{i}{2} a_0(k_x \bar{l}_1' + k_y \bar{l}_2') \right] d_{\bar{l}'}^+(k_x k_y) \right\} \tag{22}$$

The condition that this set of equations have nontrivial solutions is that the determinant of the coefficients of the $\{ \exp[(i/2)a_0(k_x \bar{l}_1 + k_y \bar{l}_2)] d_{\bar{l}}(k_x k_y) \}$ vanishes:

$$|\mathbf{I} + \mathbf{m}(k_x k_y; \omega)| = 0 \tag{23}$$

The solution of this equation gives us the desired relation between $\omega$ and the wavevector $\mathbf{k}$.

From equation (11b) and the assumption that $J(l)$ is nonzero only for interactions between nearest and next-nearest neighbor spins, we find that $\hbar\omega(\mathbf{k})$ is given by

$$\begin{aligned}
\hbar\omega(\mathbf{k}) = {}& Hg\beta + 4SJ_1(3 - \cos a_0 k_x - \cos a_0 k_y - \cos a_0 k_z) \\
& + 8SJ_2(3 - \cos a_0 k_x \cos a_0 k_y - \cos a_0 k_y \cos a_0 k_z \\
& - \cos a_0 k_z \cos a_0 k_x)
\end{aligned} \tag{24}$$

where $J_1$ and $J_2$ are the nearest and next-nearest neighbor exchange integrals, respectively. In all that follows it will be convenient to rewrite equation (24) as

$$\hbar\omega(\mathbf{k}) = f(k_x, k_y) - g(k_x, k_y) \cos a_0 k_z \tag{25a}$$

with

$$\begin{aligned}
f(k_x, k_y) = {}& Hg\beta + 4SJ_1(3 - \cos a_0 k_x - \cos a_0 k_y) \\
& + 8SJ_2(3 - \cos a_0 k_x \cos a_0 k_y)
\end{aligned} \tag{25b}$$

$$g(k_x, k_y) = 4SJ_1 + 8SJ_2 (\cos a_0 k_x + \cos a_0 k_y) \tag{25c}$$

We will be interested in those solutions of equation (23) that are characterized by the conditions

$$f(k_x, k_y) - \hbar\omega > g(k_x, k_y) > 0 \tag{26}$$

The second of these two conditions together with equation (25c) requires that

$$J_1 > 4J_2 \tag{27}$$

When the conditions (26) are satisfied, $\hbar\omega - \hbar\omega(\mathbf{k})$ is negative-definite, and the summation over $k_z$ on the right-hand side of equation (21) can be replaced by integration according to

$$\sum_{k_z} \longrightarrow \frac{La_0}{2\pi} \int_{-\pi/a_0}^{\pi/a_0} dk_z \tag{28}$$

We therefore find that

$$m_{II'}(k_x k_y; \omega) =$$

$$-8S[J(\bar{l})J(\bar{l}')]^{1/2} \frac{a_0}{2\pi} \int_{-\pi/a_0}^{\pi/a_0} dk_z \frac{1}{(f - \hbar\omega) - g\cos a_0 k_z}$$

$$\times \left[ \sin \frac{a_0}{2}(k_x \bar{l}_1 + k_y \bar{l}_2) \sin \frac{a_0}{2}(k_x \bar{l}'_1 + k_y \bar{l}'_2) \cos^2 \frac{a_0}{2} k_z \right.$$

$$\left. + \cos \frac{a_0}{2}(k_x \bar{l}_1 + k_y \bar{l}_2) \cos \frac{a_0}{2}(k_x \bar{l}'_1 + k_y \bar{l}'_2) \sin^2 \frac{a_0}{2} k_z \right] \tag{29}$$

where only terms which are even in $k_z$ have been retained in the integrand. With the change of variables

$$a_0 k_x = x, \text{ etc.} \tag{30}$$

and some rearrangement of terms, we can rewrite $m_{II'}(k_x k_y; \omega)$ as

$$m_{II'}(k_x k_y; \omega) = -4S[J(\bar{l})J(\bar{l}')]^{1/2}$$

$$\times \left\{ \cos \tfrac{1}{2}[x(\bar{l}_1 - \bar{l}'_1) + y(\bar{l}_2 - \bar{l}'_2)] \frac{1}{\pi} \int_0^\pi \frac{dz}{(f - \hbar\omega) - g\cos z} \right.$$

$$\left. - \cos \tfrac{1}{2}[x(\bar{l}_1 + \bar{l}'_1) + y(\bar{l}_2 + \bar{l}'_2)] \frac{1}{\pi} \int_0^\pi \frac{\cos z \, dz}{(f - \hbar\omega) - g\cos z} \right\}$$

$$= -\frac{4S[J(l)J(l')]^{1/2}}{\alpha} \{ \cos \tfrac{1}{2}[x(\bar{l}_1 - \bar{l}'_1) + y(\bar{l}_2 + \bar{l}'_2)]$$

$$- \beta \cos \tfrac{1}{2}[x(\bar{l}_1 + \bar{l}'_1) + y(\bar{l}_2 + \bar{l}'_2)] \} \tag{31a}$$

where

$$\alpha = [(f - \hbar\omega)^2 - g^2]^{1/2} \tag{31b}$$

$$\beta = \frac{(f - \hbar\omega) - [(f - \hbar\omega)^2 - g^2]^{1/2}}{g} \tag{31c}$$

From the results given by equation (31), or perhaps more easily from equation (29), we see that we can weaken the first condition in (26) to

$$f(k_x, k_y) - \hbar\omega \geq g(k_x, k_y) \tag{26a}$$

provided that the equality occurs only for $k_x = k_y = 0$, because at this point the integral on the right-hand side of (29) is still convergent.

We can now write down the explicit expressions for the elements of the matrix $\mathbf{m}(k_x k_y; \omega)$, using the labeling convention given in equation (12):

$$m_{11} = -\frac{4SJ_1}{\alpha}(1 - \beta) = -A$$

$$m_{12} = m_{14} = -\frac{4S(J_1 J_2)^{1/2}}{\alpha}(1 - \beta) \cos \tfrac{1}{2}x = -D$$

$$m_{13} = m_{15} = -\frac{4S(J_1 J_2)^{1/2}}{\alpha}(1 - \beta) \cos \tfrac{1}{2}y = -E$$

$$m_{22} = m_{44} = -\frac{4SJ_2}{\alpha}(1 - \beta \cos x) = -B$$

$$m_{23} = m_{45} = -\frac{4SJ_2}{\alpha}[\cos \tfrac{1}{2}(x - y) - \beta \cos \tfrac{1}{2}(x + y)] = -F \tag{32}$$

$$m_{24} = -\frac{4SJ_2}{\alpha}(\cos x - \beta) = -G$$

$$m_{25} = m_{34} = -\frac{4SJ_2}{\alpha}[\cos \tfrac{1}{2}(x + y) - \beta \cos \tfrac{1}{2}(x - y)] = -H$$

$$m_{33} = m_{55} = -\frac{4SJ_2}{\alpha}(1 - \beta \cos y) = -C$$

$$m_{35} = -\frac{4SJ_2}{\alpha}(\cos y - \beta) = -J$$

It follows, therefore, that the matrix $\mathbf{I} + \mathbf{m}(k_x k_y; \omega)$ has the following form:

$$\mathbf{I} + \mathbf{m} = \begin{vmatrix} 1 - A & -D & -E & -D & -E \\ -D & 1 - B & -F & -G & -H \\ -E & -F & 1 - C & -H & -J \\ -D & -G & -H & 1 - B & -F \\ -E & -H & -J & -F & 1 - C \end{vmatrix} \tag{33}$$

It has been found ([3b]) that the matrix $\mathbf{S}$ given by

$$\mathbf{S} = \begin{vmatrix} \frac{1}{\sqrt{2}} & 0 & 0 & 0 & 0 \\ 0 & \frac{1}{2} & 0 & -1 & 0 \\ 0 & 0 & \frac{1}{2} & 0 & -1 \\ 0 & \frac{1}{2} & 0 & 1 & 0 \\ 0 & 0 & \frac{1}{2} & 0 & 1 \end{vmatrix} \quad \mathbf{S}^{-1} = \begin{vmatrix} \sqrt{2} & 0 & 0 & 0 & 0 \\ 0 & 1 & 0 & 1 & 0 \\ 0 & 0 & 1 & 0 & 1 \\ 0 & -\frac{1}{2} & 0 & \frac{1}{2} & 0 \\ 0 & 0 & -\frac{1}{2} & 0 & \frac{1}{2} \end{vmatrix} \tag{34}$$

block diagonalizes the matrix $\mathbf{I} + \mathbf{m}$ by a similarity transformation:

$$S^{-1}(\mathbf{I} + \mathbf{M})S = \begin{vmatrix} 1 - A & -\sqrt{2}D & -\sqrt{2}E & 0 & 0 \\ -\sqrt{2}D & 1 - B - G & -(F + H) & 0 & 0 \\ -\sqrt{2}E & -(F + H) & 1 - C - J & 0 & 0 \\ 0 & 0 & 0 & 1 - B + G & -(F - H) \\ 0 & 0 & 0 & -(F - H) & 1 - C + J \end{vmatrix}$$

(35)

The equation for $\omega$, equation (23), accordingly reduces to the pair of equations

$$\begin{vmatrix} 1 - A & -\sqrt{2}D & -\sqrt{2}E \\ -\sqrt{2}D & 1 - B - G & -(F + H) \\ -\sqrt{2}E & -(F + H) & 1 - C - J \end{vmatrix} = 0 \qquad (36a)$$

$$\begin{vmatrix} 1 - B + G & -(F - H) \\ -(F - H) & 1 - C + J \end{vmatrix} = 0 \qquad (36b)$$

We look first at the second of these two equations. The heuristic motivation for making this choice is that it is equation (36b) which in the lattice dynamical analog of the present problem yields the dispersion relation for Rayleigh surface waves. Comparing equations (23) and (33), we obtain the results

$$1 - B + G = 1 - \frac{8SJ_2}{\alpha}(1 + \beta) \sin^2 \tfrac{1}{2}x$$

$$H - F = -\frac{8SJ_2}{\alpha}(1 + \beta) \sin \tfrac{1}{2}x \sin \tfrac{1}{2}y$$

$$1 - C + J = 1 - \frac{8SJ_2}{\alpha}(1 + \beta) \sin^2 \tfrac{1}{2}y$$

whereupon equation (36b) becomes

$$1 - \frac{8SJ_2}{\alpha}(1 + \beta)(\sin^2 \tfrac{1}{2}x + \sin^2 \tfrac{1}{2}y) = 0 \qquad (37)$$

The solutions of equation (37) are readily found to be

$$f - \hbar\omega = -g \qquad (38a)$$

$$f - \hbar\omega = g + \frac{32SJ_2^2}{J_1 + 4J_2}(\sin^2 \tfrac{1}{2}x + \sin^2 \tfrac{1}{2}y)^2 \qquad (38b)$$

Only the second of these solutions is compatible with the conditions given by equations (26) and (27). It can be rewritten in its final form as

$$
\begin{aligned}
\hbar\omega = {} & Hg\beta + 8S(J_1 + 4J_2)(\sin^2 \tfrac{1}{2}x + \sin^2 \tfrac{1}{2}y) \\
& - 32SJ_2\{\sin^2 \tfrac{1}{2}x \sin^2 \tfrac{1}{2}y \\
& + \frac{J_2}{J_1 + 4J_2}(\sin^2 \tfrac{1}{2}x + \sin^2 \tfrac{1}{2}y)^2\}
\end{aligned} \tag{39}
$$

In the long-wavelength limit the result given by (39) reduces to

$$
\begin{aligned}
\hbar\omega = {} & Hg\beta + 2S(J_1 + 4J_2)(x^2 + y^2) \\
& - \tfrac{1}{6}S(J_1 + 4J_2)(x^4 + y^4) \\
& - 2SJ_2\left[x^2y^2 + \frac{J_2}{J_1 + 4J_2}(x^4 + 2x^2y^2 + y^4)\right] + O(k^6)
\end{aligned} \tag{40}
$$

When we compare this expansion with the long-wavelength limit of the dispersion relation for spin waves in an infinite crystal, equation (24),

$$
\begin{aligned}
\hbar\omega(\mathbf{k}) = {} & Hg\beta + 2S(J_1 + 4J_2)(x^2 + y^2 + z^2) \\
& - \tfrac{1}{6}S(J_1 + 4J_2)(x^4 + y^4 + z^4) \\
& - 2SJ_2(x^2y^2 + y^2z^2 + z^2x^2) + O(k^6)
\end{aligned} \tag{41}
$$

we seé that the coefficients of the terms of $O(k^2)$ are the same in both expansions, and that it is only in the terms of $O(k^4)$ that the two expressions differ in an essential way. This is in contrast with the dispersion relation for Rayleigh surface waves, which already differs in the coefficient of the lowest power of $\mathbf{k}$ from the dispersion relation for bulk waves.

We must now study the $3 \times 3$ determinantal equation (36a) to see whether it has solutions which satisfy the condtions (26) and (27). In terms of the parameters

$$
\gamma = \frac{J_1}{J_2} \qquad \xi = 4SJ_2\frac{1 - \beta}{\alpha} \tag{42}
$$

Equation (36a) takes the simple form

$$
\begin{vmatrix}
1 - \gamma\xi & -(2\gamma)^{1/2}\xi \cos \tfrac{1}{2}x & -(2\gamma)^{1/2}\xi \cos \tfrac{1}{2}y \\
-(2\gamma)^{1/2}\xi \cos \tfrac{1}{2}x & 1 - \xi(1 + \cos x) & -2\xi \cos \tfrac{1}{2}x \cos \tfrac{1}{2}y \\
-(2\gamma)^{1/2}\xi \cos \tfrac{1}{2}y & -2\xi \cos \tfrac{1}{2}x \cos \tfrac{1}{2}y & 1 - \xi(1 + \cos y)
\end{vmatrix} = 0 \tag{43}
$$

When the determinant is expanded, many terms cancel, and equation (43) becomes

$$1 - \gamma\xi(\cos^2 \tfrac{1}{2}x + \cos^2 \tfrac{1}{2}y) = 0$$

or

$$\frac{\alpha}{1 - \beta} = 4SJ_1 + 8SJ_2 (\cos^2 \tfrac{1}{2}x + \cos^2 \tfrac{1}{2}y) \tag{44}$$

The solutions of equation (44) are readily found to be

$$f - \hbar\omega = g \tag{45a}$$

$$f - \hbar\omega = -g + \frac{[4SJ_1 + 8SJ_2(\cos^2 \tfrac{1}{2}x + \cos^2 \tfrac{1}{2}y)]^2}{2S(J_1 + 4J_2)} \tag{45b}$$

The first solution is not acceptable for $x, y \neq 0$, as it leads to divergences in the expression for $m_{II'}(k_x k_y; \omega)$ [see equation (31)]. The second solution is acceptable, according to (26), provided that the condition

$$\frac{[4SJ_1 + 8SJ_2(\cos^2 \tfrac{1}{2}x + \cos^2 \tfrac{1}{2}y)]^2}{2S(J_1 + 4J_2)} \geq 2g$$
$$= 8SJ_1 + 16SJ_2(\cos x + \cos y) \tag{46}$$

is satsified. In view of the remark following Eq. (31c), the equality in Eq. (46) can obtain only for $x = y = 0$, which is indeed the case. The inequality (46) is equivalent to the inequality

$$[J_1 + J_2(2 + \cos x + \cos y)]^2$$
$$- (J_1 + 4J_2)[J_1 + 2J_2 (\cos x + \cos y)] \geq 0$$

which can be rearranged to read

$$J_2^2(2 - \cos x - \cos y)^2 \geq 0 \tag{47}$$

Inasmuch as this condition is always satisfied, we conclude that the solution given by equation (45b) is an acceptable one. In fact, we can rewrite (45b) in the form

$$\hbar\omega = Hg\beta + 8S(J_1 + 4J_2)(\sin^2 \tfrac{1}{2}x + \sin^2 \tfrac{1}{2}y)$$
$$- 32SJ_2\{\sin^2 \tfrac{1}{2}x \sin^2 \tfrac{1}{2}y$$
$$+ \frac{J_2}{J_1 + 4J_2} (\sin^2 \tfrac{1}{2}x + \sin^2 \tfrac{1}{2}y)^2\}$$

whereupon we see that it is exactly the dispersion relation we obtained earlier [see equation (39)] as the solution of equation (36b). The result has no counterpart in the analogous lattice dynamical problem, in that in the latter context the equation analogous to equation (36a) has no solutions in the long-wavelength limit.

Because the solution given by equation (39) gives a dispersion relation for spin waves which is not contained within the spin-wave spectrum of a perfectly periodic, infinitely extended crystal, equation (24), and, in fact, owes its existence to the presence of the pair of free surfaces on our crystal model, we would be justified in referring to the spin waves possessing this dispersion relation as "surface spin waves." However, this name ordinarily implies that the these spin waves are also localized in the vicinity of the free surfaces. A rigorous demonstration of this localization property would require the determination of the eigenvector(s) of the matrix $\mathbf{m}(k_x k_y; \omega)$ for $\omega$ given by equation (39). Rather than following this rigorous course, we give below a simpler, and more heuristic, demonstration of the localization of the spin waves described by (39) in the vicinity of the crystal surfaces.

We begin by formally expanding the operators $b_l^+$ and $b_l$ in terms of the operators $b_s^+$ and $b_s$ which diagonalize the Hamiltonian (8):

$$b_l^+ = \sum_s B_s(l) b_s^+$$

$$b_l = \sum_s B_s(l) b_s \tag{48}$$

In these equations the index $s$ takes the values $1, 2, \ldots, N$, and the coefficients $\{B_s(l)\}$ can be chosen to be real with no loss of generality. They are the components of the eigenvectors of a matrix $\mathbf{D}$ whose elements are

$$D(ll') = Hg\beta\delta_{ll'} - 2S\mathscr{J}(ll') + \delta D(ll') \tag{49a}$$

where we have introduced the matrices

$$\mathscr{J}(ll') = J(ll') \qquad l \neq l'$$

$$\mathscr{J}(ll) = -\sum_{l'(\neq l)} \mathscr{J}(ll') \tag{49b}$$

and

$$\delta D(ll') = 2S\mathscr{J}(ll')[\delta_{l_3 0}\delta_{l_3' 1} + \delta_{l_3' 0}\delta_{l_3 1}] \qquad l \neq l'$$

$$\delta D(ll) = -\sum_{l'(\neq l)} \delta D(ll') \tag{49c}$$

The matrix $\mathbf{D}$ is a real, symmetric, $3N \times 3N$ matrix, whose eigenvalue equation will be written in the form

$$\sum_{l'} D(ll') B_s(l') = \hbar\omega_s B_s(l) \qquad s = 1, 2, \cdots, N \tag{50}$$

so that the index $s$ labels the $N$ solutions of this matrix equation. The eigenvector components $\{B_s(l)\}$ can be chosen to satisfy the following orthonormality and closure conditions

$$\sum_l B_s(l)B_{s'}(l) = \delta_{ss'}$$

$$\sum_s B_s(l)B_s(l') = \delta_{ll'} \tag{51}$$

These conditions can be used to show that the new operators $b_s^+$ and $b_s$ obey phonon commutation rules:

$$[b_s, b_{s'}^+] = \delta_{ss'} \qquad [b_s, b_{s'}] = [b_s^+, b_{s'}^+] = 0 \tag{52}$$

Substitution of equations (48) into the Hamiltonian (8) together with the relations (50)—(51) yields the Hamiltonian $H$ in the form

$$H = E_0 + \sum_s \hbar\omega_s b_s^+ b_s \tag{53}$$

We can rewrite the eigenvalue equation, (50) in the form

$$\sum_{l'} [\hbar\omega_s \delta_{ll'} - D_0(ll')]B_s(l') = \sum_{l'} \delta D(ll')B_s(l') \tag{54}$$

where the elements of the matrix $\mathbf{D}_0$ are given by the first two terms on the right-hand side of equation (49a). If we now introduce a $3N \times 3N$ matrix $\mathbf{G}(\omega)$ whose elements are given by

$$G(ll'; \omega) = [\hbar\omega\mathbf{I} - \mathbf{D}_0]_{ll'}^{-1} = \frac{1}{N}\sum_{\mathbf{k}} \frac{\exp\{i\mathbf{k}\cdot[\mathbf{x}(l) - \mathbf{x}(l')]\}}{\hbar\omega - \hbar\omega(\mathbf{k})} \tag{55}$$

we can write equation (54) equivalently as

$$B_s(l) = \sum_{l'l''} G(ll'; \omega_s)\delta D(l'l'')B_s(l'') \tag{56}$$

Because the matrix element $\delta D(l'l'')$ is nonzero only if the atom $l'$ is in one of the crystal surfaces, while the atom $l''$ is in the other, the result given by (56) expresses the normal mode amplitude $B_s(l)$ for any atom in the perturbed crystal in terms of the amplitudes of the atoms in the two free surfaces. The surface spin waves can be considered to be localized in the vicinity of the crystal surfaces if the amplitudes $\{B_s(l)\}$ describing these spin waves decay with increasing distance into the crystal from the free surfaces, that is, for increasing values of $|l_3|$.

If we now write the explicit expression for $G(ll')$, equation (55), in the form

$$G(ll' \, \omega_s) = \frac{-1}{L^2} \sum_{k_x k_y} \exp\{ia_0[k_x(l_1 - l_1') + k_y(l_2 - l_2')]\}$$

$$\times \frac{1}{L} \sum_{k_z} \frac{\exp[ia_0 k_z(l_3 - l_3')]}{(f - \hbar\omega_s) - g \cos a_0 k_z} \tag{57}$$

and recall that for surface spin waves the conditions (26) and (26a) are satisfied, we can replace the sum on $k_z$ by integration to obtain

$$G(ll' ; \omega_s) = -\frac{1}{L^2} \sum_{k_x k_y} \exp\{ia_0[k_x(l_1 - l_1') + k_y(l_2 - l_2')]$$

$$\times \frac{\{\{f - \hbar\omega_s - [(f - \hbar\omega_s)^2 - g^2]^{1/2}\}/g\}^{|l_3 - l_3'|}}{[(f - \hbar\omega_s)^2 - g^2]^{1/2}} \tag{58}$$

Because, according to conditions (26),

$$\frac{f - \hbar\omega_s - [(f - \hbar\omega_s)^2 - g^2]^{1/2}}{g} \leq 1 \tag{59}$$

the summand on the right-hand side of Eq. (58) is an exponentially decreasing
function of $|l_3 - l_3'|$, which in the present context is the distance (in units of
$a_0$) from a source point in one of the two free surfaces to a sink point in the
interior of the crystal.

When the result given by equation (58) is substituted into (56), we obtain

$$B_s(l) = \frac{2S}{L^2} \sum_{k_x k_y} \sum_{l_1' l_2'} \sum_{l_1'' l_2''} \exp\{ia_0[k_x(l_1 - l_1') + k_y(l_2 - l_2')]\}$$

$$\times \left\{ \frac{\beta_s|l_3|}{\alpha_s} J(l_1'l_2'0; l_1''l_2''1)[B_s(l_1'l_2'0) - B_s(l_1''l_2''1)] \right.$$

$$\left. + \frac{\beta_s|l_3 - 1|}{\alpha_s} J(l_1'l_2'1; l_1''l_2''0)[B_s(l_1'l_2'1) - B_s(l_1''l_2''0)] \right\} \tag{60}$$

in an obvious notation. Thus, the spin-wave amplitude $B_s(l)$ in a surface wave
is a linear superposition of elementary amplitudes, each of which is characterized by the pair $(k_x k_y)$, each of which decays exponentially with increasing
distance into the crystal from the free surfaces, and each of which is wavelike
in directions parallel to the free surfaces. By analogy with the Rayleigh surface waves in the theory of elasticity we call the exponentially decaying amplitudes labeled by $(k_x, k_y)$ surface spin waves. Although the amplitudes $\{B_s(l)\}$
are linear superpositions of exponentially decaying waves, in general, they
themselves do not decay exponentially with increasing distance into the crystal from the free surfaces. However, they do decrease in some less rapid
fashion with increasing distance into the crystal.

It should not be thought that the assumption of next-nearest neighbor exchange interactions is essential for the occurrence of surface spin waves. While this assumption is essential for the (001) free surface studied so far, it is not essential, for example, for the (011) free surface. For the latter case we find that surface spin waves exist when the exchange interaction is between nearest neighbor spins only. The excitation spectrum is given by

$$\hbar\omega = Hg\beta + 4SJ_1(1 - \cos a_0 k_1) + 4SJ_1 \sin^2 a_0 k_2 \qquad (61)$$

where $k_1$ and $k_2$ are wavevector components in the [100] and [01$\bar{1}$] directions, respectively. The detailed derivation of equation (61) will be presented in a subsequent publication. The characteristic required for the existence of surface spin waves with a nearest neighbor model appears to be that the exchange interaction couple spins whose line of centers is not normal to the surface.

In conclusion we wish to point out that one possibly important ingredient has been omitted from the discussion of surface spin waves which we have given in this paper. This is the surface anisotropy field ([4]) which may lead to pinning of the surface spins. One might expect that surface spin waves will be significantly affected by such pinning of the surface spins. The detailed investigation of the influence of a surface anisotropy field on surface spin waves will be reported in a later publication.

## SUMMARY

The dispersion relation which gives the frequencies of spin waves in a semi-infinite simple cubic Heisenberg ferromagnet has been derived. The semi-infinite crystal is constructed from a perfect, infinitely extended, crystal, obeying periodic boundary conditions, by setting equal to zero all exchange interactions between spins located on atoms on opposite sides of the plane $z = \frac{1}{2}a_0$, where $a_0$ is the lattice parameter. The negatives of these interaction terms are treated as a perturbation on the Hamiltonian of the perfect, infinite crystal, and the equations of motion of the magnon creation operators are obtained. The condition that this set of equations have nontrivial solutions possessing an harmonic time dependence leads to a determinantal equation for the magnon frequencies, which is found to have an isolated root $\omega = \omega(k_x, k_y)$, which lies below the frequencies of the magnons of the infinite crystal. This isolated root, which occurs only when the crystal possesses free boundary surfaces, is identified as the dispersion relation for surface spin waves.

ACKNOWLEDGMENS

One of the authors (A. A. Maradudin) would like to thank Professor N. V. Fedorenko for extending to him the hospitality of the A. F. Ioffe Physico-Technical Institute, where the work described in this note was begun. He would also like to thank the University of California, Irvine, for a travel grant which made possible his visit to the Ioffe Institute.

## REFERENCES

1. Lord Rayleigh, *London Math. Soc. Proc.* **17**: 4 (1887).
2. T. Holstein and H. Primakoff, *Phys. Rev.* **58**: 1098 (1940).
3a. A.A. Maradudin and J. Melngailis, *Phys. Rev.* **133**: A1188 (1964).
3b. A.A. Maradudin and R.F. Wallis, *Phys. Rev.* **148**: 945 (1966).
3c. R.F. Wallis and A.A. Maradudin, *Phys. Rev.* **148**: 962 (1966).
4. C. Kittel, *Phys. Rev.* **110**: 1295 (1958).

# Relation Between Phase Shift
# and Scattering Amplitude in
# Solid-State Scattering Theory*

J. Callaway

*Department of Physics*
*Louisiana State University*
*Baton Rouge, Louisiana*

We consider the scattering of an excitation in a solid by a potential of finite range which does not have any internal degrees of freedom. For example, we might be interested in the scattering of electrons by impurity atoms in a dilute alloy, of phonons by a mass defect in a vibrating lattice, or of spin waves in a ferromagnet by a magnetic defect.

In all these cases, there are two sets of orthonomal states in terms of which the system can be described: wavelike states and localized states. The wavelike states are generalized Bloch states characterized by a wave vector and a band index and they satisfy Bloch's theorem. The localized states (e.q., Wannier functions) are related to the Bloch states by a Fourier transformation (however, $k$-space integrals are always restricted to the Brillouin zone). By the requirement that the scattering potential be of finite range we mean that, on the basis of localized states, the scattering potential may be represented, to a sufficiently good approximation, by a finite matrix.

The scattering process may be described as follows: Suppose a Bloch wave of wavevector $\mathbf{k}'$ in band $\beta$ is incident upon the scattering center. Scattered waves emerge in all other bands and all other wavevectors subject to the requirement of conservation of energy. In particular, suppose $E_\alpha(\mathbf{k}) = E_\beta(\mathbf{k}')$ Then in band $\beta$ there will be a scattered wave with wavevector $\mathbf{k}$. The scattering process is conveniently described by a $t$-matrix element $\langle \alpha\mathbf{k} \,|\, t \,|\, \beta\mathbf{k}' \rangle$, such that the transition probability for the process $|\beta\mathbf{k}'\rangle \longrightarrow |\alpha\mathbf{k}\rangle$ is proportional to the square of this quantity. The scattering amplitude $f_{\alpha\beta}(\mathbf{k}, \mathbf{k}')$ differs from the $t$-matrix element only by a constant factor ([1]).

*Supported by Air Force Office of Scientific Research.
A more complete account of this work has been published: J. Callaway, *Phys. Rev.* **154**: 515 (1967).

The $t$-matrix satisfies the equation

$$t = V + V\mathscr{G}t \tag{1}$$

where $\mathscr{G}$ is the "Green's function"

$$\mathscr{G} = \frac{1}{E^+ - H_0} \tag{2}$$

Equation (1) possesses the solution

$$t = V[1 - \mathscr{G}V]^{-1} \tag{3}$$

In the case of a potential of finite range, the $t$-matrix on the localized basis has the same range as the potential, and is easily constructed using equation (3), since only operations on finite matrices are required. Once $t$ is obtained on the localized basis, it may be easily transformed to the Bloch basis.

Let us define the quantity $D$ to be the determinant of the matrix which appears in equation (3):

$$D = \det[1 - \mathscr{G}V] \tag{4}$$

If the scattering potential has some symmetry, i.e., if it is invariant under some group (which we will take to be the point group of the crystal) then $D$ can be expressed as a product of factors coming from the irreducible representations of the group. This occurs because neither $\mathscr{G}$ nor $V$ will have non-vanishing matrix elements among functions which transform according to different irreducible representations. Thus, if the irreducible representations are denoted by an index $s$, and if we let $g_s$ denote the degeneracy of representation $s$, we have

$$D = \prod_s (D_s)^{g_s} \tag{5}$$

We can, in fact, calculate the subdeterminants $D_s$ directly if we first form symmetrized combinations of the localized basis functions; alternatively we can determine the unitary transformation which will render $\mathscr{G}$ and $V$ block-diagonal. In any event, we have

$$D_s = \det[1 - \mathscr{G}_s V_s] \tag{6}$$

where $\mathscr{G}_s$ and $V_s$ are the portions of the symmetrized Green's function and potential matrices which refer to representation $s$.

With the use of equation (3) and the foregoing arguments, the $t$-matrix elements in the Bloch representation have the form (before symmetrization)

$$\langle \alpha \mathbf{k} \,|\, t \,|\, \beta \mathbf{k}' \rangle = \frac{\Omega}{(2\pi)^3 D} \sum_{\substack{m,n, \\ j,\gamma}} e^{i[\mathbf{k}' \cdot \mathbf{R}_n - \mathbf{k} \cdot \mathbf{R}_m]} (\alpha m \,|\, V \,|\, \gamma j)(\gamma j \,|\, P \,|\, \beta n) \qquad (7)$$

in which $\Omega$ is the volume of the unit cell. We have used the convention that angular brackets $\langle \,|\,$ refer to the Bloch representation, while the round brackets $(\,|\,$ refer to the localized representation. The matrix $P$ is defined by writing

$$[1 - \mathscr{G}V]^{-1} = \frac{P}{D} \qquad (8)$$

Thus $P$ is the "adjugate" matrix. After symmetrization we express the $t$-matrix as a sum of contributions from the representations

$$\langle \alpha \mathbf{k} \,|\, t \,|\, \beta \mathbf{k}' \rangle = \sum_s \langle \alpha \mathbf{k} \,|\, t_s \,|\, \beta \mathbf{k}' \rangle \qquad (9)$$

and we have

$$\langle \alpha \mathbf{k} \,|\, t_s \,|\, \beta \mathbf{k}' \rangle = \frac{1}{D_s} \sum C_s^*(\mathbf{k}, \mathbf{R}_m) V_{s,\alpha m,\gamma j} P_{s,\gamma j,\beta n} C_s(\mathbf{k}', \mathbf{R}_n) \qquad (10)$$

in which the $C_s(\mathbf{k}, \mathbf{R}_m)$ are symmetrized linear combinations of plane waves for representation $s$. (For convenience each $C$ incorporates also a factor of $[\Omega/(2\pi)^3]^{1/2}$. The matrices $V_s$ and $P_s$ are the portions of $V$ and $P$ on the localized basis for representation $s$.

Equations (9) and (10) are as close as we can come in general to the partial wave analysis of the scattering amplitude. They appear to be quite different from the form one becomes familiar with in ordinary quantum mechanics involving, for angular momentum $l$,

$$f_l = \frac{2l + 1}{k} e^{i\delta_l} \sin \delta_l P_l (\cos \theta) \qquad (11)$$

We want to see how phase shifts may be introduced in the solid state problem and study the way in which equation (10) reduces to equation (11) in the limit of spherical symmetry.

To begin with we will suppose that the element of $V$ are real, and for simplicity we will assume we are dealing with a single energy band. For energies $E$ which lie outside this band, either below or above, $\mathscr{G}$ is real. Within the band, $\mathscr{G}$ (and hence $D$) is complex, and so we write

$$D_s = |D_s| e^{-i\delta_s} \qquad (12)$$

The phase $\delta_s$ turns out to be the phase shift of elementary scattering theory in the appropriate limit.

Since the change in the density of states produced by the imperfection can be written as ([2])

$$\Delta N(E) = \sum_s g_s\left(-\frac{1}{\pi} Im\frac{1}{D_s}\frac{dD_s}{dE}\right) \tag{13}$$

we have

$$\Delta N = \frac{1}{\pi}\sum_s g_s\frac{d\delta_s}{dE} \tag{14}$$

Thus the phase shifts are immediately related to the density of states. Let us take the lowest energy in the band to be $E_0$. For a metal, the total change in the number of states lying below some fixed energy $E'$ is then (including a factor of 2 for spin)

$$2\int_{E_0}^{E'}\Delta N(E)dE = \frac{2}{\pi}\sum_s g_s[\delta_s(E') - \delta_s(E_0)] \tag{15}$$

Equation (15) leads to the Fiedel sum rule if we assume the usual argument that in the case of an impurity atom with $Z$ excess electrons, $Z$ states must be brought below the Fermi energy to screen the impurity:

$$\frac{2}{\pi}\sum_s g_s[\delta_s(E_F) - \delta_s(E_0)] = Z \tag{16}$$

We note that, unlike the usual form of the sum rule, this experssion contains only a finite number of terms in the sum.

The phase shifts also satisfy a form of Levinson's theorem. For this case, let us suppose that the imperfection has associated with it $n_s$ bound states for representation $s$ which may be below or above the band, and that these states ane "drawn from the band." In other words, we assume that there are as many eigenstates of $H_0 + V$ as there are of $H_0$ (and no more). Then, if the integral of equation (15) is extended over the whole band (whose maximum energy $E' = E_m$), then the number of states withdrawn from the band is just $n_s$. Thus,

$$\delta_s(E_0) - \delta_s(E_m) = \pi n_s \tag{17}$$

Now let us consider the reduction of the expression for the $t$-matrix, equation (10), to the spherically symmetric limit, equation (11). We will not attempt to demonstrate this in general; instead we will consider a specific example which illustrates what is involved: $d$-wave scattering by a potential extending to first neighbors in a face-centered cubic lattice. This is particularly interesting since the five partial waves of $l = 2$ and differing values of $m$

which are treated together in a spherically symmetric situation contribute, in a cubic lattice, to two different irreducible representations: $\Gamma_{12}$ and $\Gamma_{25'}$ [in the notation of Bouckaert, Smoluchowski, and Wigner ([3])].

First let us note that if we indicate the rows of a degenerate irreducible representation explicitly with an index $v$, we can rewrite equation (10) in the form (with only one band included),

$$\langle \mathbf{k} | t_s | \mathbf{k}' \rangle = \frac{1}{D_s} \sum_{mnj} V_{s,mj} P_{s,jn} \sum_v C^*_{sv}(\mathbf{k}, \mathbf{R}_m) C_{sv}(\mathbf{k}', \mathbf{R}_n) \qquad (18)$$

We have used the fact that the symmetrized potential energy and Green's function matrix elements are independent of the row of the representation. Now we restrict ourselves to the particular situation described above. Then for both representations considered there is only a single Green's function and potential energy matrix element after symmetrization. Thus

$$D_{12} = 1 - \mathscr{G}_{12} V_{12}$$

and

$$D_{25'} = 1 - \mathscr{G}_{25'} V_{25'}$$

The sums over $m$, $n$, and $j$ contain only a single term as a result of our assumption about the range of the potential. For this term, $P = 1$ in both cases. Thus

$$\langle \mathbf{k} | t_s | \mathbf{k}' \rangle = \frac{V_s}{1 - \mathscr{G}_s V_s} \sum_v C^*_{sv}(\mathbf{k}, \mathbf{R}_1) C_{sv}(\mathbf{k}', \mathbf{R}_1) \qquad (19)$$

and $\mathbf{R}_1$ in the argument of each $C$ indicates that the first-neighbor symmetrized functions are used. The formation of the symmetrized combinations of plane waves is just the same as in the calculation of energy bands ([4]). The desired result is obtained only in the limit of small values of $ka$ (where $a$ is the lattice constant). Under these circumstances, the functions $C$ may be expanded and we obtain, for instance, in the case of the "$xy$" row of $\Gamma_{25'}$

$$C_{25',xy}(\mathbf{k}, \mathbf{R}_1) = \frac{1}{2} \left[ \frac{\Omega}{8\pi^3} \right]^{1/2} k_x k_y a^2$$

Similar results are obtained in other cases. We now introduce spherical harmonics. Let the angles $\theta$, $\phi$ specify the orientation of the vector $\mathbf{k}'$ with respect to the crystal axes, and let the angles $\alpha$, $\beta$ similarly specify the orientation of $\mathbf{k}$. Then we have after a straightforward calculation (with $|\mathbf{k}| = |\mathbf{k}'|$)

$$\sum_v C_{25',v}^*(\mathbf{k}, \mathbf{R}_1) C_{25',v}(\mathbf{k}', \mathbf{R}_1) = \frac{\Omega}{(2\pi)^2} \frac{k^4 a^4}{30} K_{25'}(\alpha, \beta; \theta, \phi) \qquad (20a)$$

$$\sum_v C_{12,v}^*(\mathbf{k}, \mathbf{R}_1) C_{12,v}(\mathbf{k}', \mathbf{R}_1) = \frac{\Omega}{(2\pi)^2} \frac{k^4 a^4}{60} K_{12}(\alpha, \beta; \theta, \phi) \qquad (20b)$$

with

$$K_{25'}(\alpha, \beta; \theta, \phi) = Y_{21}^*(\alpha, \beta) Y_{21}(\theta, \phi) + Y_{2,-1}^*(\alpha, \beta) Y_{2,-1}(\theta, \phi)$$
$$+ \tfrac{1}{2} [Y_{2,2}^*(\alpha, \beta) - Y_{2,-2}^*(\alpha, \beta)][Y_{2,2}(\theta, \phi) - Y_{2,-2}(\theta, \phi)] \qquad (20c)$$

and

$$K_{12}(\alpha, \beta; \theta, \phi) = Y_{20}^*(\alpha, \beta) Y_{2,0}(\theta, \phi)$$
$$+ \tfrac{1}{2} [Y_{2,2}^*(\alpha, \beta) + Y_{2,-2}^*(\alpha, \beta)][Y_{2,2}(\theta, \phi) + Y_{2,-2}(\theta, \phi)] \qquad (20d)$$

The $Y_{lm}$ are the usual harmonics.

Now let us bring in the phase shifts. These arise from the Green's functions. If we form the symmetrized combination of Green's functions and assume that our energy band is spherical with

$$E = \gamma k^2$$

we get in the small-$ka$ limit

$$\text{Im } \mathscr{G}_{12} = \frac{-\Omega k}{4\pi\gamma} \frac{k^4 a^4}{120} \qquad (21a)$$

$$\text{Im } \mathscr{G}_{25'} = \frac{-\Omega k}{4\pi\gamma} \frac{k^4 a^4}{60} \qquad (21b)$$

With the use of these results and the aid of equation (12) we see that both $d$-wave phase shifts are proportional to $k^5$ for small $k$, as we would expect.

Now consider $\Gamma_{12}$: With the use of equation (21a) we have

$$\text{Im } D_{12} = \frac{\Omega k}{4\pi\gamma} V_{12} \frac{k^4 a^4}{120} \qquad (22)$$

Then we get after some algebra

$$\langle \mathbf{k} | t_{12} | \mathbf{k}' \rangle = + \frac{2\gamma}{\pi k} \frac{\text{Im } D_{12}}{|D_{12}|} e^{+i\delta_{12}} K_{12}(\alpha, \beta; \theta, \phi)$$

$$= -\frac{2\gamma}{\pi k} \sin \delta_{12} \, e^{+i\delta_{12}} K_{12}(\alpha, \beta; \theta, \phi) \qquad (22a)$$

Also, for the $\Gamma_{25'}$ representation

$$\langle \mathbf{k} \,|\, t_{25'} \,|\, \mathbf{k}' \rangle = -\frac{2\gamma}{\pi k} \sin \delta_{25'} \, e^{i\delta_{25'}} K_{25'}(\alpha, \beta; \theta, \phi) \tag{22b}$$

This is as far as we can proceed unless $\delta_{12} = \delta_{25'} \equiv \delta_2$ which implies a relation among matrix elements of the potential. Let us suppose that this relation obtains. Then we add the $\Gamma_{12}$ and $\Gamma_{25'}$ elements, and use the addition theorem for spherical harmonics. We get

$$\langle \mathbf{k} \,|\, t_{12} + t_{25'} \,|\, \mathbf{k}' \rangle = -\frac{\gamma}{2\pi^2} \frac{5}{k} \sin \delta_2 \, e^{i\delta_2} P_2 (\cos \Theta) \tag{23}$$

in which $\Theta$ is the angle between $\mathbf{k}$ and $\mathbf{k}'$ and $P_2$ is a Legendre polynomial. This is the result we want, since in our notation, the scattering amplitude defined in Ref. 1 is related to the $t$-matrix element by

$$f(\mathbf{k}, \mathbf{k}') = -\frac{2\pi^2}{\gamma} \langle \mathbf{k} \,|\, t \,|\, \mathbf{k}' \rangle \tag{24}$$

Thus we get the usual free-space scattering amplitude.

The reader will note the rather detailed approximations that are necessary to arrive at equations (23). This indicates that we cannot expect equation (11) to be a good approximation in more realistic problems.

Finally, it can be shown (see Ref. 1) from an argument involving the $S$-matrix that in the limit in which the solid gradually goes away while a spherically symmetric scattering potential remains, the phase shift for each representation $s$ must becomes equal to the sum of the phase shifts for all of the spherical partial waves which, in the solid, are basis functions for that representation.


## REFERENCES

1. If band $\alpha$ is spherically symmetric with reciprocal effective mass $\gamma_\alpha$, the factor in equestion is $-2\pi^2/\gamma_\alpha$. See J. Callaway, in *Lectures in Theoretical Physics*, edited by W.E. Brittin, University of Colorado Press, 1966, Vol VIIIa, p. 295.
2. J. Callaway, *J. Math. Phys.* **5**: 783 (1964).
3. L.P. Bouckaert, R. Smoluchowski, and E. Wigner, *Phys. Rev.* **50**: 58 (1936).
4. J. Callaway, *Energy Band Theory*, Academic Press, Inc., New York, 1964, p. 56.

# The Excited States of the F Center

## G. Chiarotti

*Department of Physics*
*University of Rome*
*Rome, Italy*

## 1. INTRODUCTION

The interest in the study of the $F$ center in alkali halides has increased considerably in the last few years, since it gradually became clear that the $F$ center possesses several higher excited states at energies corresponding to those of the conduction band of the crystal ([1-5]). The existence of levels of relatively long lifetime merged into a continuum of allowed states has in fact stimulated a great deal of thoretical investigation ([6-8]).

As it is well known, the $F$ band arises from the optical transitions of an electron trapped at a negative ion vacancy ([9]). At large distances from the vacancy, the field is Coulombic, and since the Coulomb potential does not present discrete levels at positive energies, the suggestion by Lüty ([1]), that the structure in the optical absorption spectrum observed on the high-energy side of the $F$ band was due to transitions to excited states of positive energies, caused several objections; a model for the $L$ bands as due to a separate "impurity" center was even proposed ([10]).

The purpose of the present lecture is twofold: (i) to review the present evidence of the existence of the higher excited states of the $F$ center, with special emphasis on evidence obtained by the method of "modulated $F$ center absorption" ([5]); (ii) to discuss the possible models of the $L$ bands as optical transitions to resonant scattering states ([11]) or to band levels having high density in the vicinity of the center ([8]). At the end, some new results on the effect of an electric field on the $F$ and $L$ bands will be presented.

## 2. THE PROBLEM OF THE L BANDS

The absorption spectrum of additively colored ([12]) KCl is plotted in Fig. 1. The high-energy side of the $F$ band is presented in greater detail and shows

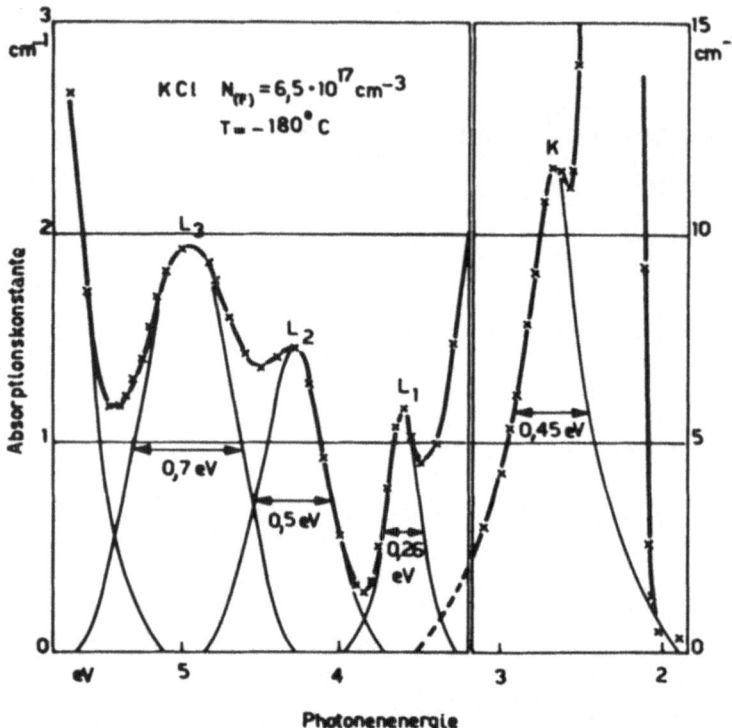

Fig. 1—The absorption spectrum of additively colored KCl at 93°K; the region of the L bands is shown with greater detail. Note a change in the ordinates scale at 3.2 eV [from F. Lüty(1)].

the presence of the $K$, $L_1$, $L_2$, and $L_3$ bands. A fourth band which behaves like the $L$ bands has been found by Hirai and Ueta [13] at higher energies and is not shown in Fig. 1. The $K$-band is thought to be associated with optical transitions to the various $np$-like states ($n > 2$) of the $F$-center below the bottom of the conduction band [14]. Accepting this interpretation of the $K$ band, and assuming that the $L$ bands are also due to transitions starting from the ground state of the $F$-center, it is evident from Fig 1 that the "$L$ levels" are located well within the conduction band.

On the other hand, the position of the bottom of the conduction band relative to the ground state of the $F$ center can be inferred from photoconductivity measurements [15]. Such measurements, in fact, show (Fig. 2) a rather well-defined threshold at an energy slightly larger than that of the $K$ band. The photoconductivity curve does not seem to follow the absorption spectrum of the $K$ and $L$ bands [16], so that it appears that the photocurrent is caused by optical transitions different from those responsible for the $K$ and $L$ bands. This result is important because it apparently rules out the hypothesis that photoconductivity arises from thermal ionization of the excited levels, which are raised close to (or within) the conduction band by the process of lattice relaxation that follows the absorption of a photon [17]. In other words. the position of the F-center ground state with respect to the conduction band, inferred from the photoconductive threshold, refers to the F-center in its unrelaxed configuration.

From the above considerations and the results of Figs. 1 and 2, it appears that the ground state of the F center of KCl lies, in its unrelaxed configuration, approximately 3 eV below the bottom of the conduction band and that, were the L-bands excited states of the F center, they would lie well within the conduction band.

The first evidence that the L bands were associated with the F center was given by Lüty [1], who discovered them and showed that their height is proportional to that of the K band over a wide range of the F center concentrations, and that, in turn, the height of the K band is proportional to that of the F band. Moreover, some of the L ands have been found also in crystals which have been X-rayed or bleached at very low temperatures [2,3], a condition severely different from that of additive coloration. However, the proportionality between the F and K bands has been at times doubted [18],

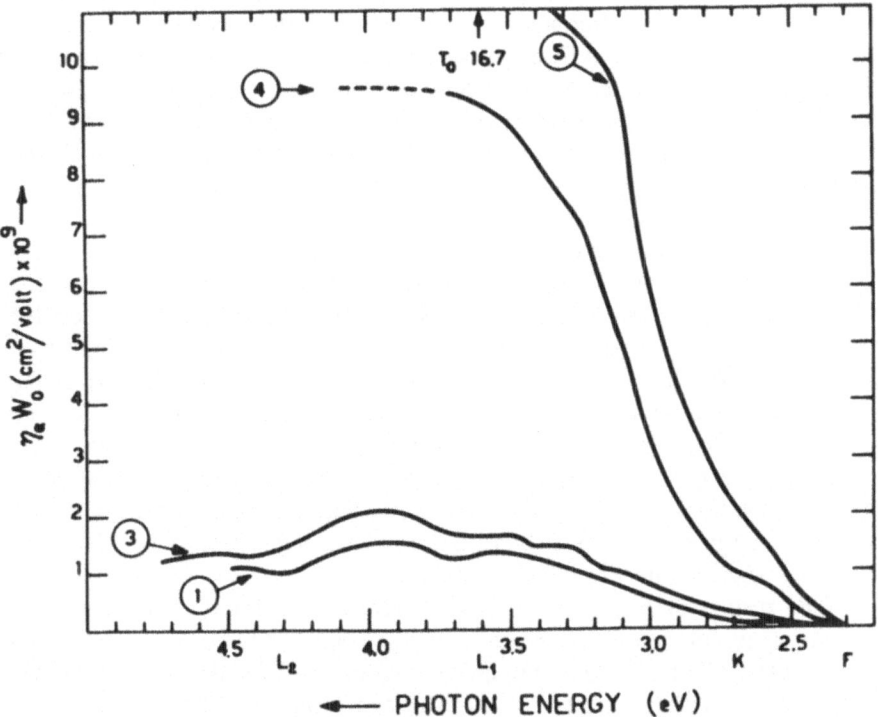

Fig. 2—Photoconductivity response of additively colored KCl at 10°K corrected for optical absorption [from R. Wild and F.C. Brown [15]].

and the ratio between the $F$ and $L$ bands found in crystals X-rayed at low temperature is in rather poor agreement with that found in additively colored crystals.

A very convincing proof (though of a rather indirect nature) has been given by Hunger and Lüty ([4]) who have shown that, under certain condittions, the quantum efficiency of the $F \rightarrow F'$ conversion obtained by irradiation into the $L_2$ band is approximately 2; i.e., each photon absorbed into the $L_2$ band destroys two $F$ centers, forming one $F'$ center (which is known to be the $H^-$-like entity formed by a negative ion vacancy which has trapped two electrons). A yield of two is not easily explained in case the $L_2$ band was due to a separate electronic center.

A more direct proof has been obtained recently by means of the results of "modulated $F$-center absorption" by U.M. Grassano and the author ([5]). The method can be applied to various other problems of current research in the field of color centers, and will be discussed in detail. The principle is the following: a crystal containing $F$ centers is exposed, at low temperature, to an intense beam of $F$ light whose intensity varies with time at a frequency $v$. As a result of this excitation, both the population of the ground state ($F^\circ$) and that of the first excited state ($F^*$) of the $F$ center vary around their average value at the same frequency. An auxiliary orthogonal beam of continuous light, which passes through the crystal inducing transitions from the $F^\circ$ and the $F^*$ levels, then shows amplitude modulation of frequency $v$, which can be detected with great sensitivity by use of synchronous amplifiers ([19]). Transitions starting from the $F^\circ$ level cause modulation in phase with the exciting $F$ light; possible transitions from the $F^*$ level would cause, in contrast, out-of-phase modulation.

The modulation index $\Delta I / I$ of the dc auxiliary light is shown in Fig. 3 as a function of energy, for a crystal of additively colored KCl ($N_F = 1.4 \times 10^{17}$ cm$^{-3}$) at liquid-nitrogen temperature. Modulation shown as positive in the figure corresponds to a signal in phase with the exciting $F$ light (the crystal becomes more transparent under $F$-light illumination). Positive peaks at energies corresponding to $F$, $K$, $L_1$, and $L_3$ bands are clearly visible. The $L_2$ peak is hidden under a much stronger negative signal, from which it can be resolved, as is shown in the inset of Fig. 3. A large negative peak at energies corresponding to the $F'$ band is also present. The shape of the curve on the high-energy side of the $F$ band is due to the superposition of the $F$, $K$, $F'$, and $L$ signals with appropriate phases.

The results of Fig. 3 prove beyond reasonable doubt that the optical transitions giving rise to the $K$ and $L$ bands originate from the $F^\circ$ level and thus that the "$K$ and $L$ levels" are indeed higher excited states of the $F$ center. They also show that, under the action of the $F$ light at 77°K, a band $F'$ is

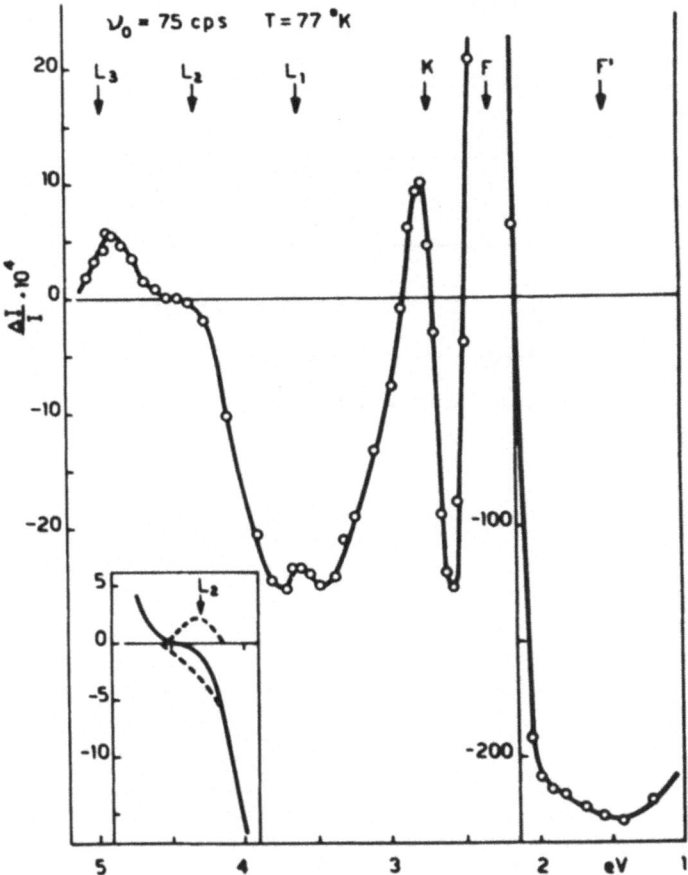

Fig. 3—Modulation index $\Delta I/I$ as a function of the energy of the photons for a crystal of KCl containing $1.4 \times 10^{17}$ F-centers/cm$^3$. In the inset the $L_2$ band is resolved [from G. Chiarotti and U.M. Grassano (5)].

developed which reconverts to the F band in the dark in times not much longer than $1/\nu$. The process can be represented by the reactions

$$F^* + F \longrightarrow F' + \text{vacancy} \qquad (1)$$

$$F' + \text{vacancy} \longrightarrow 2F \qquad (2)$$

Reactions (1) and (2) involve the tunneling of an electron from an excited $F^*$ center to a neighboring F center, followed by ionization of the F center by the electric field of the vacancy left behind.

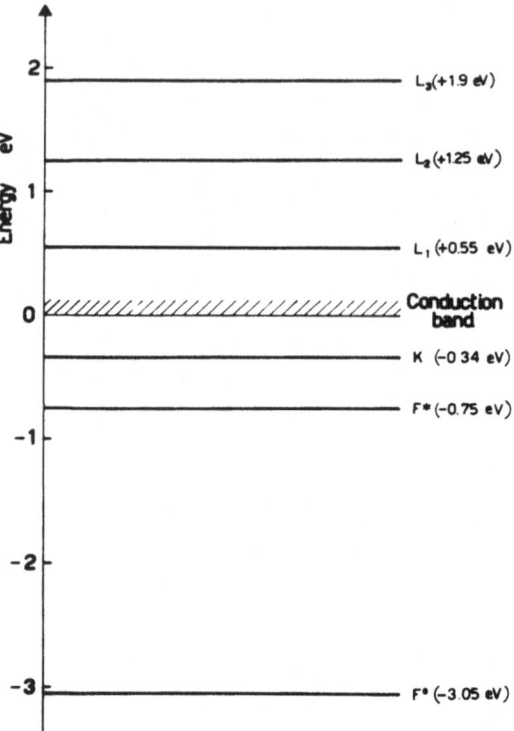

Fig. 4—The energy levels of the F-center in KCl.
The scheme corresponds to a temperature of 90°K
(though the data on photoconductivity refer to 10°K).

The analysis of the experimental results (as a function of the frequency $v$ and of the concentration of the $F$ centers) allows the characteristic times associated with reactions (1) and (2) to be determined ([5,20]). The scheme of Fig. 4 for the levels of the $F$ centers are based an the conclusions drawn from the experimental results mentioned above. The major uncertainty is in the figure giving the position of the $F$ level with respect to the bottom of the conduction band, which has been estimated from the data of Wild and Brown ([15]) as the energy corresponding to the half-value of the photoconductivity shoulder.

A great deal of theoretical investigation has recently been done in order to rationalize the existence of higher excited states of the $F$ center in the conduction band; however, the question of the lifetime of a "localized" level within a continuum does not seem to have received much attention. The presence of a gap of forbidden energy in which the localized levels could be located was proposed by Bassani and Giuliano ([21]) on the basis of band calculations done with the point-ion-model approximation; however, it has not received support from more accurate estimates ([7,22]).

Simpson, ([7]) on the other hand, proposed that the higher excited states of the F center are still below the bottom of the conduction band, and that they are raised into it during the relaxation process that follows the optical transition. Photoconductivity would then arise from the release of the electron during the latter process. It must be noted, however, that in such a case the photoconductivity spectrum should be the same as the absorption spectrum, a fact which does not receive support from the experimental evidence discussed in Section 2.

Several interesting suggestions have been recently made as to the origin of the L bands. Kojima ([8]) attributes them to transitions from the ground state of the F center directly to the levels of the conduction band. He evaluates the density of states of the conduction band for an ideal square-wave potential in the presence of a localized potential well. The wavefunction of the conduction electron is modified by the presence of the potential well (the anion vacancy) in such a way that its amplitude at the position of the vacancy increases with decreasing wavenumber of the electron. A sharp maximum of amplitude at $K = 0$ gives rise to several absorption peaks associated with transitions from the ground ([15]) state to the various $\Gamma$ points at the center of the Brillouin zone. With reference to the energy band calculations of Oyama and Miyakawa ([22]), Kojima also attempts a possible (though not entirely satisfactory) assignment to the L and K bands.

Bassani and Preziosi ([11]) on the other hand, have pointed out that three-dimensional well-like potentials show, under certain conditions, discrete scattering states of positive energy (resonant states) which can be reached by optical transitions from the ground state of the F center. In fact, it is a well-known result of elementary quantum theory that any problem of central symmetry can be handled by solving a one-dimensional Schrödinger equation with an effective potential obtained by adding to $V(r)$ the "centrifugal" potential:

$$\frac{\hbar^2}{2m} \frac{l(l+1)}{r^2} \tag{3}$$

In the case of a square well one obtains (for $l \geq 1$) the effective potential shown in Fig. 5. Although with the potential of Fig. 5, there are no stationary states of positive energy, it is well known that the escaping probability of the electron becomes small for certain energies which correspond to quasi-bound resonant levels. Transitions from the ground state of the F center to such resonant scattering states could give rise to the L bands. Bassani and Preziosi also attempt a quantitative calculation with a potential of the type used by

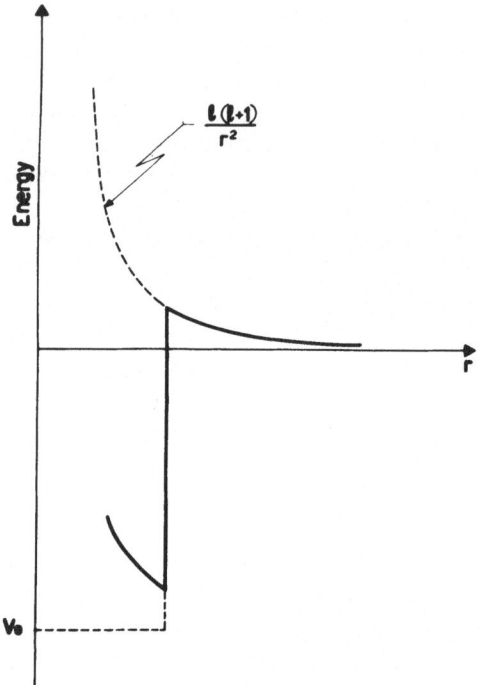

Fig. 5—Effective potential for a three-dimen-
tional potential well.

Simpson ([23]) (i.e., Coulomb-like for $r > R$ and constant $(V_0)$ for $r < R$.
$V_0$ is chosen to be intermediate between the Madelung potential $V$ and the
potential at the vacancy $(V')$ calculated under the hypotehsis that the ions
are not allowed to move from their lattice positions:

$$V_0 = V' + \beta(V - V') \qquad (4)$$

The parameter $\beta$ turns out to be the same for all alkali halides, while $R$ is
evaluated in a self-consistent way. For negative energies they obtained bound
states which are in good agreement with the experiments as well as with the
previous calculations. For positive energies, on the other hand, they found
complex solutions, which correspond to resonant states when the imaginary
part of the energy is small and of the appropriate sign. Only the lower $p$-
like state, however, has sufficiently long lifetime ($\hbar/\gamma$, $\gamma$ being the imaginary
part of the energy) to give an absorption peak of reasonable width. Table I
reports their results for negative and positive energies for various crystals,
compared with the experimental data.

**TABLE I**

| | Negative energy | | | | Positive energy | | | |
|---|---|---|---|---|---|---|---|---|
| | Theory | | Experiment | | Theory | Experiment | | |
| | $E_F$ | $E_K$ | $E_F$ | $E_K$ | $E_{resonant}$ | $E_{L_1}$ | $E_{L_2}$ | $E_{L_3}$ |
| LiF | 4.04 | 4.85 | 4.82 | | | | | |
| NaF | 3.25 | 4.12 | 3.64 | | | | | |
| KF | 2.57 | 3.24 | 2.68 | | | | | |
| NaCl | 2.53 | 3.24 | 2.68 | | | | | |
| KCl | 2.15 | 2.67 | 2.30 | 2.71 | 4.6 | 3.55 | 4.25 | 4.95 |
| RbCl | 1.97 | 2.45 | 2.04 | 2.37 | | | | |
| KBr | 2.01 | 2.45 | 2.07 | 2.36 | | | | |
| RbBr | 1.89 | 2.32 | 1.85 | 2.09 | | | | |
| KI | 1.82 | 2.15 | 1.87 | 2.12 | | | | |
| RbI | 1.68 | 2.00 | 1.96 | 1.92 | | | | |

The model used gives correct values for the energy of the bound states, but only a single resonant level which could perhaps be appropriate to the $L_3$ band. The model could certainly be improved by choosing a better potential, though it appears difficult to obtain more than one resonant state. The deficiency is most probably due to the fact that the states in the continuum are not free electron states (as assumed by Bassani and Preziosi) but Bloch states with high density at cryitical points in the Brillouin zones. Resonant states at $\mathbf{k} \neq 0$ may then be the origin of the various $L$ bands.

Most of the theoretical work on the $F$ center has been done in the scheme of the effective mass approximation. A different approach has been used by R.F. Wood[6] who carried out various calculations on the electronic structure of the $F$ center using the LCAO method in which the wavefunction of the ground and excited states were linear combinations of the $s$ and $p$ orbitals on the alkali ions nearest and next nearest to the vacancy. The perturbing Hamiltonian was the sum of the kinetic energy of the $F$ electron and of the potential energy of the field of the vacnancy. The solution of the resulting secular equation gave various levels of negative as well as of positive energy, some of which where assumed to be responsible for the $L$ bands. In this scheme the unperturbed energy of the $s + p$ state corresponds to the bottom of the conduction band; the $L$ bands would then be associated with transitions to excited states of the conduction band in the vicinity of the $F$ center.

The relationships among the various theories outlined before is not completely obvious since they start from different approximations (LCAO, effective mass, semi-continuous model); however, it appears that their refinement could provide a test for the various points of view which are commonly adopted in the theory of the electronic structure of the defects in solids.

## 3. THE EFFECT OF ELECTRIC FIELDS

If $L$ bands were due to transitions to the states of the conduction band at some critical points, one would expect a rather large effect of an applied electric field on the absorption coefficient. This effect would be similar to the well-known Franz–Keldish effect [24,25] on the band edges of semiconductors and its theory has been sketched by Tharmalingam [26], who extended Eagles [27] calculations on the transitions from acceptor levels to the conduction band of semiconductors.

A very rough estimate of the effect to be expected in the case of the $L$ bands could be done by assuming that the electric field induces tunneling states below the bottom of the conduction band with a spread in energy of the order of [26-28]

$$\Delta E \sim \left(\frac{3\hbar eF}{4m^{*1/2}}\right)^{2/3} \tag{5}$$

$F$ being the intensity of the electric field and $m^*$ an appropriate reduced mass. With a field of $5 \times 10^4$ V/cm, equation (3) gives a spread of energy of the order of $10^{-2}$ eV. On the other hand, assuming a Gaussian shape for the $L$ bands

$$\kappa = \kappa_{max} \exp\left[-\left(\frac{\omega - \omega_0}{W}\right)^2\right]$$

it is easily shown that the variation of the absorption coefficient to be expected at the point of maximum slope at lower energy would be of the order of

$$\frac{\Delta\kappa}{\kappa} \sim \sqrt{2}\,\frac{\Delta E}{W} \tag{6}$$

i.e., of the order of $10^{-2}$ for the $L$ bands in KCl containing $10^{17}$ $F$ centers/cm$^3$.

Experiments carried out on additively colored KCl at liquid-nitrogen temperature [29] have not detected any effect in the region of the $L_1, L_2$, and $L_3$ bands to the limit of $5 \times 10^{-6}$ (which was the sensitivity of the apparatus) with an external field of $5 \times 10^4$ Vp/cm and a density of color centers of $10^{17}$ cm$^{-3}$. The experiments seem to indicate that the $L$ bands are not due to direct transitions to the continuum or that the effective mass involved is extremely high. However, the qualitative considerations developed before should be taken with great caution in the absence of an explicit theory.

The Stark effect of the transitions among the bound states of the $F$ center is, on the contrary, quite detectable. Unpublished results of Grassano, Rosei, and the author [29] are plotted in Fig. 6 which shows $-\Delta I/I \simeq \Delta\kappa x$ ($x$ being the thickness of the crystal) as a function of the energy of the photons

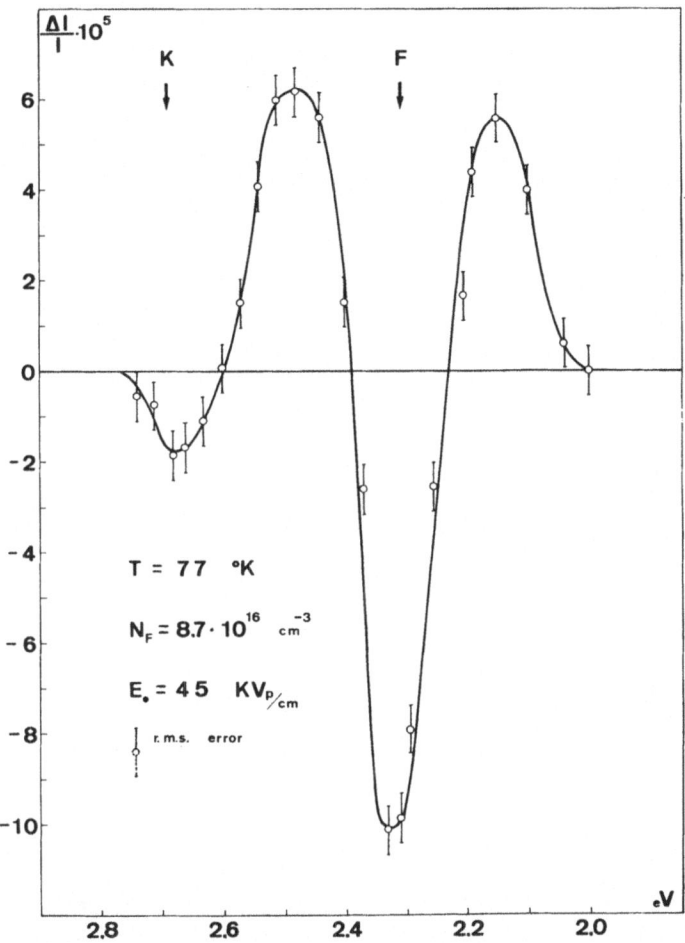

Fig. 6—The modulation index $-\Delta I/I$ at frequency $2\nu$ of a beam of light passing through a crystal of additively colored KCl to which an ac field of $4.5 \times 10^4$ Vp/cm has been applied. Measurements are shown as a function of the energy of the photons.

for a crystal of KCl containing $8.7 \times 10^{16}$ *F* centers cm³. The Stark effect has been observed by applying to the crystal an ac field of $4.5 \times 10^4$ V/cm at a frequency $\nu$ and detecting the modulation index, at frequency $2\nu$, of a beam of light passing through the crystal.

Positive values of $-\Delta I/I$ correspond to an increase of the absorption coefficient. The Stark effect appears then as a broadening of the *F* band, which roughly conserves the area. A certain amount of exchange of area between the *F* and the *K* band is also evident. Figure 7 shows the dependence of the effect on the electric field. It is seen that the *F* center shows a second-order Stark

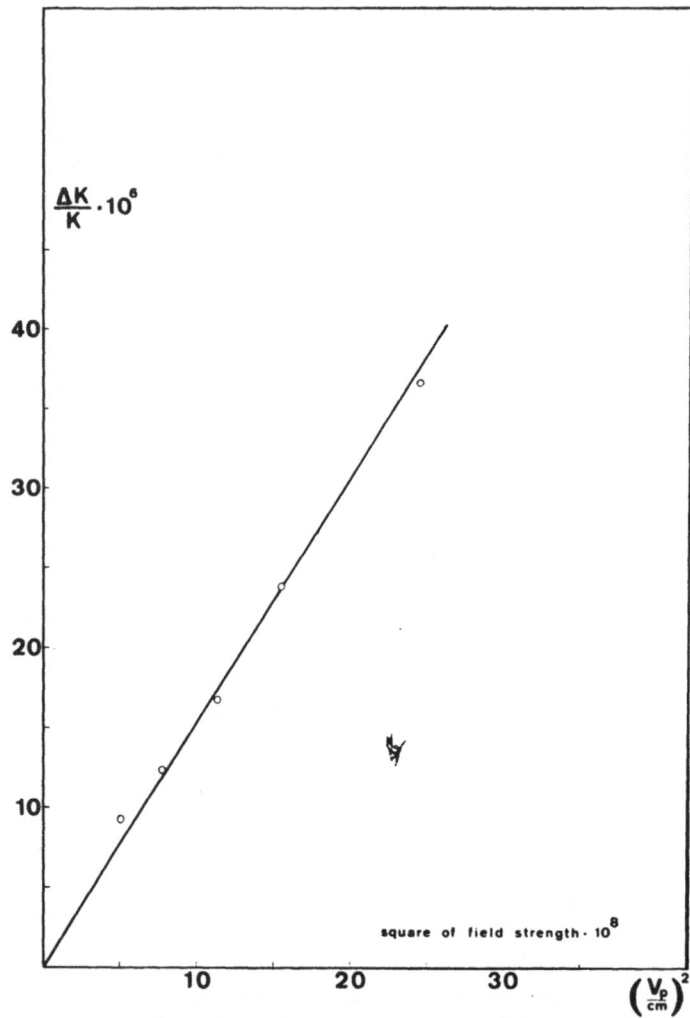

Fig. 7—The funtion $\Delta K/K$ at the peak of the $F$-band as a function of the square of the electric field. Temperature and color centers density are the same as in Fig. 6.

effect, as it should since it has a cénter of inversion. The theory of the Stark effect of the $F$ center has been worked out by Henry, Schnatterly, and Slichter ([30]). They consider the first excited state of the $F$ center as a spin-orbit doublet ($P_{1/2}$ and $P_{3/2}$) and compute the effect of the admixture with the $2S_{1/2}$ state (which cannot be detected in ordinary absorption) on the various moments of the band.

Introducing a shape function $f(E)$ proportional to the absorption coefficient divided by the energy, the various moments of the band are defined as follows:

Zero-order moment (area)

$$A = \int f(E)dE$$

First-order moment (center of the gravity of the curve)

$$\bar{E} = A^{-1} \int E f(E)dE \tag{7}$$

$n$th moment

$$\langle E^n \rangle = A^{-1} \int f(E)(E - \bar{E})^n dE \qquad (n \geq 2)$$

In presence of an external electric field in the $z$ direction, perturbation theory shows that the change of zeroth and first moments are zero and that the change of second moment is (for light polarized in the $z$ direction):

$$\langle \Delta E_x^2 \rangle = \langle \Delta E_y^2 \rangle = 0 \tag{8a}$$

$$\langle \Delta E_z^2 \rangle = (eF)^2 \, |\langle \beta \, | \, z \, | \, \gamma \rangle|^2 \tag{8b}$$

$| \beta \rangle$ and $| \gamma \rangle$ being the orbital $2P$ and $2S$ wavefunctions, respectively. In order to conserve the area, the small increase in the second moment leads to a fractional decrease in the peak height:

$$\frac{\Delta \kappa}{\kappa} \simeq -\frac{\langle \Delta E_z^2 \rangle}{2 \langle E^2 \rangle} \tag{9}$$

The results of Fig. 6 show that $\Delta \kappa / \kappa = 3 \times 10^{-5}$; however, since the measurements have been done with ordinary light, the value to be introduced in (9) is twice as high on account of 8a and 8b If the experimental value $6 \times 10^{-3} (eV)^2$ is taken for $\langle E^2 \rangle$ at liquid-nitrogen temperature [31] and $f = 5 \times 10^4$ V/cm, the matrix elemnt $\langle \beta \, | \, z \, | \, \gamma \rangle$ can be evaluated and turns out to be $1.7 \times 10^{-8}$ cm, a very reasonable value [30]. The result of the Stark effect seems moreover to indicate that the various $p$ levels for the $K$ band are mixed by the electric field with $s$ or $d$ states so that they show a smaller transition probability from the ground state of the $F$ center [32].

## 4. EFFECT ON EXCITON SPECTRA

A different approach to the study of the excited states of the $F$ center is that carried on by Fröhlich and Mahr [33] and Park [34] using laser techniques.

It is well known that excitons created in the vicinity of an $F$ center have smaller energies than in the perfect crystal. The optical absorption of alkali halides containing $F$ centers shows in fact a characteristic peak (the $\beta$ band) in the low-energy tail of the first exciton band ($^9$). The structure of the excitons in the presence of the excited $F$ centers has been studied by Fröhlich and Mahr ($^{33}$) with a method similar to that of "modulated absorption" using a laser as a source of excitation. With a pulsed ruby laser they were able to pump more than $90\%$ of the $F$ centers in their excited states. Concomitant measurements of optical absorption in the region of the $\beta$ band are plotted in

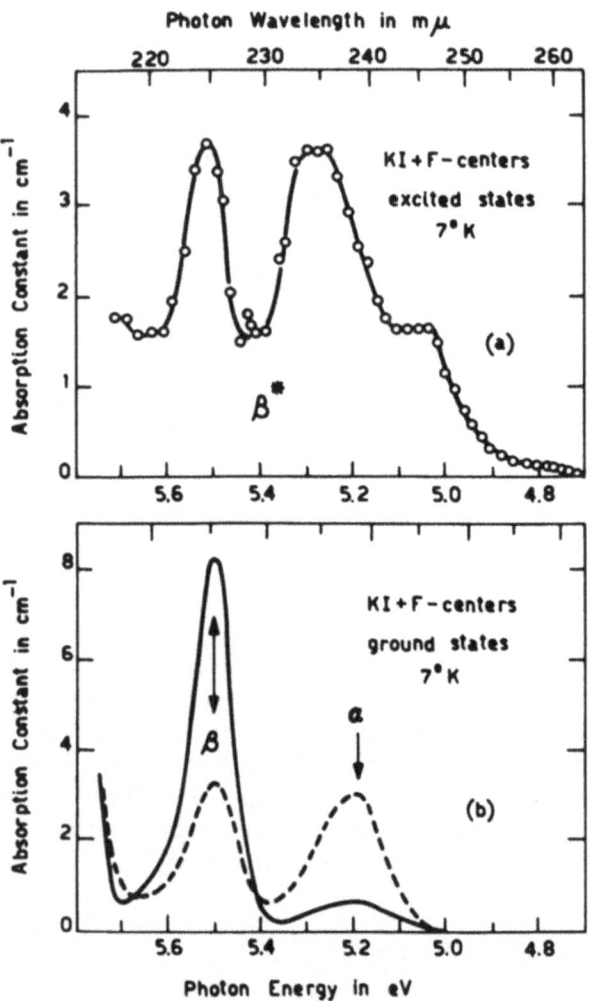

Fig. 8—(a) Exciton absorption in the presence of excited $F$ centers in KI at 7°K. (b) The well-known $\alpha$ and $\beta$ bands of KI at 7°K for comparison [from Fröhlich and Mahr ($^{33}$)].

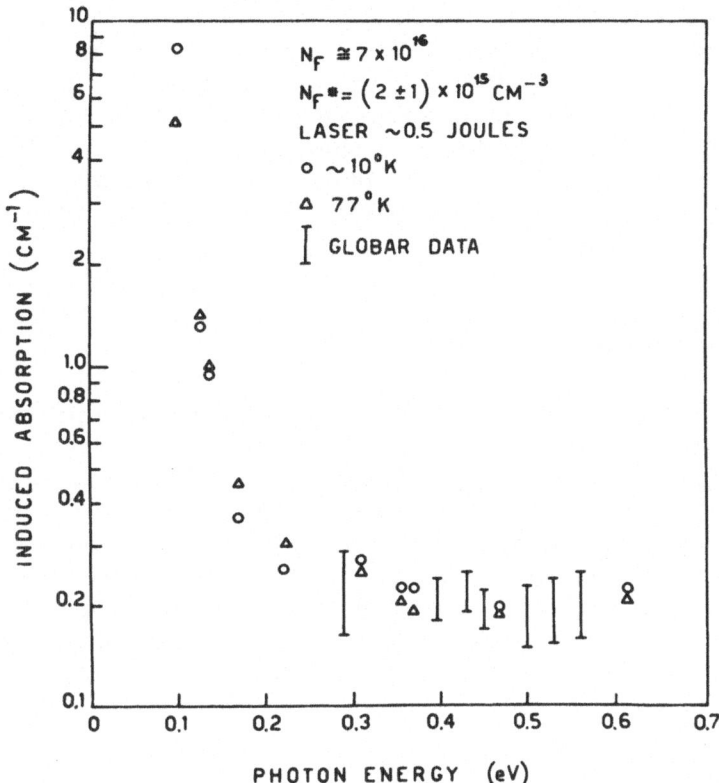

Fig. 9—Absorption spectrum of the *F*\* center [from K. Park and W.L. Faust, [36]].

Fig. 8 for a crystal of KI. The $\beta$\* spectrum shows that localized excitons are very sensitive to the nature of the defect. The wavefunction of the excited *F* center is much more extended than that of the ground state; the electron screening of the charge of the vacancy is therefore smaller and the exciton shows a larger binding energy to the defect. The $\beta$\* spectrum is displaced toward lower energies, the situation approaching that of the $\alpha$ band, which is shown for comparison in the lower section of the figure. The $\beta$\* band shows, moreover, a considerable structure, which is apparently due to the lower symmetry of the new arrangement of the ions surrounding the *F* center and to the symmetry of the excited wavefunction. Unfortunately, only the low-energy side of the $\beta$\* band is experimentally observable and, on the other hand, the number of the components of the $\beta$ and $\beta$\* excitons expected theoretically depends upon the model [34] that one assumes for the exciton [33,35]. However, a quantitative theoretical investigation of the $\beta$\* spectrum could give very valuable information on the symmetry of the localized excitons.

When most of the $F$ centers are excited, the observation of optical transitions from the excited $F^*$ level becomes possible. Tl s has been done, also with laser techniques, by Park ([36]), who was able to observe the transition of the excited electron to the conduction band (Fig. 9). From Fig. 9 one obtains a value of 0.1 eV for the optical ionization energy of the $F^*$ level ([37]), The results of Park on the transitions of the $F^*$ center in other regions of the spectrum are, however, quite complex due to the strong interaction among the excited $F$ centers. ([38])


## 5. CONCLUSIONS

The considerations of the preceding sections show that a problem of a very specialized nature, as that of the excited states of a color center, has brought about a number of experimental methods and theoretical suggestion which are of great interest for the understanding of the physics of solids. The methods of modulated absorption and of the reversal of the population with laser techniques will have fruitful applications in the study of the optical properties of solids. On the other hand, the concepts of localized density of states and of resonant scattering transitions, developed independently in several other fields, have far-reaching implications, as was stressed by almost every lecturer of this School.


## REFERENCES

1. F. Lüty, *Z. Phys.* **160**: 1 (1960).
2. C.C. Klick and M.N. Kabler, *Phys. Rev.* **131**: 1075, (1963).
3. C.C. Klick, *Phys. Rev.* **137**: A1814 (1965).
4. M. Hunger and F. Lüty, *Phys. Lett.* **15**: 112 (1965).
5. G. Chiarotti and U.M. Grassano, *Phys. Rev. Lett.* **16**: 124, (1966).
6. R.F. Wood and H.W. Joy, *Phys. Rev.* **136**: A451 (1964).
7. F. Bassani, *International Symposium on Color Centers*, Urbana 1965, page. 14. The complete text has been published in "Atti del Convegno Nazionale su proprietà ioniche ed elettroniche degli alogenuri alcalini" *Consiglio Nazionale delle Ricerche, Roma* 1966, p. 172
8. T. Kojima, *International Symposium on Color Centers*, Urbana 1965, p. 87.
9. J.H. Schulman and W.D. Compton, *Color Centers in Solids*. The Macmillan Company, Inc., New York, 1962.)
10. A. Gold, *Phys. Rev.* **123**: 1965, (1961).
11. F. Bassani and B. Preziosi, unpublished, see also Ref. 7, and G. Iadonisi and B. Preziosi, *Nuovo Cimento*, **48 B**: 92 (1967).
12. Additive coloration is a process by which a stoichiometric excess of the metal is diffused into an alkali halide crystal by means of heating it in the presence of the metal vapor. The metal ions form new layers on the surface, thus introducing into

the crystal an equal number of negative ion vacancies and of free electrons (see for example N.F. Mott and R.W. Gurney, Clarendon Press, Oxford 1953 chapter IV)

13. M. Hirai and M. Ueta, _J. Phys. Soc. Japan_, **17**: 566 (1962).

14. N.F. Mott and R.W. Gurney, _Op. Cit._ p. 113 _ff._, see also Ref. 16.

15. R.Wild and C.F. Brown, _Phys. Rev._ **121**: 1296 (1961).

16. This result which can be inferred from the curves of Fig. 2 has been stressed by D.Y. Smith and G. Spinolo (_Phys. Rev._ **140**: A2117, 1965). However, earlier work of N. Inchauspé [_Phys. Rev._ **106**: 898, (1957)] on KI and KBr points out a striking similarity between photoconduction and optical absorption.

17. J.H. Simpson, _International Symposium on Color Centers_, Urbana 1965, p. 31.

18. H.W. Hetzel and F.E. Geiger, _Phys. Rev._ **96**: 225 (1954).

19. Modulation of $5 \times 10^{-6}$ could be detected over most of the spectral range.

20. G. Chiarotti and U.M. Grassano, _Nuovo Cimento_ **46**: 78 (1966).

21. F. Bassani and S. Giuliano, Private communication—see also S. Giuliano Thesis, Università di Messina, and F. Bassani, Ref. 7.

22. S. Oyama and T. Miyakawa, _J. Phys. Soc. Japan_ **20**: 624 (1965).

23. J.H. Simpson, _Proc. Roy. Soc._ **A197**: 269 (1949).

24. W. Franz, _Z. Naturforsch_ **13a**: 484 (1958).

25. L.V. Keldish, _J. Exp. Theoret. Phys. (USSR)_ **34**: 1138 (1958).

26. K. Tharmalingam, _Phys. Rev._, **130**: 2204 (1963).

27. D.M. Eagles, _J. Phys. Chem. Solids_, **16**: 76 (1960).

28. P.H. Wendland and M. Chester, _Phys. Rev._ **140**: A138A.

29. G. Chiarotti and U.M. Grassano, and R. Rosei, _Phys. Rev. Lett._ **20**: 1043 (1966).

30. C.H. Henry, S.E. Schnatterly, and C.P. Slichter, _Phys. Rev._ **137**: A583 (1965).

31. For a Gaussian $\langle E^2 \rangle = (H^2/8 \ln 2)$, $H$ being the width at half maximum.

32. For an analogous result in the case of external uniaxial compression see: S.E. Schatterly, _Phys. Rev._ **140**: A1364 (1965).

33. D. Fröhlich and H. Mahr, _Phys. Rev. Lett._ **14**: 494 (1965); _Phys. Rev._ **141**: A692 (1966).

34. Several models have been proposed for the exciton. They differ depending on whether the excitation is obtained by transferring an electron from a halogen to a neighboring metal ion (A. von Hippel, _Z. Physik_ **101**: 680, 1936—F. Bassani and N. Inchauspé, _Phys. Rev._ **105**: 819, 1957) or by exciting one of the halogen's $3p$ electrons to bound states in the field of the vacancy (D.L. Dexter, _Phys. Rev._ **83**: 1044, 1951)

35. R. Fuchs, _Phys. Rev._ **111**: 387 (1958)

36. K. Park and W.L. Faust, _Phys. Rev. Lett._ **17**: 137 (1966).

37. This value, however, cannot be compared with that given in the scheme of Fig. 4 which refers to the _F_-center in its unrelaxed configuration.

38. K. Park, _Phys. Rev._ **140**: A1735 (1965).

# Author Index

*(The numbers in italics refer to pages on which the full citation is given)*

# Subject Index